Musculoskeletal Tissue Transplantation and Tissue Banking

Musculoskeletal Tissue Transplantation and Tissue Banking

Theodore Malinin MS MD
Professor of Orthopedics and Director Emeritus
University of Miami Tissue Bank
Miller School of Medicine
University of Miami
Miami, Florida, USA

H Thomas Temple MD
Professor of Orthopedics and Director
University of Miami Tissue Bank
Miller School of Medicine
University of Miami
Miami, Florida, USA

JAYPEE

JAYPEE BROTHERS MEDICAL PUBLISHERS (P) LTD

New Delhi • London • Philadelphia • Panama

Jaypee Brothers Medical Publishers (P) Ltd

Headquarters

Jaypee Brothers Medical Publishers (P) Ltd
4838/24, Ansari Road, Daryaganj
New Delhi 110 002, India
Phone: +91-11-43574357
Fax: +91-11-43574314
Email: jaypee@jaypeebrothers.com

Overseas Offices

J.P. Medical Ltd
83 Victoria Street, London
SW1H 0HW (UK)
Phone: +44-2031708910
Fax: +02-03-0086180
Email: info@jpmedpub.com

Jaypee-Highlights Medical Publishers Inc
City of Knowledge, Bld. 237, Clayton
Panama City, Panama
Phone: +507-301-0496
Fax: +507-301-0499
Email: cservice@jphmedical.com

Jaypee Medical Inc
The Bourse
111 South Independence Mall East
Suite 835, Philadelphia, PA 19106, USA
Phone: + 267-519-9789
Email: joe.rusko@jaypeebrothers.com

Jaypee Brothers Medical Publishers (P) Ltd
17/1-B Babar Road, Block-B, Shaymali
Mohammadpur, Dhaka-1207
Bangladesh
Mobile: +08801912003485
Email: jaypeedhaka@gmail.com

Jaypee Brothers Medical Publishers (P) Ltd
Shorakhute, Kathmandu
Nepal
Phone: +00977-9841528578
Email: jaypee.nepal@gmail.com

Website: www.jaypeebrothers.com
Website: www.jaypeedigital.com

Inquiries for bulk sales may be solicited at: jaypee@jaypeebrothers.com

Musculoskeletal Tissue Transplantation and Tissue Banking

First Edition: **2013**

ISBN 978-93-5090-506-7

Printed at: Ajanta Offset & Packagings Limited, New Delhi

Dedicated to

Our parents:

Ivan M Malinin MD and Olga A Malinin
and
Harry A Temple and Jeri M Temple

In appreciation of support and encouragement provided to us over the past years

Preface

Today's understanding and utilization of the principles of musculoskeletal tissue transplantation will lead to tomorrow's new treatment of patients with disorders of the skeletal system. Nearly all principles of biomedical research are applicable to musculoskeletal tissue transplantation. Most significant advances in the fundamental studies of the skeletal system expose how much is still unknown about bone, cartilage, tendons, ligaments and other structures comprising the musculoskeletal system. With further advances, it is hoped, unsolved orthopedic problems will become fewer. It is sad to note that the gospel of bone transplantation reached its first mass acceptance in World War I, but then declined. In World War II, the same bone grafting principles operated on a wider scale. As a consequence "amputation cases" were diminished. More and more surgeons saw their efforts justified as many men were returned to normal life.

Tissue banking, as practiced today, originated in the Korean War with the formation of the US Navy Tissue Bank which served as a prototype for civilian tissue banks in North America. One of the first successes of the US Navy Tissue Bank dealt with a Naval aviator who suffered a compound fracture of the right forearm when his plane crashed in Korea in 1952. Despite conventional treatment, an osseous ridge developed between the radius and the ulna. This resulted in his inability to rotate the forearm. This disability forced him from active flight status. Following an operation in which newly formed bone was excised, freeze-dried fascia lata was placed between the radius and ulna, full motion of the arm was restored and the aviator returned to full flight status.

In the middle of the last century, tissue banking became an important adjunct to orthopedics and other surgical specialties by promoting the concept of stored material ready for transplantation.

This book presents a summary of pertinent information on bone biology, theory and practice of tissue transplantation and of experimental results from which practical aspects of human allograft preparation, storage and transplantation have been extrapolated. The experience gained from the organization and continuous operation of a clinical tissue bank has aided us in this endeavor. It provided a philosophical and practical workshop needed for the implementation of a full service tissue bank that served, and continues to serve, a broad national and international interest. The success of the University of Miami Tissue Bank was a catalyst to the rapid growth of tissue banking in both the United States and abroad.

Bone and tissue transplantation began and evolved to satisfy patient needs. Initially, clinical success prefaced scientific discovery, but in the last several years the situation has begun to change. This book notes a momentous event in bone physiology, the isolation of several bone growth proteins, named bone morphogenetic proteins, BMP 1-7 which are members of a family of growth and differentiation factors (TGF-B). Although research involving BMPs was conducted for more than 30 years, clinical application of these powerful molecules is still in the investigational stage. Bone grafts do not simply provide support, but play an important part in osteogenesis, and to this end, growth factors play a critical role. Osteoprogenitor cells are needed as well as scaffolding to promote new bone formation.

The future is difficult to predict, but the pendulum in musculoskeletal medicine appears to be in full swing toward orthobiologics. Careful science and prudent clinical observation are needed to realize and harness the enormous potential of tissue and cell transplantation.

This book is based on cumulative knowledge as well as personal experience. Our fondest hope is that it will prove to be useful to all discriminating readers committed to the art of musculoskeletal transplantation.

Theodore Malinin
H Thomas Temple

Acknowledgments

The authors wish to express their gratitude to Mr Francis Thomas Burke, III, for photography; Ms Karen S Collins and Ms Rita M Thomas, for assistance with the manuscript; Miss Rachel Robles, for typing the manuscript, and Mrs Dorothy R Malinin, for proofreading the manuscript.

Contents

Chapter 1

John Hunter (1728–1793), British Surgeon and Physician who contributed much to medicine. His work on bone physiology was monumental and ahead of his time.

(*Reference:* Ottley D. The Life of John Hunter, FRS Haswell, Barrington and Haswell. 1839, Philadelphia.)

Introduction

The history of bone transplantation starts with a chicken. The first successful bone grafting on record was performed by John Hunter (1728–1793) who transplanted a spur of a hen into the leg of a cock. Extraordinary additions to the knowledge of bone transplantation and bone growth were made by this individual.* Despite technical limitations, compared to today, Hunter carried out some of the most significant experiments with bone that stand unique to the present. It would not be an exaggeration to say that never has a single man devoted as much energy and time to the study of bone as did John Hunter. His observations on the effects of castration on the growth of deer antlers laid the foundation to the study of the effect of endocrine secretions on bone.

At Hunter's time the prevailing doctrine of bone growth stated it took place by interstitial accretion. Hunter put this idea to test by implanting two shots 2 inches apart in the tibias of pigs and fowls. He discovered in his studies of sacrificed animals that the shots remained the same distance apart, but were much further from the ends of the tibia, thus proving the bones grew in length. Knowing the discovery that madder (a red dye obtained from the roots of *Rubia tinctorum*) is deposited in the bones, Hunter periodically fed young pigs madder and was thus able to prove that bones also grow in thickness by the addition of the new layers of bone on the outer surface as the inner layers, next to the bone marrow cavity, are resorbed.

Since Hunter's fundamental discoveries on bone growth, many scientists have contributed to the knowledge of bone growth, healing and grafting. These have led to the acceptance of bone transplantation as a clinical entity and the development of tissue banking, which allowed the development of the transplantation of allogeneic bone on a large scale.

It has been disappointing to see the gospel of bone transplantation and tissue banking preached by Hyatt, Kreutz and the present authors to be subverted to pecuniary interests propagated by commercial advertisement. The latter has been frequently based neither on documented clinical results nor on applicable laboratory studies. We have struggled to write with detachment and concentrate on the well-established facts and findings.

In 1865, NA Pirogoff, who, among other things, pioneered the use of plaster casts, noted that bone regeneration and repair were not entirely understood (Voloshin and Bernini (1998). Although much progress in the study of bone physiology and repair have been made since John Hunter's time, Pirogoff's statement of some 150 years ago still holds true, at least partially.

If a missing segment of bone cannot regenerate itself, but if closely approximated bone ends can unite, then why not replace the missing segments with other bones and have these heal to the ends of the host bone? The accounts of replacing injured and diseased bones and joints are provided by Ollier (1867), Lexer (1925), Küttner (1911) and others. In the 6th and 7th decades of the past century Ottolenghi (1966), Parrish (1973), Zatzepin (1984), Mankin et al (1983) and Mnaymneh et al (1983) have performed transplantation of massive bone allografts. However, the earliest and persistent efforts in bone transplantation can be justifiably attributed to Albee, although he worked mainly with autografts (Albee 1923).

* Ernest A Groves. Huntarian Oration. Royal College of Surgeons. February 14, 1930

The modern era of bone allograft transplantation has been ushered in large measure by the development of tissue banks through which bone allografts became readily available (Malinin 1976; Kreutz 1951). Initial experience with transplantation of large bone allografts was gained from tumor surgery (Mnaymneh et al 1985). Later, encouraged by the results with reconstructions of the extremities following resection of tumors, massive allografts started being used with regularity for non-tumorous conditions, particularly for the revision of failed hip arthroplasties.

As with all surgical procedures, the more operations are performed and the longer the patients are observed, the more complications are recorded. Some of these include infections. Some can be attributed to technical errors, the correction of which resulted in the reduction of complications. However, some complications, such as late fractures of large grafts are inherent and are undoubtedly related either to the allografts themselves or to the interactions between the allograft and the host. To this end, it is important to be fully aware of what happens to transplanted bone.

A large body of information has been collected on bone, cartilage and tissue transplantation through experimental studies, although not all of it might be applicable to man. Despite this limitation, the success of clinical transplantation depends on the biological behavior of musculoskeletal allografts and the host-graft interactions. Therefore, it is worthwhile to apply information gathered on animals in the laboratory to humans, but this must be done with caution.

Because of the relative paucity of information on the long-term behavior of bone allografts initially transplanted into humans, the development of methods for bone allograft preparation proceeded largely along empirical lines. Virtually everything was tried. Bone allografts were boiled, steam autoclaved, immersed in mercurial solutions, washed with a garden hose in the yard then sterilized with ethylene oxide, immersed in beta-propiolactone, extracted with ether, methanol and chloroform, irradiated and sterilized by repeated freeze-thawing, to name but a few. On the other hand, bone allografts were likewise excised under aseptic conditions and preserved by freezing or freeze-drying. For clinical transplantation, aseptically excised frozen or freeze-dried allografts have over the years emerged as the allografts of choice. The immunogenicity of these grafts is reduced, more so in freeze-dried allografts than frozen ones, as compared to fresh bone allografts. The reason for this is most likely the destruction of the antigen presenting cells that reside in the bone marrow. Because of the general acceptance of the allografts prepared by these methods, in this book, we will deal primarily with allografts which were either freeze-dried, rapidly frozen, or frozen at a controlled cooling velocity after exposure to cryoprotective agents. In addition, because of the magnitude of the subject and the need to focus on specific aspects of musculoskeletal tissue transplantation, the results as well as the techniques of clinical transplantation will be emphasized.

Bone grafting arose from clinical need more than 100 years ago. Paradoxically, the practice led to the science. This is largely still the case, but the field has expanded so rapidly that it is difficult to synthesize all findings. The purpose of this book is to bring together information on bone grafting, cartilage and soft tissue transplantation as well as tissue banking in a systematic and comprehensive manner. This has been aided by the authors' long-standing endeavors in tissue banking and musculoskeletal transplantation. It is hoped that the information presented herein will aid the surgeon in the treatment of disorders and diseases of bone and other musculoskeletal structures.

The publication of this book follows two important events in bone physiology. First, the isolation and encoding of several bone growth proteins named bone morphogenetic proteins, BMP 1–7. These are members of a large family of growth and differentiation factors (TGF-B). Second, the synthesis of bone growth factors by genetic engineering that stimulates osteogenesis. The function of the latter, however, is still under scrutiny. In addition, the work with pluripotent stem cells has captured the imagination of scientists and clinicians concerned with bone regeneration. These efforts, in turn, resulted in a new field, tissue engineering. Various technologies described in this field may well have a lasting impact on the treatment of musculoskeletal diseases in the future, but so far their impact has been limited.

Any new developments in medicine have a potential to become controversial. Stem cell research and therapy are no exception. However, it can be hoped the imagination and skill of new researchers will result in a new means of treating patients with diseases of the skeletal system. At present, cell therapies and tissue banking share common ground as banked bone and other tissues can provide scaffolding and matrix for growth of new tissues from cells. Thus, even with the anticipation of the new developments, tissue banking will retain its place in regenerative medicine.

Bone grafting and tissue transplantation is performed by several surgical specialties. Orthopedic surgeons have always pioneered bone grafting. However, plastic surgeons, neurologic surgeons, otolaryngologists, oral and maxillofacial surgeons and dentists have also performed musculoskeletal tissue transplantation. Since allogeneic bone transplantation is performed by many specialties, members of these specialties should be adept in excising bone grafts and knowledgeable of bone allograft biology. This knowledge will be helpful in assessing the final outcomes.

The repair of musculoskeletal tissues and bone healing are of vital concern to all surgeons. While the majority of surgical techniques now practiced are based on methods developed long ago, the techniques cumulatively referred to as "tissue engineering" have lately made their appearance. The subject is still controversial, as is usually the advent of any new therapeutic modality. Although it can be believed the developments in this field may result in significant advances, in practice these still have not materialized to a significant degree. Therefore, it seems too early to abandon conventional methods of musculoskeletal tissue transplantation and tissue banking, subjects to which this book is dedicated.

Despite numerous publications on bone transplantation and bone banking, few books combining the topics have appeared. The present book deals with both topics in a historical perspective.

The history of bone transplantation, autologous and allogeneic, is long and complicated, but one could not expect something simple in an endeavor that spans over 100 years. Bone transplantation elicited many controversies resulting both from laboratory investigation and surgical practice. Some of these have not been resolved. Research on skeletal grafting did not proceed at a uniform rate. Instead, periods of intense activity were intermittent with periods of relative dormancy. In the last decade, research on skeletal grafting has once again gathered momentum.

Organized tissue banking is of a more recent vintage. The US Navy Tissue Bank and several scientists and clinicians who gained experience in this institution provided the impetus for its development. The Office of Naval Research had also a long period of supporting endeavors in musculoskeletal transplantation and tissue preservation.

The University of Miami Tissue Bank was established in 1970 to fill the void left by the US Navy Tissue Bank when it successfully completed the demonstration of practicality of tissue banking and discontinued supplying civilian surgeons with tissue allografts. The experience of over 40 years gained by the authors in this institution allowed for the gathering and organizing of the material on which this book is based. Gathering and going over the accumulated data and experience has reinforced our feeling that long-term success with musculoskeletal tissue transplantation warrants not only an incentive for further investigation, but also for continued use of these allografts in reconstruction of the skeletal system.

BIBLIOGRAPHY

1. Albee FH. A Surgeon's Fight to Rebuild Men. EP Dutton & Co, Inc. New York, 1923.
2. Kreutz FP, Hyatt GW, Turner TC, Bassett AL. The preservation and clinical use of freeze-dried bone. J Bone & Joint Surg (Am), 1951;33:836-72.
3. Küttner H. Einige Dauerresultete der Transplantation aus der Leiche und aus der Affe. Verhandlungen Deutsche Gesselschaft für Chirurgie, 1913;11:353-65.
4. Lexer, Erich. Joint transplantation and arthroplasty. Surg. Gyn. & Obst. , 1925;40:782.
5. Malinin TI. University of Miami Tissue Bank: Collection of postmortem tissues for clinical use and laboratory investigation. Transpl Proc, 1976; 8 (Suppl 1): 53-58.
6. Mankin HJ, Doppelt S, Tomford WW. Clinical experience with allograft transplantation: The first ten years. Clinical Orthopaedics and Related Research, 1983;174:69-86.
7. Mnaymneh W, Malinin TI, Makley JT, Dick HM. Massive osteoarticular allografts in the reconstruction of extremities following resection of tumors not requiring chemotherapy and radiation. Clin Orthopaedics & Related Research, 1985;187:76-87.
8. Ollier L. Traite Experimental et Clinique de la Regeneration des os et de la Production Artificielle du Tissu Osseux, Victor Masson et Fils, Paris, 1867.
9. Ottolenghi CE. Massive osteoarticular bone grafts. Transplantation of the whole femur. J Bone & Joint Surg (Br), 1966;48:646-59.
10. Parrish FF. Allograft replacement of all or part of the end of a long bone following excision of a tumor. J Bone & Joint Surg (Am), 1973;55:1-22.
11. Voloshin I, Berninin PM. Nikolay Ivanovich Pirogoff: Innovative scientist and clinician. Spine, 1998;23: 2143-46.
12. Zatzepin ST. Limb sparing operations in bone tumors. Medizina, Moscow, 1984.

Chapter 2

Marshall R Urist MD (Rt) (1914–2001), shown with one of the authors (T Malinin), Dr Urist was a professor of Orthopaedics, University of California, Los Angeles. He was a prolific scientist, but the contribution for which he is recognized the most is the discovery of bone morphogenetic protein (BMP).

(*Reference:* Urist MR. Bone formation by autoinduction. Science, 1965; 150:899.
Urist MR, Sato K, Brownell A, et al. Human bone morphogenetic protein (hBMP). Proc Soc Exp Bio Med. 1983;173:194-9.)

Form, Structure and Peculiarities of Bone

INTRODUCTION

Transplantation of bone requires an understanding of the complex nature of this tissue. Bone is the only tissue which can undergo complete, though limited, repair. In doing so, it calls into play a host of morphologic and physiologic mechanisms specific and peculiar unto itself. Much has been written about bone morphogenesis, structure and physiology. It is not our intent to review volumes of scientific studies dealing with the subject, but rather to provide the reader with a synopsis of information that is pertinent to the ensuing chapters. In doing so, it is our hope to save the readers formidable time and effort of having to consult many other references and sources.

The mechanical properties of bone are well documented. Bone is three times lighter than iron, but its strength is approximately that of cast iron. It is also many times more flexible than cast iron (Buckwalter et al. 1996). However, it is not an inert homogeneous material like metal from which most orthepaedic devices are made. Bone, both compact and spongy, is an active living tissue; it is not static. When placed into a cell culture environment, bone fragments produce cell outgrowth.

Consideration of bone repair and bone transplantation calls for familiarity with the structure, chemistry and physiology of the components of bone. Form begets function. In order to predict bone behavior, knowing the structure and orientation of inorganic crystals and organic macromolecules is important, as it has direct bearing on the physicochemical mechanism of calcification and associated phenomena (Glimcher 1987).

The intercellular matrix of bone is composed of inorganic and organic components within which are embedded specialized cells that reside in specialized spaces, called lacunae. The latter are supplied by elaborate channels and a unique and specialized canalicular system. Bone is constantly changing in response to mechanical stresses and hormonal influences. Many studies have been devoted to understanding how bone responds to signals to stimulate or suppress its formation, but our knowledge of these complex interactions is incomplete. Although not completely understood, bone repair and regeneration depend entirely on inherent physiologic properties of this unique tissue.

PERIOSTEUM

Bones vary in size and shape, ranging from the small ossicles to long weight-bearing bones of the appendicular and axial skeleton. These mineralized structures are invested in a fibrous sheath, the periosteum, richly supplied with blood vessels and nerves (Matthews 1981).

Periosteum is a membranous structure that lines the outer surface of all bones, except at the joints of long bones. Endosteum lines the inner surfaces of all bones. Periosteum varies in thickness with age, becoming progressively thinner with time. It consists of irregular dense connective tissue, and is divided into an outer fibrous layer and an inner, osteogenic or cambium layer. The fibrous layer consists of fibroblasts while the cambium layer consists of mesenchymal progenitor cells that develop into osteoblasts, but can be induced to develop into chondroblasts and chondrocytes. Cells of the osteogenic layer of periosteum grow readily in tissue culture (**Fig. 1**).

This attribute of periosteal cell biology is consistent with the formation of osteoblasts which are responsible for increasing and maintaining the width of long bones.

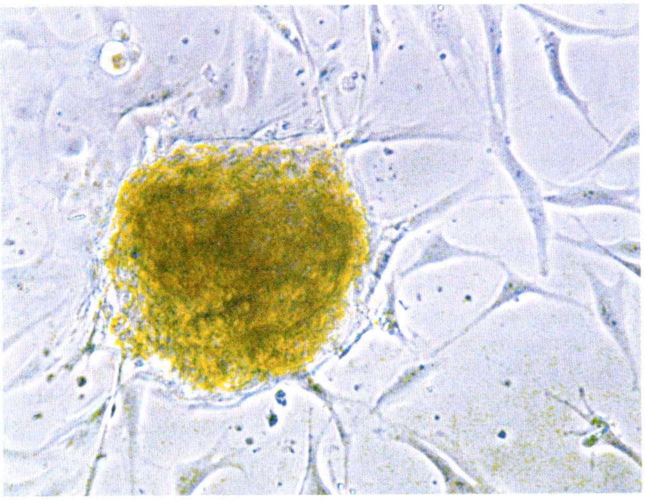

Figure 1 An explant of cortical bone periosteum cell culture. Cells originating from the explants migrate on the surface of the culture vessel. These cells have the attributes of osteoprogenitor cells (Phase contrast × 100)

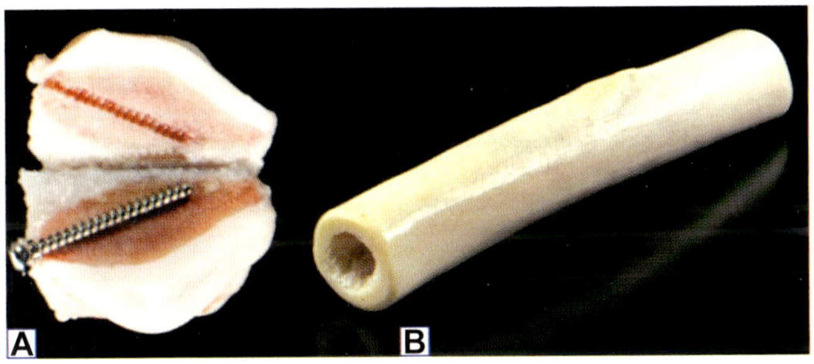

Figures 2A and B (A) Cut section of a patella with a screw in it. Patella contains compact cancellous bone which responds to the screw by osseointegration, thus demonstrating active remodeling; (B) Long bone shaft from a humerus with bone marrow removed

Experimental studies demonstrate that periosteal cells grown in monolayer culture contain cells which can be induced to follow either osteogenic or chondrogenic lineage (Gruber 2001). One of the principal functions of the periosteum is related to the cortical blood supply of bones.

Striated muscles are attached to the skeletal support system by collagen fibrils of the muscle tendons being interwoven into the bone through the fibrillar and ground substance elements of the periosteum. The latter are referred to as Sharpey's fibers (Taylor 1971). Force is applied through Sharpey's fibers to the bone and the bone responds by remodeling to accommodate these forces. Thus, bone is not a static, but a dynamic structure whose architecture is continuously modified by the activity of osseous cells. Wolf recognized the dynamic nature of bone and established the so-called Wolf's law (Wolf 1899), that simply stated: bone forms and remodels in response to the forces applied to it. The law is equally applicable to transplanted bone as it is to normal bone.

TUBULAR BONES AND FLAT BONES

It is a common practice to divide bones into tubular bones and flat bones, although it is readily apparent that there are many different shapes of bones. Chemically, bone is complex. It is made up of 30–35% organic substances and 65–70% inorganic mineral salts. The former are responsible for its toughness and the latter for its hardness. The percentages of both organic and inorganic substances vary not only between individuals but also between bones of the same individual. Flat bones are typically those of the pelvis, and tubular bones are those of the extremities (**Figs 2A and B**).

Anatomically, long bones are divided into the epiphysis, metaphysis, diaphysis and articulating surfaces. At times, this classification becomes somewhat confusing as there are no well-demarcated borders between the regions. The epiphysis is at the end of the bone that lies between the joint surface and the growth (epiphyseal) plate. However, in an adult after the growth has stopped, the division becomes somewhat obscured but is still delineated by the physical scar. The metaphysis is the widened end of the long bone shaft and diaphysis is the central part of the long bone.

MEMBRANOUS BONE AND ENDOCHONDRAL BONE

Based on embryologic development, two primary types of bone formation are recognized. These are membrane (membranous) bone formed by intramembranous ossification, woven (primary) bone and lamellar bone, both formed by endochondral ossification. In some bone transplants, as in some fractures, direct ossification may occur.

The bones of the face, cranium (except for the base) and clavicle are membranous bones. The bones of the appendicular and the axial skeleton as well as those at the base of the skull are termed endochondral bones.

Endochondral bone forms in sequence beginning with hyaline cartilage formation which ostensibly is a bone. The cartilage is replaced by bone as the organism matures. The exceptions are the articular surfaces and the growth plate, the latter becoming completely ossified only with skeletal maturity.

The traditional classification of bone formation does not take into account differences within different parts of bones. The first morphologic changes prior to cartilage replacement by bone, a process termed "ossification", are hypertrophy and proliferation of chondrocytes, followed by vascular ingrowth into normally avascular cartilage.

Mature bones contain bone marrow that can be either hematopoietic or fatty, and are surrounded by periosteum. Although these components differ in appearance and structure, they are interdependent. Marrow contains primitive mesenchymal cells that, with appropriate signaling, can transform into osteoprogenitor cells.

Visual inspection of the bone shows there are two types of bone tissue: compact (cortical) bone and spongy (trabecular) bone (**Figs 3A and B**) (Martin and Burr 1989). Spongy bone is confined primarily to the epiphysis and metaphysis of long bones, to the cuboid bones such as vertebrae, to the pelvic bones and to the mandible and maxilla (Sigh 1978).

Cortical bone is found on the outside of cancellous bones. It also makes up the diaphysis of long bones. Cortical bone can withstand the compressive forces as well as limited bending and torsional forces. Cortical bone forms approximately 80% of the mature skeleton (Kaplan et al 1994).

In the developmental sequence two types of bone are recognized: woven or primary bone and the lamellar bone. Woven bone is an immature or primitive bone. It contains more cells per unit volume than mature lamellar bone. Woven bone is coarse and fibered; these fibers have no uniform orientation. Woven bone is found in the newborn, in the metaphysis of growing bone and in the fracture callus.

Lamellar bone actively replaces woven bone in a growing organism. It is formed in place of the woven bone, as the latter is resorbed. Lamellar bone forms several structural units which make up cortical bone. These units are outer and inner circumferential lamellae, trabecular lamellae, interstitial lamellae and concentric lamellae found in osteons.

ORGANIZATION OF BONE

Grossly, woven and lamellar bones can be either trabecular, spongy, cancellous or dense compact (cortical) bones. Since bone metabolism is a surface dependent event, trabecular bone is metabolically more active than cortical bone. It has a greater surface area per unit of bone.

Cortical bone architecture is usually used to illustrate intricacies of bone structure, although both cancellous and cortical bones share the same matrix. Cortical bone accounts for approximately 80% of the skeleton in an adult (Buckwalter et al 1996). Cortical bone rims surround cancellous bone which houses either red or yellow (fatty) marrow. This is true for long, flat and cuboidal bones. The densest cortical bone is found in the diaphysis of long bones. For convenience, microscopic structure of cortical bone is depicted from this anatomical region. In a mature individual the cortical bone is composed of lamellae which are further subdivided into inner and outer circumferential lamellae, interstitial

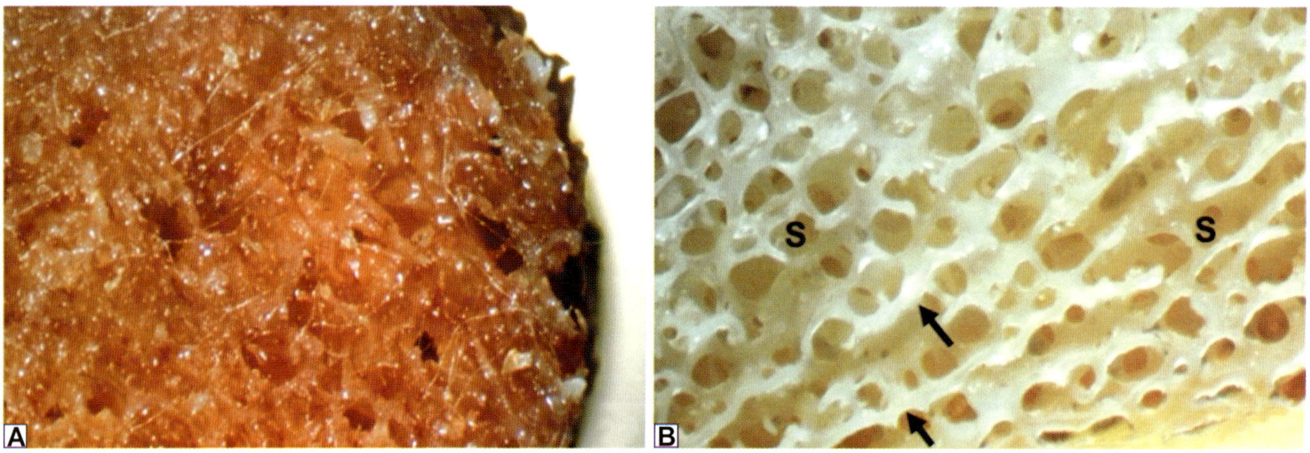

Figures 3A and B (A) Photograph of cancellous bone in its native state. The spaces between the trabeculae are filled with vascularized bone marrow. Cancellous bone bleeds readily. (B) Cancellous (trabecular) bone with bone marrow washed out. The three-dimensional nature of trabecular bone is clearly demonstrated in this preparation. The internal beams of spicules are the building blocks of trabecular bone. Large beam like structures are the trabeculae (arrows). The smaller interconnecting strut-like pieces are the spicules. The trabecular surfaces surrounding the cavities in the cancellous bone are covered with a thin layer of non-mineralized osteoid lamellae, as well as the lamellae of osteons. The organic portion of the lamellae consists of highly organized, dense collagen fibrils. Like wood grain in a tree, the lamellae are oriented in different directions. The collagen fibrils interconnect between lamellae, thus increasing their strength.

Osteonal structure, which makes up most of the diaphyseal cortex, deserves special attention. This basic structural unit of compact bone is formed around the Haversian canals surrounded by concentrically arranged lamellae. Osteons consist of branching structures longitudinally interconnected. They are oriented cylinders which spiral around the diaphysis (Cohen and Harris 1958). The Haversian canals (center of the osteons) contain blood vessels, lymphatic vessels and sometimes nerves. Canaliculi extend from the Haversian canals in a radial pattern, but not in the same plane **(Fig. 4)**.

Osteons are oriented longitudinally in long bones, having formed cylindrical structures which vary in cross-section from circular to oval shapes. With age, there is an increase in the area of compact bone occupied by osteons (Amprino 1948). The configuration of osteons is spiral and this arrangement is significant both mechanically and biologically. The importance of the continuity of osteons is that they provide an axial pathway for the diffusion of nutrient fluids through the compact bone. Each osteon is defined by a cement line (Cooper et al 1966), made up of organic matrix. Cement lines indicate where the resorption of the bone had stopped and new bone formation commenced. Although these structures were first recognized in histologic cross-sections appearing as lines, they are actually the structures enveloping the osteons.

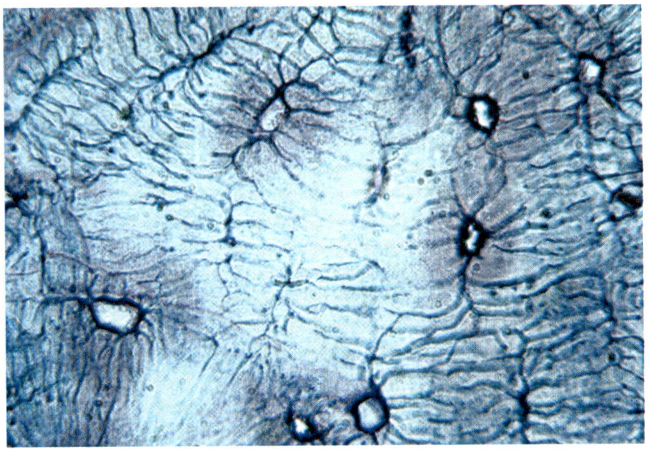

Figure 4 Haversian system with canaliculi (Romanowsky-Giemsa × 900)

The mature lamellar cortical bone contains peculiar and very intricate structures not observed in any other tissues. These are made up of a network of canals, canaliculi and lacunae. The latter house osteocytes, which are terminal cells in bone development. Osteocytes are bathed in nutrient fluid which fills the Haversian system. Osteocytes receive their nutrition through in the bone lacunar-canalicular system, as demonstrated by direct measurements by real-time solute movement in intact bone, accomplished by using fluorescence recovery after photo-bleaching and visualization of sodium fluorescein among individual osteocyte lacunae (Wang et al 2005).

Osteocytes are very sensitive to disruption of the fluid that bathes them. Even a brief interruption of the continuity and function of the canalicular system results in osteocyte death (Lecroix 1951).

Osteonal central canals branch and interconnect. They also join the Volkmann's canals, which are lateral branches of the Haversian canals through which blood vessels communicate between osteons (Coopers 1966).

Not only does the network of intraosseous canals run the entire length of long bones, but it also allows for communication between periosteum and endosteum. The remarkable feature of this system of channels and canaliculi is that it can be reproduced in the newly formed regenerated bone.

BONE CELLS

Osteocytes

Osteocytes are the terminal cells in the sequence of bone cell development. These are mature cells derived from osteoblasts. Osteocytes do not proliferate. They are confined to lacunae embedded in the calcified bone matrix. Osteocytes communicate with each other through cellular processes contained within the network of channels present in mature bone.

Lacunar spaces are incompletely filled by osteocytes. A thin layer of nonmineralized osteoid is found between osteocytes and mineralized lacunar walls **(Figs 5A and B)**.

Osteocytes form from entrapped osteoblasts. Therefore, they share many cytological features of the cells of their parent lineage. Ultrastructurally they contain large amounts of rough endoplasmic reticulum, a pronounced Golgi complex and abundant mitochondria. The nuclei are large and contain euchromatin. The cytoplasm is relatively scanty. The function of osteocytes is primarily that of maintenance, manifested by the production of the new matrix components that must be

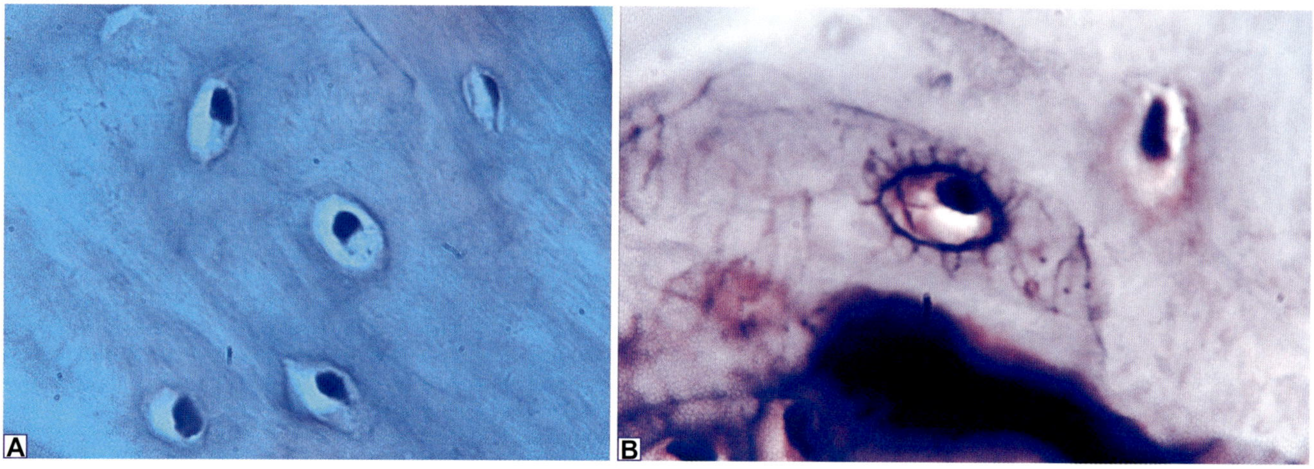

Figures 5A and B (A) Lacunae containing osteocytes. Note osteoid on the rims of the lacunae; (B) Osteocyte in a lacuna. Note strongly basophilic membrane and canaliculi originating from the same (Romanowsky-Giemsa × 1,400)

replaced. The osteocytes act through several mechanosensory mechanisms. They destroy bone by osteocytic osteolysis, but this function is a limited one. The notion that osteocytes function as mechanosensors is a logical one as the cells best situated to sense strain in bone would be those found throughout it (Skerry 2008). Since osteocytes are not able to make or remove, to a significant degree, the matrix surrounding them, their ability to alter bone architecture involves communication to the bone surface where bone resorption and formation can be initiated. Gap junctions between osteocytes and osteoblasts allow such a communication (Taylor et al 2007). Osteocytes are not active throughout the life of an individual. In older individuals, the cytoplasm is devoid of organelles as compared to osteocytes from younger individuals. The nuclei are condensed and occasionally pyknotic. These features represent various stages of cell death (apoptosis) associated with aging (Matthews 1981).

Osteoblasts

Osteoblasts lie directly on the osteoid of bone surface. These cells vary in size and shape depending on the metabolic activity. Osteoblasts are usually present in a single layer on top of nonmineralized osteoid which separates them from mineralized bone. Active osteoblasts are usually cuboid, pyramidal or rounded. The cytoplasm of these cells is strongly basophilic (**Fig. 6**).

The ultrastructure of osteoblasts has characteristic features. The cytoplasm consists primarily of rough endoplasmic reticulum made up of membranous tubes studded with ribosomes. Ribosomes are also present in a free state as is the Golgi apparatus and mitochondria. In active osteoblasts, cisternae of the rough endoplasmic reticulum are dilated with secreted protein. This accounts for the basophilia of the cytoplasm.

The process of differentiation of osteoblasts can be roughly divided in two parts (Ebisawa et al 1999). First is the conversion of undifferentiated mesenchymal cells into osteoblasts progenitor cells. The second is the maturation of the osteoblasts progenitors into mature osteoblasts (Boyant et al 1989). Experimentally osteoprogenitor cells differentiate into osteoblasts by stimulation with BMPs. However, osteoblasts do not respond to the dark reaction of 8-methoxypsoralen by increasing cell replication and RNA and protein synthesis, but stimulation with 8-methoxypsoralen results in a sharp increase in the lipid synthesis suggesting the plasma membrane

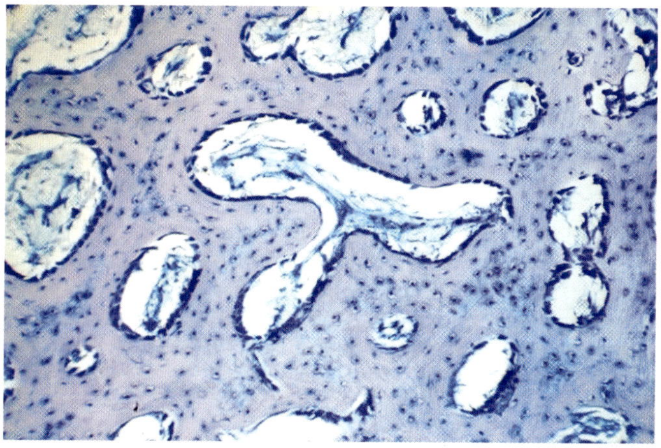

Figure 6 Newly formed bone. The cavities with vascular channels are lined with multiple osteoblasts which will be transformed into osteocytes (Romanowsky-Giemsa × 250)

of osteoblasts is the target of stimulation (Hornicek et al 1994). The response of osteoblasts and osteoprogenitor cells is multifaceted and involves, among other things, dietary factors and photosensitization (Malinin et al 1995). Interrelations with some of these factors, particularly psoralens, may cause a steatotic response in osteoblasts, but does not permanently impair their function. Thus osteoblasts are very resistant to chemical interactions, and maintain homeostasis under normal conditions (Malinin et al 1996).

The most prominent function of osteoblasts is the secretion of the collagen to form the organic matrix of bone. This requires intense metabolic activity manifested by the presence of alkaline phosphatase which is easily demonstrable histochemically. Type I collagen is also present in abundance in the cytoplasm of osteoblasts.

In bone mineralization, osteoblast activity is regulated by vitamin D and is associated with increased alkaline phosphatase production (Boyant et al 1989). Active osteoblasts may assume three functions. Some may remain on the surface and assume the flat form of a living cell bone. This coincides with reduced metabolic activity. Others may become surrounded by matrix and become transformed into osteocytes. Alternatively some osteoblasts simply disappear from the site of bone formation. This is accomplished by an unknown mechanism (Recker 1992).

Osteoclasts

Osteoclasts do not form new bone but only resorb it. These are multinucleated cells, usually with more than six nuclei. Their appearance is characteristic, although they may be sometimes confused with foreign body-type giant cells (**Figs 7A and B**). Osteoclasts are in direct contact with bone surface. The serrated border of osteoclasts provides an extensive area of specialized plasma membrane through which acid hydrolytic enzymes are released (Roodman 1999). Multinucleated osteoclasts are derived from fusion of hematopoietic cells of the macrophage-monocyte lineage.

Osteoclasts, when active in bone resorption, are found on the surface in indentations called Howship's lacunae. Osteoclasts attach to the bone surface by cell wall proteins called integrins. They isolate the area of bone under cell attachment resorbing it. This is accomplished by acidifying the area by the carbonic anhydrase system.

BONE REMODELING AND SKELETAL HOMEOSTASIS

The skeletal system performs several functions. One of these is the maintenance of homeostasis in the entire organism. For this reason, bone is in the process of continuous renewal and remodeling.

All bone surfaces are lined by dormant osteoblasts commonly referred to as bone lining cells. These cells are located directly against bone matrix and are elongated with cytoplasmic extension in contact with osteocytes (Matthews 1981). These osteoblasts are sometimes called resting osteoblasts or surface osteocytes. These cells have few organelles, and less cytoplasm than active osteoblasts. Flattened bone lining cells are most likely inactive osteoblasts that not only

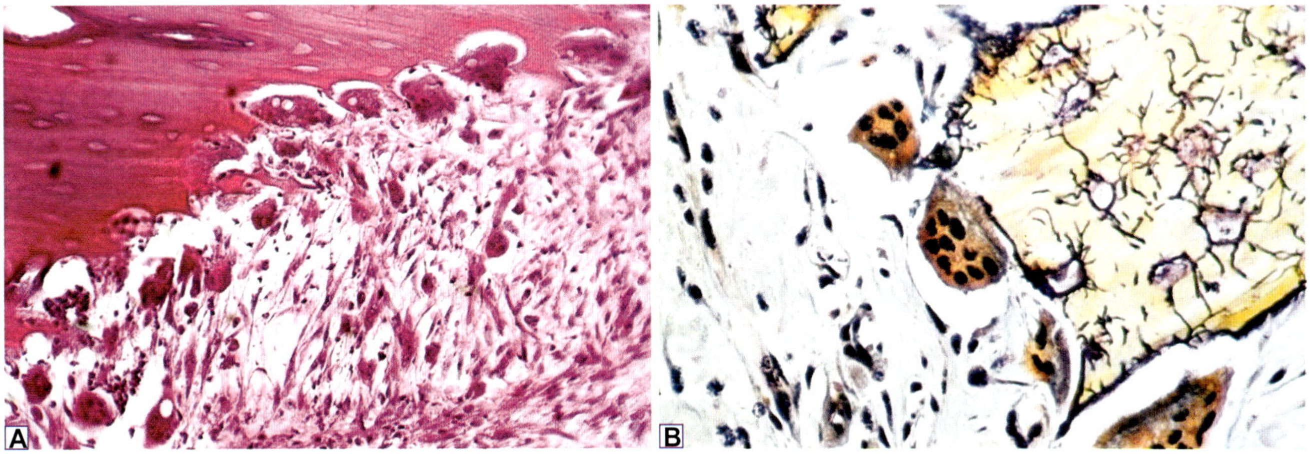

Figures 7A and B (A) Osteoclasts resorbing cortical bone and forming Howship's lacunae (H&E × 400); (B) Details of multinucleated osteoclasts and newly forming canalicular system in a bone undergoing remodeling (Silver impregnation with Bismarck Brown and Azure II counterstain × 1,400)

lie beneath the periosteum, but also form the endosteum on the trabecular and endosteal surfaces of spongy bone. Osteoblasts and lining cells are joined by adherens junctions containing cadherins. The latter are calcium dependent transmembrane proteins which aid tight junctions and desmosomes to link cytoskeletons (Shin et al 2000).

Osteoblast cells originate from the primitive mesenchymal stem cells. Commitment of mesenchymal cells to osteoblastic lineage occurs with the aid of the WnT/B-catenin pathway and associated proteins (Logan and Nusse 2004). However, all osteoblasts are not identical. They are heterogeneous with variable gene expression. This may explain the morphologic heterogeneity of the trabecular microarchitecture at different skeletal sites (Clarke 2008).

Multiple attributes of osteoblasts, combined with physiologic properties of the periosteum, are major contributors to bone remodeling, a process of bone renewal that maintains mineral homeostasis. The process involves concomitant removal of discrete portions of old bone and their replacement with new bone matrix which is subsequently mineralized. Remodeling begins at birth and continues until death.

The remodeling cycle is made up of several sequential phases. Activation becomes resorption, followed by new bone formation. Remodeling sites are targeted to the areas requiring repair as well as randomly selected sites **(Fig. 8)**.

Activation of remodeling involves conversion of osteoclast precursors, presumably from the circulation, altering of the living cell layer and fusion of mononuclear osteoclast precursors into multinucleated osteoclasts (Roodman 1999).

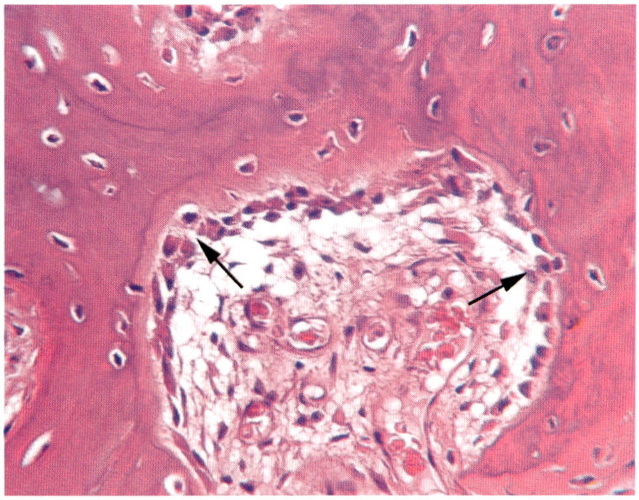

Figure 8 Perforation (targeted area) in a transplanted cortical allograft undergoing repair and remodelling. Arrows point to osteoblasts (H & E × 100)

Osteoclasts attach to the bone surface by activating integrin receptors in their cell membranes which bind with peptides of the matrix to form Howship's lacunae. Osteoclast-induced bone resorption lasts only 2–4 weeks during each remodeling cycle. Osteoclast activation and resorption is regulated by the ratio of the receptor activator of NF-KB ligand (RANKL) to osteoprotegerin (OPe) IL-1 and IL-6, colony stimulating factor (CSF), parathyroid hormone, 1, 25 dihydroxy vitamin D, and calcitonin (Boyle et al 2003, Blair and Athanasou 2004). Resorbing osteoclasts lower the pH between them and subjacent bone to as low as 4.5. This mobilizes and solubilizes bone mineral. Active, resorbing osteoclasts secrete tartrate resistant acid phosphatase, a hallmark of their activity. They also produce cathepsin K, matrix metalloproteinase 9, and gelatinase released from cytoplasmic lysosomes (Delaisse et al 2003).

During the reversal phase, bone resorption ceases and bone formation begins. At the end of bone resorption, cavities contain multiple types of mononuclear cells. Once bone resorption stops, these cells disappear. The mechanism by which this is accomplished is unknown. The process of bone resorption is quite rapid compared to subsequent bone formation. Around 4–6 months are required to complete new bone formation. Osteoblasts synthesize new collagenous matrix, and induce mineralization of new matrix by concentrating calcium and phosphate and enzymatically destroying mineralization inhibitors such as proteoglycans and pyrophosphates (Anderson 2003).

In conclusion, bone remodeling preserves skeletal function in its many forms throughout life. Under normal conditions, bone remodeling units assure concomitant bone resorption and bone formation that results in optimal form and function. To understand and to predict behavior of bone regeneration it is important to keep in mind the scheme of bone remodeling.

BONE MINERALIZATION

Bone is formed in two stages. Matrix is formed first. Later it begins to mineralize. The process of adding mineral to matrix is termed "bone mineralization". Mineralization of bone is essential to maintain its hardness and strength. It is a well regulated process in which crystals of calcium phosphate are produced by bone forming cells and are laid down in precise amounts within the bones' fibrous matrix. The regulation of this process is mitigated largely by inorganic pyrophosphate (Millan 2006).

Normal bone is composed of 50–70% mineral, 20–40% organic matrix, 5–10% water and less than 3% lipids. The mineral in bone is primarily hydroxyapatite [$Ca_{10}(PO_4)_6(OH)_2$]. However, studies of bone mineral crystals have shown that the hydroxyapatite of bone is not a pure chemical compound (Buckwalter et al 1996) but instead, contains both carbonate ions and acid phosphate groups. The acid phosphate groups in bone crystals are unique and are not the same as those found in other calcium crystals such as precursors of apatite and apatite of sedimentary rocks. Unlike pure hydroxyapatite, bone crystals do not contain hydroxy groups. Therefore, bone mineral crystal should be classified as apatite and not hydroxyapatite.

The formation of solid calcium phosphate from soluble calcium and phosphate in bone matrix is a phase transformation, a process analogous to the formation of ice from water (Aronson et al 1990). As the process of mineralization proceeds, the crystallinity of apatite increases but never approaches the highly crystalline structure of geological hydroxyapatite. Mineralization of bone matrix occurs in a progressive manner. Mineral first appears in specific zones of the collagen fibrils, separated by unmineralized regions. Mineralization usually proceeds rapidly once it begins.

Bone serves not only as a support structure, but also a calcium reservoir that is sequestered as apatite. In an adult, only a few grams of the total (1,200) circulate in extracellular and intracellular fluids. In these fluids about one-half of the total calcium is ionized. On the other hand, only 0.1% or less of the calcium is ionized within cells (de Bernard 1992). The concentration of free calcium in plasma reflects its movement in and out of bone deposition. The concentration of free calcium in plasma is about 1000 times greater than that of intracellular calcium which is about 2.4 to 4 mM.

To understand calcium metabolism and its relation to bone, it is useful to visualize an adult man with an average calcium intake and in zero balance (Nordin 1976). Since the subject is in zero balance and the rate of bone formation is equal to that of bone resorption, the net amount of calcium absorbed is equivalent to that lost in the urine. This example shows that calcium homeostasis reflects the rate of intestinal calcium absorption, the rate of urinary calcium excretion and that of calcium exchange between blood and bone. As the only organ which acts as either a source or sink of calcium, bone is essential in the regulation of calcium in the extracellular fluid. As long as normal homeostasis prevails, bone will continue to be remodeled constantly in a predictable manner. For clarification it should be noted that some older studies postulate that the matrix affects the formation of hydroxyapatite by transformation of amorphous calcium phosphate. This view is incorrect. It has been conclusively demonstrated that calcium phosphate is not present in bone (Grynpas et al 1984).

OSSIFICATION AND BONE DEVELOPMENT

On the basis of its development, bone is classified either as intramembranous or endochondral. The distinction is based on the manner in which bone is formed. Primary membranous bone formation involves the cranial vault, facial bones, and, in part, the mandible and the clavicle. The appendicular and the axial bones become involved in membranous bone formation only secondarily (Ogden 1980). The membrane derived bones are formed from proliferative ectomesenchymal tissue. At the site of primary ossification, cells form short and irregular strands which run in random directions. These cells form fibrous intercellular matrix which quickly ossifies and calcifies, forming new bone directly.

Endochondral bone formation is more complex and is dependent on an inexorably close relationship between cartilage and bone. Basic cell types produce each of the skeletal tissues. These undifferentiated cells bear several names including osteoprogenitor cells and scleroblasts. These cells of mesodermal origin give rise to tissues with close affinities for each other.

The endochondral ossification leading to the formation of long bones in the embryo begins with accumulations of undifferentiated cells which secrete cartilaginous matrix and eventually differentiate into chondrocytes (Buckwalter et al 1996). After the appearance of embryonic hyaline cartilage, a periosteal covering forms around the diaphysis. This in turn forms a thin layer of bone. Two distinctly different forms of endochondral ossification and calcification occur during bone development and growth (Hunter 1992). These are epiphyseal cartilage ossification of the growth plate of long bones, and osteoid calcification in periosteal bone formation, also referred to as the "calcification front", "linge frontiera" (Lacroix 1955) and "zone of demarcation" (Frost 1972). The organic matrices of these tissues are different. Osteoid consists primarily of Type I collagen with lesser amounts of non-collagenous proteins. Epiphyseal cartilage matrix consists primarily of proteoglycans and Type II collagen (**Fig. 9**). Calcification patterns are also different between these tissues. In osteoid, mineral deposits are made along

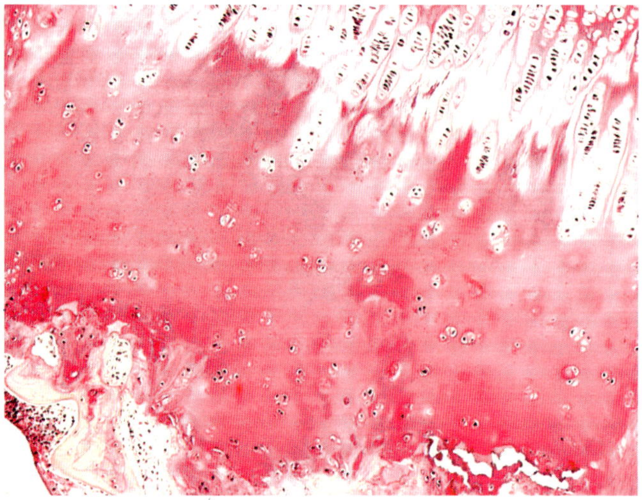

Figure 9 Epiphyseal plate showing abundant proteoglycans (PAS-hematoxylin × 250)

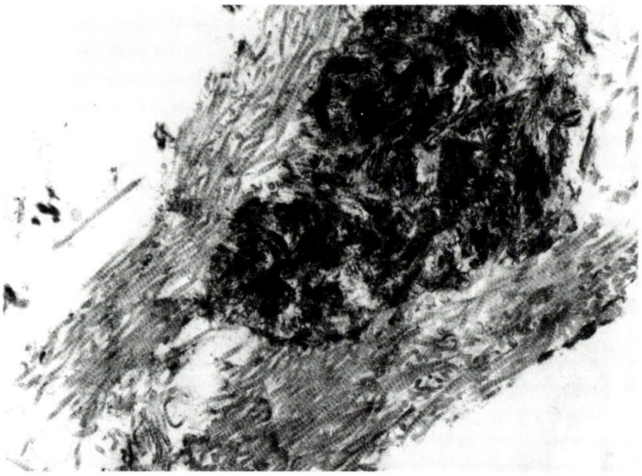

Figure 10 Electron photomicrograph showing direct ossification of collagen fibers (× 20,000)

collagen fibers (**Fig. 10**). In epiphyseal, cartilage calcification appears to be at random throughout the process of ossification across the growth plate and has a predictable temporal and special sequence. It is also associated with proteoglycans which bind calcium.

The physis is organized in several zones: resting, proliferation, hypertrophy and provisional calcification. As the process continues throughout growth and development, the bone lengthens. The cells in the hypertrophic zone undergo scheduled apoptosis and, in the process of ossification, some regions of the cartilage mineralize, chondrocytes enlarge and vascular outgrowths invade the cartilage. Invading cells resorb the central portion of the cartilage and transform it to the bone marrow cavity. This occurs in the primary zone of ossification. Osteoprogenitor cells, which are in intimate contact with the vascular sprouts, differentiate into osteoblasts to form bone matrix on the mineralized cartilage. This process which is the reverse of bone remodeling then takes place. Osteoclasts follow osteoblasts and resorb calcified cartilage and immature bone. The osteoblasts then again become active and lay down lamellar bone which replaces the mixture of calcified cartilage and immature bone. This is referred to as the secondary zone of ossification.

After the formation of the long bones of the extremities and epiphyseal centers, endochondral ossification continues until skeletal maturity is reached. It then re-emerges during healing of fractures (Muthukumaran and Reddi 1985).

APPOSITIONAL BONE FORMATION

This occurs during periosteal enlargement of bone and during bone modeling and remodeling. Unlike membranous or endochondral bone formation, appositional bone formation is initiated by osteoblasts which have aligned on the existing bone surface (Buckwalter et al 1996, Arey 1965).

The activated osteoblasts synthesize osteoid that forms bone lamellae (**Fig. 11**). Periosteal osteoblasts secrete osteoid on the outer layer of the bone and by doing so enlarge the bone layer by layer. The appositional bone formation receives relatively scant attention, but the process is a continuous and important one. Before puberty, as the length of the long bones increase, the formation of bone beneath the periosteal covering widens the shaft of the bone. The local removal and

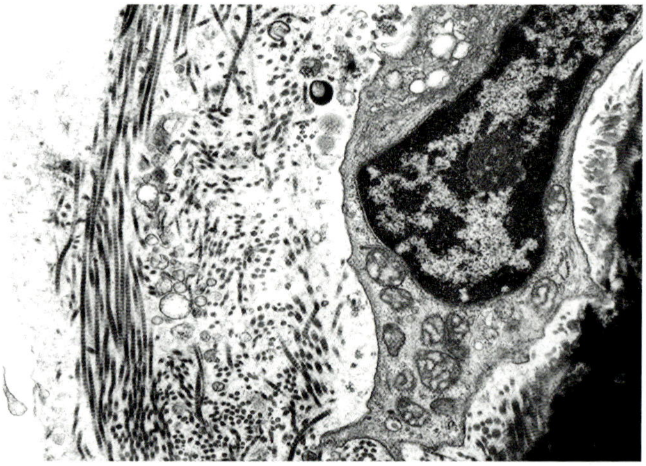

Figure 11 Activated osteoblast showing active organelles, surrounded by type II collagen fibers (Electron photomicrograph × 30,000)

replacement of a small volume of bone in the endosteum contributes to the formation of the medullary canal. As periosteal opposition exceeds endosteal resorption, the bone develops a progressively thicker cortex (Seeman 2003). The appositional bone growth ratio can be reliably measured with fluorescent markers (Lee 1964). The process is important since it has direct bearing on bone graft incorporation.

BONE MORPHOGENETIC PROTEINS AND OTHER GROWTH FACTORS

Bone morphogenetic proteins known also as bone morphogenic proteins (BMPs) are group of growth factors which can induce heterotopic bone and cartilage formation in rodents (Reddi and Reddi 2009).

These compounds also known as cytokines and metabologens are now considered constitute a group of morphogenetic signals influencing skeletal tissue architecture and bone formation throughout the body (Bleuming et al 2007). Originally isolated form bovine bone by Urist (Urist and Strates 1971), these compounds were found to be present in microscopic quantities in bone; one kg of bone yieldsing around 100 ug of BMPs.

Human BMPs were not produced until 1983 (Urist et al 1983).

Originally seven BMPs were discovered; of these BMP 2 through BMP 7 belong to the transforming growth factor Beta superfamily of proteins. Recently additional BMPs were discovered, bringing the total number to 20.

BMPs are metalloproteinases and can be produced by recombinant DNA technology, BMPs interact with specific receptors on the cell surface referred to as bone morphogenetic protein receptors (BHPRs). Signal transduction through BMPRs results in mobilization of SMAD family of proteins and results in stimulation of mesenchymal cells to proteins that transduce extracellular signals form transforming growth factor beta ligands to the nucleus where they activate downstream TGT-3 gene transcription.

SUMMARY

Bone is a highly organized and specialized tissue which is subject to continuous changes. Although extensively studied, many aspects of bone physiology are still poorly understood. Therefore, current concepts are subject to continuous revisions.

The review of the structure and function of bone provided in this chapter was intended to aid the reader in dealing with the subsequent chapters. The review was not comprehensive, but meant to focus only on highlights of osseous structure and function.

Recent advances in the understanding of the bone matrix and the regulation of cellular activity improve our understanding of this complex process but often evoke more questions. Some of the factors that control formation and resorption of bone may lead to changing approaches in bone transplantation and tissue engineering.

Many students of bone came to recognize the differences in various types of bone. However, most investigators have selected compact diaphyseal bone as the material which typically represents bone structure. We have followed the same route, realizing diaphyseal cortical bone represents the least heterogeneous sample of bone tissue. Knowing this, allowances for oversimplification have had to be made, but not to the extent that the information is inaccurate. When the study of bone was in its infancy, much emphasis was placed on calcium and phosphorus metabolism. Now cell signaling and differentiation are the focus of research with ossified tissues. Now, in the study of skeletal tissue, laboratory is closer to the bedside than it was in the past. To make full use of laboratory findings a working summary of skeletal physiology and biology is a prerequisite. To this end we hope we have provided a useful summary of the subject.

It was our aim to devote the major portion of this chapter to a description of cortical and cancellous bones as they appear in an adult human, as either tubular or flat bones. It would be, of course, natural to recall how the bones are formed before considering their final organization.

BIBLIOGRAPHY

1. Amprino R. A contribution to the functional meaning of the substitution of primary by secondary bone tissue. Acta Anatomica, 1948;5(3):291-300.
2. Anderson HC.Matrix vesicles and calcification. Current Rheumatology Reports, 2003;5(3):222-6.
3. Arey LB. Developmental Anatomy. Philadelphia: WB Saunders; 1965. pp. 396-425.
4. Aronson T, Good B, Stewart G. Preliminary studies of mineralization during distraction osteogenesis. Clinical Orthopaedics and Related Research, 1990;43-9.
5. Blair HC, Athanasou NA. Recent advances in osteoclasts biology and pathological bone resorption. Histology and Histopathology, 2004;19(1):189-99.
6. Bleuming SA, He XC, Kodach LL, et al. Bone morphogenetic protein signaling suppress tumorigenesis at gastric epithelial transition zones in mice. Cancer Research 2007; 67(17):8149-55.

7. Boyant BD, Schwartz Z, Bonewald LF, et al. Localization of 1, 25-(OH)2D3-responsive alkaline phosphatase in osteoblasts-like cells (Ros17/2.8, MG-63, and MC3T3) and growth cartilage cells in culture. The Journal of Biological Chemistry, 1989;264(20):11879-86.
8. Boyle WJ, Simonet WS, Lacey DL. Osteoclast differentiation and activation. Nature, 2003;423(6937):337-42.
9. Buckwalter JA, Glimcher MJ, Cooper RR, et al. Bone Biology I: structure, blood supply, cells, matrix and mineralization. American Academy of Orthopedic Surgeons. Instructional Course Lectures, 1996;45:371-99.
10. Clarke B. Normal bone anatomy and physiology. Clinical Journal of the American Society of Nephrology, 2008;3(Suppl 3):S131-9.
11. Cohen J, Harris WH. Three dimensional anatomy of Haversian System. The Journal of Bone and Joint Surgery (Am), 1958;40(2):419-34.
12. Cooper RR, Milgram JW, Robinson RA. Morphology of the osteons: an electron microscopic study. Journal of Bone and Joint Surgery (Am), 1966;48(7):1239-71.
13. DeBernard B. Calcium metabolism and bone mineralization. In: BK Hall (Ed). Bone Mineralization and Metabolism. Boca Raton: CRC Press; 1992. pp. 73-98.
14. Delaisse JM, Andersen TL, Engsig MT, et al. Matrix metalloproteinase (MMP) and cathepsin K contribute differently to osteoclasts activities. Microscopy Research and Technique, 2003;61(6):504-13.
15. Ebisawa T, Tada K, Kitajima I, et al. Characterization of bone morphogenetic progein-6 signaling pathways in osteoblasts differentiation. Journal of Cell Science, 1999;112(Pt 20): 3519-27.
16. Frost HM. The physiology of cartilaginous, fibrous and bony tissue. Charles C Thomas: Springfield; 1972.pp. 59-60.
17. Glimcher MJ. Chemistry, structure and organization of bone and their influence on bone healing. In: JM Lane (Ed). Fracture Healing. New York: Churchill Livingston; 1987. pp. 39-48.
18. Gruber R, Mayer C, Bobaczk K et al. Effect of cartilage derived morphogenic protein and osteogenic protein-1 on osteochonddrogenic differentiation of periosteum-derived cells. Endocrinology, 2001;142:2087-94.
19. Grynpas MD, Bonar LC, Glimcher MJ. X-ray diffraction radial distribution function studies on bone mineral and synthetic calcium phosphate, J Mat Sci. 1984;19:723-36.
20. Hornicek FJ, Malinin GI, Banovac K, et al. Morphologic and metabolic changes in rat osteoblasts cultures during the dark reaction with 8-methoxypsoralen. Chemico-Biological Interactions, 1994;90(2):185-93.
21. Hunter GK. In vitro studies on matrix mediated mineralization. In: Bone Metabolism and Mineralization. CRC Press, Inc. Boca Raton, FL, 1992.
22. Kaplan FS, Hayes WC, Keaveny TM, et al. Form and function of bone. Orthopaedic Basic Science. AAOS, 1994;127-84.
23. Lacroix P. The organization of bones. London: J&A Churchill; 1951.
24. Lee WR. Appositional bone formation in canine bone: a quantitative microscopic study using tetracycline markers. Anatomy, 1964;98:665-7.
25. Logan CY, Nusse R. The Wnt signaling pathway in development and disease. Annual Review of Cell and Developmental Biology, 2004;20:781-810.
26. Malinin GI, Hornicek FJ, Banovac K, et al. Interaction of cultured rat osteoblasts with 8-methoxypsoral during irradiation with long-wave ultraviolet light. Biochemical and Biophysical Research Communications, 1995;207(2):877-81.
27. Malinin GI, Vornberger WJ, Hornicek FJ, et al. Effects of psoralen on the structural integrity of osteoblasts. Archives of Toxicology, 1996;70(3-4):182-8.
28. Martin RB, Burr DB (Eds). Structure, function and adaptation of compact bone. NY: Raven Press; 1989.
29. Mathews JL. Bone structure and ultrastructure. In: Urist MR (Ed). Fundamental and Clinical Physiology. Philadelphia, Toronto: JB Lippincott Co; 1981. pp. 4-44.
30. Millan JL. Spotlight on Research NIAMS, NIH. 2006.
31. Muthukumaran N, Reddi AH. Bone matrix induced local bone induction. Clinical Orthopaedics and Related Research, 1985;200:159-64.
32. Nordin BEC. Calcium, phosphate and magnesium metabolism. Edinburgh: Churchill Livingstone; 1976. p. 41.
33. Ogden JA. Chondro-osseous development and growth. In: Urist M R (Ed). Fundamental and Clinical Bone Physiology. Philadelphia and Toronto: JB Lippincott Co; 1980. pp. 108-71.
34. Recker RR. Embryology, anatomy and microstructure of bone. In: Coe FL, Favas MJ (Ed). Disorders of Bone and Mineral Metabolism. New York: Raven Press; 1992. pp. 219-40.
35. Reddi AH, Reddi A. Bone morphogenetic proteins (BMPs): from morphogens to metabologens. Cytokine and growth factor reviews. 2009;20(5-6):341-2.
36. Roodman GD. Cell biology of the osteoclasts. Experimental Hematology, 1999;27(8):1229-41.
37. Seeman E. Periosteal bone formation-a neglected determinant of bone strength. New England Journal of Medicine, 2003;349(4):320-3.
38. Shin CS, Lecanda F, Sheikh S, et al. Relative abundance of different cadherins defines differentiation of mesenchymal precursors into osteogenic, myogenic, or adipogenic pathways. Journal of Cellular Biochemistry, 2000;78(4):566-77.
39. Sigh I. The architecture of cancellous bone. Journal of Anatomy, 1978;127(Pt. 2):305-10.
40. Skerry TM. The response of bone to mechanical loading and disuse: fundamental principles and influences on osteoblasts/osteocytes homeostasis. Archives of Biochemistry and Biophysics, 2008;473(20):117-23.
41. Taylor AF, Saunders MM, Shingle DL, et al. Cell Physiology. American Journal of Physiology, 2007;292(1):C545-52.
42. Taylor DW. The life and teaching of William Sharpey (1802-1880) 'Father of Modern Physiology in Britain'. Medical History, 1971;15(3):241-59.
43. Urist BS, Strates BS. Bone Morphogenetic Protein Journal of Dental Research 1971;50(6):1392-1406.
44. Wang L, Wang Y, Han Y, et al. In situ measurement of solute transport in the bone lacunar-canalicular system. Proceedings of the National Academy of Sciences of the United States of America, 2005;102(33):11911-6.
45. Wolf J. Das Gesetz der Knochen. August Hirschwald, Berlin, 1892.
46. Wolf J. Die Lehre von den functionellen Knochengestalt. Virchows Archiv, 1899;155:226.

Chapter 3

Professor Dr. Med. Maurice E Müller (1918–2008), Professor of Orthopaedics, University of Bern, Bern, Switzerland. Dr Müller together with Dr Martin Allgower developed dynamic compression plating for internal fixation of fractures. The technique advanced allograft transplantation.

(*Reference:* Müller ME, Allogower M, Schneider R, Willnegger N. *Manual of Internal Fixation Techniques Recommended by the A-O group,* 2nd ed. New York, Springer-Verlag, 1979.)

Fracture Healing: Relation to Bone Grafting

INTRODUCTION

Fracture healing is a specialized attribute of bone, the only structure in the body capable of self-repair, leading to a complete restoration of the skeletal integrity. Bone loss that exceeds a critical (length segmental bone loss) does not spontaneously heal and frequently requires a bone graft and internal fixation.

In most patients, fracture healing proceeds smoothly, without complications. However, in 5–10% of an estimated six million plus fractures sustained annually in the United States, fracture healing is either delayed or impaired (Einhorn 1996). This is caused by a number of factors that include injury severity, host factors and inadequate or inappropriate treatment. In this chapter, we will not deal with fracture healing failures as there are numerous papers and textbooks devoted to the subject. Our purpose is to review fracture healing as it proceeds under normal circumstances because the mechanisms brought into play during the healing of fractures are the same as those involved in bone graft healing.

A break in the bone or osteotomy through bone will initiate a process known generally as fracture healing, fracture repair or bone repair. Suggestions to refer to this process as "regeneration", are neither accurate nor appropriate. Regeneration means replacement of a missing part (Malinin and Deck 1958). Bone in mammals is not capable of replacing a missing part of significant length, as the repair process is limited to defects where the fragments of fractured bone are in close proximity to each other. If the fracture fragments are not aligned in perfect and rigid anatomic juxtaposition, the repair is accomplished by forming an external callus. If, in the long bone, the length of the defect exceeds the diameter of the affected bone, the gap will not be replaced with new bone, but will be filled with fibrous connective tissue. For these reasons the terms "fracture healing", "fracture repair" or "bone healing", rather than bone regeneration will be used in this chapter.

The patterns of healing of bone fractures have been studied by scores of "bone healers", surgeons and laboratory scientists since early antiquity. A number of fracture treatment modalities have been advocated, discarded and adopted again. Most of these were developed as the result of empirical observations and several have endured the test of time. Among these can be listed: fracture reductions, splinting, plaster casting, traction, bracing, tension band wiring, pinning, intramedullary nail fixation and compression plating (**Fig. 1**). The efficacy of many other modalities, to date, remains unproven (Novikoff et al 2008).

A huge amount of information dealing with fracture repair is reported in the literature and thus a comprehensive review of these studies and observations would exceed the scope of this chapter. However, understanding the events leading up to complete fracture healing is pertinent to bone transplant healing and incorporation. This process is a complex one which contains a coordinated transcriptional program initiated by a mechanical stimulus that leads to signaling angiogenesis, granulation tissue formation, recruitment of primitive mesenchymal cells and finally osteogenesis. The processes are not independent, but temporally, biologically and spatially interdependent. Since different phases of the process are difficult to isolate as independent phenomena, contribution of each phase is poorly understood.

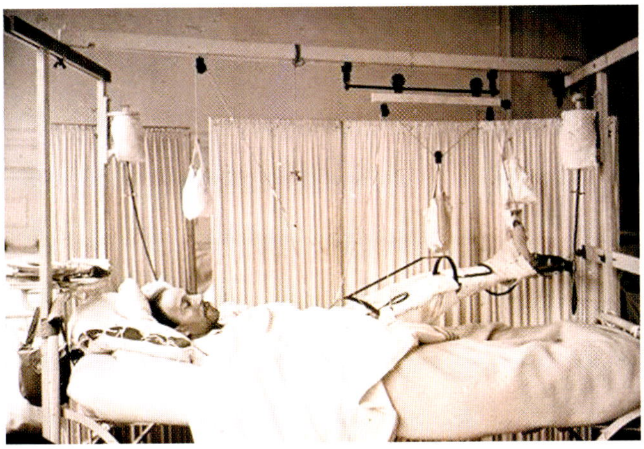

Figure 1 Plaster casting and traction were fracture treatment modalities generally accepted as the best options for healing broken bones. The notion was accepted as a result of the study of the fracture callus. The photograph shows a combination of plaster casting and traction applied to a soldier at the War Demonstration Hospital, Rockefeller Institute for Medical Research, New York during World War I. Little had changed in the management of fractures until the introduction of compression plating and intramedullary nail fixation, popularized by Kunchner during World War II

The comments in this chapter are based on general experience with fracture healing. They encompass both clinical observations and laboratory studies. It is hoped the material will be helpful to those dealing with bone grafting.

By convention, the process of bone fracture healing has been divided into several phases: reactive, reparative and remodeling. Most of the work, both clinical and laboratory, is based on the recognition of these phases as separate biologic entities. However, as stated earlier, the distinction between phases cannot be delineated accurately because these processes are interrelated and inseparable. To complicate the matter further, the process of bone healing is subject to modifications by factors such as the angle of displacement, physical attributes of the fracture itself and the general condition of the individual sustaining the fracture.

Realization that the early phase of bone healing can be modified by the mechanical state of the fracture site has come as a result of the work of the Swiss AO* group. The AO method is aimed at achieving anatomical reduction by compressing the fracture site and stabilizing the fracture so as to obviate the need for external splinting (Allgower and Spiegel 1979). However, the AO method does not constitute the first attempt at internal fixation of fractures with plates. Fixation of fractures with metal plates and screws, albeit without compression, has been previously described (Scudder 1938). In fractures reduced by compression plates, the amount of periosteal callus is negligible as compared to fractures without rigid fixation (**Figs 2A to D**).

Consideration of compression plating is important in allograft transplantation. Without rigid fixation, allograft reconstruction of the extremities, pelvic and shoulder girdles would be very difficult. The biology of fracture healing fixed with compression plates is altered as compared with the biology of fractures maintained in position by splinting, plaster casts, non-compression plates, and intramedullary nails. Therefore, healing of fractures reduced by compression plates deserves special consideration. The general biology of callus formation and bone fracture healing is based on the events in non-compressed fractures subjected to micromotion. This so-called "secondary" fracture healing takes place by creating a large mass of callus. The description of fracture healing that follow refers to "secondary" fracture healing (**Fig. 3**).

"Primary" fracture healing requires very high stability and in practice occurs only rarely, although much has been made of the concept (Phillips 2005). If the fracture is completely reduced, "osteonal" healing takes place. Osteoclasts create "cutting cones" and cross the fracture line, thus initiating direct bone repair. Compression plating exploits this phenomenon.

AN OVERVIEW OF EXISTING CONCEPTS

Laboratory Animal Studies

It is, of course, neither desirable nor advisable to experiment on humans. Therefore, to fill voids in our knowledge, data has to be obtained from experimental animals. Although some animal data might not be directly applicable to humans, the hierarchy of testing credibility is in place (Boden 2003). Data from cell culture studies and the mouse and rat ectopic bone induction assays are most likely the least stringent assays, as this data may be difficult to apply to humans since ectopic bone formation in humans is pathologic, not physiologic. The rabbit long bone defects as well as long bone defects in other

* Arbeitsgemeinschaft für Osteosynthesefragen.

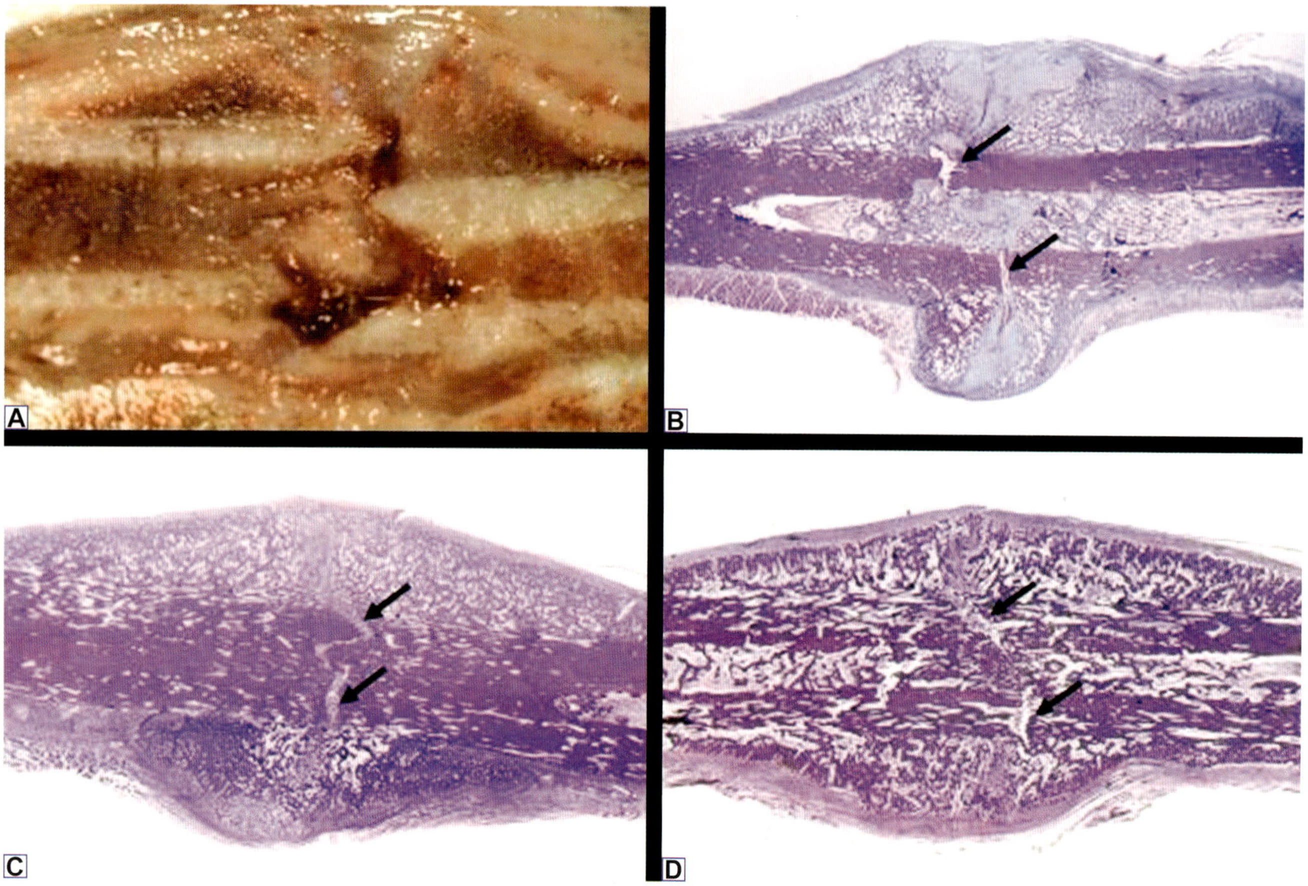

Figures 2A to D Evolution of the callus formation. (A) Acute fracture with hematoma; (B) Stage 2 of callus formation, soft callus is made up primarily of cartilage; (C) Stage 3, hard callus. The tissue has converted from chondral to osseous. The new bone is of woven type; (D) Stage 4, during the last stage of fracture repair, remodeling takes place. Woven bone is changed into lamellar bone. Medullary circulation is re-established. The stage of remodeling may take several years in a human (Hematoxylin and Eosin × 10)

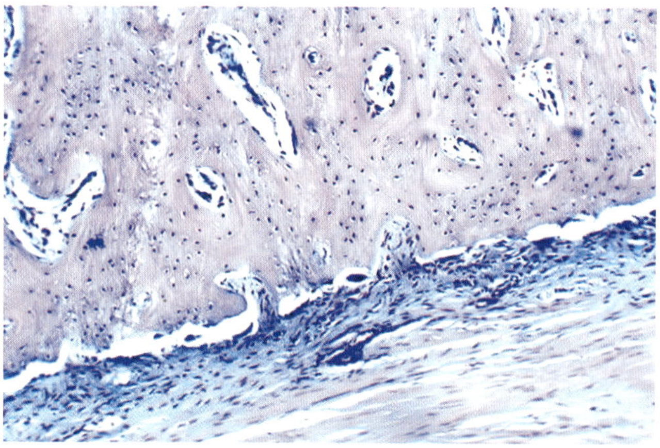

Figure 3 The relationship between the muscle and connective tissue surrounding the fracture site is of great importance, as it provides the cells and the new vasculature which participate in callus formation (Romanowski-Giemsa × 100)

larger animals are of greater stringency. Among these animals, dogs are the most commonly used laboratory models. Non-human primate long bone experiments probably correlate with similar human applications. However, non-human primate models are expensive, require sophisticated animal facilities and are not readily accessible (Malinin et al 2007).

Periosteum and Hematoma

Periosteum is an important contributor to fracture repair in uncompressed fractures. Proliferation of periosteal cells, primarily from the cambium layer, form a callus, and their invasion of the hematoma initially surrounding the fracture is an early event in fracture repair (Ozaki et al 2000, Phillips 2005). However, rigidly fixed and compressed bone may heal through endosteal proliferation without forming an external callus. Thus, participation of the periosteum in the fracture healing process varies with the degree of displacement of fractured bone ends. If these are held rigidly together, external callus is minimal, if not completely non-existent. Under these circumstances, periosteal proliferation is not marked.

The notion of the importance of the periosteum in fracture healing dates to the first part of the last century, when both Lexer (McKinley 2009) and Kolodny (1923) emphasized the importance of the periosteal blood supply. If periosteum was the foremost osteogenic factor in bone repair, its blood supply must also be of importance. Kolodny thought periosteal callus plays a far greater role in the union of fractures than does the endosteal callus. The idea was embraced, until the introduction of compression plating became generally accepted dogma in orthopedic surgery.

When bone is broken, hemorrhage follows. The amount of hemorrhage is proportional to the severity of the injury and the extent of blood vessel disruption. Bleeding stops with time and the extravasated blood forms a blood clot which within several days is invaded by capillaries and proliferating fibroblasts (Potts 1933, Rhinelander 1965, 1968, 1974). The size of the blood clot varies. An ideal size of the blood clot in relationship to the fracture is unknown. Were it known, it would have practical relevance. If a large blood clot might be inhibitory to fracture healing, it would be best to evacuate it. If, on the other hand, a small blood clot is inadequate to promote fracture healing, perhaps an addition of blood to the fracture site might be a good idea (Sarmiento and Latta 2006).

The relationship between the periosteum and the fracture hematoma has been the subject of a number of studies, (McKinley 2009, Aro et al 1990) as were the studies of callus formation in general. If one compares the healing patterns in the experimental animals in which the periosteum or the bone marrow had been removed with that of untreated animals, some striking differences become apparent. In the bone marrow deprived animals, oddly enough, the periosteal proliferation is decreased while in the animals with removed periosteum, chondrogenesis is inhibited. This suggests periosteum does indeed contribute to the early chondrogenesis and endochondral ossification in uncomplicated fracture healing. The process is not a simple one. Several growth promoting factors, (Schmidmayer et al 2002) bone morphogenetic proteins, (Szpalski and Gunzburg 2005) acidic and basic fibroblast promoting factors as well as platelet derived growth factors, have been identified at the fracture sites (Nash et al 1994). Endogenous transforming growth factor-beta has been localized at the periosteum that underwent membranous bone formation. All of these factors are present in fracture healing sites, and are the normal components of the fracture healing process. However, it is not clear whether or not addition of any of these biological entities to the fracture sites will result in the acceleration of the healing process (Novikoff et al 2008). Optimal mechanical and biochemical conditions for fracture healing are still unknown. Quantification of bone formation and bone resorption is essential to the understanding of the bone repair process, but it is difficult to achieve since the callus has an ever changing cell population (Dunham 1992).

Fracture healing is dependent not only on the periosteum and endosteum, but also on the surrounding muscle mass (Utvag et al 1999, Sarmiento et al 1974). Isolation of the bone periosteum complex from the surrounding muscle mass in the early stages of healing reduces the capacity to form osseous tissue between the bone ends.

Callus forms as the result of complex interactions between skeletal muscle and periosteum. The integrity of the muscle periosteum interface in the early stages following fractures is an important prerequisite for the formation of the callus (**Fig. 3**). A large blood clot separating the bone periosteum complex from the surrounding muscle envelope might have an inhibitory rather than a beneficial effect on the early callus formation. Healing of uncomplicated fractures of long bones might actually be delayed by an extensive hematoma, as it is by the soft tissue injury.

Investigators overlook muscle tissue as a significant source of osteoprogenitor cells and focus principally on periosteum, oversimplifying the complexity of the fracture healing mechanism. Long bone fractures are always associated with some injury of the adjacent soft tissues, even if the fractured area is non-displaced (Sarmiento et al 1974).

The primary goal of the treatment of fractures is the re-establishment of the structural integrity of the fractured bone, and thus restoration of the function of the affected limb. This is best accomplished with the full understanding of the pathophysiology of the fracture healing process. Surgical intervention disrupts the normal healing process, as does an excessive hematoma. Bearing this in mind, nonoperative treatment of uncomplicated bone fractures is still a very viable option (Sarmieto and Latta 2006). However, it is not a viable option for bone grafting other than in confined spaces such as bone cavities, spine and as supplementation of arthroplasties.

Metal Plates and Cortical Bone Allograft Plates

In dealing with fracture healing, it is impossible to separate the fracture from the subject in whom it occurs. The overall state of the patient's health, nutrition and the bone health before the injury are of importance. The present overview deals with the biology of fracture healing in healthy average individuals and experimental animals not compromised by osteopenia or osteoporosis. Although biological fracture fixation is gaining popularity, metal compression plating is still used extensively in fracture treatment. In either case, fracture sites are reduced anatomically. This calls for surgical intervention, and most investigators agree that trauma caused by surgical procedures amplifies the existing injury (McKinley 2009). Rigid internal fixation with metal plates alters the normal sequence of bone fracture healing. Dynamic compression plates effect healing of fractures (Sarmiento et al 1980). However, when these plates are used, atrophy of bone beneath the plate is frequent. This may predispose the bone to subsequent fracture. When the plate is removed, fractures may also occur through the screw holes (Uhthoff and Finnegan 1983). With anatomical reduction effected by the compression plates, healing takes place directly across the cortices and the medullary canal. There is no external callus formation. This is most likely due to significant reduction or complete inhibition of periosteal activity by rigid immobilization.

Because of problems with metal plates, studies have been performed on fracture fixation with semi-rigid biological plates. Cortical bone allograft plates have also been used clinically, albeit in a small number of patients (Hornicek et al 2001, Head et al 1999). Although initially not as strong as metal plates, these provide not only fixation but also biological stimulation of healing. In experiments with non-displaced fractures of radii and ulnas, freeze dried cortical bone allograft contributed to the fracture-healing by immobilization, and by not preventing external callus formation on the opposite side of the fracture. Bone plate allografts were incorporated into the host bone by 12 weeks postfracture. At this point, the fractured limbs regained two-thirds of the strength of the contralateral control limbs. By 24 weeks biomechanical strength of control and fractured limbs was equal. These observations suggest that the disadvantages of fixation with rigid metal plates might be overcome by the use of biologically active allograft plates. Therefore, an investigative effort in the clinical use of these allografts should be encouraged. However, a word of caution is that these allografts must be biologically active. Thus, the method of their preparation is important (Malinin et al 2007).

Non-displaced fractures in experimental animals show a sequential pattern of fracture healing which involves callus formation. The callus itself has many attributes, some of which have been the subject of intensive investigations. Metabolism of fracture callus was one of the subjects investigated (Dunham 1992). These studies were carried out with the aim of obtaining information necessary to positively influence the healing process. Although much basic information was obtained, in practice little could be applied to the overall improvement of fracture healing. A number of studies on the composition of the fracture callus have been carried out. Qualitative and quantitative variations in fracture callus composition of proteoglycans have been analyzed.The characteristic differences between the glycosaminoglycans of bone and cartilage set apart precartilage from pre-osseous cells within the fracture callus (Dunham et al 1983). Although it is widely believed that ossification within the callus is strictly endochondral, experimental data shows fracture callus undergoes both endochondral and to a lesser, but measureable extent, membranous ossification.

One of the hallmarks of the callus is the presence of alkaline phosphatase in the tissues of the callus. Although undifferentiated cells of the callus show little enzymatic activity, with cellular differentiation, alkaline phosphatase activity is increased until it reaches its peak coinciding with chondrocyte proliferation in the most active stage of the fracture callus formation. As cartilage becomes calcified, the alkaline phosphatase activity diminishes. The activity of 6-phosphate dehydrogenase follows an opposite route. In the proliferative stage its activity is high, while the activity of alkaline phosphatase is negligible. In the later stages of callus development activity of both enzymes increases, but the loci of activity are different. Six-phosphate dehydrogenase activity is high in granulation tissue, while alkaline phosphatase accumulates in the cartilage and its ossifying component.

Initially, studies of calcification and ossification of the fracture callus were performed on decalcified histologic sections. These studies, performed at the beginning of the last century, provided an accurate picture of the callus in the late stages of its development (Urist and McLean 1941). However, the methods employed in these studies were not reliable enough to demonstrate time relationships and to provide details of the early stages of fracture callus development. The latter could be clarified only after the development of methods for sectioning and staining decalcified bone (McLean and Bloom 1940). Availing themselves of the latter, Urist (1941) and McLean (1940) demonstrated the formation of fibrocartilaginous callus in and around the fracture site and the subsequent invasion, removal and replacement of this fibrocartilaginous mass by new bone arising from the cambium layer of the periosteum and endosteum. Healing from the endosteum is of particular importance in bone graft incorporation. The onset and progress of calcification and mineralization has been observed by impregnation of sections with silver nitrate. The technique of silver impregnation has proven to be extremely useful and is still used today **(Fig. 4)**.

In the ensuing years a large volume of work was done on fracture callus ossification and calcification with current concepts formulated by Buckwalter, Einhorn and others. According to Glimcher (1987), tissue mineralization is an all encompassing process which includes all intra- and extracellular tissue components that become mineralized. These include intracellular organelles. It is important to bear in mind that extracellular and intracellular components are distinct entities. Mineralization of each of the tissue components proceeds independently. Calcification of each of the structures occurs from a metastable solution containing the requisite mineral ions (Glimcher and Krane 1968). Calcification of tissue components as independent events, with crystals being formed from metastable solutions, is now an accepted mechanism of calcification. This concept superceded the hypothesis of solid phase particles extruded into intervening space, initiating calcification by nucleation and multiplication.

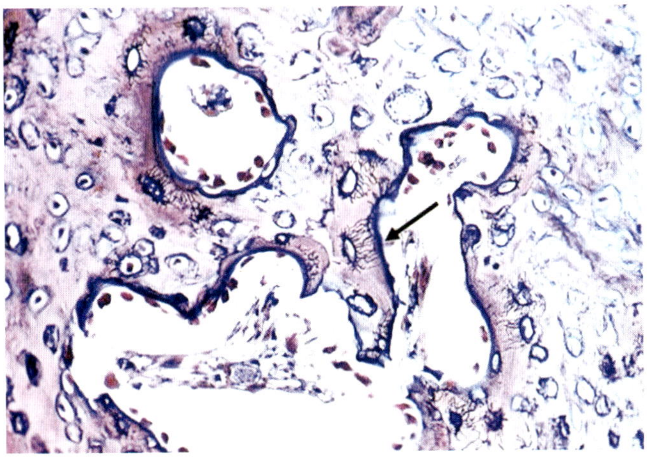

Figure 4 Silver staining of newly formed bone of the callus. Newly formed canalicular system (arrow) is clearly visible (Silver impregnation, eosin azure II counterstain × 400)

A new prevalent theory of callus calcification purports specific calcification of collagen fibrils. Accordingly the extruded mineral phase particles initiate calcification within the intervening space between mineralizing components.

Summary of Chronologic Events Which Result in Fracture Healing

In chronological order, hematoma formation is followed by an inflammatory reaction and angiogenesis. Cartilage formation then takes place. Cartilage undergoes subsequent calcification, cartilage removal and bone formation. The final event is bone remodeling. Some of the processes occur simultaneously.

The inflammatory reaction or inflammatory cascade is of paramount importance in initiating bone healing. On the molecular level, the cascade is initiated by signaling molecules. These are interleukin (IL-1), interleukin 6 (IL-6), (Einhorn 1995, Lind et al 1993) transforming growth factor, (TGF), insulin-like growth factor (IGF), (Trippel 1998) fibroblast growth factor (FGF), (Lind 1998) platelet derived growth factor (PDGF) (Canalis et al 1989) and most importantly the bone morphogenetic proteins of Urist (Schmitt et al 1999).

As mentioned earlier, the detailed studies on callus formation have been performed on animals, not humans. Therefore, extrapolation of the data from animals to humans has to be made with reservations. This is particularly true with regard to the time required for various phases of fracture healing to take place. Generally, fractures heal much slower in humans than they do in animals. Roughly, time sequences described for rodent models are doubled in humans. Nevertheless, data from animals is useful in understanding the process. The cellular and molecular events of fracture healing in rodents have been elaborated on by Einhorn (1998). Within hours of the newly formed hematoma, platelets release PDGF and TGF. Granulocytes release IL-1 and IL-6. Within 24 hours, mesenchymal cells begin to appear and apparently release bone morphogenetic proteins (BMPs) concomitant with their differentiation into osteoprogenitor cells (OPCs). In addition to a number of BMPs described in animals, recognition of human BMP is of relatively recent vintage (Urist and McLean 1941). In humans it is found only in bone. Endothelial cells lining the blood vessels in the bone marrow transform into macrophages (Maximow 1927) and osteoprogenitor cells. Bone morphogenetic proteins initiate differentiation of OPCs into osteoblasts and osteocytes. Hard callus begins to form in the subperiosteal area on days 3–5. This is accomplished by intramembranous ossification that is promoted by motion and inhibited by rigid fixation. Proliferation of OPCs and undifferentiated mesenchymal stem cells (MSCs) continues. Cartilage formation begins in the soft callus. It is accompanied by the phenotypic expression of collagen type II. In 2–3 weeks, cellular proliferation in the hard callus stops. At 2 weeks fibronectin levels go down and cartilage begins to ossify. After 3 weeks the hard callus area is made up of woven bone. In the soft callus, chondrocytes undergo apoptosis and necrosis. Solid union with woven bone occurs at 4–5 weeks. This is followed by the formation of lamellar bone, a hallmark of remodeling.

Primary Fracture Repair and Dynamic Compression Plates

The introduction of the dynamic compression plating added a new dimension to the study of primary bone healing (Müller et al 1965). The technique offered a distinct advantage over previously used rigid plates, (Bick 1968, Sherman 1912, Lane 1913) in as much as it afforded perfect approximation of the ends of fractured bones and produced healing by direct osteon formation.

The term compression plating may be somewhat misleading, but it is well entrenched through continuous usage. The term compression fixation might be more accurate because the plate itself is under tension, and only bone fragments are compressed across the osteotomy or fracture sites. At any rate the terms are used interchangeably, and mean the same thing. The goal of internal fixation is to place the fracture site between two oval holes with screws compressing the bone fragments because of the shape of the holes.This results in primary healing with interfragmentary gaps filled with bone without intermediate formation of cartilage (Schnek and Willengger 1963).

Experimental studies in animals demonstrated healing of diaphyseal osteotomies of fractures without periosteal bridging callus, providing that absolute stability with rigid plates had been achieved (Woo et al 1976). However, the method, successful as it is, has several disadvantages, primarily the slow rate of healing (McKibbin 1978). Osteonal healing is the same as remodeling in the late stages of fracture repair with callus formation. Periosteal callus healing in the closed fractures in a canine animal model has been shown to have major mechanical advantages over fractures treated by compression plating (Sarmiento et al 1980). Finally, rigid fixation with large plates significantly shields the bone from physiologic loads according to Wolf's law. Consequently, local osteopenia occurs in stress shielded areas (Uhthoff and Dubuc 1971). After removal of the plates osteopenic bones are vulnerable to refracture, a condition that persists for some time. Because of these limitations numerous studies have been carried out to overcome the deficiencies inherent in dynamic compression plating.

One such study investigated freeze-dried allograft bone plates and compared these with compression plates (Malinin et al 1984). An alternative to compression plating with metal plates are the freeze-dried cortical bone plates which provide not only fracture fixation, but also biological attributes of bone allografts. Secondary removal of the implant from the patient is not required. Previous work demonstrated that freeze-dried cortical bone grafts can effect early incorporation and healing of the fracture (Malinin 1984) (**Fig. 5**). Healing has also been recorded in several clinical cases, but concern still exists about the mechanical strength of these devices as a function of time after fracture fixation (Hornicek 2001).

When fractured dog radii fixed with freeze-dried bone plates are refractured sequentially, it shows the results. With time there is complete restoration of the biomechanical properties of fractured bone as compared to a contralateral control. The results with fractures fixed with stainless steel compression plates are different. When the plates are removed the bone becomes weaker. Biomechanical data on fractures fixed with metal and allograft plates is summarized.

At 2 weeks postfracture, the allograft plates were clearly visible radiographically and were not adherent to the underlying bone. Biomechanical support was provided primarily by bicortical screws that retained the bone plates in place. Granulation tissue had formed between the allograft plate and the host. This was gradually replaced with vascularized mesenchymal tissue which by 4–6 weeks was replaced with a typically appearing callus (**Figs 6A and B**). Periosteal callus was likewise formed on the opposite side from the bone plate.

At 8 weeks the fracture sites were healed. However, the size of the external callus varied from animal to animal. Mechanically, the fractured radius bone plate constructs are either comparable to or actually exceed the mechanical strength of the contralateral control bones if the screws are not removed from bone plates. In animals tested after screw removal, the mechanical strength of the construct reached 73% of the strength of the contralateral side. All refractures occurred through the screw holes. Histologically, the interface between the bone plate and the host bone presents a variegated appearance. Major portions of the bone plates become attached to underlying bone by osseous unions. Bone marrow cavities between the bone and the plate were also present. The plates themselves were vascularized and filled with channels accompanied by active osteogenesis. Bone resorption was noted at the edges of the bone plates.

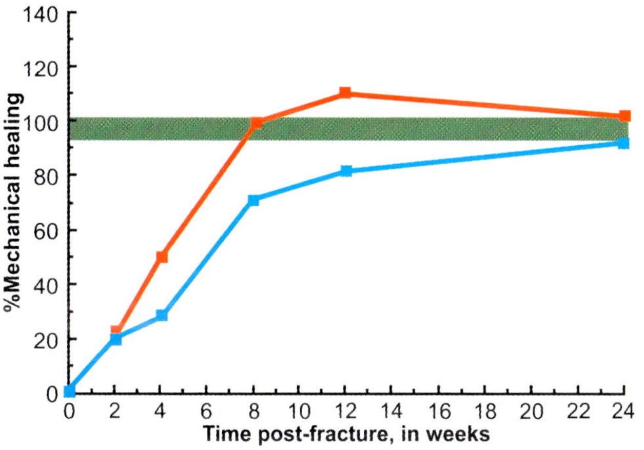

Figure 5 Graph showing progressive change in the structural strength of fractured canine radii stabilized with onlay cortical bone allografts. Green area represents structural strength of normal contralateral radii. Red line: fractured radii with bone plate allografts and screws in place. Blue line: same with screws removed

At 12 weeks postfracture there is only a relatively minor mechanical difference between the animals whose radii were refractured with or without screw removal. With intact screws, the experimented preparations consistently reached greater strength than the contralateral controls. However, to produce these fractures 16% more force was required compared to the contralateral limb.

In animals 24 weeks postfracture, the biomechanical properties of the experimental limb were almost identical to the contralateral controls when the screws were left in place. All of the refractures occurred in the center of the plate and the underlying bone, resembling the original fracture. After screw removal refracture strength reached 90% of the contralateral control strength with failure modes divided equally. Healing and incorporation of bone plates was evident both grossly (**Figs 7A and B**) and microscopically (**Figs 8A and B**).

Fractured limbs of animals treated with stainless steel compression plates reach their peak refracture strength at 8 weeks. On the other hand, fractured limbs of animals fixed

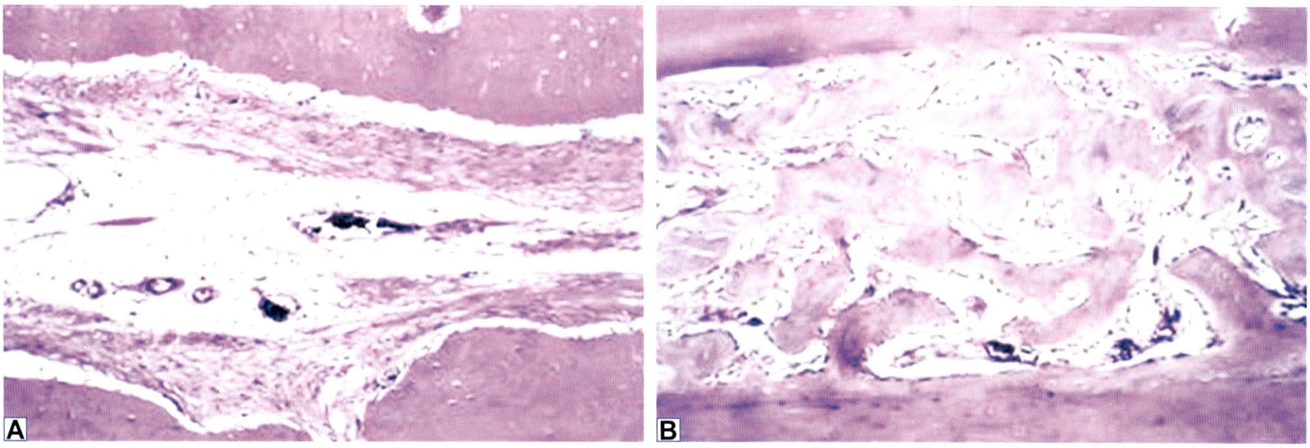

Figures 6A and B (A) Accumulation of vascularized mesenchymal tissue between the freeze-dried allogeneic bone plate and underlying fractured radius; (B) By 4–6 weeks mesenchymal tissue is replaced with well-organized callus

with bone plate allografts continue to increase in strength upto 24 weeks, at which time it levels off. These findings indicate continual activity and integration of bone plate allografts into the underlying bone. However, biological activity in the bone plates must be present if one would expect them to behave as described above. If biological principles in the allografts are inactivated by treatment such as irradiation, the plate will not integrate, nor will it stimulate repair in the bone to which it had been applied. The cited examples re-emphasize the complexity of the mechanisms of fracture repair. Unlike other tissues, bone does not repair itself by forming a scar. Instead it reactivates the process of embryonic development which reduplicates the original structures.

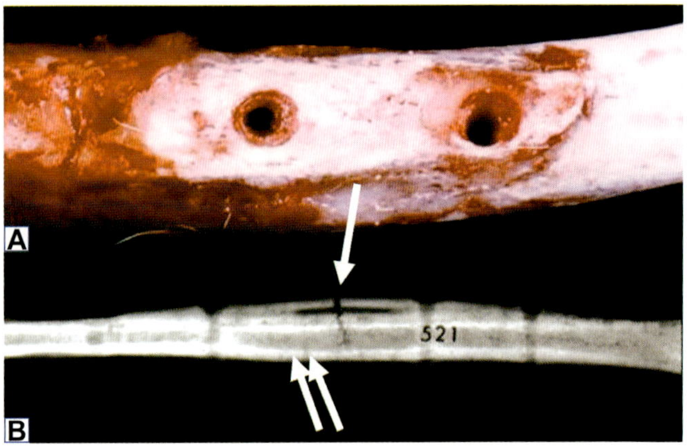

Figures 7A and B (A) By 24 weeks cortical bone plate became incorporated into fractured bone of the host. The allograft had also undergone canalization as evidenced by the appearance of blood channels on the surface; (B) X-ray showing bone incorporation distally and proximally. Refracture for biomechanical testing (arrow) occurred away from the original fracture site (double arrow). Underlying bone has also undergone transformation suggesting diffusion of bioactive substances from the bone plate

Despite repeated statements that bone can regenerate itself, it must be realized that the capacity for bone regeneration is limited. If a significant length of diaphyseal bone is lost (usually longer than the diameter of the shaft), it will not be replaced by new bone (Key 1934). Instead the defect will be filled by fibrous tissue, with the extremity remaining structurally compromised. For this reason, numerous attempts have been and are being made to stimulate bone regeneration in the true sense of the word.

Ancillary Methods for Stimulating Healing of Fractures

Many products and methods have been introduced to promote the healing of fractures. One of the most commonly encountered methods of promoting fracture healing is electrical stimulation and pulsed electromagnetic field stimulation (Bassett 1965, Abeed et al 1998). Literature reports on electrical stimulation in fracture healing have been numerous. There have been differences of opinion as to

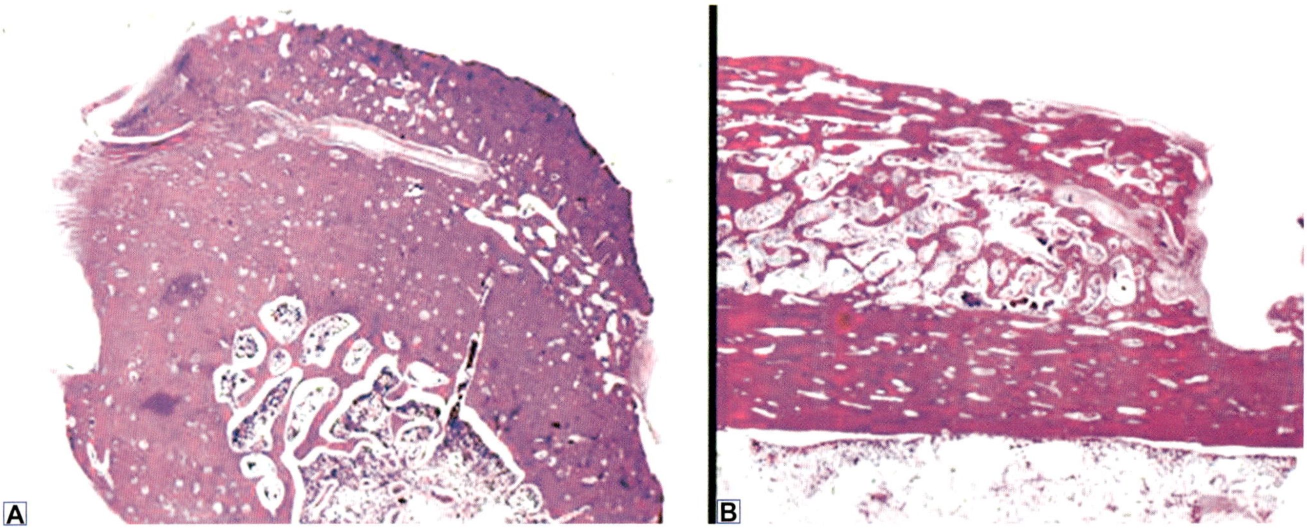

Figures 8A and B Sections of fractured radii with superimposed cortical bone plates, 8 weeks after fracture and allograft application. (A) Cross-section showing osseous union between the allograft and the underlying host bone. Center of the junction is occupied by vascularized mesenchymal tissue; (B) Longitudinal section. Bone allograft is united with underlying fractured radius through newly formed callus. Both the host bone and the allograft have undergone cancellation (H&E × 100)

whether there exists a narrow window for shape and amplitude of waveform as originally claimed by Bassett (1965). To understand the underlying mechanisms which are still obscure, further studies are needed. Since the treatment with electrical stimulation lasts from 14 to 15 weeks, and many patients subjected to this modality are also treated by other means including surgery and immobilization, the contribution of electric stimulation to bone healing is difficult to determine. Given these facts, orthopedic surgeons contemplating the use of electric stimulation devices must decide based on experience and observation as to its potential benefit or lack thereof. Currently, the Food and Drug Administration (FDA) has approved electrical bone growth stimulation, but FDA approval neither produces nor guarantees clinical results.

Ultrasound bone growth stimulator is likewise approved by the FDA. Studies suggest mechanical forces of the ultrasound waves are transformed into electrical impulses as they travel through the tissues (Rubin et al 2001). In reported studies, older patients treated with ultrasound are said to heal faster than those in the placebo groups. Another group of "fracture healing helpers" are bone substitutes made from non-human sources. These include coralline hydroxyapatite bone void filler, purified bone collagen and hydroxyapatite-tricalcium phosphate. These substitute grafts, when placed next to healthy bone, are remodeled by the body in the same manner as bone grafts. The clinical results with these substances remain uncertain. There is simply not enough clinical information to warrant advocacy for the use of these substances in clinical practice. One of the difficulties is that most of these substances are not strong enough to immobilize the fracture, and thus must be used in combination with other fixation devices (Novikoff et al 2008). Moreover, they are purely osteoconductive and have no inductive potential.

Fracture healing is influenced by numerous hormones, growth factors and cytokines (Barnes et al 1999). The systemic administration of growth hormone (GH) is said to accelerate bone regeneration. Local application of insulin-like growth factors such as growth factor-1 (IGF-1) and transforming growth factor-1 have been shown to stimulate bone metabolism, but their influence on accelerating fracture healing is inconclusive (Einhorn et al 1995).

Perhaps the best known factor which has been used in fracture healing is bone morphogenic proteins (BMPs) (Swiontkowski et al 2006). Bone morphogenic protein is a part of the transforming growth factor beta (TGF-β) super family proteins secreted by cells. These serve as signaling agents and influence tissue differentiation. There are two commercially available BMP preparations, BMP-7 (also known as osteogenic protein or OP-1) and BMP-2 (also known as "Infuse and Inductor"). These are attached to a bovine collagen carrier which releases BMP slowly over time. Clinically BMPs are used only in non-unions and delayed unions. It is clear that BMP plays a significant role in fracture healing (Swionkowski 2006, Giannoudis and Tzioupis 2005). We know BMP causes chemotaxis of cells, but this depends on its concentration and gradient. It is not known exactly what local concentrations of BMP are needed to stimulate maximum osteogenesis. It may well be that high concentrations of BMP may have little additional effect on fracture healing since there are several inhibitory proteins that influence BMP activity. It is now known from many animal studies and limited clinical trials that BMPs might be effective in promoting bone regeneration, but we must find out how they work so they can be manipulated for the desired effect.

Addition of autologous bone marrow to the fracture site as advocated by Connolly et al (1989) 20 years ago has found limited clinical application. The technique is an effective one, but its success depends on the number of osteoprogenitor cells in the aspirate. However, little information is available currently regarding the number of cells that are necessary to initiate bone repair. Randomized studies are still needed to define the use of the method in clinical settings.

In the last few years, there has been considerable interest in platelet concentrates to promote bone healing. However, since platelet concentrates do not contain osteoinductive factors they are used only in conjunction with other graft extenders (Flamini et al 2007).

DISCUSSION

In the last several decades, our knowledge of the bone healing process has expanded partially as the result of advancements made in other areas of medicine and biology. The ability to recognize important growth factors, deliver these at the fracture site and manipulate mesenchymal stem cells has led to the proliferation of studies, many performed on rodents. Although the fact is not frequently emphasized, many studies show it is more difficult to induce new bone formation in humans than in laboratory animals.

The present hierarchy of laboratory testing related to fracture healing ranges from cell culture assays to rodent bone induction assays. The availability of much of this technical information unfortunately does not elucidate fully the mechanism of fracture healing. To understand bone fractures one must remain aware of the cascade nature of the process and strive to exploit it rather than work against it. In dealing with fracture healing is it possible to surpass nature? It may be possible but not likely. The current trend to avoid surgical intervention unless absolutely necessary seems to support this contention.

The induction of accelerated bone fracture healing by chemical, physical, hormonal and other means has been a rapidly growing area which has produced a plethora of information (Lane and Wernz 1987). Some of it is overwhelming, some competitive, and some has commercial overtones. Clinical evidence of the efficacy of many of these modalities is not clear, and neither are indications, dosage timing and the mode of action. The same applies to the healing of allografts.

If information on bone induction and related topics is to be applied successfully to bone transplantation it must be clear and concise. In the meanwhile, clinical experience is of paramount importance. The bone fracture callus is still the king and immobilization the queen. The popular principle that guided fracture treatment for the past 100 years holds that rigid fixation provides an optimal environment for osteogenesis. This widely held concept is based on radiographic observations and biomechanical data obtained in large part on static materials. Although it is now evident, rigid internal fixation alters the normal sequence of callus formation and it is still the only method that can be used effectively to transplant bone segments. If it can be combined with the induction of the external callus formation, graft incorporation and retention will be enhanced. For this reason it behooves surgeons transplanting bone to be thoroughly familiar with the principles governing fracture healing.

BIBLIOGRAPHY

1. Abeed RI, Naseer M, Abel EW. Capacitively coupled electrical stimulation treatment. Results from patients with failed long bone fracture unions. Journal of Orthopaedic Trauma, 1998;12:510-3.
2. Allgower M, Spiegel PG. Internal fixation of fractures: Evolution of concepts. Clinical Orthopaedics, 1979;138: 26-9.
3. Aro HT, Wipperman BH, Hodgson SF, et al. Internal remodeling of periosteal new bone during fracture healing. Journal of Orthopaedic Research, 1990;8:218-46.
4. Barnes GL, Kostenuik PJ, Gerstenfeld LC, et al. Growth factor regulation of fracture repair. Bone and Mineral Research, 1999;14:1805-15.
5. Bassett C. Electrical effects in bone. Scientific American, 1965;213:18-25.
6. Bick EJ. Source Book of Orthopaedics. New York: Hafter Publishing; 1968.
7. Boden SW. Osteoconductive bone graft substitutes. AAOS online service. Bulletin; 2003. pp. 1-3.
8. Canalis E, McCarthy TL, Centrella. Effects of platelet-derived growth factor on bone formation in vitro. Journal of Cellular Physiology, 1989;140:530-7.
9. Connolly J, Lippiello L, Dehne R. Development of an osteogenic bone-marrow preparation. Bone and Joint Surgery (Am), 1989;71:684-91.
10. Dunham I. Metabolism of the fracture callus. Fracture Repair and Regeneration, Vol V. In: BK Hall (Ed). Boca Raton: CRC Press. 1992. pp. 1-32.
11. Dunham J, Catlerall A, Bitensky L, et al. Metabolic changes in the cells of the callus during fracture healing of the rat. Calcified Tissue International, 1983;35:56-61.
12. Einhorn IA. Enhancement of fracture healing. Bone and Joint Surgery (Am), 1995;77:940-56.
13. Einhorn TA, Majeska RJ, Rush EB, et al. The expression of cytokine activity by fracture callus. Journal of Bone and Mineral Research, 1995;10:1272-81.
14. Einhorn TA. Enhancement of fracture healing. AAOS Instructional Course Lectures, 1996;45:401-16.
15. Einhorn TA. The cell and molecular biology of fracture healing. Clinical Orthopaedics and Related Research, 1998;355S:S7-21.
16. Flamini S, Francesco A, Colafarina O, et al. Autologous platelet gel for delayed union of tibial shaft treated with intramedullary nailing. Italian Journal of Orthopaedics and Traumatology, 2007;8:1-5.
17. Giannoudis PV, Tzioupis C. Clinical applications of BMP-7: the UK perspective. Injury, 2005;(suppl 3):S47-50.
18. Glimcher MJ, Krane SM. The organization and structure of bone and the mechanism of calcification.Treatise on Collagen. In: GN Ramachandren, BS Gould (Eds). New York: Academic Press; 1968. pp. 69-72.

19. Glimcher MJ. Chemistry, structure and organization of bone and their influence on bone healing. Fracture Healing.In: JM Lane (Ed). New York: Churchill Livingstone; 1987. pp. 39-48.
20. Head WC, Emerson RH, Malinin TI. Structural bone grafting for femoral reconstruction. Clinical Orthopaedics and Related Research, 1999;369:223-9.
21. Hornicek FJ, Zych GA, Hutson J, et al. Salvage of humeral non-unions with onlay bone plate allograft augmentation. Clinical Orthopaedics and Related Research, 2001;386:203-9.
22. Key JA. The effect of local calcium deposit on osteogenesis and healing of fractures. Journal of Bone and Joint Surgery (Am), 1934;16:176-84.
23. Kolodny A. The periosteal blood supply and healing of fractures. Journal of Bone and Joint Surgery (Am), 1923;5:698-741.
24. Lane JM, Werntz JR. Biology of fracture healing. In: Fracture Healing.In: JM Lane (Ed). New York: Churchill Livingstone; 1987. pp. 49-59.
25. Lane WA. The operative treatment of fractures. London: Medical Publishing Co; 1913.
26. Lind M, Schumacher B, Soballe K, et al. Transforming growth factor beta enhances fracture healing in rabbit tibiae. Acta Orthopaedica Scandinavica, 1993;64:553-6.
27. Lind M. Growth factor stimulation of bone healing.Effects on osteoblasts, osteotomies, and implant fixation. Acta Orthopaedica Scandinavica, 1998;suppl 283:2-37.
28. Malinin T, Deck JD. The effects of implantation of embryonic and tadpole tissues into adult frog limbs. I. Regeneration after amputation. Journal of Experimental Zoology, 1958;139:307-28.
29. Malinin T, Latta LL, Wagner JL, et al. Healing of fractures with freeze-dried cortical plates. Clinical Orthopaedics, 1984;190:281-6.
30. Malinin TI, Carpenter EM, Temple HT. Particulate bone allograft incorporation in regeneration of osseous defects; importance of particle sizes. Open Orthopaedics Journal, 2007;1:19-24.
31. Malinin TI. Principles of musculoskeletal tissue banking. In: W.S. Pietrzak (Ed). Orthopaedic Biology and Medicine: Musculoskeletal Tissue Regeneration, Biological Materials and Methods. NJ, Totowa: Humana Press; 2007. pp. 483-508.
32. Maximow AA. Bindegewebe und blutbildende Gewebe. Handb d mikr Anat d Menschen (v Möllendorf). Berlin: Julius Springer; 1927. p. 232.
33. McKibbin B. The biology of fracture healing in long bones. Journal of Bone and Joint Surgery (Br), 1978;60:150-62.
34. McKinley T. Principles of fracture healing. Surgery, 2009;21: 209-12.
35. McLean FC, Bloom W. Calcification and ossification. Calcification of normal growing bone. Anatomical Record, 1940;78: 333-8.
36. Muller ME, Allgower M, Willenegger H. Technique of Internal Fixation Fractures. New York: Springer-Verlag; 1965.
37. Nash TJ, Howlett CR, Martin C, et al. Effect of platelet-derived growth factor on tibial osteotomies in rabbits. Bone, 1994;15:203-8.
38. Novikoff WM, Manawasi A, Hogan MV, et al. Critical analysis of the evidence for current technologies in bone healing and repair. Journal of Bone and Joint Surgery (Am), 2008;90:85-91.
39. Ozaki A, Tsunoda M, Kinoshita S, et al. Role of fracture hematoma and periosteum during fracture healing in rats: interaction of fracture hematoma and periosteum in the initial step of healing process. Journal of Orthopaedic Science, 2000;5:64-70.
40. Phillips AM. Overview of the fractures. Healing cascade. Injury, 2005;36(suppl 1):S5-7.
41. Potts WJ. The role of hematoma in fracture healing. Surgery, Gynaecology and Obstetrics, 1933;57:318-31.
42. Rhinelander FW, Phillips RS, Steel WM, et al. Microangiography in bone healing. II Displaced closed fractures. Journal of Bone and Joint Surgery (Am), 1968;50(4):643-62.
43. Rhinelander FW. Some aspects of the microcirculation of healing bone. Clinical Orthopaedics and Related Research, 1965;40:12-6.
44. Rhinelander FW. Tibial blood supply in relation to fracture healing. Clinical Orthopaedics and Related Research, 1974;105:34-81.
45. Rubin C, Bolander M, Ryaby JP, et al. The use of low-intensity ultrasound to accelerate healing of fractures. Journal of Bone and Joint Surgery (Am), 2001;83:259-70.
46. Sarmiento A, Latta LL, Zilioti A, et al. The role of soft tissues in the stabilization of tibial fractures. Clinical Orthopaedics and Related Research, 1974;105:116-29.
47. Sarmiento A, Latta LL. The evolution of fracture bracing. Journal of Bone and Joint Surgery (Br), 2006;88:141-8.
48. Sarmiento A, Mullie DL, Latta LL. A quantitative comparison analysis of fracture healing under the influence of compression plating vs. closed weight bearing treatment. Clinical Orthopaedics, 1980;149:232-9.
49. Schmidmayer G, Wildermann B, Heeger G, et al. Improvement of fracture healing by systemic administration of growth and local application of insulin-like growth factor-1 and transforming growth factor et al. Bone, 2002;31:165-72.
50. Schmitt JM, Hwang K, Winn SR, et al. Bone morphogenetic proteins: an update on basic biology and clinical relevance. Journal of Orthopaedic Research, 1999;17:269-78.
51. Schnek R, Willenegger H. Histologische Bild der sogenauten Primerheilung der Knochenkompaktanach experimentallen Osteotomien am Hund. Experientia, 1963;19;593-5.
52. Scudder CL. The treatment of fractures. Philadelphia and London: WB Saunders Co, 1938.
53. Sherman WON. Vanadium steel bone plates and screws. Surgery, Gynecology and Obstetrics, 1912;14:629-36.
54. Swiontkowski MF, Aro HT, Donell S, et al. Recombinant human bone morphogenetic protein 2 in open tibial fracture

subgroup analysis of data combined from two prospective randomized studies. Journal of Bone and Joint Surgery (Am), 2006;88:1258-65.

55. Szpalski M, Gunzburg R. Recombinant human bone morphogenetic protein 2: a novel ostoinductive alternative to autogenous bone graft? Acta Orthopaedica Belgica, 2005;71(2):133-48.
56. Trippel SB. Potential role of insulin like growth factors in fracture healing. Clinical Orthopaedic and Related Research Supplement, 1998;(355):S301-13.
57. Uhthoff HK, Finnegan M. The effect of metal plates on post-traumatic remodeling and bone mass. Journal of Bone and Joint Surgery (Br), 1983;65:66-71.
58. Uhthoff HK, Dubuc FL. Bone structure changes in the dog under rigid internal fixation. Clinical Orthopaedics and Related Research, 1971;81:165-70.
59. Urist MR, McLean FC. Calcification and ossification. Journal of Bone and Joint Surgery (Am), 1941;23:1-16.
60. Utvag SE, Grundues O, Reikezas O. Early muscle-periosteal lesion inhibits fracture healing in rats. Acta Orthopaedica Scandinavica, 1999;70:62-6.
61. Woo SL-Y, Akeson WH, Couts RD. A comparison of cortical bone atrophy secondary to fixation with plates with large differences in bending stiffness. Journal of Bone and Joint Surgery (Am), 1976;58:190-201.

Chapter 4

Professor Dr. Med. Alfred Benninghoff MD (1890–1953), Professor of Anatomy, University of Freiburg, Frieburg, Germany. His contribution to the study of cartilage was most significant. He is best remembered for "Benninghoff's Arcades" and recognition of the articular cartilage chondron as an anatomical unit. (*Courtesy:* University of Marburg and Professor Dr of Med G Ulrich Exner).

(*Reference:* Benninghoff A. Form und Bau der Gelenkknorpel in ihren Beziehungen zur Funktion. Zweiter Teil: der Aufbau des Gelenkknorpels in seinen Beziehungen zur Funktion. Zeitschrift für Zellforschung und mikroscopische Anatomie. 1926;2: 783-862.)

Cartilage

INTRODUCTION

Cartilage is a dense, resilient fibrous substance consisting of cells, matrix water, and macromolecular intercellular substances. Hyaline (articular) cartilage lines the large as well as small synovial joints which produce rapid, controlled and smooth movements required for locomotion and other musculoskeletal functions.

Cartilage is a firm, dense, but elastic tissue having many features in common with other connective tissue in keeping with their common embryologic origin from the mesodermal cells (Bergman and Affi 1974). Cartilage contains ground substance, so-called matrix, and fibers which may be either collagenous or elastic. The amount and the character of these fibers determine the type of cartilage, of which three distinct types are recognized (Bloom and Fawcett 1962). Three varieties of cartilage are distinguished; hyaline, fibrous, and elastic.

Hyaline cartilage is the most abundant of the three varieties of cartilage in existence. It is present in the respiratory system, forming the cartilage of the nose, larynx, trachea and the bronchi; in the costal cartilage of the ribs; as articular cartilage covering the surfaces of the joints, and in the fetus where it forms the osseous precursors (Sobotta and Piersol 1930). In most locations cartilage is surrounded by a thin membrane called the perichondrium, a vascular membrane containing many small cells, presumably those responsible for the formation of cartilage (Jordan 1952).

Since hyaline cartilages in different anatomic locations are considered as the same tissue, these are studied concomitantly with attributes found in one being assigned to all. This may be misleading as hyaline cartilages respond to the demands placed on them differently in different anatomic locations. For instance, articular cartilage responds to axial loads while tracheal cartilage does not.

Fibrocartilage is found in the articulation of the mandible, the clavicle, semilunar cartilages of the knee, intervertebral discs and other cartilaginous symphyses of the body. It also lines tendon grooves of the bones and forms the glenoid ligament of the shoulder as well as the labrum of the hip joint. Like hyaline cartilage, fibrocartilage is not vascularized and it is devoid of nerves.

Elastic cartilage occurs in the epiglottis, in the cuneiform and corniculate cartilage and in the vocal processes of the arytenoid cartilage of the larynx. It also lines the auditory canal and forms the skeleton of the external ear. Elastic cartilage is essentially hyaline cartilage permeated with elastic fibers.

Hyaline Cartilage

Hyaline cartilage is composed of specialized cells, chondrocytes embedded in extracellular matrix which they produce and which is composed primarily of collagen Type II, proteoglycans and elastin fibers. Articular cartilage, a specialized hyaline cartilage lining the joints is of particular interest. The structure of articular cartilage varies according to the distance between its surface and subchondral bone. From the subchondral bone to the articular surface the differences lie in the shape and volume of chondrocytes, collagen fiber orientation and diameter, and proteoglycan concentration. The latter is reflected in the intensity of staining of the matrix in histological sections. For convenience, articular cartilage is divided into four zones which are not distinct, but blend

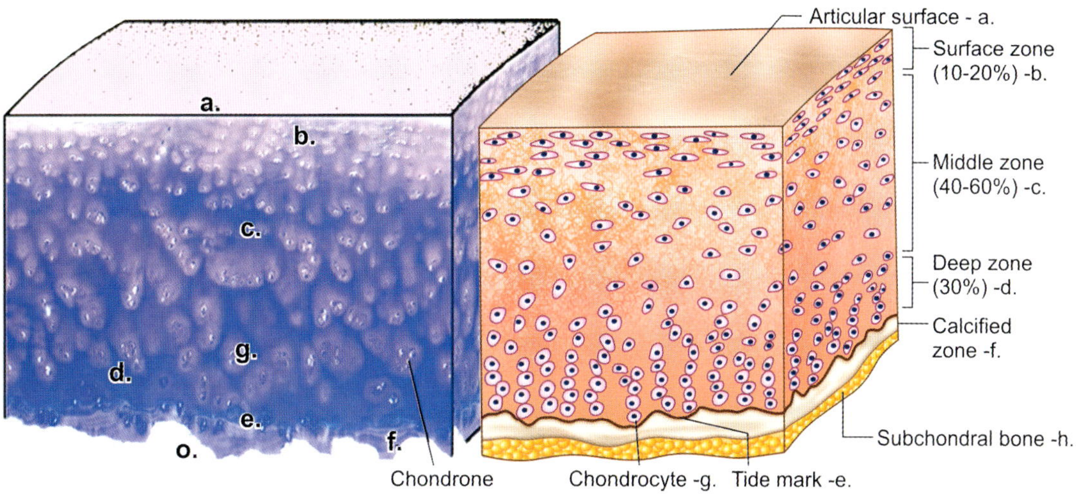

Figure 1 Normal articular cartilage: Left-stained cartilage section. (a) articular surface; (b) surface zone; (c) middle zone; (d) deep zone; (e) tidemark; (f) calcified zone; (g) chondrocyte; (h) subchondral bone. Chondral architecture is well-defined. Proteoglycan in the matrix stains blue. With the exception of the tidemark and the calcified zone, borders between other zones are indistinct, and blend into each other (Romanowsky-Giemsa stain × 250). On the right is a schematic representation depicting various zones of articular cartilage

into each other. These are the superficial zone, middle zone, deep zone and the calcified zone (**Fig. 1**).

The middle zone is the largest, accounting for 40–60% of the cartilage thickness. The uppermost, superficial zone of the cartilage constitutes its gliding surface, providing it is lubricated with synovial fluid. In this relatively thin zone, chondrocytes are elongated with their long axes parallel to the surface. Collagen fibrils are likewise arranged parallel to the surface. In this zone, proteoglycan content is the lowest among all zones. Therefore, staining of the matrix in this zone is the least intense. The surface of the articular cartilage is covered with a thin membrane which can be demonstrated only by special techniques (**Fig. 2**).

The transitional or the middle zone contains collagen fibers with larger diameters, but less organization; here chondrocytes are rounded and spheroid in shape. Each chondrocyte is enclosed within a clearly defined small space or lacuna (**Fig. 3**).

The deep zone contains the highest concentration of proteoglycans, hence in histologic sections it stains more

Figure 2 Fixed section of normal articular cartilage stained with acridine orange and viewed in ultraviolet light. A distinct line underlying the synovial surface is plainly visible (× 250)

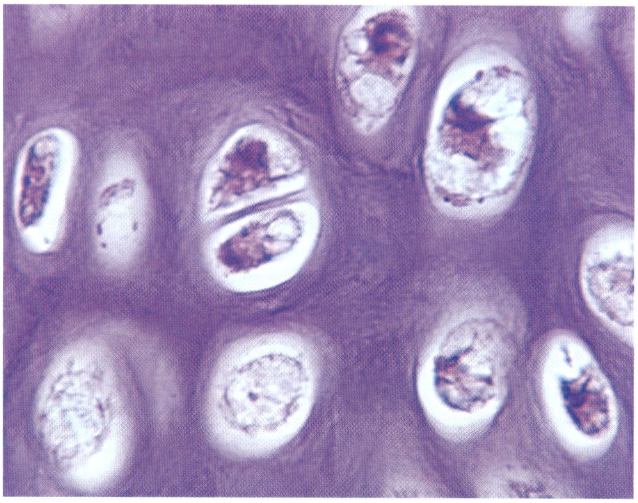

Figure 3 Chondrocytes are encased in tight small spaces termed lacunae (Periodic acid-Schiff hematoxylin-staining × 900)

intensely (basophilic) than other zones. The collagen fibers are oriented perpendicular to the surface and are of larger diameter than those in the upper zones.

The zone of calcified cartilage (the deepest zone) separates hyaline cartilage from subchondral bone. It contains small cells embedded in calcified tissue. A wavy, basophilic line, called the "tidemark" separates the deep and calcified zones **(Fig. 1)**.

In addition to the division of articular cartilage into zones, the extracellular matrices (ECM) are further divided into pericellular, territorial and inter territorial regions depending on the proximity of chondrocytes containing lacunae. The regions differ in the content of proteoglycans, collagen and other matrix components (Benninghoff 1925, Malinin and Malinin1999).

Articular cartilage is a vestigial part of the cartilaginous epiphysis, except for its superficial layers, which develop from embryonic perichondrium. With completion of growth, cartilage, which covers the epiphysis, migrates toward the joint and forms articular surface margins which are continuous with periosteum (Jordan and Kindred 1948).

Adult articular cartilage rests on the subchondral bone plate which is penetrated by vascular channels. Cartilage is attached to subchondral bone by fibrils and interdigitations (Collins et al 1973). While immature articular cartilage is bluish and translucent, mature cartilage is yellowish-white presumably due to more widely dispersed chondrocytes. With aging, non-enzymatic modification of tissue proteins by reducing sugars, the so-called Maillard reaction, produces accumulation of pentosidine. This, along with the accumulation of a variety of other Maillard reaction products produces more brittle and stiffer cartilage encountered with advancing age (Verzijl et al 2000 and Armstrong and Mow 1982). Maturation of the subchondral bone is accompanied by the formation of a thin continuous bone plate which was presumed by many students of cartilage to present a barrier impermeable to metabolic exchange between cartilage and subchondral bone. The zone of calcification of the deepest layer of the cartilage which varies in thickness is demarcated by a wavy line staining blue with Romanowsky type stains and vividly red with Masson's trichrome stain (Malinin and Malinin 1999) **(Figs 4A and B)**.

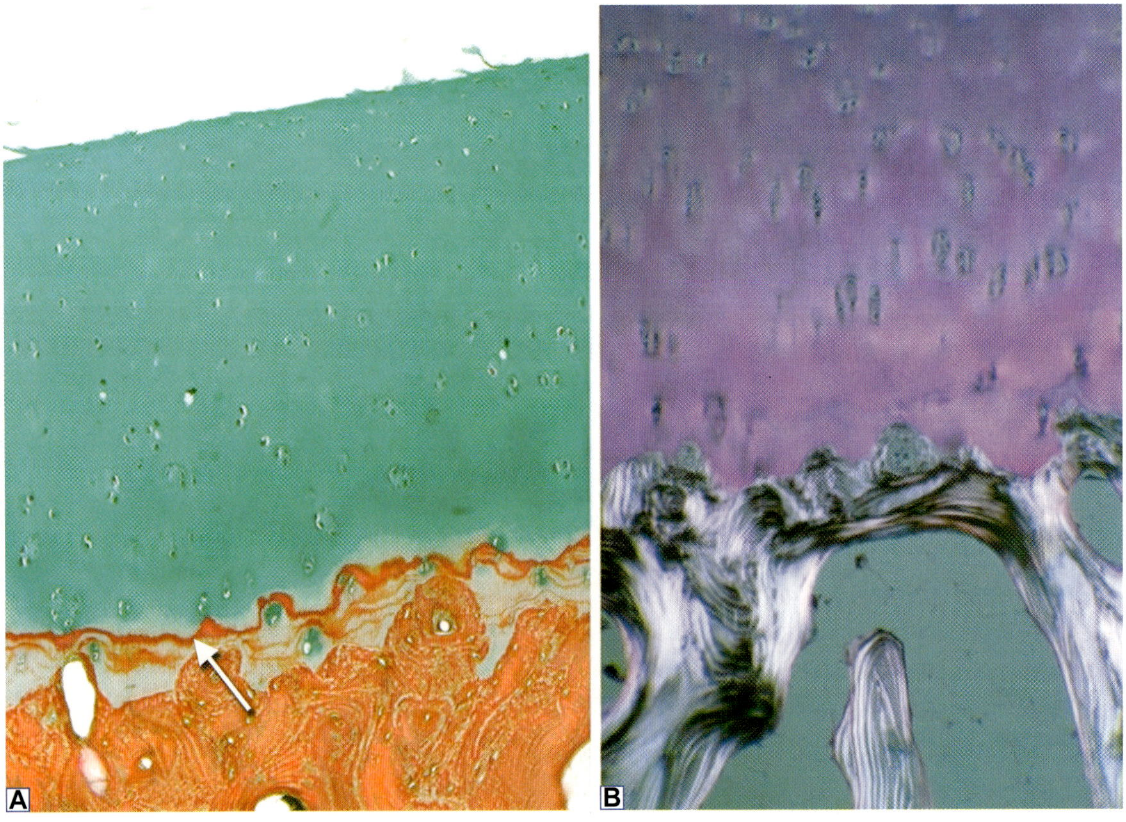

Figures 4A and B (A) Normal articular cartilage showing the tidemark (arrow), and underlying calcified zone and subchondral bone, staining orange red. Masson trichome stain x100; (B) Collagen in calcified zone differs from that in subchondral bone as shown in polarized light (Romanowsky-Giemsa x250)

When articular cartilage reaches maturity, the calcified cartilage zone as well as the subchondral bone plate are said to form a continuous impermeable barrier between cartilage and the subchondral bone marrow spaces. However, it may be recalled the calcified zone was perforated by various vascular channels. Apparent prevention by this lamina of blood vessel contact with the articular cartilage led to a notion that the nutrition of the latter is provided exclusively by the synovial fluid. This notion is incorrect.

Nutrition of Articular Cartilage

The presence of a seemingly impermeable plate between the cartilage and the subchondral bone influenced the concepts regarding articular cartilage maintenance and nutrition. As a consequence of this anatomic structure, many investigators ascribed cartilage nutrition solely to the synovial fluid (Honner and Thompson 1971, Hodge and Mckibbin 1969). However, a number of studies, including our own, show that articular cartilage receives nutrition both from the synovial fluid and the underlying bone (Serafini-Fracassini and Smith 1974; Malinin et al 1998). Thus a theory that cartilage receives no nutrition from subchondral bone is subject to revision (Muller-Gerbl 1998). The notion that synovial fluid contributes to the nutrition of articular cartilage dates back to Leidy (1849) who demonstrated that the cartilage surface was able to imbibe fluid. A description by Virchow in 1863 of loose joint bodies with normal cartilage supported this idea which was later reinforced by Strangeway (1920), and Phemister (1924). The evidence to suggest that articular cartilage nutrition is supplied by synovial fluid is very strong, as loss of synovial fluid leads to a rapid deterioration of articular cartilage. Normal articular cartilage cannot survive in the absence of synovial fluid, but the latter probably provides more than nutrition alone. Synovial fluid also lubricates the joint and provides a specific protective effect (Hodge 1969). Moreover, synovial cells are thought to release both cytoprotective and destructive molecules in response to disease and injury. Despite its obvious importance, cartilage is not sustained by the synovial fluid alone; subchondral bone is also essential for maintaining cartilage on a long-term basis.

Support for long-term dependence of articular cartilage on subchondral bone is demonstrated in studies using adult animal models (Malinin and Ouellette 2000). Interruption of contact between viable subchondral bone and articular cartilage produces injury that requires extended time to manifest itself. Nevertheless, the cited experiments demonstrate the importance of subchondral bone in the maintenance of articular cartilage homeostasis as shown in **Figures 5A to D**.

Collagen in Hyaline Cartilage

Collagen content and organization of articular cartilage has been an object of intense study for over 100 years (Hultkranz 1898, Benninghoff 1925, Speer and Dahners 1979). Benninghoff performed systematic and meticulous analysis of collagen organization in articular cartilage primarily by the use of polarized light microscopy. He concluded that collagen fibers are oriented horizontally in superficial layers, form arches in intermediate layers and are placed vertically in deep layers. This arrangement appears to have functional implications and suggests that cartilage fibers orient along lines of stress. Transmission electron microscopy and scanning electron microscopy provided further elaboration on the collagen architecture of articular cartilage. The collagen fibers of hyaline cartilage consist of collagens type II, IX, and XI. However, the majority of cartilage collagen (90–95%) is made up of collagen type II (**Figs 6A and B**).

In addition to some 29 genetically distinct collagen types, cartilage matrix also contains small amounts of collagens type V, VI, and X. The functions of these are not known.

The function of collagen type II in articular cartilage is to provide the tensile, and perhaps more important shear strength, and to retain proteoglycans within the extracellular matrix (ECM). The articular cartilage collagen fibers are smaller than those of bone and tendons. They present a cross-linked network. This arrangement probably adds stability to the network and contributes to the tensile properties of articular cartilage.

In articular cartilage, cell-matrix interactions are essential for the maintenance of extracellular matrix. The network of collagen fibers provides the capacity of the tissue to withstand the swelling pressure of the embedded proteoglycans.

Proteoglycans of Hyaline Cartilage

A variety of proteoglycans essential for its normal function are found within the cartilage. These include aggrecan, decorin, biglycan, fibromodulin, and lumican. Each proteoglycan serves functions that are determined by both its protein and its glycosaminoglycan chains (Rougley 2006).

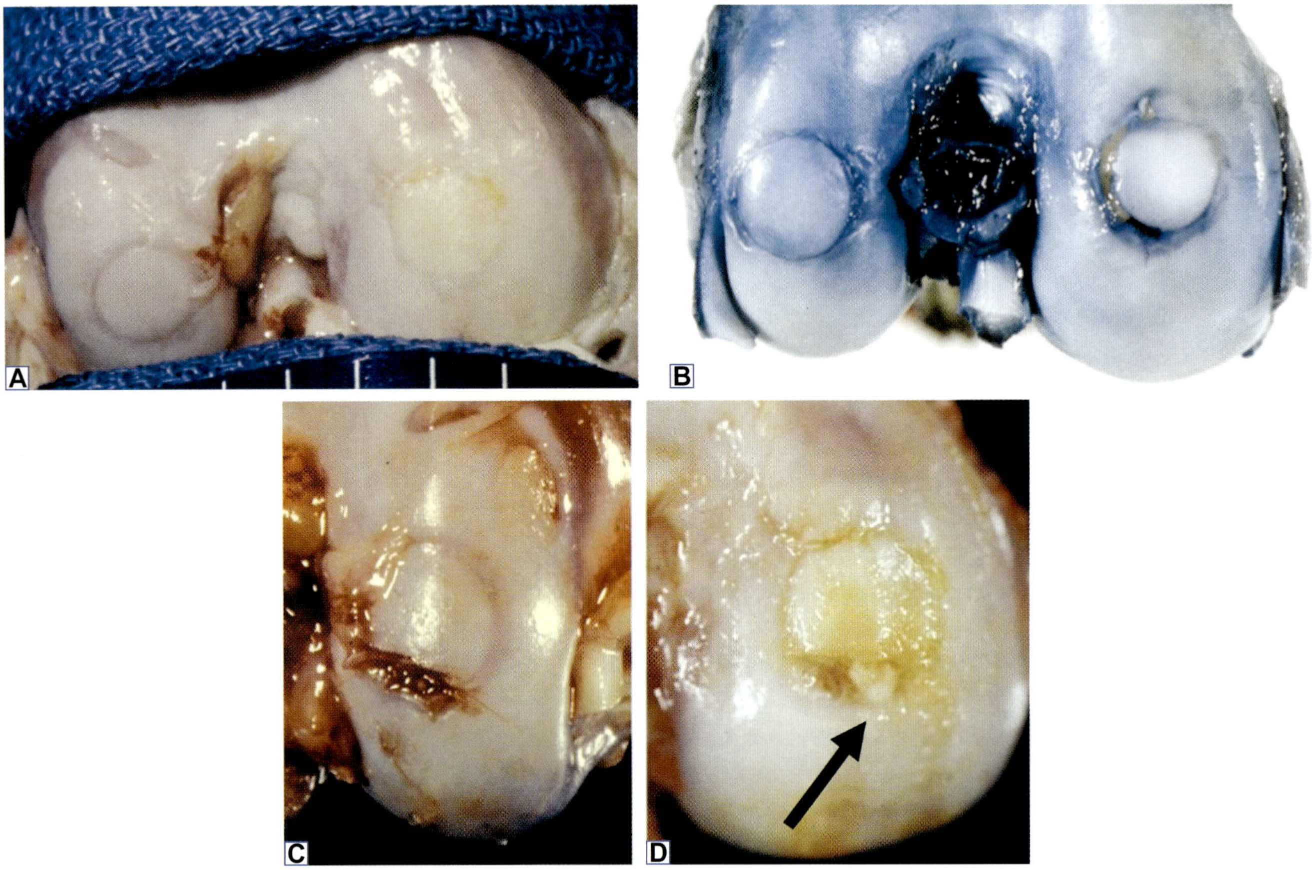

Figures 5A to D (A) Cartilage in contact with subchondral bone maintains its normal gross and microscopic appearance, years post-transplantation (left). Cartilage transplant deprived of contact with subchondral bone appears discolored (right); (B) Specimen with normal transplant and transplant deprived of contact with subchondral bone in a specimen injected with India ink. Vascularized bone beneath normal cartilage is perfused with India ink (left). The cartilage on the right, deprived of contact with subchondral bone, is pale and appears non-viable; (C) Three years post-transplantation, cartilage on the graft in contact with subchondral bone appears normal. Biopsy site shows active bleeding; (D) Cartilage on the transplant deprived of contact with subchondral bone is completely degenerated, three years post-transplantation; there is no subchondral bleeding at the biopsy site (arrow)

The reason for separate consideration of aggrecans relates to the uniqueness of cartilage as a load bearing tissue (Knudsen and Knudsen 2001). Cartilage is particularly rich in extracellular matrix, which accounts for over 90% of the dry weight of the tissue (Hardingham and Fosang 1992). Aggrecan exhibits a unique feature in that the core protein has the capacity to interact with hyaluronan, thus establishing a relationship with chondroitin sulfate. Although aggrecan was originally thought to be a specific proteoglycan of cartilage, it was subsequently found to be present in other tissues (intervertebral disc, aorta and tendon). Nevertheless, aggrecan is a hallmark of chondrogenesis and is essential for its studies (Schwarz et al 1999).

Aggrecan, a large chondroitin sulfate proteoglycan, is the predominant proteoglycan (PG) in cartilage. After its secretion, aggrecan self assembles into a structure with 50 monomers bound to a filament of hyaluronan, also known as hyaluronic acid or hyaluronate. Aggrecan provides osmotic resistance for articular cartilage to withstand compressive loads. Other proteoglycans formed during chondrogenesis are cell surface syndecans and glypicans, decorins (small leucine rich proteoglycans), biglycans, fibromodulins, lumicans and epiphycans and pelicans (basement membrane proteoglycans). The exact functions of all of these proteoglycans are still obscure. Cartilage is made up primarily of extracellular matrix which constitutes

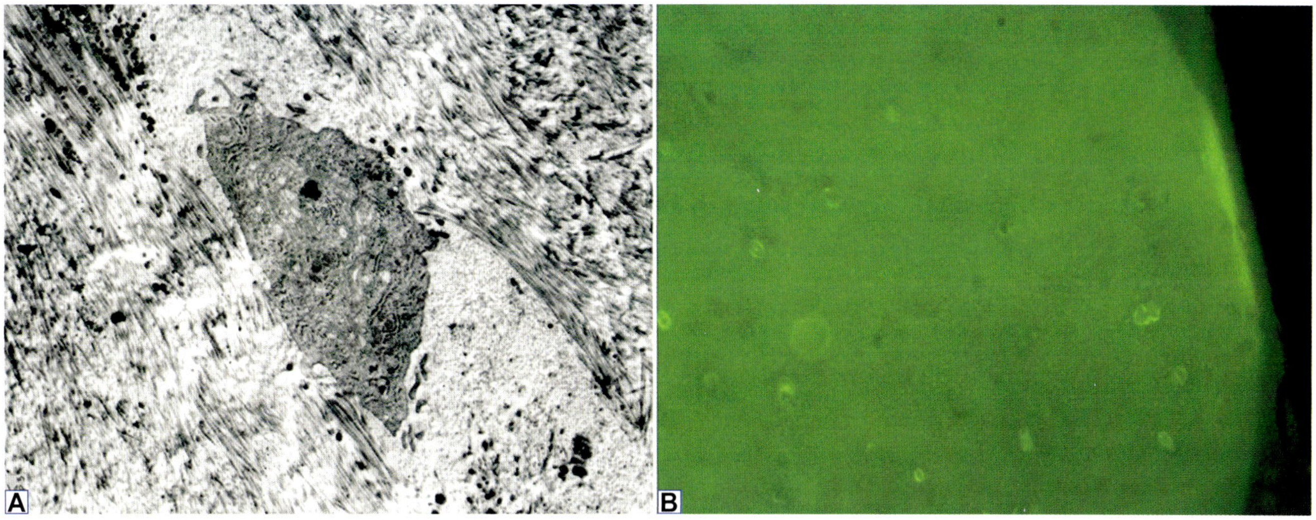

Figures 6A and B (A) Electron photomicrograph of normal cartilage from the medial condyle of the distal femur. Tangentially sectioned chondrocyte is embedded in the matrix rich in collagen II, × 16,800; (B) Section of normal human cartilage with collagen type II demonstrated by immunohistochemical technique

about 90% of its dry weight (Hardingham et al 1992). The main component of the matrix is glycosaminoglycan which was known for a long time to be chondroitin sulfate (Mathews et al 1968). Further elucidation of the nature of this compound showed it to be chondroitin sulfate proteoglycans (Hascall 1986), later termed "aggrecan".

The recognition of aggrecan as an entity came as a result of cloning and sequencing of its core protein (Doege et al 1991). Further investigations showed a variety of proteoglycans including aggrecan, decorin, biglycan, fibromodulin and lumican, to be essential for the normal function of hyaline cartilage, characterized by high concentration of aggrecan—which exists as proteoglycans aggregates in association with hyaluronan and link protein (Roughley 2006). Owing to the quantitative abundance of ECM in cartilage, analytical and morphologic characterization of PG's was natural. Among these, visualization was most attractive. Glycosaminoglycans of cartilage are negatively charged. This made possible the application of anionic dyes and oxidation reactions to the study of cartilage very attractive. The subject had been reviewed in detail by several investigators (Grauman 1964, Schrevel et al 1981, Hyatt 1993). For practical reasons, only selected cationic reactions and a few oxidative reactions need to be considered. Of these, the acetic anhydride-thiocarbohydrazide-silver proteinate reaction for structural coloration of cartilage is particularly useful (Malinin et al 1998), because it allows for detailed study of chondrocyte structures as well as those of the cartilage matrix.

Many publications rely on the histological/histochemical properties of the cartilage matrix to demonstrate the integrity of the cartilage or lack thereof. Since these represent single attributes of hyaline cartilage, it is important to know the exact significance of the given histochemical reactions and/or staining characteristics. In the study of hyaline cartilage histology, Masson's trichrome stain provides even coloration of calcified and uncalcified cartilage matrix, while vividly coloring the tidemark red and hypertrophic deep zone orange. This form of staining, although advantageous, lacks histochemical specificity.

In periodic acid Schiff (PAS) preparations, dark red coloration is present in deep layers of the cartilage and its calcified zone. In the upper layers this reaction is weak. The most intense reactivity is present in the interterritorial matrix in the deep zones of the cartilage (**Fig. 7**).

The most popular and frequently used stain for demonstrating proteoglycans in the hyaline cartilage matrix is safranin-O. When bound to polyionic proteoglycans, safranin-O is metachromatic in aqueous medium and orthochromatic in dehydrated, mounted tissue sections. There are several techniques for staining and counterstaining cartilage with safranin-O. In solutions, chondroitin and keratin sulfates bind safranin-O stoichiometrically (Rosenberg 1971), producing vividly red coloration. However, in fixed and mounted sections, the coloration is orange (**Fig. 8**).

At very low levels, glycosaminoglycans may elude detection with safranin-O. Therefore, absence of staining with

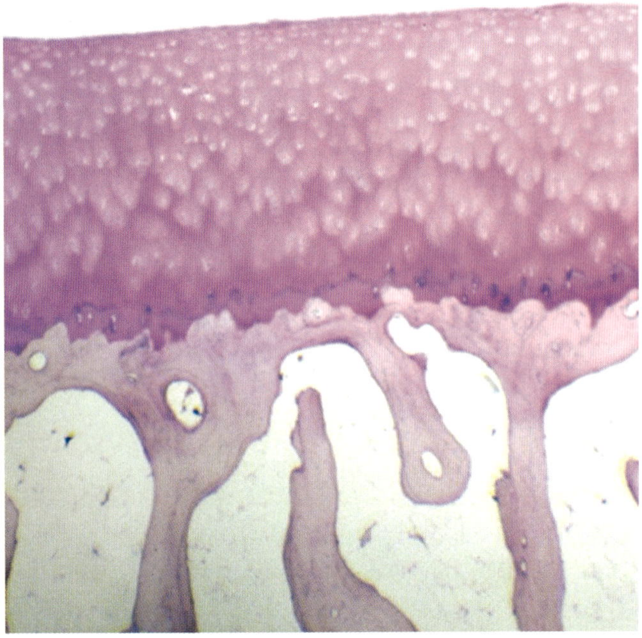

Figure 7 Normal articular cartilage stained with periodic acid Schiff technique. The deep intercellular matrix stains dark red-purple, indicating presence of neutral polysaccharides including chondroitin sulfate and hyaluronic acid (PAS × 100)

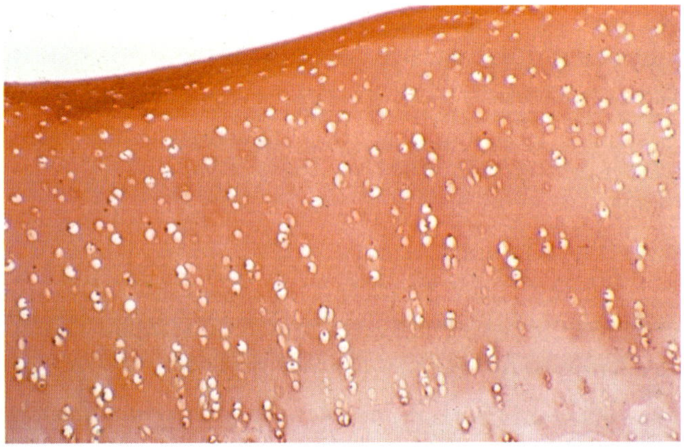

Figure 8 Normal articular cartilage stains uniformly with safranin-O (× 250)

safranin-O may not necessarily indicate complete absence of proteoglycans in the matrix (Cample & Allard 1988).

Romanowsky-Giemsa stain, a mixture of methylene blue and eosin is a differential stain used to supplement hematoxylin and eosin, stain. It stains cartilage matrix blue in the depth, and reddish in the upper zones. The tidemark is strongly basophilic. Toluidine blue, a basic dye, is used in two variants. First, conventional technique employs toluidine blue for quick staining, and in cartilage for the demonstration of metachromasia which depicts glycosaminoglycans and other non-neutral polysaccharides. Another method employs periodic acid-sodium bisulfite toluidine blue stain (Malinin 1970). This stain colors pericellular and inter-territorial matrix blue and produces a distinct pattern of staining in the deep zones of articular cartilage. It also stains chondrocyte nuclei.

In summary, results with safranin-O stained sections of articular cartilage are almost identical to those stained with PAS. Staining with toluidine blue and Romanowsky-Giemsa stains demonstrate metachromasia attributable to the presence of proteoglycans. Staining with toluidine blue of sodium bisulfite addition derivatives produces a distinct pattern of coloration in the deep zones of articular cartilage. In normal human articular cartilage, there is a diffuse metachromatic staining of the matrix with diminution of intensity of staining in the most superficial zone of the cartilage. Alcian blue stains acid mucopolysaccharides and acetic mucin. It is specific for sulfated groups when the staining solution is below pH 2.

Incorporation of radioactive sulfur corresponds closely to alcian blue staining (Curran and Kennedy 1955). Alcian blue is a water-soluble polyvalent acid dye (Steedman 1950). The blue color is due to the presence of copper in the molecule. Alcian blue is, in some ways, similar to hematoxylin and eosin in which basophilic elements stain blue, while eosinophilic elements do not stain. In the intervertebral disc, staining with alcian blue corresponds to metachromasia in toluidine blue preparations (Butler 1968).

The distribution of the chondroitin sulfate and related substances can be also demonstrated by the immunofluorescent techniques (Barland et al 1966). With rabbit antisera prepared against protein-polysaccharides of human articular cartilage, the fluorescence is restrictively localized to perichondrial regions of the matrix. This corresponds roughly to staining with toluidine blue of sodium bisulfate addition derivatives of periodic acid-labile substances (Malinin 1970). More recent studies with immunofluorescence allow for localization of different protein epitopes. In addition, specific histochemical stains help to visualize nuclear DNA, mitochondria and other cellular elements. However, as with many methodologies, technical knowledge and experience are important for the interpretation of results (Soeder et al 2004).

Glycogen in Articular Cartilage

Since the discovery of glycogen in cartilage by Rouget in 1859, many studies have dealt with this topic (Stockwell and

Meachim 1979). However, systematic studies of cartilage glycogen have not been performed. There is an entrenched view of glycogen being composed entirely of glucose in self-contained glycogen depositions in tissues. Glycogen in cartilage increases in the areas to be calcified and disappears on calcification, suggesting glycogen accumulation is an attribute of cartilage matrix alone (Schajowicz and Cabrinin 1954). Glycogen is a polysaccharide, but it is doubtful articular cartilage glycogen can be mobilized to release glucose into circulation. The function of glycogen in articular cartilage is specialized. Because of low oxygen tension in the cartilage, anaerobic glycolysis is regarded as a primary pathway of ATP production by chondrocytes (Brighton 1978). In this respect the function of glycogen is understood, but there is also evidence for concomitant aerobic glycolysis (Yamamoto and Gay 1988). Since glycogen has been found in close association with lipid inclusions, participation of glycogen in lipid synthesis is possible.

The presence of glycogen in articular cartilage is clear, but its function is multifaceted. It is also present in chondrocytes. As chondrocytes undergo apoptosis, glycogen is converted to glucose. In the ECM (extracellular matrix), glycogen forms proteoglycans. In addition, because histochemical reactions used in the study of the cartilage matrix depend in a large measure on the Schiff reaction, glycogen assumes through its vivid coloration a prominent role in the studies of articular cartilage.

Factors affecting mobility of fibrillar proteins and other ECM constituents determine the properties of cartilaginous tissues. A very important process leading to immobilization of proteins and their complexes is the reaction of saccharides with the amino groups of proteins. These nonenzymatic reactions, for which glycogen is essential, proceed under physiologic conditions, and do not require heat (Zyzak et al 1995). Collectively referred to as the Maillard reaction, the phenomenon is actually a series of independent events (Malinin and Malinin 1999) with numerous ramifications. The first step in the Maillard reaction is the condensation reaction of the amino groups of proteins with the aldehydes of glucose. This yields a Schiff base, a compound with a functional group containing a carbon-nitrogen double bond. The nitrogen atom is connected to an aryl or alkyl group, not hydrogen. Schiff base is not to be confused with a Schiff reagent (Liggins and Furth 1997). Another, much faster interaction between aldoses and amino groups of proteins also produces Schiff bases. Aldose and ketose derived Schiff bases spontaneously undergo a series of changes known respectively as Amadori and Heyns rearrangements (Malinin & Malinin 1999). These changes are important because they contribute to loss of elasticity and changes in coloration of cartilage associated with aging.

Condensation of glucose with the amino groups of lysine forms collagen bound Schiff base-hexosyl lysine. Amadori rearrangement of the latter forms ketoimine. The initially formed Schiff bases and ketoimines can undergo further reaction to produce advanced glycation end products. The latter may enter into a series of polymerization reactions yielding colored and fluorescent Schiff bases. Schiff base is a stable imine. These reactions lead to the oxidation of lipids which in turn undergo degradation to aldehydes again resulting in changes associated with aging.

Cartilage Lipids

Lipids and lipid containing compounds constitute important components of cartilage matrix and chondrocytes. These were first identified in cartilage in 1849 by Leidy (Leidy 1849). Neutral fats were identified in chondrocytes 50 years later (Sacerdoti 1900). Neutral fats were not identified in the matrix until 1931 (Putschar 1931) when large osmiophilic fat inclusions up to 45 μm in diameter were noted in the developing cartilage (Clark and Clark 1942). These inclusions filled the cytoplasm and displaced the nucleus to the periphery of the cell. Lipid inclusions were observed during all stages of cartilage development. Close association of lipid inclusions with glycogen inclusion led to a speculation that lipids in chondrocytes were derived from glycogen (Montagna 1949). Subsequently synthesis of lipids from glucose became established (Howard 1977). Many of light microscopy studies of lipids were directed towards histochemical characterization and detection of matrix lipids (Goldberg and Boskey 1996). Surprisingly, lipids in normal chondrocytes attracted relatively little attention, but their presence is important in the aging process, and the maintenance of cartilage metabolism.

A sharp melting point can be an indication of purity of tested compounds and has been used for characterization of cell lipids *in situ* (Small 1970). For this purpose, polarization of conventional microscopes with a heating stage can be utilized (Malinin 1980).

Ultrastructural localization of cell and matrix lipids depends on osmic acid-lipid interactions. Reduction of osmic acid by unsaturated lipids makes visualization of lipids possible both by light and electron microscopy (Collins et al 1973). In cell culture inclusion of saturated triglycerides and fatty acids, do not reduce osmic acid and therefore remained unstrained in osmium treated electron microscopic preparations (Geyer 1967, Mackenzie et al 1967). At the ultrastructural level, osmiophilic and non-osmiophilic lipid inclusions are found in human and animal cartilage cells. In chondrocytes, the size of lipid inclusions varies from very small structures to very large aggregates which deform surrounding cytoplasm and displace cell nuclei. Although lipids and proteoglycans are found in the intra chondrocytic inclusions, their mode of formation and their precise composition await further elucidation. Lipids change with age (Bonner et al 1975). Arachidonic acid, a precursor of PGE-2 and PGF alpha, levels increases in the surface of cartilage and decreases in deep layers. Since PGE-2 is an inflammatory mediator, fats are indirectly involved in the inflammatory process (de Grauw et al 2006).

Water in Articular Cartilage

Water makes up 65–80% of the total composition of articular cartilage. However, water in cartilage does not exist in free form. Some 30% is bound to collagen, and the remainder is bound to ECM. Water content is not distributed homogeneously, producing a viscous gel. A large portion of articular cartilage water can be displaced by a pressure gradient, but the frictional resistance against pressure induced flow is high. This frictional resistance against movement of water due to pressurization constitutes the mechanism to support high loads on the joints. Water content of articular cartilage decreases with age and disease (arthrosis).

The flow of water and tissue, and the drag it creates, are major determinants of the compressive behavior of cartilage. Thus, the mechanical behavior of cartilage is dependent on its permeability, i.e., how easy it is for the water to move in and out of the cartilage. Flow of fluid through solid, permeable matrix is governed by Darcy's law, which states that the rate of volume discharge through a porous solid is related to the pressure gradient applied to the solid and the hydraulic permeability coefficient (Lu & Mow 2008).

By considering cartilage a porous, permeable solid filled with fluid, a determination of the intrinsic equilibrium modulus of the cartilage matrix and its permeability to fluid flow can be made (Armstrong and Mow 1982). The interaction of water and the macromolecular matrix of the cartilage provide it with its mechanical properties and function. Water accounts for up to 65–80% of the wet weight of the cartilage (**Fig. 9**). The cartilage matrix becomes softer and more permeable with the increase of its water content.

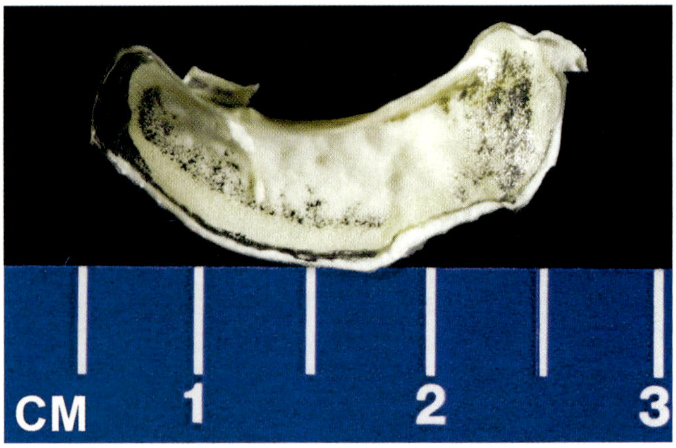

Figure 9 Section of dehydrated articular cartilage showing non-aqueous matrix

Water in articular cartilage has been the subject of several recent studies. New testing techniques have demonstrated that articular cartilage can be studied in terms of its nonlinear and viscoelastic behavior, mechano-electrochemical properties and anisotropy characteristics. New theories have been formulated based on new findings. The evidence obtained from experiments and theoretical studies provides for in-depth understanding of the structure-function relationship of articular cartilage hydration. "There is now no doubt that each phase (the charged solid matrix, water and ions) of the cartilage contribute to its compressive, tensile, electrokinetic and transport behaviors" (Armstrong and Mow 1982).

Chondrocytes

The chondrocyte is the sole architect of cartilage (**Fig. 10**).

Chondrocytes reside in smooth walled lacunae which join with their surrounding tissues to form chondrones. Each chondrocyte has a centrally located round or spherical nucleus containing DNA (**Fig. 11**). The word "chondron", denoting a cluster of chondrocytes with pericellular and territorial matrices, has been reintroduced recently (Poole 1997, Malinin and Malinin 1999). Although chondrons

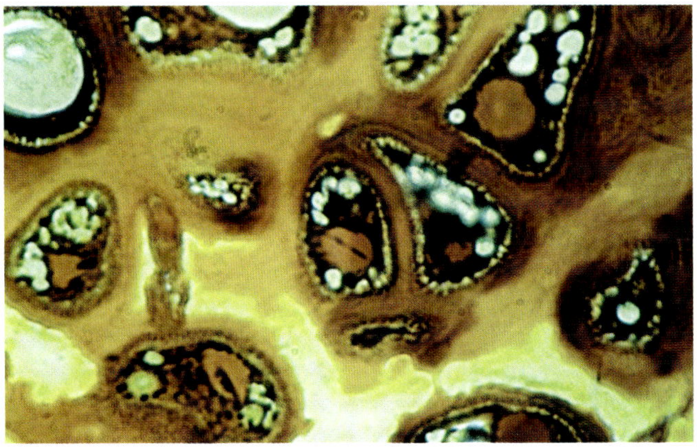

Figure 10 Chondrocytes in normal hyaline cartilage show intranuclear lipid and glycogen inclusions in the nuclei as well as along the nuclear edge. These inclusions are thought to be in the form of liquid crystals (Structural coloration technique × 1,600)

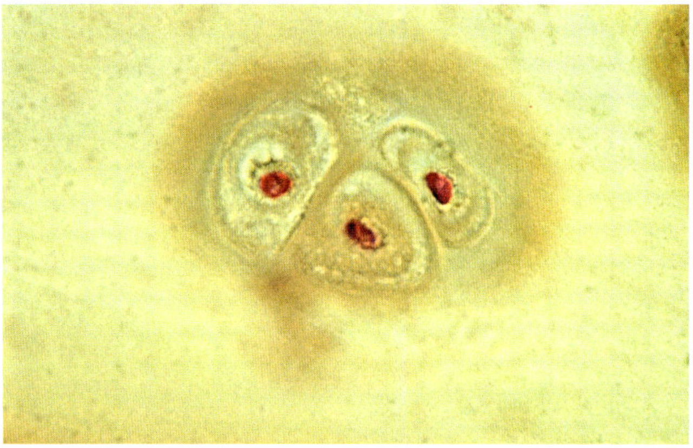

Figure 11 Chondron containing 3 chondrocytes surrounded by a rim of tannophilic protein. Red staining nuclei indicate DNA (FOTA × 900)

were described by Benninghoff in 1925, (Benninghoff 1925 a, b) his findings were largely ignored for some 40 years until Szirmai (1969) reintroduced the chondron concept. Benninghoff used polarized light microscopy to demonstrate a structural unit of a chondrocyte or a group of chondrocytes in their pericellular microenvironment. The concept of fluid-filled chondron representing a defined unit did not receive recognition until recently when Pool and co-workers established chondron as the microanatomical unit of adult articular cartilage (Poole et al 1988).

The cytoplasm of chondrocytes is granular and contains glycogen and lipid droplets as well as other inclusions. *In vivo* the cells fill the lacunae completely. However, in fixed histological preparations, with some exceptions, there is distortion which leaves cells separated from the walls of the lacunae. This is explained by the presence of lipid and glycogen which become dissolved during routine processing for histological examination, thus creating this artifact. Chondrocytes assume different shapes according to the zones of articular cartilage in which they reside; those close to the surface are flattened and elongated with long axes parallel to the surface. In the central and deep zones, cells are generally spherical and ovoid. In the deep zone, chondrocytes are in columns oriented perpendicular to the surface. Vacuoles are present in the cytoplasm in all but the superficial zone chondrocytes. Chondrocyte properties are unique and are clearly different than other mesenchymal cells. Ultrastructurally, chondrocytes have been studied primarily in experimental animals (Palfrey and Davis 1966) whereas similar studies in humans are scant (**Fig. 12**).

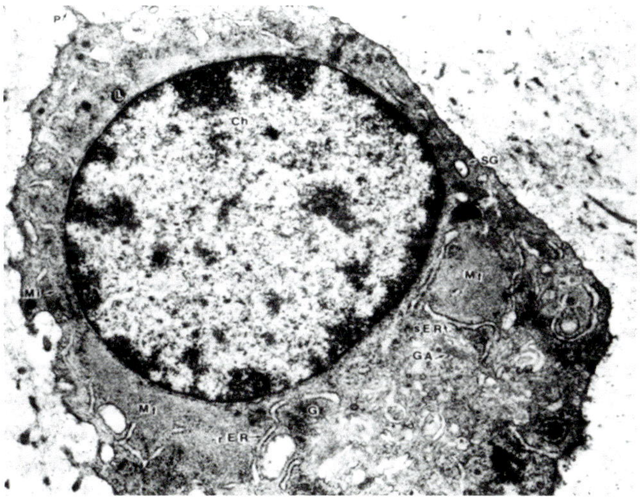

Figure 12 Human chondrocyte from the deep zone of articular cartilage. Nucleus has euchromatin at the margin. Cytoplasm is scant in comparison to the nucleus. The presence of organelles and glycogen granules indicate high metabolic activity (× 16, 800)

In chondrocytes from the middle and deep zones, the nucleus is oval with a smooth contour and a complete nuclear membrane made up of two layers. Nuclear pores are rare. Chromatin is distributed evenly throughout the nucleus, but there is some peripheral clumping. Nucleoli are seen frequently in serial sections.

The plasma membranes of chondrocytes exhibit short branching processes (**Fig. 13**). These processes are sometimes seen in sections without the nucleus. In some sections, pinocytotic vesicles are present beneath the cell membrane.

Endoplasmic reticulum is abundant. Many cells contain glycogen accumulations in the cytoplasm (Ghadially 1975).

The Golgi apparatus is readily identifiable but is not present in cells from all zones, suggesting there are differential levels of metabolic activity in the different layers of cartilage. It is usually absent in the cells from the superficial zone. Mitochondria are numerous. Apart from the Golgi apparatus, cytoplasmic vesicles are also present in chondrocytes. They are invested with a typical unit membrane and can be either oval or round. Their content varies; some may be empty while others may be filled with electron-dense material. Whirl bodies may also be present. These are made up of irregular concentric lamellae. Dense bodies are also present in the cytoplasm. This suggests the presence of degenerative changes (Hyatt 1981).

Microtubules are found only in some of the chondrocytes, usually in the middle zone of cartilage. Under compression, the Golgi apparatus may contribute to the inhibition of proteoglycan synthesis (Parkkinen et al 1993). Many chondrocytes contain the so-called cytoplasmic crystals, also called inclusion crystals (Malinin and Malinin 1999). These are distinct from the accumulation of glycogen or lipids. They are blue-colored in transmitted light and strongly birefringent in polarized light. These inclusion crystals appear in various shapes, usually aggregating at the nuclear poles. The most likely composition of the inclusion crystals is liquid crystal configuration (**Fig. 10**). The significance of these structures is unknown at this time.

Reaction to Injury

Articular cartilage has a very limited capacity for regeneration. If a defect in the cartilage is confined to the cartilage plate alone, it will fail to heal spontaneously (Hunziker and Rosenberg 1996; Kim et al 1991). If, on the other hand, injury involves both the cartilage and the subchondral bone, repair may take place. However, the extent of the repair process is limited in both quantity and quality (De Palma et al 1966). The degree of its unpredictability may be either partial or it might be completely lacking (**Fig. 14**). Complete repair has not been demonstrated in clinical or experimental conditions (Hunziker 1999).

There are several possible explanations as to why spontaneous articular cartilage repair does not take place. Most obvious is the fact that cartilage is avascular and thus, movement of inflammatory cells to the site of injury does not take place. In addition to the lack of vascular supply, cartilage does not harbor a repository of undifferentiated mesenchymal cells that can participate in the repair response that is observed in other tissues. Chondrocytes in cartilage may also have a limited capacity to produce and deliver matrix at distances away from the chondrons in which they reside.

Injury to the articular cartilage does not produce hemorrhage, or an inflammatory response and thus granulation tissue formation. Chondrocytes near the site of injury proliferate and form clusters, but they do not migrate into the lesion. Any new matrix produced remains in the vicinity of chondrones, and although cartilage maintains malleability and the potential for restructuring, it does not regenerate although the mechanism of cartilage flow exists (Calandruccio and Gilmer 1962). The mechanism of cartilage repair has been a subject of much attention (Hunziker and Rosenberg 1996).

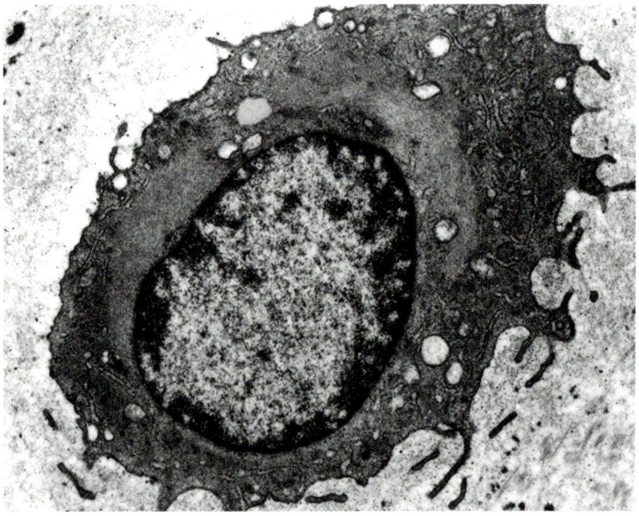

Figure 13 Electronphotomicrograph of a human chondrocyte showing short branching processes × 12,000

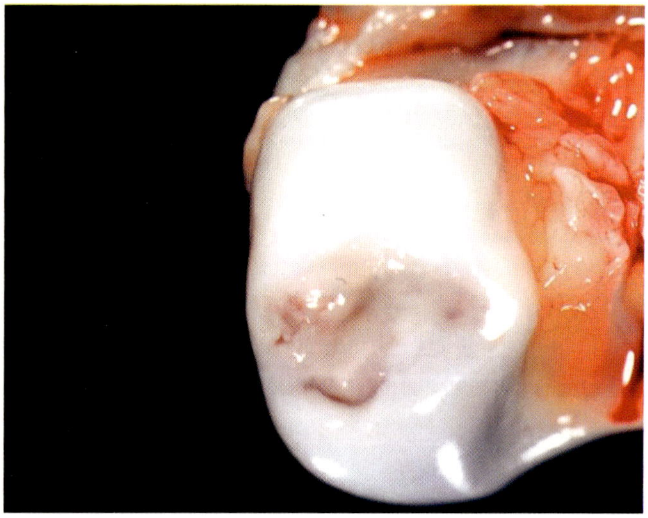

Figure 14 Cartilage repair and regeneration are limited, as seen in a knee of an experimental animal 12 weeks post injury

The phenomenon of cartilage flow constitutes the reduction in the size of cartilage defect by a rapid flow of surrounding matrix. Cartilage flow can be observed in defects produced by drilling (Bruns et al 1997). Exuberant cartilage flow was also observed in transplanted cartilage after prolonged storage at hypothermic temperatures in cell culture media **(Fig. 15)** (Malinin et al 1985). In the latter case cartilage flow was accompanied by extensive chondrocyte proliferation **(Fig. 16)**.

The mass of the "flowing" cartilage is characterized by arciform bending toward the defect of horizontally aligned superficial collagen fibers (Burns 1997). Ultrastructurally, numerous matrix-derived morphologically diverse elongated structures constitute another feature of cartilage flow mass (Ghadially et al 1977). Like collagen fibers, spindle-shaped cells presumably of perichondral origin, form punctuate chains transitioning into defects. The cell density of the flow mass is variable. Some observers report cell mass lower than that which is observed in normal cartilage; however, our observations on experimental animal flow mass show the cell density to be higher than that in normal cartilage (Malinin et al 1985).

Concomitantly, with the onset of cartilage flow, the native cells at the edge of the defect become necrotic and lose metachromasia as demonstrated in toluidine blue stained preparations. Mitotic activity is reported to be absent in the proximity of the defect and in the mass of the cartilage flow (Fuller 1972).

The repair of cartilage defect involving subchondral bone depends on several factors. One of these is the severity of the injury and the extent of tissue loss. Repair of larger osteochondral defects is less complete and less predictable than healing of smaller defects.

Mechanical injury to the cartilage involving subchondral bone results in hemorrhage, followed by a fibrin clot formation and an accumulation of inflammatory cells. This is the same sequence of events as in a fracture healing with recruitment of progenitor cells and the temporal release of cytokines. Hemorrhage or the lack thereof is not the only difference between injuries involving only cartilage and those involving subchondral bones as well. Injury to only cartilage evokes no response. On the other hand, involvement of subchondral bone evokes a definite and predictable reaction. Bone osteoid and bone marrow contain several growth factors. Bone marrow contains cells which release with cytokines and signal other cells to participate in injury and repair. In addition, platelets release platelet derived growth factor (PDGF) and transforming growth factor (TGF). Although the role of growth factors in cartilage repair has not been defined with precision, it is likely that platelet released factors stimulate proliferation of chondrocytes and

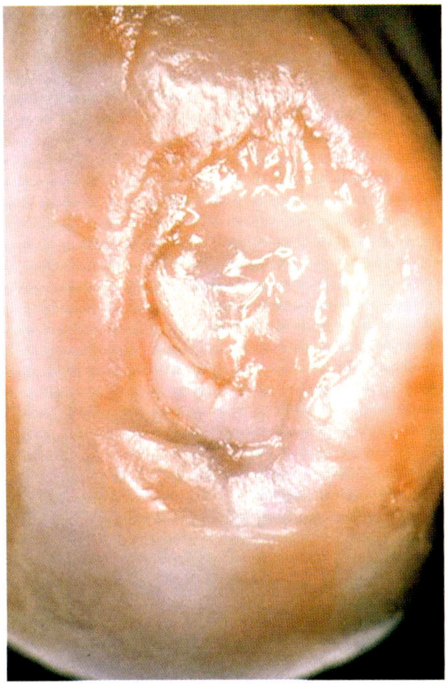

Figure 15 Example of "cartilage flow", a process which sometimes occurs during cartilage repair. The phenomenon is of interest because it demonstrates cartilage malleability and massive restructuring. Unfortunately, the process of cartilage flow does not culminate in complete remodeling and restoration of normal cartilage structure

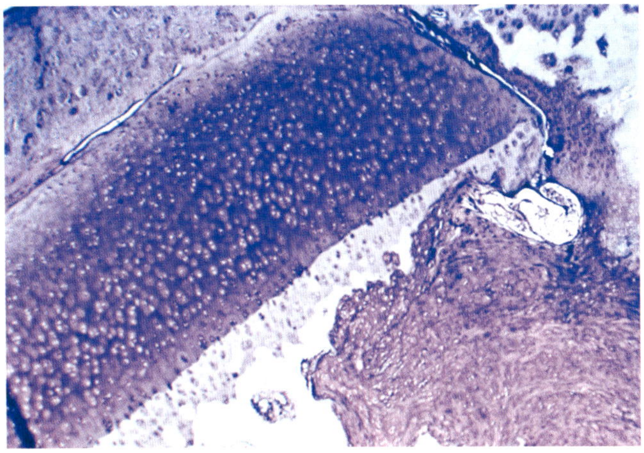

Figure 16 "Cartilage flow" is accompanied by translocation and formation of cartilage plates in subchondral spaces (Romanowsky-Giemsa × 35)

undifferentiated mesenchymal cells that may either directly or indirectly participate in repair. Within weeks, some of the mesenchymal cells assume morphologic characteristics of chondrocytes and produce collagen II and proteoglycan containing matrix. Within the subchondral zone of injury, immature bone is formed. The pattern or repair here is identical to that encountered in fracture callus formation. The bone repair restores the original level of subchondral bone. The chondral defect is rarely repaired completely and usually contains, in addition to chondrocytes and chondroid matrix, a large amount of fibrous tissue. Thus, the cartilage repair is incomplete, the defect being filled with a combination of hyaline cartilage and fibrocartilage. The difference between bone and cartilage in the formation of the regenerating tissue points to the discrepancies between these tissues. In the chondral portion of the defect, primitive mesenchymal cells are not converted into chondrocytes capable of forming normal hyaline cartilage, while in the osseous portion of the defect these are converted into osteoprogenitor cells which effect complete replacement of the lost osseous tissue. For completion, repair of cartilage requires appropriate matrix, cell signaling and progenitor cell differentiation from the periphery inwards. Our observations suggest that under optimal, induced conditions, repair can occur over time and produces cartilage that is morphologically and physiologically identical to articular cartilage. However, it is unclear whether or not this regenerated cartilage is sustainable over time.

Chondrocyte In Vitro

The enthusiasm for clinical transplantation of autologous chondrocytes has generated many investigative efforts dealing with chondrocyte cultures. To date, transplantation of autologous chondrocytes constitutes a promising modality in the treatment of isolated chondral defects (Brittberg et al 2003) when compared to the gold standard, transplantation of fresh osteochondral allografts (Bugbee and Convery 1999, Fox et al 2002).

Autologous chondrocyte transplantation is made possible by cell culture techniques that expand the number of cells needed to fill a local defect. Since the source of autologous cartilage is limited to chondrocytes that are isolated from small slices of cartilage harvested arthroscopically, the number of cells available is limited. In addition to the time needed to expand the cultures, cost is significant.

Under these circumstances, cartilage cell culture has clinical relevance. A brief description of *in vitro* chondrocyte cultivation is provided. Two methods are in existence for the propagation of chondrocytes *in vitro*. One depends on the enzymatic dispersion of tissues and isolation of chondrocytes. The other one employs an explant technique in which cells grow directly from the normal cartilage without disruption of the cartilage matrix.

The procedures for isolating chondrocytes became feasible because of the availability of degrading enzymes such as trypsin, collagenase, pronase, deoxyribonuclease, and hyaluronidase which disrupt the cartilage matrix, releasing single and clumped chondrocytes (Nevo et al 1992). Enzymatic digestion of tissues for culturing cells is not applied to cartilage alone, but has been used with many other tissues. It is a standard cell culture technique.

Many attempts have been made in the past to grow chondrocytes *in vitro* in a manner which would allow for retention of chondrocyte phenotypic expression indefinitely. None of these met with unqualified success. Chondrocytes isolated by enzymatic digestion measure in diameter 14.9 +/- 2.2 microns, those in histological sections 14.2 +/- 2.9 microns. Thus the cells in culture are comparable to cells in the native state (Wang et al 2007). When cultured in monolayers, cells tend to adhere to the culture vessel surface and go through a dedifferentiation process acquiring fibroblast-like appearances and characteristics (Mandl et al 2004). This is the usual pattern of cell growth in cell culture. Substitutions of media seem to have little effect. However, bovine fetal serum appears to be essential for chondrocyte propagation. In order to produce cell cultures capable of forming organized hyaline cartilage, it is essential to prevent chondrocyte dedifferentiation (De Camargo and Bittencourt 2009). Several reports in the literature state chondrocytes in culture require 3D scaffolding in order to produce cartilage matrix. To this end, several strategies have been devised. The most popular of these are growing chondrocytes on hydrogel, and synthetic polymer matrices. According to some investigators, chondrocytes cultured in alginate maintain their phenotypes for as long as 8 months (Masuda et al 2003), while others claimed shorter periods of about 4 months (DeCamargo 2009). The other method of initiating chondrocyte cultures is by an explant technique (Hornicek et al 1988). By this method, fragments of cartilage are placed into tissue culture flasks

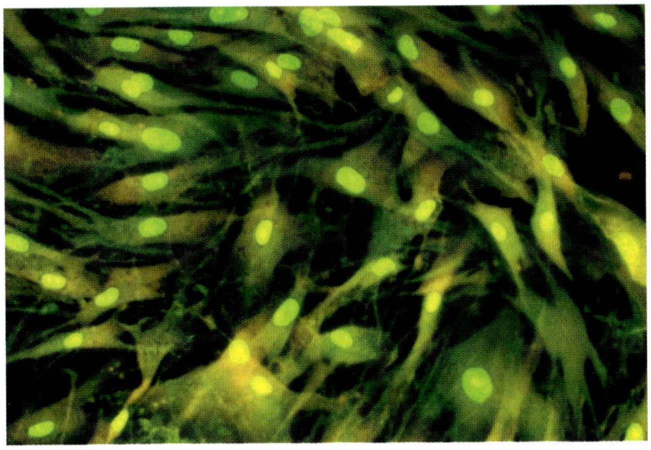

Figure 17 Chondrocytes in a monolayer culture stained supravitally with acridine orange (x100)

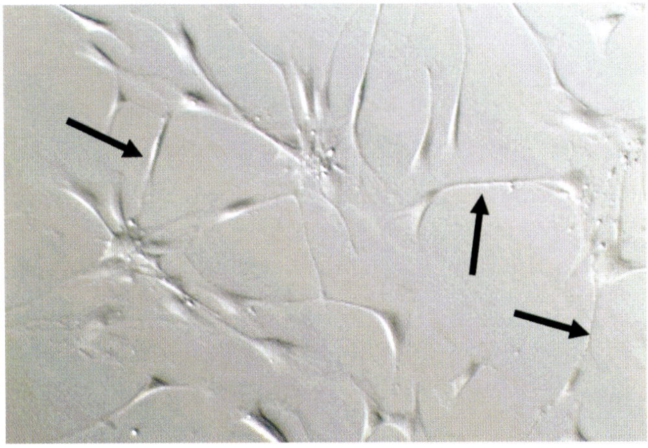

Figure 18 Large cells and numerous intercellular bridges (arrows) are typical in chondrocyte cultures (Hoffman modulation microscopy x50)

and cells migrate directly from these onto the surface of the culture vessel. Like cultures derived from the enzymatically dispersed chondrocytes, explant-initiated monolayer cultures will eventually dedifferentiate. Cell density is one of the critical factors in maintaining chondrocyte phenotypes. When chondrocytes are plated at high density in culture flasks their phenotypes do not change (Ruggiero et al 1993). Placement of a large number of cartilage explants into a single culture flask produces a high density monolayer culture. Chondrocytes in culture maintain a characteristic stellate appearance **(Fig. 17)**. There is also continuous interaction between chondrocytes as revealed by time-lapse cinematography. Intercellular bridges are numerous suggesting, in culture at least, that chondrocytes are dynamic and physiologically robust, not seemingly asocial, asexual and isolated in dense matrix **(Fig. 18)**.

Despite initial enthusiasm, cell therapies alone have been discouraging. Cells devoid of matrix and hence appropriate signaling are doomed to fail. The combination of an appropriate matrix, cells and perfectly timed signaling are essential to normal repair.

BIBLIOGRAPHY

1. Armstrong CG, Mow VC. Variations in the intrinsic mechanical properties of human articular cartilage with age, degeneration and water content. Journal of Bone and Joint Surgery (Am), 1982;64(1):88-94.
2. Barland P, Janis R, Sandson J. Immunofluorescent studies of human articular cartilage. Annals of the Rheumatic Diseases, 1966;25:156-64.
3. Benninghoff A. Der functionelle Bau des Hyalineknorpels. Erg Anat Entw Gesch, 1925;26:1-54.
4. Benninghoff A. Form und Bau der Gelenkknorpel in Ihren Bezeihungen zur Funktion. Zweiter Teil. Der Aufbau des Gelenkknorpels in seinen Bezeihungen zur Funktion. Zeitschrift für Zellforchung, 1925;2:783-862.
5. Bergman, Affi AK. Atlas of Microscopic Anatomy. Philadelphia: WB Saunders Co; 1974.
6. Bloom W, Fawcett DW. A Textbook of Histology. London, Philadelphia: WB Saunders Co; 1962.
7. Bonner WM, Jonsson H, Malanos C, et al. Changes in the lipids of human articular cartilage with age. Arthritis and Rheumatism, 1975;18:461-73.
8. Brighton CT. Structure and function of the growth plate. Clinical Orthopaedics and Related Research, 1978;136:22-32.
9. Brittberg M, Peterson L, Sjorgen-Johnson E, et al. Articular cartilage engine city with autologous chondrocyte transplantation. Journal of Bone and Joint Surgery (Am), 2003;85(suppl 3):109-15.
10. Bruns J, Kersten P, Silberman M, et al. Cartilage-flow phenomenon and evidence for it in perichondrial grafting. Archives of Orthopaedic and Trauma Surgery, 1997;116: 66-73.
11. Bugbee WD, Convery FR. Osteochondral allograft transplantation. Clinical Journal of Sport Medicine, 1999;18:67-75.
12. Butler WF. Metachromasia and alcian blue staining of the intervertebral disc of the calf. Journal of Anatomy, 1968;102(Pt. 2):301-10.
13. Calandruccio RA, Gilmer WS. Proliferation, regeneration, and repair of articular cartilage of immature animals. Journal of Bone and Joint Surgery (Am), 1962;44:431-55.

14. Cample JKL, Allard SA. Limitations of safranin O staining in proteoglycan-depleted cartilage demonstrated with monoclonal antibodies. Histochemistry, 1988;89:185-8.
15. Clark ER, Clark EL. Microscopic observations on new formation of cartilage and bone in the living mammal. American Journal of Anatomy, 1942;70:167-200.
16. Collins R, Griffith WP, Phillips FL, et al. Staining and fixation of unsaturated membrane lipids by osmium tetraoxide. Crystal structure of a model osmium intermediate. Biochimica et Biophysica Acta, 1973;320:745-7.
17. Curran RC, Kennedy JS. The distribution of sulfated mucopolysaccharides in the mouse. Journal of Pathology and Bacteriology, 1955;70(2):449-57.
18. De Camargo Bittencourt RA, Pereira HR, Felisbino SL, et al. Chondrocyte cultures in tridimentional scaffold: alginate hydrogel. Acta Ortopédica Brasileira, 2009;17:1-10.
19. De Grauw JC, Brama PA, Wiemer P, et al. Cartilage-derived biomarkers and lipid mediators of inflammation in horses with osteochondritis dissecans of the distal intermediate ridge of the tibia. American Journal of Veterinary Research, 2006;67:1156-62.
20. De Palma AF, McKeever CD, Subin DK. Process of repair of articular cartilage demonstrated by histology and autoradiography with tritiated thymidine. Clinical Orthopaedics and Related Research, 1966;48:229-42.
21. Doege KJ, Sasaki M, Kimura T, et al. Complete coding sequence and deduced primary structure of the human cartilage large aggregating proteoglycan, aggrecan. Journal of Biological Chemistry, 1991;266:894-902.
22. Fox JA, Freedman KB, Lee SJ, et al. Fresh osteochondral allograft transplantation for articular cartilage defects. Operative Techniques in Sports Medicine, 2002;10:168-73.
23. Geyer RP. Uptake and retention of fatty acids by tissue culture cells: Lipid Metabolism in Tissue Culture Cells. In: Rothblatt GH, Kritchevsky D (Eds). Philadelphia, PA: The Wistar Institute Press; 1967. pp. 33-47.
24. Ghadially FN, Thomas I, Oryshak AF, et al. Long-term results of superficial defects in articular cartilage: a scanning electron-microscope study. Journal of Pathology, 1977;12:213-7.
25. Ghadially FN. Ultrastructural pathology of the cell. London and Boston: Butterworths; 1975.
26. Goldberg M, Boskey AL. Lipids and biomineralizations. Progress in Histochemistry and Cytochemistry, 1996;31(2):1-87.
27. Grauman W. Knorpelgewebe. In: Graumann W, Neumann K (Eds). Handbuch der Histochemie. Stuttgart: Gustav Fischer; 1964. pp. 148-63.
28. Hardingham TE, Fosang AJ, Dudhia J. Aggrecan: The Chondroitin Sulfate/Keratin Sulfate Proteoglycans from Cartilage. In: KE Kuettner, R Schleyerbach, J Peyron, et al (Eds). Articular Cartilage and Osteoarthritis. New York: Raven Press; 1992. pp. 5-20.
29. Hardingham TE, Fosang AJ. Proteoglycans: many forms and many functions. Federation of American Societies for Experimental Biology, 1992;6:861-70.
30. Hascall VC (Ed). Articular Cartilage Biochemistry. New York: Raven Press, 1986. pp. 257-71.
31. Hodge J, Mckibbin B. The nutrition of mature and immature articular cartilage. Journal of Bone and Joint Surgery, 1969;51B:140-7.
32. Honner R, Thompson RC. The nutritional pathways of articular cartilage: an autoradiographic study in rabbits. Journal of Bone and Joint Surgery, 1971;53A:742-8.
33. Hornicek FJ, Reiss E, Malinin TI, et al. Establishment of primary cell cultures from human and canine parathyroid gland explants. Journal of Bone and Mineral Research, 1988;4:157-65.
34. Howard BV. Acetate as carbon source for lipid synthesis in cultured cells. Biochimica et Biophysica Acta, 1977; 488(1):141-51.
35. Hultkranz W. Ueber die spaetzichtung der Genelkkorpel. Verhandlungen der Anatomischea Gesselschaft. Kiel; 1898. p. 248.
36. Hunziker EB, Rosenberg LC. Repair of partial-thickness articular cartilage defects. Cell recruitment from the synovium. Journal of Bone and Joint Surgery (Am), 1996;78:721-33.
37. Hunziker EB. Biologic repair of articular cartilage. Clinical Orthopaedic and Related Research, 1999;367:5135-46.
38. Hunziker EB. Articular cartilage repair: are the intrinsic biological constraints undermining this process insuperable? Osteoarthritis and Cartilage, 1999;7:15-28.
39. Hyatt MA. Stains and Cytochemical Methods. New York: Plenum Press; 1993.
40. Hyatt MA. Principles and techniques of electron-microscopy. In: Biological Applications (3rd edn). FL, Boca Raton: CRC Press; 1981.
41. Jordan HE, Kindred JE. Textbook of Embryology. New York, London: Appleton-Century-Crofts; 1948.
42. Jordan HE. A Textbook of Histology. New York: Appleton-Century-Crofts; 1952.
43. Kim HKW, Moran ME, Salter RB. The potential for regeneration of articular cartilage in defects created by chondral shaving and subchondral abrasion: An experimental investigation in rabbits. Journal of Bone and Joint Surgery (Am), 1991;73:1301-15.
44. Knudsen CE, Knudsen W. Cartilage proteoglycans. Cell and Developmental Biology, 2001;12:69-78.
45. Leidy J. On the intimate structure of and histology of articular cartilage. American Journal of the Medical Sciences, 1849;17:227-94.
46. Liggins J, Furth AJ. Role of protein-bound carbonyl groups in the formation of advanced glycation end products. Biochimica et Biophysica Acta, 1997;136:123-30.

47. Lu XL, Mow VC. Biomechanics of articular cartilage. Medicine and Science in Sports and Exercise, 2008;40:195-9.
48. Mackenzie CG, Mackenzie JB, Reiss OK. Regulation of cell lipid metabolism. In: Rothblatt, GH, Kritchevsky D, (Eds). Lipid Metabolism in Tissue Culture Cells. Philadelphia, PA: The Wistar Institute Press; 1967. pp. 63-83.
49. Malinin GI, Lo HK, Malinin TI. Transmission light microscopy of structurally colored semi thin cartilage sections. Histochemistry and Cell Biology, 1998;109:1-10.
50. Malinin GI, Malinin TI. Microscopic and histochemical manifestations of hyaline cartilage dynamics. Progress in Histochemistry and Cytochemistry, 1999;34:163-242.
51. Malinin GI. Metachromatic staining of sodium bisulfate addition derivatives of glycogen. Journal of Histochemistry and Cytochemistry, 1970;18:834-41.
52. Malinin GI. The in situ determination of melting-solidification points of liquid inclusions in fixed cultured cells. Journal of Histochemistry and Cytochemistry, 1980;28:708-9.
53. Malinin TI, Ouellette EA. Articular cartilage nutrition is mediated by subchondral bone: a long-term autograft study in baboons. Osteoarthritis and Cartilage, 2000;8:483-91.
54. Malinin TI, Wagner JL, Pita JC, et al. Hypothermic storage and cryopreservation of cartilage. Clinical Orthopaedics and Related Research, 1985;197:15-26.
55. Mandl EW, Van Der Ween SW, Verhaar JAN, et al. Multiplication of human chondrocytes with low seeding densities accelerates cell yield without losing redifferentiation capacity. Tissue Engineering, 2004;10(1-2):109-18.
56. Masuda K, Sah RL, Hejna MJ, et al. A novel two-step method for the formation of tissue eugenllized cartilage by mature bovine chondrocytes. The alginate-recovered-chondrocyte (ARC) method. Journal of Orthopaedic Research, 2003;1(1):139-48.
57. Matthews JJ, Martin JH, Collins EJ. Metabolism of radioactive calcium by cartilage. Clinical Orthopaedics and Related Research, 1968;58:213-24.
58. Montagna W. Glycogen and lipids in human cartilage with some cytochemical observations on the cartilage of the dog, cat and rabbit. Anatomical Record, 1949;103(1):77-92.
59. Muller-Gerbl M. The subchondral bone plate. Advances in Anatomy, Embryology and Cell Biology, 1998;141:1-134.
60. Nevo Z, Robinson D, Halperin N. The use of grafts composed of cultured cells for repair and regeneration of cartilage and bone. In: BK Halled. Bone, Fracture Repair and Regeneration. London, Boca Raton, Ann Arbor: CRC Press; 1992. pp. 123-52.
61. Palfrey AJ, Davis DV. The line structure of chondrocytes. Journal of Anatomy, 1966;100:213-26.
62. Parkkinen J, Lammi MJ, Pelttazi A, et al. Cultural Golgi Apparatus in hydrostatically loaded cartilage chondrocytes. Annals of the Rheumatic Diseases, 1993;52:192-8.
63. Phemister DB. The causes of and changes in loose bodies arising from articular surface of the joint. Journal of Bone and Joint Surgery, 1924;6:278-315.
64. Poole CA. Articular cartilage chondrons: form, function and failure. Journal of Anatomy, 1997;191:1-13.
65. Poole CA, Flint MH, Beaumont BW. Chondrons extracted from canine tibial cartilage: Preliminary report on their isolation and structure. Journal of Orthopaedic Research, 1988;6:408-19.
66. Putschar W. Über Fett in Knorpel unter normalen und pathiologischen Verhalsnissen. Beitr Path Anat, 1931;87:526-9.
67. Rosenberg L. Chemical basis for the histological use of Safranin O in the study of articular cartilage. Journal of Bone and Joint Surgery (Am), 1971;53:69-82.
68. Rouget C. Des substances amyloides de leur role dans la constitution des tissues des animaux. Journal of Physiology, 1859;2:305-25.
69. Roughley PJ. The structure and function of cartilage proteoglycans. European Cells and Materials. Journal of Histochemistry and Cytochemistry, 2006;12:92-101.
70. Ruggiero F, Petit B, Ronziere M, et al. Composition and organization of the Collagen Network produced by fetal bovine chondrocytes cultured at high density. Journal of Histochemistry and Cytochemistry, 1993;41:867-75.
71. Sacerdotti C. Über das Knorpeltett. Virchows Archiv Für Pathologische Anatomie Und Physiologie, 1900;159:152-72.
72. Schajowicz F, Cabrinin RL. Histochemical studies of bone in normal and pathological conditions. Journal of Bone and Joint Surgery, 1954;36:474-89.
73. Schwarz NB, Pirok EW III, Mensch JR, et al. Domain organization, genomic structure, evolution, and regulation of expression of the aggrecan gene family. Progress in Nucleic Acid Research and Molecular Biology, 1999;62:177-225.
74. Serafini-Fracassini A, Smith JW. The Structure and Biochemistry of Cartilage. Edinburgh-London: Churchill-Livingstone; 1974. pp. 206-19.
75. Schrevel J, Gross D, Monsigny M. Cytochemistry of cell glycoconjugates. Progress in Histochemistry and Cytochemistry, 1981;14(II):1-269.
76. Small DM. The physical state of lipids of biological importance: cholesterol esters, cholesterol triglycerides. In: Blank M (Ed). Surface Chemistry of Biological Systems. New York: Plenum; 1970.
77. Sobotta J, Piersol WH. Atlas of Human Histology and Microscopic Anatomy. New York, G.E: Stechert and Co; 1930.
78. Soeder S, Kuhlmann A, Aigner T. Analysis of protein distribution in cartilage using immunofluorescence and laser confocal scanning microscopy. Methods in Molecular Medicine, 2004;101:107-25.
79. Speer DP, Dahners L. The collagenous architecture of articular cartilage. Clinical Orthopaedics and Related Research, 1979;139:267-75.
80. Steedman HF. Alciar blue 8Gs: A new stain for mucin. Quarterly Journal of Microscopical Science, 1950;3-91:477-9.
81. Stockwell RA, Meachim G. The Chondrocytes. In: Freeman MAR (Ed). Adult Articular Cartilage. Kent: Pitman Medical Publishing Co. Ltd; 1979. pp. 69-144.

82. Strangeway ISP. The nutrition of articular cartilage. British Medical Journal, 1920;1:661-3.
83. Szirmai JA. Structure of cartilage. In: Engel A & Carsson T (Eds). Aging of Connective and Skeletal Tissue. Stockholm: Nordiska Bokhanhandelns; 1969. pp. 163-84.
84. Verzijl N, De Groot J, Oldehiukel E, et al. Age-related accumulation of Maillard reaction products in human articular cartilage collagen. Biochemical Journal, 2000;350 (pt. 2):381-7.
85. Virchow R. Die Krankheiten Geschwulste. Berlin. 1863.
86. Wang XH, Weix C, Chen WY. Mechanical properties of chondrocytes isolated from normal articular cartilage: Experiment with rabbit knees. Journal of Orthopaedics and Related Research. 2007;87(13):916-20.
87. Yamamoto T, Gay CV. Ultrastructural analysis of cytochrome oxidase in chick epiphyseal growth plate cartilage. Journal of Histochemistry and Cytochemistry, 1988;36:1161-6.
88. Zyzak DV, Richardson JM, Thorpe SR, et al. Formation of reactive intermediates from Amadori compounds under physiological conditions. Archives of Biochemistry and Biophysics, 1995;316:547-54.

Chapter 5

Professor Dr. Med. Peter Ivanovich Buchmann (1872–1948), Dr Buchmann was the first to successfully transplant whole joint autografts. Dr Buchmann was Professor of Operative Surgery and Topographic Anatomy, Don Medical Institute, Rostov-on-Don, Russia.

(*Reference:* Buchmann P. Behandlung knocherner Ellbogengelenksankylosen mittels Überpfanzung von ganzen Gelenken. Zentralblatt für Chirurgie. 1908;35:582-89.)

Transplantation of Autologous Bone

INTRODUCTION

Transplantation of autologous (autogenous) bone, often referred to as bone grafting, is one of the most common transplantation procedures. Both cortical and cancellous, as well as corticocancellous bone grafts are used. The principles and techniques of bone grafting procedures were established long before the advent of the modern, metal oriented orthopedic surgery (Albee 1915). Bone grafting brought with it not only the concept of internal fixation and stabilization of the skeletal structures, but also the concept of osteogenesis. Presently bone autografts are routinely used to fill defects, to promote union, to bridge gaps in long bones and to effect arthrodesis.

The fate of autologous bone transplants has been studied in detail for over a century. The biologic behavior of these grafts is predictable. Autografts are the gold standard of bone transplantation. Behavior of autologous bone grafts serve as a yardstick for gauging the performance of all other grafts. However, osteogenesis, the process through which transplanted autogenous bone functions, still remains a subject of some controversy, particularly with respect to the origin of cells participating in it. The only safe statement one can currently make with regard to the exact mechanism of osteogenesis is that it is largely obscure (Burwell 1961).

The factual data on the behavior of autogenous bone grafts in humans, when viewed in historical perspectives, at first appear confusing. About every decade, one or several surgeons showing interest in the subject recorded observations on the biology and techniques of bone grafting. This generated advocates as well as adversaries of particular theories and techniques. Consequently, the publications on the subject are voluminous. All of these could not be cited without making this chapter monotonous, repetitious, and unduly long. However, the review of these writings reveals that with the exception of bone morphogenetic protein of Urist (Urist 1952, 1979), the present concepts regarding the behavior of bone autografts had been established by 1930 (Albee 1923; Scudder 1939). These concepts have been proven to be correct, and the subsequent findings on the physiology of the transplanted autogenous bone have been simply amplifications of previous findings.

CURRENT CONCEPTS

The "discipline" of bone grafting has produced terms such as incorporation, revascularization, osteoinduction (bone induction principle), substitution, remodeling, osteoconduction and accretion. These terms are deeply ingrained in the terminology of bone transplantation (Anderson et al 1964). Incorporation means integration of the graft into the skeletal system or union of the graft with the recipient. It does not mean complete replacement of the graft with osseous tissue of the host. Urist conveniently divided incorporation of an autograft into several phases (Urist 1980). These include mobilization of mesenchyme and vascular system. Revascularization denotes the extension of the host's blood vessels into the interstices of the cancellous bone and of the macrocanalicular system of cortical bone. Stringa (Stringa 1957) and Herndon (Herndon 1960) related graft incorporation with the speed of vascular ingrowth into the graft. The terms substitution, replacement and accretion describe the process by which the bone graft is replaced by new tissue from the surrounding host bed. This is accomplished by the resorbtion of the graft and replacement of it with new bone. Penetration of irregular layers of new bone into the bone of the graft has been termed "creeping replacement" or "creeping substitution". Remodeling is a continuous process of alteration of the graft architecture until it conforms to that of the host.

Cancellous bone autografts are generally held to promote osteogenesis, while cortical autografts are used primarily for fixation and biomechanical support. This common belief may not be entirely correct. Since bone morphogenetic protein (BMP), the substance responsible for promoting osteogenesis, is more abundant in cortical than cancellous bone (Urist et al 1983), cortical bone may be actually more osteogenic than cancellous bone. However, to be incorporated, cortical bone must undergo cancellation. What places cancellous bone at an advantage is its morphology which allows for early revascularization and rapid incorporation into the skeleton.

Bone cells (osteocytes) die when bone is deprived of its circulation. After a bone graft is removed from the donor site, osteocytes in its depth disappear and the bony matrix becomes non-viable. However, the surface cells, or at least some of them, survive for short periods of time provided they are protected from dessication, heating or exposure to toxic substances. The surface cells which survive are usually deep cells of the periosteum, those in the endosteum and cells on the surfaces of the trabeculae.

Once a bone autograft is placed in a bed with adequate vascular supply, it initiates osteogenesis from surrounding mesenchymal cells by turning these into osteoprogenitor cells. A few of the remaining cells of the graft may participate in the new bone formation, but their contribution is a minor one. The osteoid of the graft itself is gradually resorbed and replaced with new bone. Capillaries from the recipient bed enter the graft and the primitive mesenchymal-stem cells of the host differentiate into osteoblasts and osteoclasts. The process initiated by the graft is in many ways similar to the callus formation about the fracture. Cortical bone, like cancellous bone, becomes revascularized and is temporarily transformed into a cancellous-like structure, before reverting to its original form.

In humans, for autografts to become osteogenic, they must be placed adjacent to or into the recipient's bone. Autologous grafts placed in the soft tissue away from bone become resorbed. However, such is not the case in some animals, particularly rodents. In these, grafts placed heterotopically survive and form new bone.

When placed in contact with bone, osseous autografts heal and become incorporated without complications (Haas 1922). Once the graft unites with the recipient's bone, it undergoes remodeling, commensurate with the function of the bone into which it is placed (**Fig. 1**).

Histological sequence of bone transplant repair and incorporation is well documented (Enneking et al 1975). Both cancellous and cortical autogenous transplants share many similar features during the first 2 weeks post-transplantation. Osteocytes of the transplanted bone die and osteoid becomes necrotic. Transplanted bone becomes surrounded by coagulated blood. During the first week the blood clot is changed into granulation tissue with characteristic vascular buds (Abbott et al 1947). Lymphocytes, plasma cells, polymorphonuclear granulocytes, monocytes, histiocytes, mesenchymal cells, and apparently osteoclasts accumulate in the granulation tissue surrounding the graft. By the second week connective tissue cells and fibers increase, while the number of inflammatory cells decreases. Osteoclastic activity becomes apparent and proceeds along the vacant lacunae. Revascularization of small cancellous autografts may be completed in 2 weeks (Burchard 1978), and is accomplished by gradual host tissue and blood vessel ingrowth into the marrow spaces (Zeiss et al 1960). Ingrowth of blood vessels into marrow spaces is accompanied by migration of mesenchymal cells, some of which differentiate into

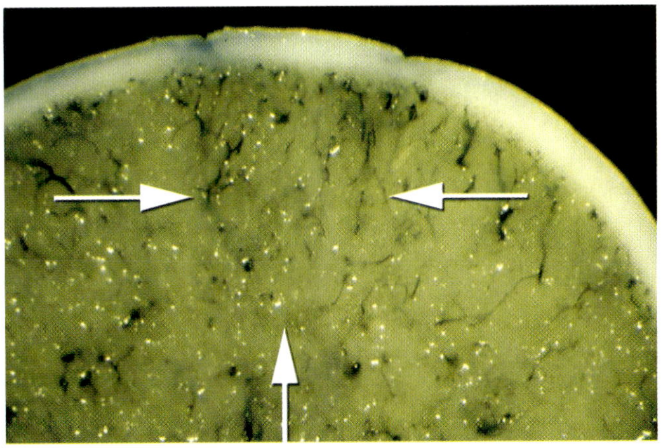

Figure 1 Autograft in an animal model 2 years post-transplantation (arrows). The specimen has been injected with India ink showing revascularization of the graft

osteoprogenitor cells. Although the exact genesis of osteogenic cells is still a subject of controversy, it is clear that these originate from the host tissue surrounding the transplant. In the marrow spaces of cancellous autografts, the osteogenic cells first differentiate into osteoblasts which line the dead bone and deposit new bone at its edges (Urist 1952). At the same time empty marrow spaces of the transplant are repopulated with new hematopoietic cells. Osteoclastic activity is present simultaneously with osteoblastic activity, and this accounts for the removal of dead compact bone of the transplant.

The most striking difference between the cancellous and cortical autografts is the rate of revascularization. Roughly, the cortical grafts become revascularized at half the rate of the cancellous grafts. This can be explained by the structure of cortical bone. Blood vessel ingrowth occurs through the Haversian system, Volkman's canals and pre-existing vascular channels. Some investigators hold that vascular penetration of cortical graft can occur only subsequent to osteoclastic resorption at the periphery of the graft (Abbott 1947, Enneking et al 1975, Deleu and Trueta 1965). The second difference between the cortical and cancellous autografts is in the rate of resorption. In cortical grafts, osteoclastic activity precedes osteoblastic activity and revascularization, while in the cancellous grafts, deposition of new bone proceeds concomitantly with revascularization and resorption. However, once the reparative process is well underway, the Haversian canals and other pre-existing channels become enlarged and lined with osteoblasts. The cavities of new bone become filled with active bone marrow. Eventually the bone is completely remodeled (**Figs 2A and B**).

The incorporation of the transplanted autogenous bone graft is primarily the function of host cells and tissues. As noted, the contribution of the viable cells from the graft itself is not very large (Adamkeiwicz 1889). In studying small bone fragments, de Souza-Ramos (1980) estimated degeneration of almost 94% of osteocytes. Similar findings were reported by Frame et al (1982). Gray and Elves actually measured donor cells' contribution to osteogenesis in experimental cancellous grafts in rats by the uptake strontium-85 Sr (Gray and Elves 1982) and noted that endosteal lining cells, together with the marrow stroma cells outnumbered periosteal cells 2:1. No significant contribution was noted from hematopoietic cells and osteocytes. When these findings were compared to those for cortical bone grafts it appeared that the rate of graft cell participation in osteogenesis in cancellous grafts was 1.7 times higher than in cortical grafts.

Using intravital microscopy, Albrektsson and Linder followed vascular changes and bone remodeling in the same bone autograft from the time of transplantation to 1 year post-transplantation. Autografts which received new blood vessels early (5–9 days post-transplantation) contained viable bone and maintained circulation 1 year

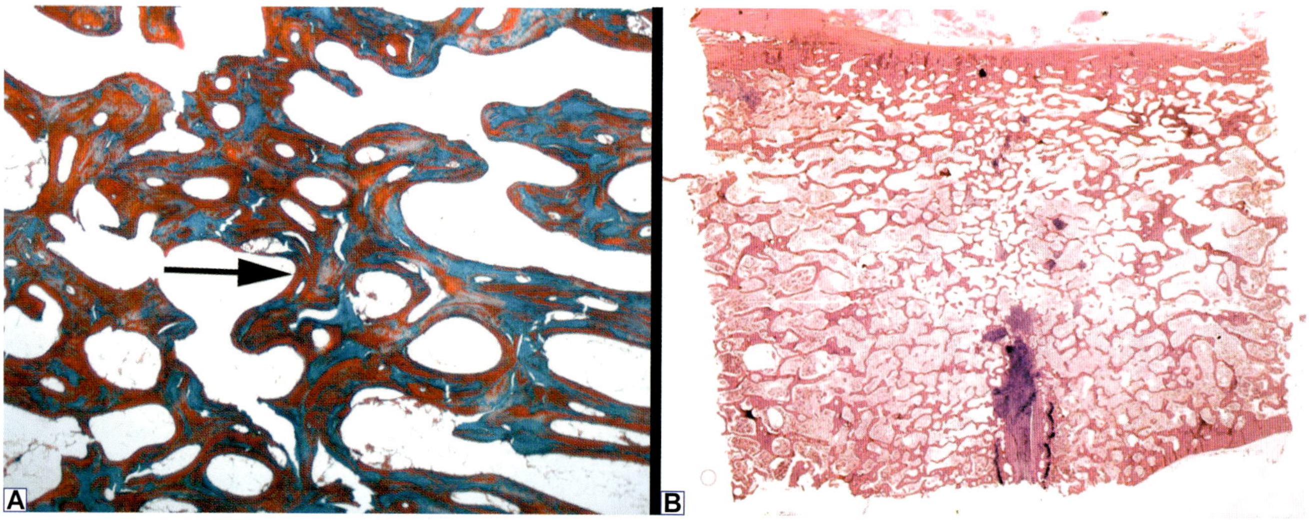

Figures 2A and B (A) Junction of the autograft with the host 2 years post-transplantation (arrow). The autograft has been completely incorporated (Masson-trichrome stain × 25); (B) Spine of an experimental animal (baboon) 3 years post discectomy and autologous bone transplantation. The graft has been completely replaced with host bone (H&E × 10)

after the transplantation (Albrektsson and Linder 1981). The rate of revascularization was related to the amount of trauma suffered by the graft and the host bed. In autografts receiving minimal trauma, blood vessels appeared on the average of 7 days post-grafting. On the other hand, in severely traumatized graft, blood vessel ingrowth occurred only at 15 days post-grafting. Bone remodeling started when the vascular density resembled the normal pattern. This took place earlier and more rapidly in the minimally injured autografts and host bed (Albrektsson 1980). These findings strongly suggest the advisability of exercising due care when bone autografts are excised and transplanted. Avoidance of undue trauma to the graft and the bed into which it is implanted will result in a quicker incorporation and healing of the graft.

Onlay grafts in contact with bone diminish in size over a period of time, i.e., they undergo remodeling. Bone grafts subjected to compression and strain either retain their general space or become larger. "Thus functional use appears to influence the retention of the calcified matrix of the graft" (Peer 1955).

Bone autografts in humans transplanted in contact with the host bone undergo revascularization, incorporation, replacement and remodeling. Periosteum is not needed for a successful transplantation of a bone autograft. However, when the same grafts are transplanted into soft tissues, not in contact with bone, they are eventually replaced with fibrous tissue. An exception to this is the survival of the corticocancellous bone implanted in the nose. Unlike other bones, nasal bones likewise retain their calcified structure when implanted into muscle or fat.

The various stages of bone graft healing are similar to those of healing of fractures except that in fractures there are usually only two opposing surfaces. With grafts there are several surfaces which must heal. Good vascular supply is needed to bring this process about (Buxton 1951).

The general health of the graft recipient must be satisfactory to allow for adequate healing. Bone graft incorporation and osteogenesis are obviously more rapid in the young and middle aged, but bone autografting is not contraindicated in patients in advancing age. The local conditions are also of importance. Tissues other than bone must be adequate. A bone graft will not survive without good nutrition, coverage with muscle, subcutaneous tissue and skin.

History of the Autologous Bone Transplantation

The acceptance of bone grafting techniques did not take place overnight. Its development spanned several decades and was not without controversy. In the United States, the surgeon who was probably most influential in promoting the use of autologous bone grafts was Fred H Albee. In 1908 he published a report on arthodesing the hip joint with a bone graft (Albee 1908). He then developed the technique of spinal fusion, and by 1923 based his experience on 1,000 clinical cases (Albee 1915, 1940), and laid down some well defined surgical principles. He stated the graft should be so placed as to receive sufficient blood supply and whenever possible should be fixed internally (Albee 1923). He believed the bone graft acted as a stimulus to osteogenesis to the bone into which it was grafted or with which it was placed in contact. He presented his findings before the Congress of German Surgeons on April 14, 1914. Finally, he showed that when a bone graft is "well contacted" it becomes adherent to the recipient bone by newly formed tissue which changes to solid bone within a few weeks (Albee 1914).

Ollier, a French surgeon, is generally credited with being the first to seriously deal with bone grafts (Ollier 1867). However, he considered the procedure dangerous in humans and his entire clinical experience consisted of transplanting periosteum in only one patient. Since he performed his operations in the pre-antiseptic era and without benefit of anesthesia, his reluctance to use bone grafts clinically is hardly surprising. The operation involving human graft was performed by him on May 13, 1835. He transplanted a piece of tibial periosteum under the skin of the forehead to remodel the nose. The periosteum was removed from the bone with a chisel, the skin of the forehead elevated and the graft placed underneath. The wound became infected and because of the fear of erysipelas the graft was removed. When this was being done Ollier noted that the periosteal graft was adherent to underlying bone and was beginning to grow (Delangeniere 1921).

John Hunter provided an early impetus to tissue transplantation by showing general interest in it (Hunter 1837), but one of the first recorded transfers of autogenous bone was performed by von Walther in 1820 (von Walther 1821) and Wolf in 1869 (Wolf 1869). However, it was not until the publication of the experimental work by Ollier (Ollier 1867),

Radzimowski (Radzimowski 1881), Jakimowitsch (Jakimowitsch 1881), Bonome (Bonome 1885), Adamkiewicz (Adamkiewicz 1889), and others, as well as the advent to aseptic surgery that transplantation of bone in patients began to be performed in increasing numbers. It is interesting that Adamkiewicz stressed the importance of antisepsis, stating he had not encountered failures with aseptically transplanted bone. He also said the healing of the graft proceeded via the connective tissue cells which invaded the transplanted bone and subsequently became ossified.

Chase and Herndon, in reviewing literature, found forty-five reasonably documented clinical reports of autologous bone transplantation during the last decade of the 19th century (Chase and Herndon 1955). Curtis, in 1892, described 4 patients which received bone grafts. One had a 5 cm post-traumatic loss of the tibia. The gap was filled with the fibular graft inserted into the medullary canal of the tibia. Perfect union resulted, with the patient walking long distances 8 months post-grafting (Curtis 1892). Bardenheuer, in 1896, transplanted the scapular spine to the humerus for osteomyelitis, following the resection of a tuberculous shoulder joint (Bardenheuer 1896). Good results with bone autografts were also reported by Poncet (Poncet 1887), von Eiselberg (von Eiselberg 1897) and others.

The first decade of the 20th century witnessed an explosive development in autologous bone transplantation. No less than 162 reports dealing with autologous bone grafts appeared during this period. Hibbs achieved a remarkable success with bone transplants (Curtis 1892). He denuded cortical bone of the vertebrae, excised cartilage from facet joints and used transverse processes as well as bone obtained from vertebral bodies to fuse the spine. He likewise fused the knee and the hip with local bone as graft material.

Experimental work by Saltykow (1909), Mayer and Wehrner (1914), and particularly Axhausen (1907 and 1911) further stimulated the interest in bone grafts by clinical investigators such as Lexer (1908 and 1911), Murphy (1912), Tietze (1902), Müller (1900) and Albee (1923) to name but a few. By this time Lexer became convinced that autologous transplanted bone died but the cells of the endosteum and periosteum survived and gave rise to new bone.

The idea of periosteum being necessary for the success of the graft was persistent. Delangeniere (1920), a student of Ollier, used osteoperiosteal grafts in an attempt to reconstruct the noses destroyed by lupus. In the first patient the graft was prepared in two stages. Osteoperiosteal graft 9 × 2 cm was removed from the medial aspect of the tibia, with one end left attached to the tibia. The skin over the tibia was then sutured around the base of the free end of the graft, thus allowing it to protrude through the wound. A tunnel was then made in the forearm, the graft inserted into the tunnel and the forearm attached to the tibia with a plaster cast. Two weeks later the pedicle was cut and the wounds of the forearm and the tibia closed. The pedicle was transferred to the nose 4 months later. The graft took well and 10 years later the grafted bone was solid but the reconstructed nose was irregular in appearance. Osteoperiosteal graft from the tibia used to repair a pseudarthrosis of the radius resulted in a solid union with perfect function.

In 1908, Buchmann performed the first joint transplants by transferring the first metatarsal-phalangeal joints into 2 patients following resection of the elbow joint (Buchmann 1908). He resected the ankylosed joint first and after preparing the bed for the graft, excised the metatarsalphalangeal joint from the contralateral foot. The resection was extra-capsular. The joint was placed into the elbow so that the plantar surface faced posteriorly with metatarsal bone next to the humerus and the phalanx abutting on the ulna. No internal fixation was used. The elbow was immobilized in a plaster cast in an extended position. Twenty-one days after surgery the cast was changed with the elbow in flexion. Roentgenograms showed healing of the host graft junctions. Long term results were good.

Lexer was not enthusiastic about the operation, stating that a homograft would be more suitable for the purpose (Lexer 1925). However, the idea of autologous joint transplantation was appealing and several attempts in this direction were made. Successful replacement of the elbow joint with tibiotarsal joint was performed by Duval (Tuflier 1913). Goebel reported on whole joint transfer in the hand (Goebel 1913), as did McKeever many years later (McKeever 1952). Behrend transplanted the head and shaft of the fibula to replace proximal humerus (Behrend 1930).

Petrow replaced metacarpal bone resected because of a tumor with a distal phalanx from the toe (Petrow 1933). He examined the autograft 25 years after the operation and noted that the graft became completely remodeled.

Excellent results were obtained by Vorschutz in 2 patients with autografts from the tibia to the mandible (Vorschutz and Klinische 1911). Carter transplanted ribs into the noses of 9 patients with saddle deformities. The grafts healed and remained in place 18 months post surgery (Carter 1911). McWilliams transplanted a rib to bridge the defect in the mandible which resulted from resection of a sarcoma (McWilliams 1912).

By 1915 the use of autologous intramedullary grafts or inlay grafts for the treatment of fractures and non-unions became rather common (Dyas 1915). Their employment for trauma obviously increased during World War I. Delangeniere and Lewin described 273 cases in which osteoperiosteal grafts from the tibia were used for reconstruction of osseous defects from war wounds and injuries (Delangeniere 1920).

In 1917, Murphy summarized his experience with bone grafts, emphasizing that the clinical success with bone grafts depended entirely on the understanding of bone regeneration. He also evolved a method for using intramedullary dowels which furnished effective internal fixation, and being entirely intraosseous, favored stimulation of osteogenesis by an intimate contact of the graft with the bone of the recipient. He described several patients in whom resected portions of bone were replaced with autografts from the tibia (Murphy 1917). Bloodgood, commenting on Murphy's work, stated that "the function of the limb can be restored after removal of large pieces of bone, even involving joint or joints so that amputation is never necessary for a disease of the extremity unless the soft parts are so involved that their removal leaves the limb without function; that bone transplantation and joint transplantation are not at all difficult" (Bloodgood 1915).

Gallie and Robertson wrote in 1919 "during the past few years many of the views on inflammation and repair in bone have been subjected to searching investigation. The result has been that some of the old beliefs have been rudely shaken and wide differences of opinion developed on points of fundamental importance (Gallie and Robertson 1919). Since periosteum contributed to the circulation of bone, it was important to them to include it in the graft. They used autologous bone plates for the treatment of non-unions, securing these to the underlying bone with steel screws, wire or pegs. The best results were obtained when an inlay method of Albee was used.

Initially osteoperiosteal grafts were taken from the tibia. Later the ribs were used as the source of these grafts. Lewin in 1924 repaired a skull defect with osteoperiosteal grafts from two ribs, with the cranial defect filling entirely with bone (Lewin 1924). Carter considered the rib, with its layer of periosteum, the bone best suited for transplantation in the face (Carter 1932). He used costochondral cartilage with a portion of a rib to reconstruct the saddle nose deformities and noted growth of the transplant at points of contact with the frontal nasal bone.

The popularity of osteoperiosteal grafts continued throughout the 1920's, but by the early 1930's they were largely replaced with full thickness grafts. It was noted it took a longer time for the osteoperiosteal graft to become solid as compared to a full thickness graft. Dorrance wrote in 1930 that full thickness grafts were now generally considered to be the best, but osteoperiosteal grafts might still be indicated in select cases (Dorrance 1930). He cited an example of a "pull through" osteoperiosteal graft used for arthrodesis of the knee, but held that to be successful. Osteoperiosteal grafts should contain a good supply of bone, and one must be as careful to fasten the graft to the bone as in a full thickness graft. He also noted it took longer for the osteoperiosteal graft to become solid as compared to a full thickness graft. By the mid-thirties discussion of the role of periosteum in the incorporation of bone autograft seemed to be no longer of paramount importance. From the practical standpoint it had been amply demonstrated that bone defects could be repaired by "transplanting or grafting periosteum, osteum and periosteum, osteum and endosteum, complete sections to include all layers, complete bone segments and even the entire joints" (Harbin and Liber 1934). The use of autografts for the treatment of non-united fractures became routine. The grafts were also used to promote healing and for internal fixation. In 1936, Henderson reported on 247 non-united fractures to which autografts were applied. A high percentage of these patients healed their non-unions. Osteoperiosteal grafts were by this time largely abandoned, with Henderson stating, "Osteoperiosteal grafts give no support and we have not been satisfied with results from their use" (Henderson 1936).

The departure from the osteoperiosteal graft did not mean abandonment of tibia as the primary source of the graft. Orthopedic surgeons still preferred it as the source of autograft, although according to Campbell, the shaft of the fibula provided a much stronger graft (Campbell 1939). Finally it was realized that removal of full thickness cortical bone plate from the tibia necessitated prolonged external support from one-half to a full year to prevent fracture, and that unless considerable biomechanical support was required, cancellous bone grafts were as good as anything. In a biopsy and radiologic study of human iliac bone autografts, Mowlem (Mowlem 1941) showed that the fate of the graft depended upon the nourishment the graft received from the host bed, something Albee noted to begin with (Albee 1923). When the grafts were placed in an avascular bed the bone was replaced by fibrous tissue. The chips of cancellous bone eventually proved to be very effective in promoting osteogenesis and their use became widespread (Muller 1900). However, the shift to the ilium as a primary source of autograft was not sudden (Mrazik et al 1980). Autografts from the ilium were apparently first introduced by Linderman in 1916 (Linderman 1916). Matti said he likewise started using cancellous autografts during World War I (Matti 1931). Henry and Geist in 1933 reported satisfactory results with multiple chips from the ilium in spinal fusions. Arthrodeses were obtained in over 75 cases (Henry and Geist 1933). Henry further noted their utility in the filling of curetted cavities from cysts, giant cell tumors and "benign types of chondrosarcoma" (Henry 1937). It followed that while the strength and stability of solid cortical bone were of great importance, this kind of bone graft might not have been the best for promotion of osteogenesis. Curiously, chips of cortical bone apparently had not been tried. Thus it followed that if mechanical stability could be maintained by other means while the bone was healing, chips from the ilium would be the preferable autograft in the arthrodesis of joints, in the repair of bone defects and in some forms of non-union or delayed union of fractures (Dick 1946). It was not until 1976 that Spence et al compared cancellous and cortical bone chips and demonstrated the utility of the latter, but the study was done with allografts, not autografts.

Several publications confirming the value of cancellous chip grafts in reconstructive surgery appeared during and after World War II (Converse 1945, Blocker and Weiss 1946). Horwitz advocated the use of cancellous bone as a graft to bridge or fill osseous defects at the ends of long bones (Horwitz 1949). However, he noted cancellous chips or sliver grafts became rapidly incorporated and were therefore not advisable alone in the repair of the long bone defects at or near the center of the shaft. In this location, massive iliac grafts with one or two cortical surfaces served very well. He based these opinions on the analysis of 85 patients with un-united fractures and large defects of long bones who received iliac bone autografts (Horwitz et al 1947). The principal advantage of iliac bone grafts in these circumstances has been their rapid revascularization and strong osteogenic potential. Failure of union with these autografts, and/or reabsorption occurred 4–5 months postoperatively in only three cases.

LABORATORY INVESTIGATIONS

Transplantation of bone in animals preceded that in humans. The results of animal experiments are largely responsible for the development of clinical bone transplantation. The first reported instance of successful autogenous bone transplant in experimental animals dated to 1809 and the description by Merren of reimplantation of trephined cranial bones (Merren 1952). From the very beginning the problem of bone transplantation was interconnected with the investigative endeavors dealing with periosteum. Duhamel, who is credited with initiating scientific studies of osteogenesis, placed silver wires subperiosteally and discovered they became encased in bone (Duhamel 1742). He believed new bone formation resulted from the osteogenic property of the periosteum. This view conflicted with that of those who regarded the periosteum as an inactive vascular support structure and thought the arteries were solely responsible for osteogenesis observed in the healing of fractures. Two divergent schools of thought arose based on the views of these men. The controversy was not settled until Flourens in 1840 showed conclusively that periosteum was osteogenic and active in the healing of fractures (Flourens 1840).

The first truly scientific study was made by Julius Wolf who, in 1863, with great exactness, described difficulties with transplantation of osseous tissues. Since the surface of the bone did not necessarily reflect on the viability of the bone itself, he studied the growth of the transplant with the vital dye, employing madder dye for the purpose. He found with the use of vital stain that autogenous bone graft can be maintained alive, but only in the flat bones.

Madder, a red vegetable dye containing anthraquinone, is selectively deposited in bone if taken internally. Madder is fixed in the entire skeleton, but greater concentration of the dye is required to stain old bone. The story of the discovery relating to madder staining of bones is curious. Sir Arthur Keith (Keith 1918) tells that in the summer of 1736 a London surgeon, John Belchier, was dining with a friend who was a calico painter. He was served a leg of pork in which the bone was red. On inquiry, he found his host had fed the pigs madder, a vegetable dye used in calico dyeing. Belchier guessed this might have been the cause of the bone staining red, and having performed an experiment to prove the point, reported his findings to the Royal Society the same year (Belchier and Goodsir 1738). Belchier subsequently did use madder to stain skeletons, but it was Duhamel who recognized its importance for the study of bone development (Lecroix 1951).

Ollier, having little success with a periosteal graft in a patient, undertook a systematic study of periosteal transplants in experimental animals. He concluded the periosteal grafts and transplanted bone remain alive and could become either osteogenic or chondrogenic. He believed the best graft to be an osteoperiosteal graft, and that both endosteum and the content of the Haversian system participated in bone regeneration (Ollier 1867). Buccholz confirmed by microscopic observation chondrogenicity of periosteum from the tibia in an orthotopic position but not in the cranium (Buchholz 1863). Physiological concepts developed by Ollier remained unchallenged until 1881 when Radzimowski showed that cells in bone transplanted with or without periosteum remained alive only temporarily (Radzimowski 1881). Boheme examined bone implanted into the back muscle of rats and reported the death of autogenous bone, but survival of the periosteum (Boheme 1885). Transplanted periosteum formed cartilage, most of which underwent endochondral ossification. The reader must be reminded at this point that in the rat, bone transplanted into muscle will differentiate, grow and will not be absorbed. Such is, of course, not the case in man nor in some other animals. Boheme equated this finding with osteoid modification of chondroid such as was observed by Kassowitz in a fracture callus (Kassowitz 1881).

At the end of the century, the equilibrium of bone transplantation physiologists was disrupted by Barth, who in a series of papers, stated that all elements of autografts died regardless of whether bone was transplanted with or without periosteum. According to Barth bone tissue itself became necrotic and was replaced by osteoblasts from the adjacent living bone. Barth's work was based on careful histologic observation of trephined bone discs removed from the skull (Barth 1893, 1895, and 1898). It attracted several supporters which included Darkschewitsch and Weidenhammer, who presented similar findings (Darkschewitsch and Weidenhammer 1892). On the other hand, Barth's contentions precipitated opposition from the proponents of the living cell transfer. Notable among these was David who published several polemical papers on the subject (David 1896, 1897).

Much was contributed to the understanding of osteogenesis and bone transplantation by Axhausen who performed his exhaustive studies between 1907 and 1909 (Axhausen 1907, 1908, 1909). His experimental work with rabbits was supplemented with microscopic examination of human biopsy material. Axhausen's studies conclusively showed transplanted bone died in an autograft, but most cells in the periosteum survived and participated in osteogenesis. These findings have been substantiated by many subsequent investigators. However, at the time one contradictory view was held by Macewan who, on the basis of experiments on young dogs, reaffirmed his previous conviction, based on clinical experience, that periosteum per se did not form bone. Macewan did not consider the cambium layer as part of periosteum and hence this might have been the basis for his disagreement with Axhausen. Davis and Hunnicutt supported Macewan's findings, but on reading their paper one finds the only thing their experiments showed conclusively was the disappearance with time of autologous periosteum placed in soft tissues of dogs (Davis and Hunnicutt 1915).

McWilliams observed perfect healing of autologous bone transplants, without periosteum, implanted into long bones (McWilliams 1912). In 15 transplants of the dog fibula with periosteum intact implanted in contact with bone, over 90% were successful. Transplanted human fibula likewise healed. Of 25 transplants without periosteum only about one-half produced good results (McWilliams 1914). Thus, experiments led him to believe that periosteum was needed to maintain nutrition of the graft, but that by itself it was not osteogenic.

Phemister removed one-third of the ulna of dogs and replaced it with bone with periosteum and endosteum, bone without periosteum and endosteum, as well as with periosteum alone. When periosteum and endosteum were left on the graft, osteogenesis proceeded at a faster rate than when these structures were absent. Periosteum transplanted into the ulnar defect by itself failed to produce new bone (Phemister 1913). Baschkirzew and Petrow showed that large bones transplanted with periosteum into soft tissue in animals were capable of regeneration (Baschkirzew and Petrow 1912).

Brown and Brown in 1913 observed resorption of bone transplanted into subcutaneous tissue regardless of whether it contained periosteum or not. However, when the same autografts were transplanted in contact with host bone, and particularly when the grafts were under stress, they healed. The healing took place in the presence or the absence of the periosteum. Thus, they concluded, periosteum was not essential to the healing and incorporation of autografts (Brown and Brown 1915).

Similar findings were reported by Haas some 10 years later (Haas 1922). Finally the evidence in favor of Axhausen's hypothesis became so overwhelming it was accepted even by Barth.

In autografts of dog radii, little influence on the osteoblastic activity by periosteum was demonstrated by Gallie and Robertson (Gallie and Robertson 1919). They felt that if a piece of bone has been cut free from its circulation and transplanted in the same animal with its surfaces bathed in lymph, the cells on the surfaces and in the open mouths of the Haversian canals would live and undergo rapid proliferation. The cells deep in the graft would die and disappear in 2–3 weeks. "The proliferating cells on the surface, whether derived from the graft or from neighboring living bone, immediately commence the absorption of the graft, and do not cease their activity until every vestige of the old bone has disappeared." In the meantime, circulation was re-established and the new bone was laid down, particularly on the surfaces. Initially the process of absorption progressed more rapidly than that of replacement, and at the end of 2 months the autografts became rarefied. If the graft was not subjected to strain the absorptive process outstripped the replacement and the new bone gradually disappeared. If the graft was implanted so that it was subjected to physical forces, after 2 months the process of bone replacement exceeded that of bone resorption and the new bone thickened and became hard.

Brooks and Hutson in 1920 established a relationship between the ages of the experimental animals and the rate of healing of the graft (Brooks and Hutson 1920). Willich in 1926 concluded from experiments with free auto-transplants in dogs that bone marrow and endosteum were as important in new bone formation as other portions of the graft (Willich 1926). Ely, in an experimental study on cats, showed that when a piece of bone was removed and immediately buried in the soft tissue of the animal from which it had been taken the bone and bone marrow within the graft died. About a week later vascularization of bone marrow began and was followed by new bone formation (Ely 1924). Kartaschew in 1930 transplanted small autologous bone segments in contact with bone and noted transplanted bone to be completely replaced by new bone.

Osteogenesis was stimulated by bone marrow cells as well as by periosteum and endosteum (Kartaschew 1930). In free autogenous bone marrow and endosteal grafts placed in the resected portions of the fibula in dogs, radiologic evidence of the formation of a new medullary canal was noted by McGaw and Harbin (McGaw and Harbin 1934). The defect was filled with newly formed bone which apparently developed *in situ*. The authors believed this to be specific evidence of the osteogenic activity of bone marrow and endosteum.

Definitive work on the transfer of autologous fresh pieces of bone was performed by Lecroix (Lecroix 1951). To simplify the experiments he dealt only with extra-skeletal grafts of bone, implanting them under the renal capsule. Young rabbits were used and the fragments of bone for transplantation were obtained from the tibia. Precautions were taken in preparing the subcapsular bed before removing the graft so that the latter would not be out of the organism for more than a few seconds. It would be difficult to imagine conditions more favorable to graft survival than those created in these experiments. Nevertheless, serial histological studies showed that during the first few days the cells of transplanted bone died and the bone lacunae were emptied. However, few cells at the surface of the graft maintained normal appearance. By 45 days the grafted fragments had disappeared. They were replaced with bony plaques on the inner surface of the renal capsule. A thick layer of hematopoietic tissue was interposed between the bony plaque and the kidney. Lamellar bone was distinguished in the plaque.

After 7 months bone marrow was active and new bone was still being formed. Since the number of cells retained on the surface of the graft was very small, their contribution would not be great. *In vitro* cultivation of the fragments of bone was performed as early as 1916 when Dobrovolskaja observed cell outgrowth from such fragments cultivated in homologous plasma using hanging drop technique (Dobrovolskaja 1916).

In tissue culture, cell growth can be initiated by a few superficial cells of the explants. When a bone fragment is placed in tissue culture medium, the cells of the Haversian system and the intertrabecular spaces migrate into the culture medium and are seeded there (Fell 1933, Gaillard 1942). Only a few are needed to initiate the growth. Similar circumstances prevail in the anterior chamber of the eye. This technique, originated by Connheim (Connheim 1882) in 1887 and later popularized by Greene (Greene 1952), was adapted to the study of bone growth. The anterior chamber of the rabbit eye allows for direct observation of the transplant which is not in direct contact with the blood supply. Hutchinson (Hutchinson 1949) in 1949 studied a series of bone autotransplants from the ilium into the anterior eye chamber. Bone transplants examined 180 days after being placed into the eye chamber resembled normal bone with periosteum. However, bone grafts used in these experiments were very small and were vascularized by the time they were examined. But be it as it may, these observations as well as studies in tissue culture were important because they showed that some cells in an autograft are capable of survival and replication *in vitro* despite devitalization of the predominant portion of the explant.

The great amount of experimental work with animal bone autografts implanted into soft tissues is repetitive and in certain aspects confusing. Different interpretations and theories are often presented with factual data, but the factual data itself is rather consistent, particularly if one keeps in mind the differences in experimental animals used. Bone grafts in rabbits and rats behave differently from those in guinea pigs, dogs, monkeys and certainly man. For this reason determination of the fate of autografts in humans is very pertinent. Fortunately, most valuable information on the microscopic appearance of human autogenous bone transplant has been accumulated by Peer (Peer 1951) who studied 36 bone grafts from the tibia, ilium and rib implanted into soft tissues. All bone grafts, whether cancellous or cortical, were replaced by fibrous tissue 8–12 months post-transplantation. The bone cells of the graft disappeared when bone matrix was absorbed. The same thing happened with or without periosteum. Thickness of the autograft did not prevent its absorption. Bone grafts placed in the vascular tissues of the neck and in the muscle were absorbed in the same manner as those placed into the subcutaneous fat of the abdominal wall. On the other hand, sixteen autogenous grafts attached rigidly to host bone retained their osseous structure and were joined to the underlying bone by an osseous union. These findings still hold true today and re-emphasize the often repeated dictum—in humans, bone autografts must be placed either in bone or in intimate contact with bone.

Excision and Preparation of Bone Autografts

Optimum results can be achieved by placing the autografts into the recipient bed as soon as possible after their removal in as much as the exposure to the operating room lights and desiccation damages the superficial cells of the graft. Therefore, it is advisable to remove the grafts after the recipient bed has been adequately prepared. If there is a delay, the grafts can be conveniently wrapped in a gauze sponge soaked with the patient's own blood, saline or Ringer's lactate which facilitates the survival of the superficial cells of the grafts. The grafts must be kept moist at all times. Although prolonged exposure of tissues to saline may result in a loss of an appreciable number of cells, Marx and Snyder showed that bone marrow and cancellous bone cells placed in saline for 1 hour survived as well as those placed into autologous blood or tissue culture media (Marx 1980).

The choice donor site is based on the quantity and quality of bone needed, age of patient, recipient site location and potential donor site morbidity. The donor sites should be readily accessible and the removal of the bone autograft ideally should not cause significant cosmetic and/or functional disability. Obviously this is not always possible and many authors have reported patients complaining more of about the sites from which the autografts were taken than the sites of the operations for which the grafts were used (Finkemeier 2002, Arrington 1996, Heneghan and McCabe 2009).

The usual donor sites are the iliac crest, tibia, fibula, and ribs (Ebzaheim 2001). Rarely, the greater trochanter, distal radius and proximal ulna can be used for obtaining a small amount of cancellous bone. However, autologous grafts can also be obtained from the local source, i.e., the area where primary surgery is performed. Ilium is by far the most common source of bone autografts not only because it can provide a considerable amount of cancellous bone, but also because it is easily accessible.

Ilium

The crest of the ilium is convex and it varies in thickness. It is thinner in the center than at the ends, the anterior and posterior superior iliac spines. The crest is subdivided into internal and external lips, separated by an intermediate line. The tubercle of the crest, an elevation in the outer lip, some 5–6 cm from the anterior superior iliac spine, is a basic anatomical orientation landmark. The anterior superior spine, the most readily palpable landmark, gives origin to the fascia lata, the tensor fascia lata, inguinal ligament, and the sartorius muscles. The posterior superior spine is more difficult to palpate and is of less surgical significance.

The crest of the ilium contains abundant cancellous bone between the inner and outer cortical bone tables, whereas in the center of the ilium the cortex is thin and little cancellous bone exists. Iliac crest functions as an epiphysis until relatively late. It fuses between the ages of 20–25 years. For this reason, in younger patients lateral sub-epiphyseal approach is advisable (Crockford & Converse 1972).

Iliac crest can provide cancellous bone chips, corticocancellous strips, full thickness and unicortical corticocancellous blocks, tri-cortical wedge grafts and dowel grafts. If the patient is placed in a prone position, the grafts are taken from the posterior third of the crest where the ilium is somewhat thicker.

The iliac crest is exposed through a transverse skin incision which is made along the lateral border of the crest and is carried down to bone. The ilium is covered with iliac fascia, incision of which lateral to the crest and subsequent stripping from the crest facilitates closure of the wound. Cutaneous sensory innervations of the lateral thigh and hip region are by the lateral femoral cutaneous nerve. It usually emerges at the lateral border of the psoas muscle, crosses the iliac muscle deep to the iliac fascia, which penetrates medial to the anterior superior iliac spine and posterior to the attachment of the inguinal ligament. The position of this nerve varies and this must be kept in mind during the lateral dissection of the ilium (Marx 1980). With the exception of this nerve, there are no major nerves in the area of the lateral ilium. However, cutaneous branches of the posterior primary rami of the upper sacral nerves pass by the posterior superior iliac spine. These can be easily identified and retracted off the ilium when the posterior superior iliac spine is approached **(Figs 3A and B)**.

Exposure of the ilium is simple if the anatomical landmarks are identified but it can cause considerable bleeding (Canale 2002). Hemorrhage can be minimized by subperiosteal dissection. If corticocancellous chips are needed, only the outer cortex of the ilium is exposed by reflecting the gluteal muscles subperiosteally. Long corticocancellous grafts, 5–10 mm wide and 2–5 mm thick are obtained using straight osteotomes to make longitudinal parallel osteotomy

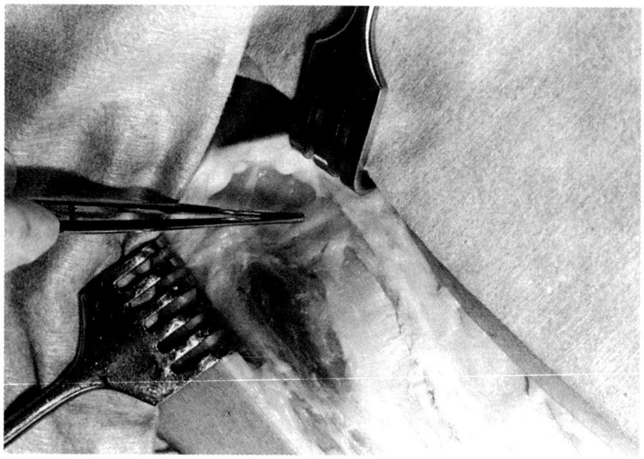

Figure 3A The lateral femoral cutaneous nerve emerges from the lateral portion of the psoas major muscle, runs toward the crest of the ilium, crossing it near anterior superior iliac spine. It is embedded in the fascia, but can be easily identified

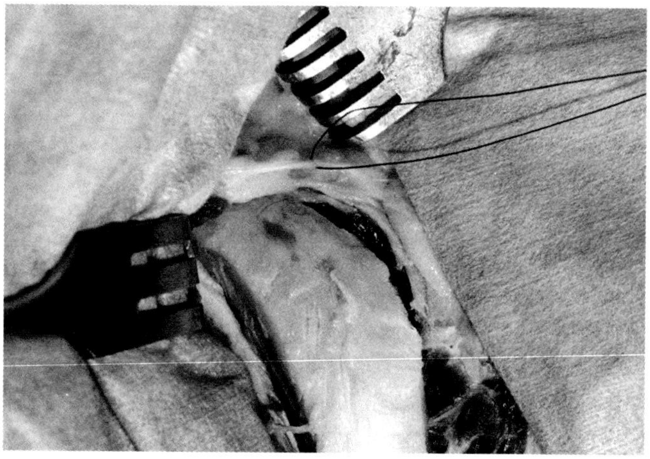

Figure 3B To avoid injury to the nerve it can be dissected and retracted

cuts followed by curved osteotomies to divide the inner cortex in a superior to inferior direction. Additional cancellous strips can be easily removed from the exposed cancellous bone surface using curved gouges or curettes. Again, care must be taken not to violate the inner iliac cortex. Bleeding from the large cancellous bone surface can be minimized by the use of a coagulating substance such as thrombin soaked gelfoam, bone wax, or microcrystalline collagen. Obliteration of the dead space is achieved by accurate replacement of the musculoperiosteal flap with interrupted sutures and by the closure of the iliac fascia (**Fig. 4**). Wound suction drainage is used for 24–48 hours.

If only cancellous chips are needed, these can be obtained by inserting gouges or curettes through the exposed superior surface of the iliac crest into the cancellous bone between the two intact tables of the ilium.

To obtain full-thickness bi-cortical blocks of ilium, muscles are stripped sub-periosteally from the inner and outer cortices. The grafts are removed with straight osteotomes or with power saws. Dowel grafts are removed with special circular cutting drills. The anterior superior iliac spine is always spared and preserved.

Wolf and Kawamoto described a technique for removing a full thickness graft from the anterior ilium which allows for reapproximation of the edges of the ilium after excision of the graft (Wolf and Kawamoto 1978). The crest is split obliquely both medially and laterally and each segment is reflected subperiosteally. The anterior superior spine is left intact. After removal of the graft, the two fragments of the crest are then sutured together and to the intact anterior spine. Cosmetic results are good and the patient cannot detect absence of bone. This method however, is not applicable in children because in children the iliac crest is not ossified. In children the apophysis of the iliac crest can be preserved by making a cut parallel to and below it. The apophysis can be replaced and sutured securely in place.

In patients with compact bone, i.e., in the absence of significant osteopenia or osteoporosis, one can often obtain unicortical blocks of ilium which will function as full-thickness coronal segments of ilium. The technique for obtaining such grafts is illustrated in **Figures 5 and 6**. This technique offers an advantage of preserving the contour of the iliac crest and facilitates obliteration of the dead space and closure.

If grafts are removed from the posterior iliac crest, care must be taken not to violate the sacroiliac joint.

In addition to the commonly used lateral subcrestal approaches to the ilium, several other techniques for obtaining bone grafts from the ilium have been described (Marx 1980). When small amounts of cancellous bone are needed a simple anterior or posterior crestal trephine curettage technique can be employed (Scott 1966). This requires only a 3 cm long incision, stripping of a small section of the crest, and excision of a piece of bone sufficient to fill a small defect. Kazanjian and Converse (Kazanjian and Converse 1974) advocated sagittal splitting of the crest, fracturing both

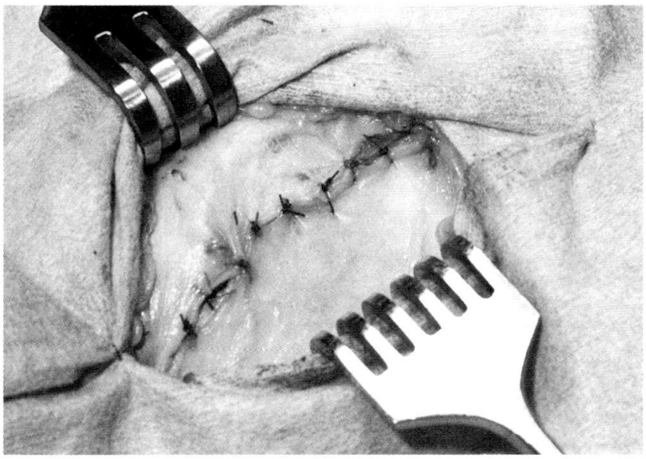

Figure 4 After removal of the graft the incision which is best made lateral to the iliac crest is closed with strong interrupted sutures. The sutures approximate the periosteum muscular origins and the fascia

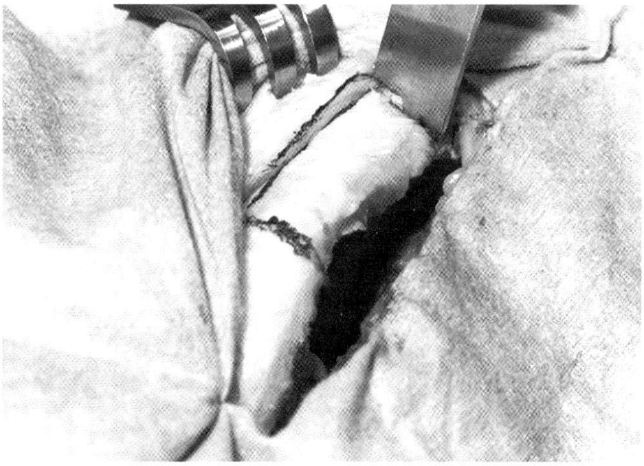

Figure 5 In a healthy patient partial thickness iliac graft will provide enough bone for most grafting procedures. The graft to be removed is outlined, cut with a power saw and then paired with an osteotome

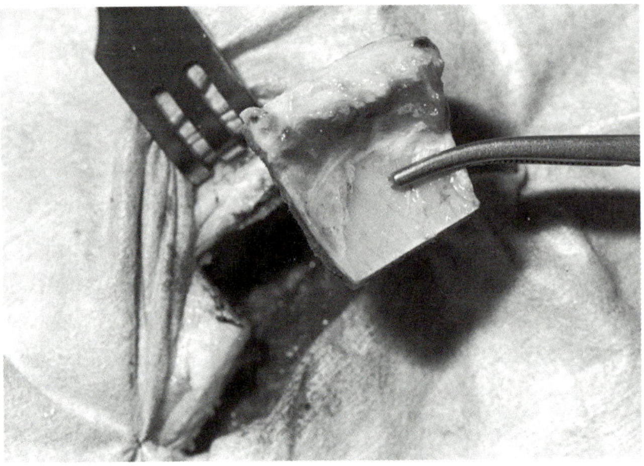

Figure 6A The outer table of the ilium provides thick cortical plate and a thick portion of the iliac crest

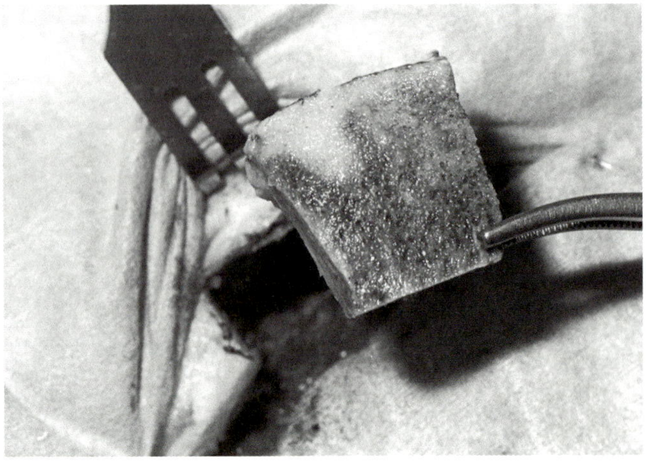

Figure 6B The smooth inner portion of the graft contains compact trabecular bone and abundant bone marrow

inner and outer tables to expose the cancellous bone, and once this is removed, compressing the fractured bone plates towards each other to eliminate the dead space.

Fibula

Although ilium is the most common site for obtaining bone autografts, morbidity, particularly pain associated with removal of such grafts is not inconsiderable. Lipscomb noted the morbidity accompanying removal of grafts from the ilium can be avoided by using corticocancellous bone grafts from the fibula (Lipscomb 1974). Clinically these grafts do as well as the grafts from the ilium.

The middle one-half to two-thirds of the fibula may be removed without causing any deleterious effects to the function of the leg (**Fig. 7**). However, the distal fourth of the bone must be left intact to maintain ankle stability. The proximal portion of the fibula including its head may be removed to replace a resected distal radius or a proximal humerus. The middle half of the fibula is easily removed through a Henry approach (Henry 1927, 1950). After application of a tourniquet, the dissection is deepened along the natural plane between the peroneus longus anterior and soleus muscles posteriorly. The peroneal muscles are reflected anteriorly by subperiosteal dissection which begins distally and progresses proximally so that oblique muscle fibers originating from the bone will tend to keep the periosteal elevator against the fibula. The bone is cut proximally and distally using either a power saw or a gigli saw rather than an osteotome, which may split the bone. The nutrient artery which usually enters

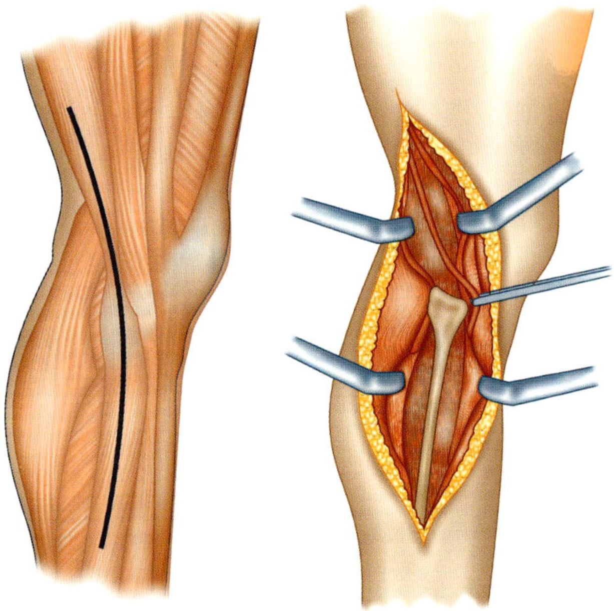

Figure 7 Resection of the fibula can be accomplished through a curved cutaneous incision running along the posterior margin of the fibula. After dividing the superficial and the deep fascia, the common peroneal nerve is identified and retracted as is the biceps tendon. After detaching the part of the peroneus longus muscle originating from the head of the fibula and further retracting the common peroneal nerve, the fibula becomes exposed for either extra-periosteal or sub-periosteal dissection, sectioning and detachment from other structures

the bone near its middle over the posterior surface may have to be ligated.

If the proximal fibula is needed it can be removed through proximal extension of the Henry approach, taking care to avoid injury to the peroneal nerve and the adjacent blood vessels. Ideally the nerve should be exposed proximally at the posterior medial aspect of the distal end of the biceps femoris tendon and then traced distally to where it winds around the neck of the fibula. Here the origin of the peroneus longus muscle which covers the nerve is divided and the nerve is retracted anteriorly and protected. Following removal of the proximal fibula, the stump of the biceps tendon and the fibular collateral ligament are sutured to the adjacent soft tissue and fascia.

Tibia

After applying a tourniquet to mid-thigh, a long slightly curved longitudinal incision is made over the anteromedial surface of the tibia. The periosteum is incised and reflected medially and laterally with a periosteal elevator to expose the whole anteromedial surface. A full thickness cortical graft can then be removed. When using an osteotome, multiple drill holes are made along the outline of the graft. Alternatively, the graft can be removed with a power saw. The cortical graft is then removed from its bed. Additional cancellous bone grafts can be removed from the proximal end of the tibia with a curette. The graft is wider and thinner proximally than distally. In order to fill a defect, several grafts must be used as the grafts are usually thin. Care must be taken to avoid the articular surface of the tibia or the epiphyseal plate in children. In children the periosteum is thick and can be sutured as a separate layer. If the amount of bone removed is small, there is little residual effect (Longaere 1964). If, on the other hand, the amount of bone removed is large the defect produced is disfiguring and painful. Secondary fractures are likewise frequent.

Rib

Although rib autografts were utilized by Macewen as early as 1912 (Macewen 1912), they never became too popular. This can be explained by the manner in which the rib autografts have been used. When a whole rib is transplanted it becomes vascularized only through cut ends and Volkmann's canals. The facts that the rib bone is highly osteogenic and is capable of regenerating itself repeatedly were not appreciated until 1957 when the split rib autografts were subjected to an extensive study (Longaere and De Stafano 1957). Rib grafts are now used extensively for reconstruction of mandibles, cranioplasties and spinal fusions (Knight and Wood 1945).

If only a small portion of a rib is required, an anterior incision directly over the seventh rib provides an adequate access for the removal of the graft. If a large portion of the rib is needed, posterolateral thoracoplasty approach over the seventh rib is recommended. The incision is made parallel to the seventh rib, from the angle of the scapula to the anterior axillary line. The fascial plane is opened after elevating the scapula, and by dissecting through this plane, a full view of the second to tenth rib is obtained. By utilizing this approach, several ribs can be taken. Obviously only alternate ribs should be removed and each rib bed closed individually. Periosteum is incised and elevated off the ribs. After removal of the rib it is carefully closed to facilitate rib regeneration.

After removal of the rib, it can be split lengthwise if it is desirable to contour the graft.

Local Sources of Autografts

The greater trochanteric region of the femur is a good source for a small amount of cancellous chips which can be used in the surgery of the hip. Other sites used to obtain bone grafts for regional use include the distal radius for corticocancellous blocks for wrist fusions and small amounts of cancellous bone for grafting non-unions of metacarpal, carpal or phalangeal bones (McGrath and Watson 1981). Femoral condyles are occasionally used to obtain small amounts of cancellous bone used to graft a depressed tibial plateau fracture. The proximal ulnas are also occasionally used as a source of a small amount of cancellous bone grafts (Clendenin and Green 1981). Local autografts were also used to fill cages for anterior cervical fusions. However, to date there are no reports in the literature comparing results with local autograft and iliac crest graft. Lack of donor site morbidity is, of course, obvious and this alone might be enough to justify the employment of the local autograft.

Types of Grafts and Bone Grafting Techniques

Single onlay cortical grafts: The single onlay graft is used like a metallic plate. It is placed across the site of a defect and is fixed to the underlying bone with two or three screws on either side of the defect. The graft may or may not be supplemented with cancellous bone grafts. The tibia and the fibula are the most common donor sites for this autograft. Application of this autograft was the simplest and most effective treatment for most ununited diaphyseal fractures prior to the advent of and the acceptance of metallic internal fixation devices. At present the importance of the dual function of the onlay cortical grafts, i.e., fixation and osteogenesis has diminished. It has been largely replaced by the more secure metallic internal or external fixation devices supplemented by cancellous bone grafts placed circumferentially around the defect. However, this graft is still used occasionally in non-unions not responding to conventional surgical treatment.

Dual onlay grafts: These are occasionally used in bridging of large defects or treating difficult non-unions, especially those near a joint, as the fragment nearest to the joint is usually small, osteopenic and mostly cancellous. Stability is achieved by the viselike grip of the dual onlay grafts. Supplementary cancellous bone chips are also added to fill the trough-like space between bone plates and to further stimulate osteogenesis.

These grafts were initially advocated by Boyd in 1941 (Boyd 1941) for the management of congenital pseudarthrosis of the tibia because they provided both structural stability and stimulus for osteogenesis. In bridging bone defects, the advantages of the dual cortical grafts include improved mechanical fixation and added strength and stability.

Like single onlay cortical grafts, these dual grafts have become less popular with the advent of stronger metallic fixation devices.

Inlay bone grafts: In this technique a rectangular slot is made in the cortex of the bone with defect usually across a non-union or pseudarthrosis. A graft of the same size is then fitted into the slot. Inlay grafts at present have a limited application but are still occasionally used in arthrodeses, particularly at the ankle. The onlay graft is more efficient and simpler than the inlay graft.

Dowel grafts: This type of bone graft has been used in the management of non-union of the carpal navicular, depressed tibial plateau fracture and non-union of femoral neck fractures. It is also used in anterior cervical spine fusion, as well as in lumbar fusions. These grafts are usually taken from the ilium, and are bicortical. Special instruments for taking dowel-shaped grafts have been described by Cloward (Cloward 1958).

Muscle pedicle grafts: This technique is useful in the management of non-unions of the femoral neck in young patients (Meyers et al 1973).

Conceptually, this graft is a vascularized graft where the bone continues to survive through the vascularization provided by the muscle pedicle. In this technique a rectangular corticocancellous graft is harvested from the posterior-lateral aspect of the intertrochanteric crest including the insertion of the quadratus femoris muscle. A slot is cut at the base of the femoral head and the proximal end of the pedicle graft is inserted and impacted into this slot. One or two screws are used to fix the distal portion of the pedicle graft to the trochanteric region of the femur. During the procedure the quadrates femoris muscle is not twisted or put under tension so as not to compromise the blood supply to the bone graft. This procedure is technically difficult, but good results in terms of decreased incidence of non-union and avascular necrosis of the femoral head are reported by its advocates.

Strut grafts and peg grafts: These grafts serve as internal fixation devices and are used in stabilization of the long bone segments and arthrodesis of the spine. The fibula, the rib, or the iliac crest are the most common donor sites for these grafts. These grafts provide structural stability and stimulate osteogenesis.

Clothespin or H-grafts: These grafts are used in spine fusions. The graft is a large rectangular iliac crest graft with a slot cut out in both the proximal and the distal portion (H-shape) so that the graft may be wedged in place between adjacent spinous processes.

Whole bone grafts: The fibula is the most common "whole bone graft" although in actuality the entire fibula is seldom used, if ever. Fibular shafts are quite suitable for bridging large defects in long bone diaphysis of the upper extremity. Fibular graft is stronger than a tibial cortical plate graft and the morbidity and disability created by the removal of the fibular graft is less than that rendered by removing a large tibial graft. The proximal fibula has been used to replace a resected proximal humerus, a distal radius, and as a substitute for a resected distal fibula (Edmonson 1930 Campbell 1980).

Multiple cancellous bone grafts: Cancellous bone chip grafts provide the most readily available osteogenetic material. Unless biomechanical support is required, these grafts fulfill almost all requirements. They are used in a wide variety of orthopedic conditions. They are particularly useful in filling cavities or defects of cysts, tumors, or tumor-like lesions. Being soft, they can be packed into small cavities or crevices. They are also the most commonly used grafts in the treatment of delayed union, non-union and pseudarthrosis. In non-unions, the bone is fixed either with a compression plate or a tightly snug intramedullary nail and then the cancellous bone grafts are packed around the side of the non-union to stimulate osteogenesis. Cancellous bone chips are also commonly used in arthrodesis of the spine, with or without spinal instrumentation.

Free Vascularized Bone Grafts

The grafting of bone into tissues with reduced vascularity, which may result from previous trauma, infection, surgery or irradiation, has not met with success and constitutes a difficult clinical problem. To overcome this, autologous bone with its own blood supply may be transplanted, as first demonstrated by Taylor in 1975 (Taylor et al 1975). This is made possible by relatively recent developments in the techniques of microvascular anastomoses. This technique allows for the detachment and a transfer to the distal site of large bone segments with their blood vessels. The vascularized bone autografts actively participate in the healing process of the graft-host junctions, rather than serving merely as a framework for inductive creeping substitution as is the case with non-vascularized bone autografts.

Experimental studies on animals demonstrated complete survival of corticocancellous grafts maintained by microvascular anastomosis and the importance of the reconstruction of the nutrient artery circulation (Osterman and Bora 1984). The experimental work was paralleled clinically both in the reconstruction of the craniofacial structures and the long bones. Taylor and Watson (1978) reported the first clinically successful microvascular transfer of a 22 cm long fibula to bridge a traumatic defect of the tibia. The use of a vascularized rib to reconstruct the mandible and of the osteocutaneous rib graft for combined skin-bone defects followed shortly thereafter (Osterman and Bora 1984).

Free vascularized grafts are of value in bridging intercalary defects, but are not applicable in osteoarticular defects as they contain no articular surface. The operation is long and tedious and demands expertise in microvascular surgery. It also entails a separate major operative procedure to harvest the graft. Nevertheless, if properly executed, this technique seems to be very promising. Weiland (Weiland 1981) reported a success rate of 87% of which 9% needed supplementary grafting procedures. In massive diaphyseal intercalary bone loss with intact overlying skin, the revascularized fibular grafts proved ideal. If both skin and bone defects exist, the free osteocutaneous groin flap may be used (Taylor 1978). However, the curvature of the iliac crest limits its applicability to defects of 10 cm or less, whereas the fibular graft is used in segmental bone defects greater than 6 cm. The technique for the preparation of vascularized fibular graft is described by Taylor (Taylor 1977). The fibula is approached through a long straight lateral skin incision extending from the neck of the fibula distally as needed. Dissection is deepened through the natural plane between the peroneous longus and soleus muscles. Using an extra periosteal dissection technique, these muscles are reflected from the fibula anteriorly and posteriorly respectively. The lateral border of the fibula is exposed and both the peroneal longus and brevis muscles are elevated from the lateral border of the fibula until the anterior crural septum is reached. The septum is divided and the anterior compartment muscles are dissected from the interosseous membrane until the anterior tibial artery and nerve are identified and protected. The interosseous membrane is divided along the entire length of the graft. The soleus and flexor hallucis longus are reflected, using extra periosteal dissection from the posterior border of the fibula until the peroneal vessels are encountered. The fibula is cut proximally and distally using a gigli saw. It is highly recommended to leave the distal 6 cm of the fibula in place to maintain the integrity of the ankle joint. The distal limb of the peroneal vessels are ligated at the site of the distal fibular osteotome. The resected graft is then retracted and the interosseous membrane is divided and the posterior tibial muscle is dissected from the graft. The fibular graft is now isolated on its vascular pedicle proximally. Proximally the peroneal vessels are cut close to the junction with the posterior tibial artery. The free fibular graft is then transplanted to the recipient bed and microvascular anastomosis of the peroneal artery and vein to the recipient vessels are performed after bone stable fixation is achieved.

Free vascularized iliac crest graft provides an excellent method of bridging smaller gaps. Here the main feeding artery is the deep circumflex iliac artery. This artery is isolated by exposing the external iliac artery and then the deep circumflex artery is usually found opposite the origin of the inferior epigastric artery. A segment of the iliac crest measuring 3–10 cm long may easily be raised with its vascular pedicle intact. The eternal surface of the ilium is exposed subperiosteally. The periosteum and a strip of abdominal muscle is left attached to the inner cortex of the ilium to avoid damaging the perforator vessels. The deep circumflex vessels are then divided from the external iliac vessels, thus completing the free bone graft isolation.

In the last decade, free-vascularized bone transfer has become an established surgical technique with high success rates reported in most series (Taylor 1983). The donor sites now include iliac crest, fibula, metatarsal bone, rib, radius and scapula. However, size, shape and vascular supply limit the utility of iliac crest and rib grafts. The clinical indications for these grafts include traumatic long-bone defects, composite defects of bone, muscle and skin, mandibular reconstruction, reconstruction of defects secondary to tumor ablation, non-unions resistant to conventional treatment, and congenital pseudarthroses (Osterman and Bora 1984).

In general, operations with free-vascularized bone grafts are reserved for the circumstances when the recipient bed is poorly vascularized and there is an associated soft tissue loss. Vascularized cortical grafts heal rapidly, but the advantage over nonvascularized grafts is transient as revealed by the comparison of the biological features of repair of vascularized graft with that in non-vascularized graft. Non-vascularized grafts contained necrotic bone marrow and empty lacunae at 2 weeks, but by 6–12 weeks osteoid replacement took place and necrotic osteons were either replaced with new bone or resorbed. However, at 24 weeks the interstitial lamellae still remained necrotic. Vascularized cortical autografts, on the other hand, contained viable osteocytes and bone marrow at all time intervals. At 24 weeks the difference between vascularized and non-vascularized grafts became less distinct (Dell and Sheppard 1985).

Donor Site Complications

Although relatively rare, a wide variety of complications at the donor sites have been reported. Most of these have occurred in the ilium probably because now it is the most common site for autografts. Usually surgical complications at the donor site are relatively minor and easy to manage, however major complications with significant morbidity and prolongation of postoperative rehabilitation do occur (Seiler et al 2000, Silber et al 2003). Cockin reported 6% minor complications and 3.4% major complications with donor sites in the ilium (Cockin 1971). The majority of the complications are avoidable if, in removing the graft, good surgical techniques are adhered to. Unfortunately, quite often excision of bone autografts is felt to be of minor importance in relation to the primary operative procedures and is consequently delegated to the least experienced member of the surgical team.

The reported complications of the iliac crest donor site include:

- *Hematoma and seroma formation*: This can be avoided with good hemostasis and closed wound suction drainage. The average postoperative blood loss in the suction canister is about 250 cc (Laurie et al 1984).
- *Delayed healing and wound breakdown*: This complication is reported in some 20% of incisions placed directly over the iliac crest (Laurie et al 1984). When secondarily healed, these incisions result in depressed and adherent scars that are painful under belts or clothing.
- *Infection*: Infections can be minimized with proper preoperative preparation, intraoperative prepping and draping of the iliac crest region, as well as prevention of hematoma formation and perioperative antibiotic prophylaxis, but as with all surgical procedures they will occur from time to time.
- *Postoperative wound pain*: Postoperative wound discomfort as well as buttock anesthesia following posterior iliac crest graft removal are common and usually last from 2 weeks to 2 months or longer. The pain at the donor site is often more severe than the pain at the primary operative site (Laurie et al. 1984). In some patients pain may persist for years. Painful neuromas can be prevented by making the incision within 8 cm of the posterior superior iliac spine and thus avoiding the superior cluneal nerves.
- *Iliac hernia*: Iliac hernia at the donor site (Cowley and Anderson 1983, Reid 1968) occurs only following the removal of full thickness large bone grafts particularly from the middle part of the crest. Reconstruction of the large iliac defect with methylmethacrylate is an easy technique to prevent the formation of these hernias (Lubicky and Dewald 1982). However, methylmethacrylate will fail eventually, but by this time new bone growth and scar tissue usually prevent herniation of abdominal viscera. Lumbar hernia in

the region of the inferior lumbar triangle of Petit was also reported (Challis et al 1975, Castelein and Sauter 1985).

- *Meralgia paresthetica*: (Weikel and Habal 1977, Massey 1980): Patients usually complain of burning, tingling and numbness of the anterolateral thigh, in the region supplied by the lateral femoral cutaneous nerve. The sensory disturbances vary from slight hyperesthesia to total anesthesia. The lateral femoral cutaneous nerve is a branch of the second and third lumbar nerves and it emerges from the lateral border of the psoas major muscle. It crosses the ilium obliquely, running toward the anterior superior iliac spine where it passes medial to it and behind the inguinal ligament and then over the sartorius muscle, and proceeds laterally to innervate the skin of the anterolateral thigh. During excisions of the iliac grafts the nerve is vulnerable to injury in the region of the anterior superior spine. The nerve can also be entrapped in the sutures or later in the operative scar (Weikel 1977). To avoid injury to the nerve, anterior iliac crest incisions should be placed below and lateral to the crest as incisions made medial to the crest carry a 10% risk of injury to the nerve (Laurie et al 1984).
- *Fractures of the iliac bone*: (Blackman 1983, Guba 1983, Ubhi and Morris 1984, Reale et al 1979): These fractures can be complicated by large hematomas (Guba 1983) or hernia (Ubhi and Morris 1984). Fracture of the anterior iliac spine (Reynolds et al 1978, Kuhn and Moreland 1986) can be avoided by taking the graft at least 3 cm posterior to the anterior-superior spine. If a full thickness bicortical graft is taken too close to the anterior superior iliac spine, the latter becomes rather weak and could be avulsed by strong contraction of the tensor fascia lata and sartorius muscles which originate from it.
- Lacerations of the superior gluteal artery during removal of bone grafts from the posterior iliac crest have occurred (Kahn 1979). The artery is injured as it exits the greater sciatic notch. If divided, bleeding can be alarming due to the retraction of the vessel into the pelvis. To avoid this injury, bone graft removal should not be extended caudal to the inferior margin of the bony origin of the gluteus maximus. By following this safety landmark, one can avoid entering the greater sciatic notch.
- Subluxation of the hip following extensive removal of the iliac crest **(Fig. 8)** (Cocklin 1971).
- *Acetabular fracture*: This very rare complication may occur if excessive force or leverage is used when freeing the graft.
- Contour abnormality of the crest can occur if the crest is split sagittally (Wolf and Kawamoto 1978). Although this technique gives good access and good early reconstruction over a longer period, the remodeled crest may become irregular and depressed **(Fig. 9)**. In some patients regenerating bone hypertrophies and produces undesirable exostoses.

Figure 8 An autopsy specimen from a deformed iliac crest following removal of bone graft a number of years back

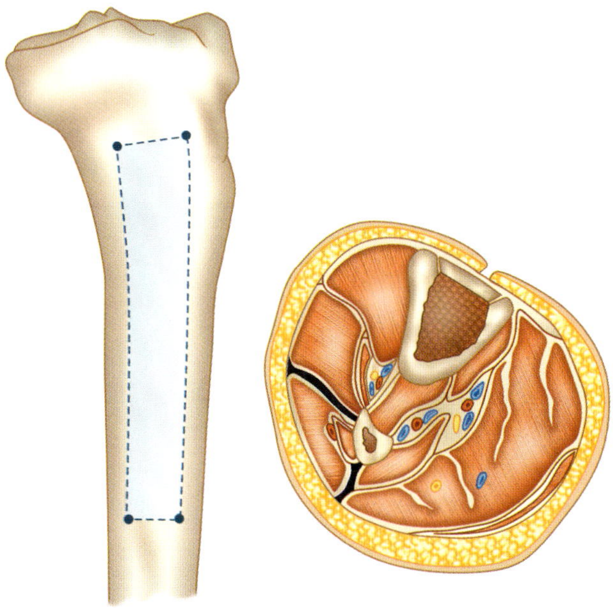

Figure 9 Removal of the cortical plate from a tibia. A small hole can be drilled in each corner, and a cut made with an oscillating saw connecting the holes. Care is taken not to violate the underlying lesion. Graft is little wider proximally than distally. After making the cuts the graft is pried from its bed. Press fitting assures good sealing of the graft as shown in the cross-sectional drawing

- Gait disturbance is very rarely seen following posterior iliac crest incisions (Longaere 1957). It is caused by massive detachment of gluteal muscles (Cockin 1971) or the quadratus lumborum and erector spinae muscles.
- Heterotopic bone formation and neuroma formation have been noted in rare instances. Sagittal splitting of the crest and rewiring of the crest can give good access, but about one-third of the patients show significant irregularity or spreading of the crest (Laurie et al 1984). Raising a trapdoor on the crest or removing the graft below an intact iliac crest would avoid these problems (Jackson et al 1982). In children this approach minimizes damage to the epiphysis (Crockford and Converse 1972, Mrazik 1980).
- Intestinal iliac hernia at the site of a massive donor site defect: (Cowley 1983, Challis 1975, Reid 1968, Kave and Silverstein 1986). This occurs only following the removal of large full-thickness bone grafts, particularly from the middle part of the crest. Reconstruction of the large iliac defect with methyl methacrylate is an easy technique to prevent the formation of these hernias (Lubicky 1982).
- Meralgia paresthetica with symptoms of burning, tingling and numbness of the anterolateral thigh. This is caused by injury to the lateral femoral cutaneous nerve with anterior iliac crest incision (Massey 1980, Walker 1977). The nerve is vulnerable to injury in the region of the anterior superior iliac spine. Therefore, to prevent this complication, the incision should remain 2 cm lateral to the anterior superior iliac spine.
- Sacroiliac and pelvic instability may occur (Coventry and Tapper 1972, Lichtbau 1962) but is easily prevented if when taking full thickness grafts from the posterior iliac crest, the ligamentous structures are preserved. Compromising the posterior-superior supporting ligaments of the sacroiliac joint may cause instability of this joint and sometimes may render the pelvic ring unstable.
- Injury to the peritoneum may happen when grafts are taken from the inner table of the anterior iliac crest. Careful retraction of the abdominal wall and the iliacus muscle will prevent this injury.

Comments on Complications

All in all, biological properties of bone autografts make these transplants the most suitable grafts for the restoration of the bone defects. No other materials including allografts, growth factors, inorganic materials, etc. can match the biologic properties of autografts. Therefore, whenever possible, autologous bone must be considered as the primary and the most desirable source of bone grafts. However, at times the price for an ideal graft might be too high. The price paid is in the currency of morbidity, and especially patient's pain and discomfort. Many studies have shown that iliac crest autograft removal is not risk free (Banwart 1995, Coventry 1972, Younger 1989). Overall, complication rate ranges from 9.4% to 49%. However, major complications are reported to be below 2.5% in some series and as high as 5.8% in others (Arrington 1996). This level of major complications, with minor complications considerably higher, has turned some surgeons to abandon autografting altogether, particularly in children (Dudd 1988). Such a recommendation might be too severe, but it correctly points to the need for a careful and thoughtful approach to autografting.

Clinical Indications for Autografting

The orthopedic literature is replete with information on the different clinical indications and techniques of autografting. This is readily available to an interested reader. Therefore, in this chapter, only brief discussion of the most common clinical conditions in which bone autografts are used is presented. These include:

- Benign bone tumors and tumor-like lesions such as bone cysts, enchondroma, chondroblastoma, fibrous dysplasia, etc. The tumor or cyst is curetted and the cavity is packed with autogenous bone chips. Details of this procedure are given in Chapter 12. In the same connection the only truly orthotopic autograft may be used. This consists of removing a rectangular cortical graft to create a "bone window" through which access can be gained to the underlying lesion. The bone segment is then replaced exactly in the same location from which it was removed. The technique is illustrated in Figure 9. In removing the cortical plate, it is important not to violate the underlying lesion. In replacing the autograft, one must assure the absence of the tumor on the medullary surface of the graft.
- Malignant bone tumors in which the surgical treatment entails the sacrifice of large segments of long bones. Massive autografts are therefore needed. These can be remote free grafts such as the proximal fibula to replace

a resected proximal humerus or distal radius, or the turn-up hemi-tibia or turn-down hemi-femur local grafts to replace a resected distal femur of proximal tibia, respectively, with arthrodesis of the knee (Enneking et al 1977). This procedure, although relatively popular at one time, is now largely abandoned. A resected acetabular portion of the innominate bone can be replaced and the hip joint reconstructed using the ipsilateral proximal femur combined with a total hip prosthesis as reported by Puget and Utheza (Puget and Utheza 1986). A vascularized fibula can also be used to bridge a large diaphyseal defect of a long bone (Weiland et al 1977). A pediçled vascularized patella can be used to reconstruct a resected femoral or tibial condyle as reported by D'Aubigne (D'Aubigne 1973).

- *Fresh fractures*: Severely comminuted fractures often contain several devascularized small bone fragments, which increase the risk of non-union. Grafting the fracture site with autogenous cancellous chips enhances union of the fracture. Moreover, occasionally there may be loss of large bone segments. In such cases large autogenous grafts, either from the ilium or fibula (free or vascularized), are used to substitute for the lost bone.
- *Delayed and non-united fractures*: By and large, atrophic non-unions require bone grafting in addition to rigid stabilization, whereas hypertrophic non-union does not require grafting. This is because hypertrophic non-unions with an "elephant's foot" or "exuberant mushroom callus" retain substantial biologic capability of healing. Conversely, atrophic non-unions, which have minimal callus with atrophic bone ends as well as sealed medullary canal with dense sclerotic bone, have minimal biologic potential for healing. The fibrocartilaginous tissue at the site of the non-union does not need to be excised as it has osteogenic potential and does provide fibrous stability. However, this tissue has to be excised in some unusual circumstances, such as the presence of a deformity that needs to be corrected, or existing infected necrotic bone or synovial pseudarthrosis that should be resected. Autogenous cancellous bone grafts are used, and every attempt is made to obtain the best possible vascularized soft tissue coverage and bed for the grafts. In the tibia, where non-unions are most common, the technique of posterolateral approach and cortico-cancellous bone grafting have given good results (McMaster and Hohl 1965, Freeland and Mutz 1976, Reckling and Waters 1980, Hanson and Eppright 1966). The grafts are laid across the site of the non-union of the tibia and fibula, the interosseous membrane and the fibrous defect. Healing of this posterior tibial graft to the fracture fragments is dependent on the rich extra-osseous blood supply provided by the calf muscles, the interosseous membrane and the periosteum. This method is particularly useful when there is a coexisting infection with anterior drainage from the anterior compartment, as this compartment is not entered during the procedure. After union of the fracture is achieved, the anterior drainage often ceases. Occasionally, in open infected non-unions with bad anterior skin and excessive drainage, a staged operative procedure is recommended. The first stage generally consists of radical debridement and excision of the devitalized osseous and soft tissue and skeletal stabilization by external fixation device. The second stage is an open cancellous bone grafting as popularized by Papineau (Papineau 1973, 1976; Green and Dlabal 1983; Gaines and Reid 1976). The grafts, obtained from the iliac crest, are laid on the well vascularized bed of healthy granulation tissue covering the cavity. Granulation tissue will then grow through the bone grafts and cover the entire grafted area. The third stage is split thickness grafting of this newly formed layer of granulation tissue. In severely infected non-unions with chronic osteomyelitis that requires radical bone resection for control of sepsis, a large skeletal defect is created. If this defect is greater than 6 cm, the vascularized bone transfer procedure is quite successful. A vascularized fibula or iliac crest is the usual donor sites (Wood and Cooney 1984; Weiland 1984, 1979).
- *Spine fusions*: see Chapter on: bone grafts in spine stabilization.
- *Joint arthrodesis*: Autografts, preferably cancellous, are used to obtain more rapid and certain fusion. These grafts could be local or remote grafts (from ilium or fibula). They could be free or muscle pedicle grafts such as the hip arthrodesis described by Davis (Davis 1954) where an anterior iliac crest graft is transferred with its gluteal muscle attachment left intact. Grafts are most often used to enhance intra-articular type of arthrodesis, where either internal fixation or compression external fixation devices are used (Charnley 1951, Charnley and Lowe 1958; Lucas and Murray 1961). The once popular extra-

articular arthrodesis is very rarely used nowadays (Lucas and Murray 1961).

- *Limb lengthening*: The tibia and femur could be lengthened to equalize limb lengths in patients with short lower extremity. The most frequently used method is the Wagner tibial or femoral lengthening technique (Wagner 1971, 1977). This procedure involves two stages. The first stage is a transverse osteotomy of the diaphysis of the tibia or femur and application of a special external distraction apparatus. Lengthening is carried out daily in the postoperative period until the desired length is obtained. The second stage is then carried out. Here, the distracted fragments are fixed with a plate and screws and the distraction apparatus is removed. If the defect between the two bone fragments contains insufficient callus, it is filled with autogenous iliac grafts.
- *Congenital pseudarthrosis*: The tibia is the most common site of this congenital non-union. More rarely, other ones including the fibula, radius, ulna and clavicle are involved. Osseous union is probably more difficult to obtain in this condition than in any other. Several types of bone grafting procedures have been described for tibial pseudarthrosis. All include complete excision of the dense tissue in the area of pseudarthrosis. Either a single massive onlay autogenous graft or more frequently dual cortical grafts, as described by Boyd, are used (Boyd 1941, Boyd and Sage 1958).
- *Hip and knee arthroplasties*: Autogenous bone grafting is used to augment a deficient acetabulum in patients with dysplastic acetabulum or congenital hip dislocation. The patient's excised femoral head or his iliac crest provides the source for the graft. Iliac bone grafts are also used to augment a deficient acetabulum in some revision arthroplasties. In knee arthroplasty, the patient's excised portion of femoral condyles (in primary procedures) or his iliac crest (in revision) could be used to augment a deficient tibial condyle to provide structural support for the prosthesis. In a primary procedure the distal or posterior portion of the femoral condyle usually fit the defect in the tibial condyle. In a revision procedure, the iliac bone graft is obtained from the thick portion between the anterior superior spine and the high point of the crest. The graft could be either half thickness or full thickness tricortical graft. The graft is sculptured to fit the tibial condyle defect. The graft is usually fixed to the underlying tibial condyle with two screws. The graft should be in complete opposition with the exposed surfaces of the tibial defect, and should be at the same level as the opposite tibial condyle to maintain optimum component and leg alignment. The tibial prosthesis is then cemented in over the graft.
- Other infrequent clinical indications for autogenous bone grafting included open wedge osteotomies, bone blocks to limit joint motion such as the posterior bone block operations of Campbell (Campbell 1923), Gill (Gill 1933) or Inclan (Inclan 1949), to prevent plantar flexion of the ankle in paralytic talipes equinus in poliomyelitis. To prevent shoulder dislocation, the Eden-Hybbinette bone graft is placed against the anterior aspect of the neck of the scapula and rim of the glenoid cavity so that it blocks anterior displacement of the humeral head (Eden 1920, Hybbinette 1932).

Autologous Bone Grafts in Infections

In the treatment of bone defects associated with infections and the loss of the soft tissue, open cancellous bone grafting can be utilized with predictable results. Neither sophisticated technology nor prolonged antimicrobial therapy is required in supplementation of bone grafting. However, the method requires prolonged external fixation and avoidance of weight bearing in the lower extremity. Patient cooperation is essential as hospitalization is usually protracted (Cabanella 1984).

The open bone grafting technique for infected bone defects evolved largely during World War II. Several surgeons have used cancellous bone chips to obliterate bone cavities following traumatic osteomyelitis (Knight 1945, Prigge 1946, Robertson and Barron 1946). The technique, which remained largely unchanged, was divided into three stages. At first the infected bone cavity was thoroughly debrided. In the second stage, after granulation tissue had formed, the cavity was covered with split thickness skin graft. In the third stage the skin graft was lifted off and the epithelialized cavity filled with bone chips. Coleman et al thought preliminary split thickness skin grafting unnecessary and performed debridement and bone grafting in one sitting. Bickel, in 1953, reported on the treatment of chronic hematogenous osteomyelitis by marsupialization and bone grafting, while others described various modifications of Coleman's technique (Cabanella 1984).

The technique which is now most commonly used is the one originally described by Rhinelander (Cabanella 1984) and later popularized by Papineau (Papineau 1973). It consists of excision and stabilization, cancellous bone grafting and skin coverage (Green and Dlabal 1983). According to VanDerwerken and Marti, cancellous bone graft has the ability to fill the space in the bone providing there is an adequate granulation tissue bed, even in poorly vascularized areas (VanDerwerken and Marti 1982). However, the majority of authors now advocate the use of free vascularized fibular or iliac bone grafts for avascular areas (Dell 1984).

Segmental defects of more than 4 cm long should not be treated exclusively with chipped cancellous bone autografts because of the problems with stabilization.

In summary, cancellous bone grafting provides a useful addition to the armamentarium of a surgeon treating bone infections. Primary indication for the grafting are infected non-unions, bone cavities with osteitis and certain infected compound fractures (Cabanella 1984).

CONCLUSION

Autogenous bone grafts remain the best biological grafts for the reconstructive surgery of the skeletal system. However, the morbidity associated with the excision of these grafts is an important clinical complication. This not withstanding, the biologic properties of autogenous bone transplants have not been matched by any other grafts or graft substitutes. Such being the case the autografts still remain the "gold standard" of bone transplantation.

BIBLIOGRAPHY

1. Abbott LC, Schottstaedt ER, Saunders JB, et al.The evaluation of cortical and cancellous bone as grafting material. Journal of Bone and Joint Surgery, 1947;29(2):381-414.
2. Adamkiewicz A. Uber Knochen transplantation. Wiener Medicine. Blatter. 1889;12, 7:355.
3. Albee FH. Arthritis deformans of the hip joint: Report of a new operation. The Journal of the American Medical Association, 1908;50:1553.
4. Albee FH. Original surgical uses of the bone-graft. Surgery Gynecology and Obstetrics, 1914;18:699.
5. Albee FH. Bone graft surgery. Philadelphia: WB Saunders Co.; 1915.
6. Albee FH. Fundamentals in bone transplantation. The Journal of the American Medical Association, 1923;81:1429-32.
7. Albee FH. Bone graft surgery in disease, injury and deformity. New York: D. Appleton Century Co.; 1940.
8. Albrektsson T. The healing of autologous bone grafts after varying degrees of surgical trauma. Journal of Bone and Joint Surgery, 1980;62(3):403-10.
9. Albrektsson T, Linder L. Intravital, long-term of autologous experimental bone grafts. Archives of Orthopaedic and Trauma Surgery, 1981;98:189.
10. Anderson KJ, LeCocq JF, Akeson WH, et al. End-point results of processed heterogenous, autogenous and homogenous bone transplants in the human: a histologic study. Clinical Orthopaedics Related Research, 1964;33:220.
11. Axhausen G. Histologische Untersuchungen uber Knochen-transplantation am Menschen. Deutsch Zeitschrift L Chirurgie, 1907;91:388-428.
12. Axhausen G. Die pathologisch anatomischen Grundlagen der Lehre von der freien Knochen transplantation bein Menschen and bein Tier. Medical Clinic, 1908;2:23.
13. Axhausen G. Uber den Vorgang partieller sequesteirung transplantierten Knochengewebes, nebst neuen histologischen Untersuchungen uber Knochen transplantation am Menschen. Arch F Klin Chir, 1909;89:281.
14. Axhausen G. Arbeiten aus dem Gebiet der Knochenpathologie und Knochenchirurgie. I. Kritische Bemerkugen un neue Beitrage zur freien Knochen transplantation. Arch F Klin Chir, 1911;94:241.
15. Bardenheuer. Transplantation der Spina Scapula Zum Ersatz des Oberen Humeruushalfte. Verhandl. Deutsch Gesellschaft Kinderchir, 1896;25:295.
16. Barth A. Uber histologische Befunde nach Knochen implantation en. Arch F Klin Chir, 1893;46:409.
17. Barth A. Histologische Untersuchungen uber Knochen implantationen. Beitz z Path. Auch u z Alig Pathol, 1895;17:65.
18. Barth A. Die Enstehung und das Wachsthum der freien Gelenkkorper. Eine histologischeklinische Studie. Arch F Klin Chir, 1898;56:507.
19. Baschkirzew NJ, Petrow NN. Betrage zur freien Knochenuberpflanzung. Deutsch Zeitschr F Chir, 1912;113:32.
20. Behrend M. Transplantation of the head and shaft of the fibula to the humerus. Surgery, Gynaecology and Obstetrics, 1930;51:717.
21. Belchier J, Goodsir J. An account of the bones of animals being changed to a red color by ailment only. Philosophical Transactions of the Royal Society of London, 1738;39:287.
22. Blackmore ME. Fractures at cancellous bone graft donor sites. Injury, 1983;14:519.
23. Blocker TG, Weiss LR. Use of cancellous bone in the repair of defects about the jaw. Annals of Surgery, 1946;123:622.
24. Bloodgood JC. Discussion of Dr Murphy's paper. Southern Medical Journal, 1915;15:235-6.

25. Bonome A. Zur Histogenese der Knochenregeneration. Arch L Path Anat u Physiol u Klin Med, 1885;100:293.
26. Boyd HB. Congenital pseudarthrosis, treatment by dual bone grafts. Journal of Bone and Joint Surgery, 1941;23:497.
27. Boyd HB and Sage FP. Congenital pseudarthrosis of the tibia. Journal of Bone and Joint Surgery, 1958;40(A):1245.
28. Brooks B, Hutson WA. Studies in bone transplantation. Archives of Surgery, 1920;1:284.
29. Brown WL, Brown CP. Further observations on the fate of free bone transplants. Journal of the American Medical Association, 1915;45:1007.
30. Buchholz R. Einige Versuche uber kunstlische Knochenbildung. Virchows Arth Path Anat, 1863;26:78.
31. Buchmann P. Behandlung knocherner Ellbogengelenksankylosen mittels Überpfanzung von ganzen Gelenken. Zentralbl f Chir, 1908;35:582.
32. Burchard H, Enneking WF. Transplantation of bone. Surgical Clinics of North America, 1978;58:405.
33. Burwell RG. Studies in transplantation of bone. VIII Treated composite homografts. Journal of Bone and Joint Surgery, 1961;43B:820.
34. Buxton SJD. Bone grafting. Medicine III, 1951;6:541.
35. Cabanella ME. Open cancellous bone grafting of infected bone defects. Orthopedics Clinics of North America, 1984;15:427.
36. Campbell WC. An operation for the correction of "drop-foot". Journal of Bone and Joint Surgery, 1923;5:815.
37. Campbell WC. The autogenous bone graft. Journal of Bone and Joint Surgery, 1939;21:694.
38. Campbell's Operative Orthopedics. In: Allen S Edmonson, A Hernshaw (Eds). The C.V. London, Toronto, St. Louis: Mosby Co; 1980. p. 20.
39. Carter WW. The correction of nasal deformities by mechanical means and by transplantation of bone. M Rec, 1911;80:1159.
40. Carter WW. The ultimate fate of bone when transplanted into the nose. Archives of Otolaryngology, 1932;15:563.
41. Castelein RM, Sauter AJ. Lumbar hernia in an iliac bone graft defect: a case report. Acta Orthopaedica Scandinavica, 1985;56(3):273.
42. Challis, JH, Lythe JA, Stuart AE. Strangulated lumbar hernia and volvulus following removal of iliac crest bone graft: a case report. Acta Orthopaedica Scandinavica, 1975;46:230.
43. Charnley J. Compression arthrodesis of the ankle and shoulder. Journal of Bone and Joint Surgery, 1951;33(B):180.
44. Charnley J, Lowe HG. A study of the end results of compression arthrodesis of the knee. Journal of Bone and Joint Surgery, 1958;40(B):633.
45. Chase SW and Herndon CH. The fate of autologous and homologous bone grafts. Journal of Bone and Joint Surgery, 1955;37(A):809.
46. Clendenin MB, Green DP. Arthrodesis of the wrist complications and their management. Journal of Hand Surgery, 1981;6:253.
47. Cloward RB. An anterior approach for removal of ruptured cervical disks. Journal of Neurosurgery, 1958;15:602.
48. Cockin J. Autologous bone grafting complications at the donor site. Journal of Bone and Joint Surgery, 1971;53(B): 153.
49. Connheim J. Vorlesungen uber allgemeine Pathologie. Berlin: August Hirschwald; 1882.
50. Converse JM. Early and late treatment of gunshot wounds of the jaw in French battle casualties in North Africa and Italy. Journal of Oral and Maxillofacial Surgery, 1945;3:112.
51. Coventry MB, Tapper EM. Pelvic instability, a consequence of removing iliac bone for grafting. Journal of Bone and Joint Surgery, 1972;54(A):83.
52. Cowley SP, Anderson LD. Hernias through donor sites for iliac bone grafts. Journal of Bone and Joint Surgery, 1983;65(A):1023.
53. Crockford D, Converse JM. The ilium as a source of bone grafts in children. Plastic and Reconstructive Surgery, 1972;50:270.
54. Curtis B Farquhar. Case of bone implantation and transplantation for cysts of tibia, osteomyelitic cavities and ununited fracture. American Journal of the Medical Sciences, 1892;106:30.
55. Darkschewitzch LO, Weidenhammer WW. Uber den Ersatz von Trepanationstucke des Schadels. Centralbl F Chir, 1892;19:835.
56. D'Aubigne RN. Reconstruction of the knee by pedicled patellar transplant. Acta Orthopaedica Scandinavica, 1973;44:550.
57. David M. Uber die histologischen Befunde nach Replantation trepantirter Knochenstucke des Schadels. Arch F Klin Chir, 1896;53:740.
58. David M. Zur Frage des Verhaltens austrepanirter und weider eingeheilter Schadelstucke. Deutsch Med Wochenschr, 1897;23:751.
59. Davis JB. The muscle pedicle bone graft in hip fusion. Journal of Bone and Joint Surgery, 1954;36(A):790.
60. Davis JS, Hunnicutt JA. The osteogenic power of periosteum with a note on bone transplantation: an experimental study. Annal Surgery, 1915;61(6):672-85.
61. Delageniere H. Repair of loss of bone substance and reconstruction of bones by osteo-periosteal grafts taken from the tibia. American Journal of Surgery, 1921;35:281.
62. Deleu J, Trueta J. Vascularization of bone grafts in the anterior chamber of the eye. Journal of Bone and Joint Surgery, 1965;47:319.
63. Dell PC, Sheppard JE. Vascularized bone grafts in the treatment of infected forearm non-unions. Journal of Hand Surgery (Am), 1984;9:653.
64. de Souza-Ramos R. Osteogenetische Induktion. Orthopaedics, 1980;118:781.
65. Dick IL. Iliac-bone transplantation. Journal of Bone and Joint Surgery, 1946;28:1.

66. Dobrovolskaja NA. On the regeneration of bone in its relation to cultivation of bone tissue. British Journal of Surgery, 1916;4:332.
67. Doi K, Tominaga S, Shibata T. Bone grafts with micro vascular anastomosis of vascular pedicles. Journal of Bone and Joint Surgery, 1977;59(A):809.
68. Dorrance GM. Osteoperiosteal bone grafts. Annals of Surgery, 1930;42:161.
69. dos Reis GB, Fallopa F, Alvachian Fernandes HJ, et al. Outcome of diaphyseal forearm fracture non-unions treated by autogenous bone grafting and compression plating. Annal Surgery of Innovation and Research, 2009;3:5.
70. Duhamel HL. Sur Ie developement et la crue des os des animaux. Mem Acad soy de Sc, 1742;55:354.
71. Dyas FG. Treatment of fractures by autogenous bone transplants. Surgery in Gynaecology and Obstetrics, 1915; 21:115.
72. Eden R. Zur operativen behandlung der habituelien schulterluxation. Zentralbl Chir, 1920;47:1002.
73. Ely LW. Bone growth in transplanted bone. Archives of Surgery, 1924;9:215.
74. Enneking WE, Burchard H, PuhlJJ. Physical and biological aspects of repair in cortical dog transplants. Journal of Bone and Joint Surgery, 1975;57(A):232.
75. Enneking WF, Shirley PD. Resection-arthrodesis for malignant and potentially malignant lesions about the knee using an intramedullary rod and local bone grafts. Journal of Bone and Joint Surgery, 1977;59(A):223.
76. Fell HB. Chondrogenesis in cultures of endosteum. Proceedings of the Royal Society, 1933;112:412.
77. Flourens P. Nouvellas recherches concernant l'action de la garance sur les os. CR. Academy of Sciences, 1840;10:305.
78. Frame JW, Browne RM, Brady CL. Biologic basis for interpositional autogenous bone grafts to the mandible. Journal of Oral and Maxillofacial Surgery, 1982;40:407.
79. Freeland AE, Mutz SB. Posterior bone grafting for infected ununited fracture of the tibia. Journal of Bone and Joint Surgery, 1976;58(A):653.
80. Gaillard PJ. Hormones regulating growth and differentiation in embryonic explants. Paris: Hermann; 1942.
81. Gaines JH, Reid RL. Salvage of the infected tibial non-union. Proceedings of the American Academy of Orthopaedic Surgeons. Journal of Bone and Joint Surgery, 1976;58(A):732.
82. Gallie WE, Robertson DE. The repair of bone. British Journal of Surgery, 1919;7:211.
83. Gill AB. An operation to make a posterior bone block at the ankle to limit foot drop. Journal of Bone and Joint Surgery, 1933;15:166.
84. Goebel R. Ersatz von Fingergelenken durch Zehengelenke. Münchener Medicine Wochenschr, 1913;60:1598.
85. Gray JC, Elves MW. Donor cells' contribution to osteogenesis in experimental cancellous bone grafts. Clinical Orthopaedics and Related Research, 1982;163:261-81.
86. Green SA, Dlabal JA. The open bone graft for septic non-union. Clinical Orthopaedics and Related Research, 1983; 180:117.
87. Greene HSN. The significance of the heterologous transplantation of human cancer. Cancer, 1952;5:24.
88. Guba L, Pool MD. Stress fracture of the iliac bone with subfacial femoral neuropathy. British Journal of Plastic Surgery, 1983;36(3):305.
89. Haas SL. Spontaneous healing inherent in transplanted bone. Journal of Bone and Joint Surgery, 1922;4:209.
90. Hammack BL, Enneking WF. Comparative vascularization of autogenous and homogenous bone transplants. Journal of Bone and Joint Surgery, 1960;42(A):811.
91. Hanson LW, Eppright RH. Posterior bone grafting of the tibia for non-union. Journal of Bone and Joint Surgery, 1966;48(A):27.
92. Harbin M, Liber KE. The behavior of transplanted bone. Surgery in Gynaecology and Obstetrics, 1934;59:149.
93. Henderson MS. The massive bone graft in ununited fractures. Journal of the American Medical Association, 1936; 107:1104.
94. Heneghan HM, McCabe JP. Use of autologous bone graft in anterior cervical decompression: morbidity and quality of life analysis. Musculoskeletal Disorders, 2009;10:158.
95. Henry AK. Exposure of long bones and other surgical methods. Briston: John Wright and Sons; 1927.
96. Henry AK. Extensile exposure applied to limb surgery. Edinburgh: E and S Livingstone Ltd; 1950.
97. Henry MO, Geist ES. Spinal fusion by simplified technique. Journal of Bone and Joint Surgery, 1933;14:622.
98. Henry MO. Chip grafts in orthopedic surgery. Journal of Bone and Joint Surgery, 1937;19:1057.
99. Herndon CH. Principles of bone graft surgery different methods of operative procedure and indications for each. Instructional Course Lectures, 1960;17:149-64.
100. Higgs SH. The use of cancellous chips in bone grafting. Journal of Bone and Joint Surgery, 1946;28:15.
101. Horwitz T, Lambert RG. Massive iliac bone grafts in the treatment of ununited fractures and large defects of long bones; the combined bone graft-metallic plate technique, Surgery, Gynaecology and Obstetrics, 1947;84:435.
102. Horwitz T. The behavior of bone grafts. Surgery, in Gynaecology and Obstetrics, 1949;89:310.
103. Hunter J. "The Works of John Hunter". In: JF Palmer (Ed). London: Longman; 1837.
104. Hutchinson J. Observation of bone transplants in the anterior chamber of the eye. Glasgow Medical Journal, 1949;30:357.
105. Hybbinette S. De la transplantation d'un fragment osseux an luxations recidivantes de l'epaule; constatationns et resultats operatoires. Acta Chirurgica Scandinavica, 1932; 71:411.
106. Inclan A. End results in physiological blocking of flail joints. Journal of Bone and Joint Surgery, 1949;31(A):748.
107. Jackson IT, Munro IR, Salgev KE, Whitaker LA. Atlas of craniomaxillofacial surgery. Mosby, St. Louis; 1982. pp. 17.

108. Jakimowitsch DR, Versuche uber das Wiederanheilen volekommen getrennter Knochensplitter. Deutsche Zeitschen F Chir, 1881;15:201.
109. Kahn B. Superior gluteal artery laceration, a complication of iliac bone graft surgery. Clinical Orthopaedics and Related Research and Related Research, 1979;140:204.
110. Kartaschew SI. Beitrage zur Frage der reien autoplastischen Knochen transplantation. Arch F Klin Chir, 1930;156:758,
111. Kassowitz M. Die normale Ossification. Wien: L Braumuller. 1881;p. 37.
112. Kave VG and Silverstein GS. CT demonstration of hernia through an iliac crest defect. Journal of Computer Assisted Tomography, 1986;10(3):432.
113. Kazanjian VH, Converse JM. Surgical Treatment of Facial Injuries, 3rd edition. Baltimore: Williams and Wilkins; 1972. p. 572.
114. Keith A. Bone growth and bone repair. British Journal of Surgery, 1918;5:685.
115. Knight MP, Wood GO. Surgical obliteration of bone cavities following traumatic osteomyelitis. Journal of Bone and Joint Surgery, 1945;27:547.
116. Kuhn DA, Moreland MS. Complications following iliac crest bone grafting. Clinical Orthopaedics and Related Research, 1986;209:224.
117. Laurie SWS, Kahan LB, Mulliken JB, et al. Donor site morbidity after harvesting rib and iliac bone. Plastic and Reconstruction Surgery, 1984;73:933.
118. Lecroix P. The organization of bones. London: J and A Churchill; 1951. p. 23, 190-4.
119. Lewin JL. Case of successful grafting of ribs into skull for cranial defect. Medical Journal of South Africa, 1924;20:61.
120. Lexer E. Freie Knochenplastic und Gelengimplantation. Medical Clinics, 1908;4:686.
121. Lexer E. Uber freie. Transplantationen. Arch F Klin Chir, 1911;95:827.
122. Lexer E. Zwanzig Jahre Transplantation forschung in der Chirurgie. Arch Klin Chir, 1925;138-251.
123. Lexer E. Joint transplantation and arthroplasty. Surgery, Gynaecology and Obstetrics, 1925;40:782.
124. Lichtbau S. Dislocation of the sacroiliac joint. A complication of bone grafting. Journal of Bone and Joint Surgery, 1962;44(A):193.
125. Lipscomb PR. When to use cortical bone grafts. Texas Medical, 1974;70:76.
126. Linderman A. Die gegenwartigen Behandlungswege der Kieferschussverlentzungen. Wiesbaden. 1916.
127. Longaere JJ, de Stafano GA. Further observations on the behavior of autogenous split-rib grafts in reconstruction of extensive defects of the cranium and face. Plastic and Reconstruction Surgery, 1957;20:281.
128. Longaere JJ. Transplantation of bone. In: JM Converse (Ed). Reconstructive and Plastic Surgery. London, Philadelphia: W B Saunders co; 1964. p. 140.
129. Lubicky JP, Dewald RL. Methyl methacrylate reconstruction of large iliac crest bone graft donor sites. Clinical Orthopaedics and Related Research, 1982;164:252.
130. Lucas DB, Murray WR. Arthrodesis of the knee by double plating. Journal of Bone and Joint Surgery, 1961;43(A):795.
131. Macewen W. Intrahuman bone grafting and reimplantation of bone. Annal Surgery, 1909;50:959.
132. Macewen W. The Growth of bone. Observations on osteogenesis. The experimental inquiry into the development and reproduction of diaphyseal bone. Glasgow: James Mac Pehose and Sons; 1912.
133. Massey EW. Meralgia paresthetica secondary to trauma of bone graft. Journal of Trauma, 1980;20(4):342.
134. Mayer L, Wehrner E. An experimental study of osteogenesis. American Journal of Orthopedics, 1914;12:213.
135. Mrazik J, Amato C, Leban S, et al. The ilium as a source of autogenous bone for grafting. Clinical considerations. Journal of Oral Surgery, 1980;38:29.
136. McFarland B. Pseudarthrosis of the tibia in childhood. Journal of Bone and Joint Surgery, 1951;33(B):36.
137. McGaw WH, Harbin H. The role of bone marrow and endosteum in bone regeneration. Journal of Bone and Joint Surgery, 1934;16:816.
138. McGrath MH, Watson HK. Late results with local bone graft donor sites in hand surgery. Journal of Hand Surgery, 1981;6:234.
139. McKeever F. Discussion of experimental studies in the transplantation of whole joints, by C.H. Herndon and S.W. Chase. Journal of Bone and Joint Surgery, 1952;34(A):578.
140. McMaster PE, Hohl M. Tibiofibular cross-peg grafting. Journal of Bone and Joint Surgery, 1965;47(A):1146.
141. McWilliams CA. A discussion of bone transplantation and the use of the rib as a graft. Annal Surgery, 1912;56:377.
142. McWilliams CA. The periosteum in bone transplantation. Journal of the American Medical Association, 1914;62:346.
143. Merren. Animadversiones guaedan chirurgical experimentes in animalibus factus illustratae, Giessae. Cited in: Hutchinnson. Journal of British Surgery, 1952;39:552.
144. Meyers MH, Harvey JP, Moore TM. The treatment of displaced subcapital and transcervical fractures of the femoral neck by muscle-pedicle bone graft and internal fixation. Journal of Bone and Joint Surgery, 1973;55(A):257.
145. Mowlem R. Bone and cartilage transplants: their use and behaviour. British Journal of Surgery, 1941;29:182.
146. Müller W. Demonstration zur Frage Osteopoastik. Münch Med Worchenschr, 1900;47:1472.
147. Murphy JB. Contributions to the surgery of bones, joints and tendons. Journal of the American Medical Association, 1912;58:985, 1094.
148. Murphy JB. Limitations of bone regeneration and reconstruction. Southern Medical Journal, 1917;10:223.
149. Ollier L. Traite experimental et clinique de la regeneration des os et de la production artificielle du tissu osseux, Paris: Victor Masson et Fils; 1867.

150. Osterman AL, Bora IM. Free vascularized bone grafting for large gap non-union of long bones. Orthopaedics Clinic of North America, 1984;15:131.
151. Papineau L. L'excon-greffe avec fermeture retardee deliberee dans l'osteomyelite chronique. Nouv Presse Med, 1973;2: 1753.
152. Papineau, LJ, Alfagme A, Delcourt JP, et al. Chronic osteomyelitis of long bone – resection and bone grafting with delayed skin closure. Journal of Bone and Joint Surgery, 1976;52(B):138.
153. Peer LA. The fate of autogenous human bone grafts. British Journal of Plastic Surgery, 1951;3:237.
154. Peer LA. Transplantation of Tissues. Baltimore: Williams and Wilkins Co.; 1955. p. 196.
155. Petrow N. Ein 25 Jahre altes Knochentransplantat. Arch F Klin Chir, 1933;175, 176.
156. Phemister DB. The fate of transplanted bone and regenerative power of its various constituents. Surgery in Gynaecology and Obstetrics, 1913;19:303.
157. Poncet A. Transplantation osseuse interhumaine (greffe massive) dans un cas de pseudarthrose du tibia gaulhe chez un jeunne homme de dix-neuf ans. Comp. Rend. Paris: Academy of Science; 1887. p. 929.
158. Prigge EK. The treatment of chronic osteomyelitis by use of muscle transplant or iliac graft. Journal of Bone and Joint surgery, 1946;28:576.
159. Puget J, Utheza G. Reconstruction de L'os iliaque aL'aide du femur homolateral après resection pour tumeur pelvienne. Revue de Chirurgie Orthopédique, 1986;72:151.
160. Radzimowski A. Concerning explantation and transplantation of bone. Kiew: Universitats Nachrichten; 1881.
161. Reale F, Gambacorta D, Mencattini G. Iliac crest fracture after removal of two bone plugs for anterior cervical fusion. Journal of Neurosurgery, 1979;51:560.
162. Reckling FW, Waters CH III. Treatment of non-unions of fractures of the tibial diaphysis by posterolateral cortical Cancellous bone grafting. Journal of Bone and Joint Surgery, 1980;62(A):936.
163. Reid RL. Hernia through an iliac bone graft donor site: a case report. Journal of Bone and Joint Surgery, 1968;50(A):757.
164. Reynolds AF, Turner PT, Loeser JD. Fracture of the anterior superior iliac spine following anterior cervical fusion using iliac crest. Journal of Neurosurgery, 1978;48:809.
165. Robertson JM, Barron JN. A method of treatment of chronic infective osteitis. Journal of Bone and Joint Surgery, 1946;28:19.
166. Saltykow S. Uber Replantation lebender Knochen.Beiträge. ZU Path Anat U z Allgemeine Patho, 1909;45:440.
167. Scott W, Petersen RC, Grant S. A method for procuring iliac bone by trephine curettage. Journal of Bone and Joint Surgery, 1966;50(A):757.
168. Scudder CL. The treatment of fractures. London, Philadelphia: WB Saunders Co.; 1933.
169. Silber JS, Anderson DG, Dafner S. Donor site morbidity after anterior iliac crest bone harvest for single level anterior cervical discectomy and fusion. Spine, 2003;28:134.
170. Seiler JG, Johnson J. Iliac crest autogenous bone grafting: donor site complications. Journal of the Southern Orthopaedic Association, 2000;9:91.
171. Stringa G. Studies of vascularization of bone grafts. Journal of Bone and Joint Surgery, 1957;39(B):395.
172. Taylor GI, Miller GDH, Ham FJ. The free vascularized bone graft. A clinical extension of micro vascular technique. Plastic Reconstruction Surgery, 1975;55:533.
173. Taylor GI. Microvascular free bone transfer. A clinical technique. Orthopaedics Clinic of North America, 1977;8:425.
174. Taylor, GI, Watson N. One stage repair of compound leg defect with a free re-vascularized flap of groin skin and iliac bone. Plastic and Reconstruction Surgery, 1978;61:494.
175. Taylor, GI. The current status of free vascularized bone grafts. Clinics of Plastic Surgery, 1983;10:185.
176. Tietze A. Erstaz des resecierten untered Radiusendes durch eine Grosszehenphalange. Verhandl d. Deutschen Ges L Chir. 1902;31 Congr, 1: 77.
177. Tuflier T. Sur les greffes osteo-articulares. Apropos d'une observation de M. Pierre Duval. Bull. et Mem. Soc. de Chir. de Paris, 1913;39:1078.
178. Ubhi CS, Morris DL. Fracture and herniation of bowel at bone graft donor site in the iliac crest. Injury, 1984;16(3):202.
179. Urist MR. Bone Transplants and Implants. In MR Urist (ed.), Fundamental and Clinical Bone Physiology. Philadelphia: Lippincott, 1980; pp. 331-68.
180. Urist MR, McLean FC. Osteogenic potency and new-bone formation by induction in transplants in the anterior chamber of the eye. Journal of Bone and Joint Surgery, 1952;34(A):443.
181. Urist MR, Mikulsky A, Leitze A. Solubilized and insolubilized bone morphogenic protein. Proceedings of the National Academy of Sciences, 1979;76:1828.
182. Urist MR, Sato K, Bronnell AG, et al. Human bone morphogenetic protein (BMP). Proceedings of the Society for Experimental Biology and Medicine, 1983;173:194.
183. VanDerwerken C, Marti RF. Bone transplantation, Injury, 1982;13:271.
184. Von Eiselberg A. Zur Heilung grosserer Defecte der tibia durch gestielte Haat-Periost-Knochenlappen. Arch F Klin Chir, 1897;55:443.
185. Von Walther P. Wiedereinheilung der bei der Trepanation ausgebohrten Knochenscheibe JD. Chiru Augen-Heilk, 1921;2:571.
186. Vorschutz, Klinische Beiträge. zur Frage der freien Knochens transplantation bei Defekten des Unterkiefers. Deutsche Atschr Chir, 1911;11:591.
187. Wagner H. Operative beinverlangerung. Der Chirurg, 1971;42:260.

188. Wagner H. Surgical lengthening or shortening of femur and tibia: Techniques and Indications. In: DS Hungerford (Ed). Progress in Orthopaedic Surgery. Berlin: Springer-Verlag; 1977.
189. Weikel AM, Habal MB. Meralgia paraesthetica: a complication of bone procurement. Plastic and Reconstruction Surgery, 1977;60:572.
190. Weiland AJ, Daniel RK, Riley LH. Application of the free vascularized bone graft in the treatment of malignant or aggressive bone tumors. Johns Hopkins Medical Journal, 1977;140:85.
191. Weiland AJ, Daniel RK. Micro vascular anastomosis for bone grafts in the treatment of massive defects in bone. Journal of Bone and Joint Surgery, 1979;61(A):98.
192. Weiland AJ. Current concepts review: Vascularized free bone transplant. Journal of Bone and Joint Surgery, 1981;63(A):166.
193. Willich CT. Significance of bone marrow for regeneration in free autoplastic bone transplantation in animal experiments. Journal of the American Medical Association, 1926;86:1875.
194. Wolf J. Zur Osteoplastic. Berlin Klin Wochenschr, 1869;6:492.
195. Wolf J. Die Osteoplastic in ihren Beiziehungen zur Chirurgie und Physiologie. Arch F Klin Chir, 1863;4:183.
196. Wolf SA, Kawamoto HK. Taking the iliac-bone graft: a new technique. Journal of Bone and Joint Surgery, 1978;60(A):411.
197. Wood MB, Cooney WP III. Vascularized bone segment transfers for management of chronic osteomyelitis. Orthopaedic Clinic of North America, 1984;15(3):461.
198. Zeiss IM, Nisbet NW, Heslop BF. Studies on transference of bone. II. Vascularization of autologous and homologous implants in rats. British Journal of Experimental Pathology, 1960;41:345.

Chapter 6

Professor Dr. Med. Erich Lexer, MD (1867–1937), Professor of Surgery, University of Munich, Munich, Germany. Dr Lexer was truly the "Father" of clinical bone allograft transplantation. He made lasting contributions to the development of Plastic and Orthopaedic surgery as illustrated by the two volume books on "free transplantation". He was the first to reconstruct joints with allografts.

(*Reference:* Lexer E. Die Venvendung der freien Knochenplastic nebst Versuchen über Gelenkversteifung und Gelenktransplantation. Archiv für Klinische Chirurgie. 1908;86:939-54.)

Bone Allografts

INTRODUCTION

Much of the experimental work on allogeneic bone transplantation was performed by individuals cited in the previous chapter for their investigations of the biology of autologous bone transplants. Stimulus for the interests in allografts was provided by the anatomic limitations of autografts and by the donor site morbidity (Banwart et al 1995). Some of the work on allografts was performed concomitantly with that on autografts, with results of the studies described in the same papers. Thus, at times, it is difficult to avoid duplication in citing the original source material. However, unless this was done it would be difficult to maintain accuracy in recapitulating what the various investigators actually accomplished. The matter of clinical use of allografts is simpler to deal with, but even here it is difficult to avoid some duplication and recapitulation of findings (Albee 1923, 1944).

History of Bone Allograft (Homograft) Transplantation

The history of the clinical transplantation of allogeneic bone is long, as is, of course, the history of the scientific investigation of properties and biology of bone allografts (Mowlem 1941, Müller 1900, Murphy 1912, Malinin 1976). In this chapter, an attempt has been made to condense this voluminous material, but still preserve the accuracy and continuity in describing what results were obtained and what observations were made in the laboratory and in clinical practice.

It is obvious that despite numerous and continuous endeavors, as well as clinical successes (Darkschewitzch and Widenhammer 1892 Delangeniere and Lewin 1920), all problems with the transplantation of bone allografts have not been solved, but general acceptance of the technique was predicited as early as 1948 (Alldredge 1948). Although, considerable progress has been made with developing methodology of bone allograft preparation, storage and clinical use, the overall results have not been unequivocal. There still remains controversy regarding the significance of immune responses to transplanted bone and cartilage and their final integration into the host. Whatever concepts are now held might have to be revised in the future. However, in reading the accounts of past and current work dealing with transplantation of bone, one cannot help but notice the changes these have brought about not only in orthopedic surgery but in the entire broad field of tissue transplantation (Newhof 1923, Urist 1980).

Fresh Allografts

Fresh bone allograft from a phalanx of a large toe of an amputated leg was placed by Poncet in 1887 over an ununited tibia fracture. It was not subjected to rigid internal fixation and nonunion persisted (Poncet 1887). Better results were reported by Adamkiewicz (1889). Büdinger in 1900 closed a cranial defect in a child with cancellous bone plates from the calcaneus of an amputated leg. The implanted graft healed, and over the years became indistinguishable from the surrounding bone (Büdinger 1900). These two examples illustrate an understandable controversy that still surrounds transplantation of fresh allogeneic bone. Although laboratory studies showed fresh allografts to be highly antigenic (Curtis and Herndon 1956), with most antigenic stimulus possibly provided by the cells residing in the bone marrow (Czitrom et al 1985), some fresh allografts have been transplanted with success (Barth 1893, Baschkirzew and Petrow 1912). In certain circumstances they continue to be used with good results (McDermott et al 1985, Meyers 1985).

Experimentally, Axhausen transplanted fresh allogeneic bone with periosteum, as early as 1908. These transplants regenerated, but not as well as the autogenous grafts (Axhausen 1908). He also transplanted bone clinically (Axhausen 1907, 1908) and with cranial defects as did David (David 1896, 1897). Ely implanted fresh allografts intramuscularly in dogs, and noted that although most allografts were absorbed, one persisted for 3 years. Like his contemporaries who studied allografts, Barth and Ely believed both bone and the marrow cells died in the transplanted allografts. They thought periosteum of the allografts did not form bone (Barth 1895, 1898, Ely 1919). Some years later, a clear difference between fresh autografts and fresh allografts was demonstrated *in vitro* by Hutchison (Hutchison 1949). Autologous bone transplanted into the anterior eye chamber of rabbits maintained normal appearance and yielded living cells into the surrounding medium. There was well recognized periosteum in the transplant. At 180 days, the cells in the allografts became necrotic and the bone formed sequestrae. There was no formation of periosteum. On the other hand, chondrocytes which were present in the rabbit illiums used for anterior eye chamber transplants remained viable in both the auto and the allografts.

Early experience with fresh massive osteochondral bone grafts is well known. The most noted operations of this kind were performed by Lexer (Lexer 1908) followed by Deutschlander (Deutschlander 1914) and others.

The first major allogeneic transplant performed by Lexer in 1907 was that of a whole knee joint. The patient was an 18-year-old girl with an ankylosis of the knee joint due to suppurative osteomyelitis. The knee joint was fixed in flexion. Lexer excised the entire joint, but left the patella attached to the soft tissue flap. The epiphyseal portion of both the tibia and the femur were excised at about 1½ fingerbreadth. The allograft was obtained from a leg amputated simultaneously because of "senile gangrene". It consisted of both epiphyses and contained the cruciate ligaments and menisci. Both ends of the graft were secured to the bone of the recipient with nails. Three months post-transplantation joint motion was restricted by the adhesion of the patella to the femur. Lexer decided to perform a patellectomy, and at the same time examine the allograft. The host-recipient junction had healed solidly, the menisci were intact, the cruciate ligaments preserved and the articular cartilage was smooth and not fibrillated. After 16 months the tibial surface became irregular on roentgenograms, and dense bone was noted at the host-graft junction. At about 4 years, both joint surfaces were irregular and part of the transplant was replaced with dense bone. At 10 years the marginal eminences were broadened and thickened, the joint surfaces were irregular, and the motion was restricted. However, 16 years after the transplant there was still painless weight bearing and sufficient motion to allow the patient to go up and down stairs. Another patient with a total knee joint transplant followed a similar course.

The first hemi-joint transplants were likewise performed by Lexer (Lexer 1908, 1911 and 1985). The first patient had a "cystic sarcoma" of the proximal tibia. The allograft was again obtained from an amputated leg, but from the opposite side. The clinical result was good, despite the fact that joint surfaces obviously did not articulate normally. However, because of the patient's "religious considerations" the leg was amputated 1 year and 5 months post-transplantation. The specimen showed a complete union at the host-graft junction, and revascularization in the periphery. The medullary canal of the allograft was packed with iodoform paste. It was dry and the endosteum necrotic. The joint surfaces of the graft were attached to the menisci of the host as were the cruciate ligaments. There were pressure lesions in the cartilage opposite the condyles. Otherwise it appeared normal.

Lexer performed a total of 23 operations in which he reconstructed the extremities with fresh massive allografts. The follow-up of these patients exceeded 10 years. Lexer reported good results in 12 patients and poor results in eight. Information on the other three was not available. In the "good results" group, there were nine patients with allografts involving knee joints. Nine of these were operated on because of infections or trauma and only one because of a tumor. Of eight failures, one was due to postoperative hemorrhage and seven due to infection. Three of the infections were due to recurrent tuberculosis (Lexer 1925).

Remarkable success with fresh cadaver bone transplants was achieved by Küttner (Küttner 1911 and 1913). He implanted an entire proximal one-half of the femur in place of the upper femur resected for chondrosarcoma. The allograft was obtained from a cadaver 35 hours postmortem. The patient died of metastatic disease 13 months after the operation. By Küttner's observations, the allograft, on postmortem examination, was united with the host bone and was surrounded by the newly formed periosteum. The musculature was attached to the transplant. The bone cells in the transplant were dead, but there was extensive replacement of the graft with the new bone. Other early investigators likewised made many contributions to bone transplantation (Müller 1912, Nageotte 1918).

Wade, in 1920 described a patient with a sarcoma of the upper humerus and a pathologic fracture. A segment of the femur from an amputated leg was placed into the surgically created defect. Articular cartilage of the femur was placed in contact with the glenoid cavity, and the lower end of the allograft was surrounded with periosteum of the host. Six years later, the clinical result was excellent (Wade 1920).

Carrel advocated transplanting bone and other tissues from newborns (Carrel 1912) and Neuhof reported excellent results obtained with this material (Neuhof 1923). He noted the callus which united the allograft with the bone of the recipient was derived from the host. Under favorable circumstances, the dead bone of the graft became absorbed and replaced by "creeping substitution".

In the same time-framework a number of prominent surgeons recorded their experiences with bone grafts, both autologous and allogeneic (von Eiselberg 1897, Phemister 1914, Tietze 1902, Tuffier 1913, Wade 1920, Wereschinski 1924). A radius from an amputated arm was transplanted as a substitute for a resected sarcoma of the radius by Elmer and Schmincke in 1925. The allograft healed and the functional result was good until the graft fracture resulted in a nonunion (Elmer and Schmincke 1925).

Fresh allogeneic bone grafts were used successfully by Smith for the treatment of patients with osteogenesis imperfecta (Smith 1937), and for ununited fractures of the leg. In the latter case, onlay grafts were fashioned from the tibias of donors. Behrend transplanted tibia to the humerus (Behrend 1930).

Boyd performed bone grafts in 97 patients with congenital pseudarthrosis of the tibia. One-third of these had successful results. He reaffirmed the necessity for rigid internal fixation when transplanting allografts. In his series all grafts fixed with wire loops failed (Boyd 1982).

Ghormley performed fresh bone allograft transplants in 14 patients, with 10 successful outcomes. Most of these allografts were obtained from living related donors. However, incorporation of the graft did not appear to be related to the blood groups. ABO matched and unmatched allografts did equally well (Ghormley 1942).

In 1948, Henry described a series of patients who received fresh bone allografts taken from male relatives. Because of clinical successes with these patients, he felt fresh bone allografts warranted wider application in orthopedic surgery. He noted: "Objections to the use of homografts have been made by some authors who fear transmission of disease, incompatibility of blood and Rh factor, or differences in sera". "Theoretically the closer the relationship of the donor to the recipient, the better the chances of success". For this reason, Henry obtained allografts not only from close relatives, but matched the blood type in every case. He concluded this was unnecessary, as the grafts behaved identically to the unmatched grafts transplanted by Boyd and Ghormley (Henry 1948). He also noted that "World War II has stimulated interest in preserved bone grafts; and at present work is progressing in some clinics to establish banks of preserved bone. As soon as the danger of infection can be eliminated, the use of preserved grafts will undoubtedly find a place in orthopedic surgery". Henry's prediction proved correct, and soon after the publication of his paper fresh bone allografts obtained from relatives of the recipients were replaced by preserved bone allografts from unrelated donors. Likewise Lexer's work with fresh massive allografts was not continued.

Harbin and Liber listed indications for each type of graft, and reported rapid revascularization of allografts (Harbin and Liber 1949). These and other authors noted that ultimately, after revascularization, both cortical and cancellous grafts of fresh allogeneic bone were restored to normal structures.

Alldredge and Terwilliger used fresh cortical allografts from cadaver donors on 15 patients with trauma. In 10 of these the overall results were excellent. Two grafts became infected. Three patients sustained allograft fractures after healing of the graft-host junction had taken place (Alldredge 1948).

Early laboratory experiments with fresh bone allografts followed the course of those with autologous grafts. The prevailing question at the turn of the last century related to periosteum (Dorrance 1930). However, soon the differences in the behavior of autologous and allogeneic grafts were noticed. Axhausen in 1909 transplanted bone allografts with periosteum and endosteum into rats, cats, and dogs and noted that these transplants regenerated, but not as rapidly and as well as autografts (Axhausen 1909). Groves, on the other hand, observed little difference between fresh femur allograft pegs and autograft pegs that were driven into holes drilled in the tibia (Groves 1917). Microscopically, tissue of neither grafts showed surviving cells. In the periphery there was ingrowth of new bone into the graft. This ingrowth of the host bone into the allograft produced osseous union.

As mentioned earlier, Hutchison compared fresh allografts with autografts placed into the anterior chamber of the rabbit eye. During all periods of observation, the autotransplants appeared like normal rabbit bone containing viable cells. A distinct periosteal layer was present. On the other hand allografts at 180 days did not contain viable cells, and formed sequestrae. A periosteal layer was absent (Hutchison 1949). Hutchison then implanted bone grafts into the muscle of rabbits. Bone autografts, after the initial period of implantation, formed new bone regardless of the site of implantation. Allografts implanted into muscle died and either sequestered or became absorbed. However, allografts transplanted in contact with the recipient's bone stimulated new bone formation and were replaced by the same (Hutchison 1952).

Vascularization is a keystone of bone graft incorporation, both autologous and allogeneic. The time sequence of vascular ingrowth appears important (Gallie 1919, Mayer & Werner 1914).

It is possible by using vital microscopy to follow the vascular changes and bone remodeling in the same bone graft from the time of transplantation to 1 year post-transplantation. Autografts which received their first vessels early (5–9 days after grafting) contained viable bone and an active blood circulation 1 year after transplantation. Injured grafts which received their first vessel late (17–29 days after grafting) were only partly revascularized after 1 year and contained islands of dead bone. Therefore, minimal surgical trauma to both the graft and the host bed is important for early revascularization of the graft. The latter contributes to a complete incorporation of the graft (Albrektsson and Linder 1981). Likewise preservation of the osteogenic potential in the graft must be preserved. Extensive exposure to hydrogen peroxide destroys it (Carpenter et al 2006).

Presently, fresh bone allografts are used primarily in combination with cartilage for resurfacing of joints and for filling full depth articular defects. Experience with bone and cartilage allotransplantation has been described by Gross, colleagues and by others (Czitrom et al. 1986, Meyers 1985).

Osteoarticular allografts differ from other bone transplants, in that vascularization proceeds only through the osseous graft-host interface (**Fig. 1**). Experimental and clinical studies demonstrated that thin layers of cancellous bone support the cartilage head without collapse (Campbell et al 1963, De Palma et al 1963, Lance and Fisher 1970, Pap and Krompecher 1961). Gross' experience with this type of allograft extends over a 14-year period and includes observations of over 100 patients (McDermott et al 1985). Best results are obtained with traumatic arthritis of the knee. Patients with primary osteoarthritis did not benefit from this procedure, and results in patients with osteonecrosis and osteochondritis dissecans were generally poor. Experience with osteochondral plugs was different. Of biological significance were observations that allografts did not heal well in patients on steroids. Most patients receiving joint resurfacing transplants showed immune reactivity to the graft (Langer et al 1978). Meyers showed the suitability of osteoarticular allografting in the treatment of osteonecrosis with late segmental collapse of the head of the femur (Meyers et al 1983 and 1985). In this series the fresh allografts healed well except in patients receiving steroids.

Fresh osteoarticular allografts were studied by Kandel et al (Kandel et al 1985) by light and electron microscopy.

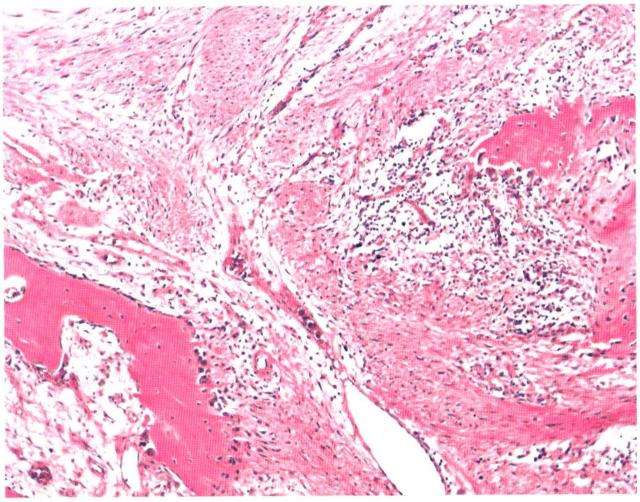

Figure 1 Photomicrograph of the junction of a fresh cancellous bone allograft and the recipient's bone. Trabecular spaces of the graft have been replaced by fibrous connective tissue. No replacement of the allograft bone is noted, but osteoblasts and osteoclasts are active in the host bone. Areas of chronic inflammatory cells are present in the graft (H&E × 100)

By necessity the observations were made only on failed allografts, but findings relating to bone healing and graft incorporation are probably applicable to fresh bone allografts in general. Some of the allografts were examined as late as 7 years post-transplantation. Ingrowth of viable host tissue and fibrovascular tissue was observed in most grafts. Bone was remodeled and some allografts were completely replaced by host tissue. Interestingly, changes suggestive of an immune response were not observed.

The host response to the allograft varied with the age of the recipient and post-transplantation time. Complete or almost complete replacement of the graft was seen as early as 17 months post-transplantation. However, in some grafts residual allograft bone was present as late as 7 years post-transplantation. Most of the changes observed in this material were comparable to those described in animals. Goldberg and Lance showed ingrowth of host bone into grafts of iliac bone in rabbits (Goldberg and Lance 1972).

Although serologically bone and cartilage transplants were shown to evoke an immune response, clinical rejection was not detected in either animals or humans receiving fresh allografts (Brown and Cruess 1982, Burchard 1983, Goldberg et al 1983). This is in concert with the clinical observations of early pioneers using fresh allografts. Empirical observations on patients receiving fresh osteoarticular allografts suggest deformity and failure of the grafts may be due to repetitive stress and mechanical factors rather than immunlogical effects resulting in structural graft deficiencies.

In conclusion, it is of interest that despite positive immunological laboratory findings, fresh bone allografts have functioned clinically much better than one might expect. However, a word of caution might be appropriate if one wished to apply the results with osteoarticular shell and cylinder allografts to other fresh bone allografts. The concept here was that narrow rims of fresh bone and relatively small amounts of bone in cylinder grafts while maintaining and securing the articular surfaces, would not be subjected quantitatively to the changes occurring in larger grafts (Meyers et al 1983). Reducing the osseous component to a minimum and washing out bone marrow, specifically cells residing in the marrow would decrease the immunologic load, as well as the amount of tissue needed to be replaced. This would result in shorter healing time (Lance 1985, Myers 1978, Paacola et al 1979, Pritzker et al 1977, Rodrigo et al 1978 and Tanaka et al 1980). However, in some experimental animals, receiving cylindrical osteoarticular allografts, histological evidence of rejection was present (Malinin et al 2000). Immunologic responses following transplantation in animals of fresh iliac cancellous bone, femoral bone and tibial bone have been shown over many years by several investigators (Burwell 1985, Chalmers 1975, Brooks et al 1963, Masse et al 1965) (**Fig. 1**). Similar changes are not observed in freeze-dried allografts in humans (**Fig. 2**).

The principal antigenic component has now been shown to be red marrow that is abundantly present in fresh bone (Burwell 1963, Czitrom et al 1985). In animals receiving fresh bone allografts, experimental studies reveal production of both humoral and cell mediated immunity (Bos et al 1983, Friedlaender 1983). According to Burwell, these lead to cessation of early osteogenesis in the grafts (Burwell 1985).

Preserved Allografts

Many varieties of preserved bone allografts have been investigated experimentally and used clinically (Malinin 1976). Since osteocytes and osteoblasts in preserved allografts are non-viable, the grafts themselves do not contribute cells to new bone formation. The biologic basis for the activity of many, but not all of these grafts is the stimulation of the recipient mesenchymal cells to become osteoprogenitor cells and to form new bone that gradually replaces the graft. For this reason, to be effective, preserved bone allografts must possess osteogenic potential of which bone morphogenic proteins (BMPs) seem to be most important. Osteogenicity

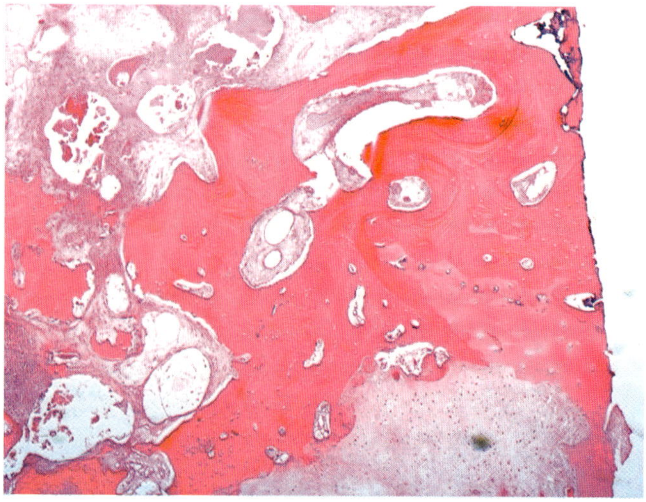

Figure 2 Biopsy of a cancellous bone graft in an acetabulum of a patient 1 year post-transplantation. The bone allograft has been replaced by the new bone (H&E × 100)

of allografts is maintained by some methods of preservation and destroyed by others.

One of the earliest methods of preserving bone allografts was by boiling. Hass, Ely, Pollock et al showed that transplanted boiled bone does not initiate new bone formation (Haas 1926, Ely 1922, Pollock et al 1922). Other investigators demonstrated that boiled bone allografts are not only non-osteogenic, but undergo very slow resorption (Baschkirzew 1912, Phemister 1914, Rohde 1924, Wereschinski 1924, Bull 1928, Engstrom 1937). In definitive experiments, Lacroix studied the properties of boiled bone allografts by implanting them renal capsules in rabbits. Seven months later the grafts were nearly completely intact. At most, the grafts had undergone scant marginal erosion by forming a minimal amount of Howship's lacunae containing osteoclasts. There was no obvious mesenchymal cell recruitment or in other words, signs of osteogenesis or hematomyelopoiesis adjacent to the grafts.

Boiling of allografts was eventually supplanted by autoclaving. This method of preservation was published in 1937 by Smith (Smith 1937). He removed bone from living donors, placed it in test tubes, and autoclaved the grafts prior to transplantation. As one might expect, in the autoclaved bone the observed biologic activity was similar to that of boiled bone.

Chemically Preserved Bone

A simple method of preserving and at the same time sterilizing bone allografts is by immersing them into chemicals such as alcohol. This method of chemical preservation and sterilization of bone started and evolved at the same time as boiling and autoclaving techniques. Early in the last century LeRiche and Christophe reported successful results with alcohol preserved bone allografts (LeRiche 1922, Christophe 1926). When alcohol-fixed bone grafts were implanted into rodents, they were absorbed, but a few layers of new bone and later hematopoietic foci appeared in and around the graft (Nageotte 1918, Polettini 1922, Rohlich 1941). Lacroix made similar observations, but noted the absence of osteoclasts at the periphery of alcohol-fixed grafts (Lacroix 1951). Thus, it was evident that, unlike boiling or autoclaving, the osteogenic potential of the graft was not completely destroyed by alcohol fixation. The "osteogenic principle" has been extracted by alcohol from normal bones and from fracture calluses. When this alcoholic extract was injected into rabbits, osteogenesis took place in about one-third of the animals (Levander and Willstead 1946, Annersten 1940). However, quantitatively osteogenesis induced by alcohol-fixed grafts (either autografts or allografts) was reduced as compared to fresh bone (Lacroix 1951).

Other procedures advocated for the storage of bone grafts were preservation in para phenol solution (Carnevale 1951), acetone, formaldehyde (Sabanas et al 1955), merthiolate (Reynolds et al 1951) and others. None of these chemical preservation techniques were generally accepted.

Frozen and Freeze-Dried Bone Allografts

The above cited experiments served to demonstrate the differences of the recipient's response to bone allografts prepared in different ways. The aforementioned chemically preserved allografts are now used rarely if ever. Most studies dealing with the biology of bone allografts have been performed on either frozen or freeze-dried bone.

Elliott and Scott showed that bone grafts implanted and then removed from patients were not dead, but surrounded by mesenchymal tissue which was undergoing metaplasia and ossification (Elliott and Scott 1951). These observations summarize the entire spectrum of bone allograft incorporation. The earlier phases of the host allograft interaction in the implantation site, including increased vascularity, are similar to those of autografts. However, the vascular ingrowth into frozen or freeze-dried allogeneic bone takes longer and the onset and progress of creeping substitution is slower. Aside from temporal considerations, no qualitative differences are noted between the process and the end results of autograft and allograft incorporation. These go through stages of revascularization, osteoclastic resorption, new bone formation and remodeling (Mankin and Friedlaender 1983, Helsop et al 1960).

Studies demonstrating frozen and freeze-dried bone allografts incorporate and undergo accretion faster than do fresh bone allografts led not only to the clinical acceptance of frozen or freeze-dried allografts, but also stimulated many investigations of the biology of bone grafts (Bonfiglio et al 1955, Curtis et al 1956, Heiple et al 1963, Herndon and Chase 1952). However, despite numerous attempts to define the exact mechanism of preserved bone allograft incorporation, the understanding of these events is still relatively incomplete (Mankin et al 1983). The most serious handicap in this regard is the lack of knowledge of the exact mechanism of osteogenesis, specifically the complex cell signaling pathways.

Cancellous Bone Allografts

Compared with autogenous controls, the most striking difference shown by allografts is the time sequence of the healing process (Heiple et al 1983). With cancellous bone, in the early healing stages, the differences are minor. The slight diminution of the size of the callus can be explained by the lack of contribution of osteogenic cells by the allograft. The differences become more pronounced by about 4 weeks after transplantation. Although by this time considerable callus is formed, the allograft is joined to host bone by fibrocartilage. Marrow replacement is slow, and usually there is no new bone formation at the margins. By 3 months the differences between the allografts and autografts are most noticeable, but by 6 months, once union with an allograft has occurred, the differences are subtle. Graft maturation and remodeling at this stage are morphologically similar (Heiple et al 1963, 1983, Blocker and Wise 1946).

Cancellous bone allografts behave differently from cortical bone grafts (Burchard 1983) although the first steps of graft incorporation are similar. These include: formation of a blood clot, accumulation of inflammatory cells, recruitment of mesenchymal cells and fibroblasts. At about 2 weeks osteoclast activity begins, and osteocyte autolysis in the graft is noticed.

Cancellous bone usually revascularizes within 2 weeks. The process depends on the size of the graft. Cancellous graft is strengthened by the new bone that replaces the graft. Finally cancellous grafts become completely resorbed and replaced with new bone. In allografts, initial accumulation of inflammatory cells is more pronounced than in autografts. This is most likely related to immunologic events. This delays the subsequent bone induction process. As mentioned earlier, by 3 months the differences between the allo and autografts are most profound, but by 6 months, if allograft-host union has occurred, the differences subside. Graft maturation and remodeling proceed along similar lines (Heiple et al 1963 and 1983).

Cortical Bone Allografts

In cortical bone allografts repair is initiated along periosteal and endosteal graft surfaces. There is mesenchymal tissue ingrowth into the Haversian system that is accompanied by resorption of the surrounding bone and its replacement with new bone (**Fig. 3**). This sequence of events is dictated by the cortical bone architecture. Space must be made within the non-viable graft before new bone formation can take place. During repair and remodeling, large canals are opened in the cortical graft, followed by vascular ingrowth into these channels from the host (**Fig. 3**). New bone is formed and channels are revascularized. As the process progresses, cortical bone may be converted into a cancellous like structure (**Fig. 4**).

Eventually new bone formation exceeds resorption and the mass of the bone is restored. This, of course, occurs only in allografts intimately attached to the recipient's bone.

The sequence of events in healing of cortical bone allografts becomes quite apparent when cortical bone plate onlay allografts rigidly fixed to the host bone are examined (Malinin et al 1985).

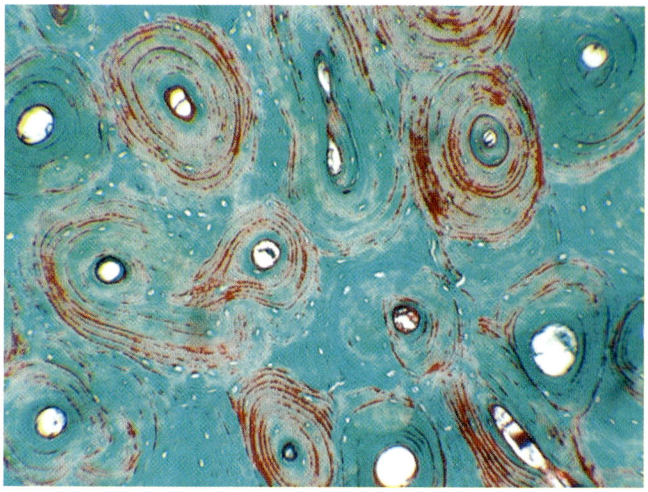

Figure 3 Biopsy of a cortical allograft in a patient 2 years post-transplantation. New lamellar bone is formed around revascularized Haversian canals (Masson trichomes stain × 250)

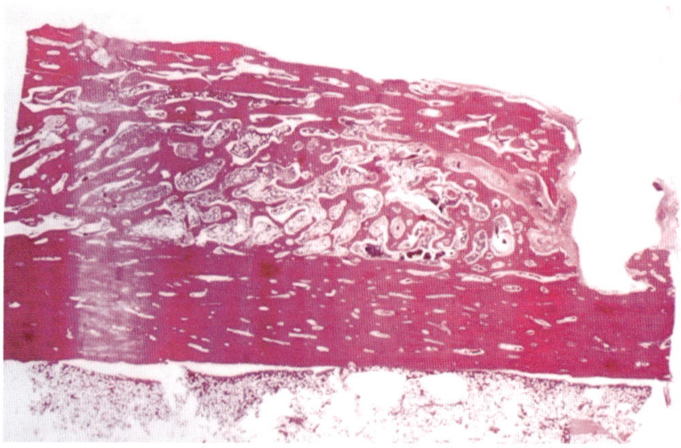

Figure 4 Onlay cortical bone allograft in a canine model 6 weeks post-transplantation. The cortical bone of the graft had been converted into a callus-like structure (H&E × 100)

Intercalary allografts show the same pattern of healing and incorporation. Clinical and laboratory experience gained over the last 3 decades indicate that freeze-dried intercalary allografts secured in place by rigid internal fixation show radiological evidence of healing at host-graft junctions and replacement of the graft with host bone. However, in laboratory experiments Burchard et al (Burchard et al 1978) reported dissolution of fibular allografts. These were not subjected to rigid internal fixation. When freeze-dried intercalary allografts are rigidly fixed, the host-graft junctions heal and the allograft bone is gradually replaced with host bone (Malinin et al 1985).

Immunology of Allogeneic Bone Transplantation

Immunology, the study of mechanisms utilized by the host to neutralize materials foreign to itself, has not produced clear-cut results with allotransplantation of bone, as it has with blood transfusion, skin and organ transplantation. The many ways by which grafts are processed, the defects in which they are placed and secured, and the actual size of the implanted materials are but a few of the reasons for disparate immunologic reactions to allograft bone (Horowitz and Friedlaender 1991).

The key observation which precipitated the development of immunology was the description of phagocytosis by Metchnikoff (Tauber 2003). Vertebrate phagocytes are derived from hematopoietic precursor cells. However, despite their common origin they developed in two distinct groups of cells; short-lived granulocytes and mononuclear cells of the blood and tissue macrophages.

Lymphocytes have a separate origin and function. Mononuclear phagocytic cells scattered throughout the body were described as the "reticuloendothelial" system by Aschoff (McCallum 1942). There is a significant difference between the phagocytic cells of the reticuloendothelial system and lymphocytes. The phagocytes of the reticuloendothelial system are non-specific cells; and engulf or phagocytize many types of particles. Lymphocytes on the other hand respond to very specific antigens via T lymphocytes. Any one lymphocyte will take part in only one specific immune reaction. In order for B lymphocytes to produce antibodies they must come in contact with T lymphocytes stimulated by an antigen (Pappenheimer 1914).

Initially, immunology was focused on infectious diseases. As surgical techniques were developed and improved, transplantation immunology came into being. Transplantation immunology includes the studies of host immune responses to transplanted bone. These were stimulated by the knowledge gained from transplantation of skin and organs, particularly kidney. Many of the concepts of transplantation immunology are derived from technical laboratory tests and procedures applicable to organ transplants. Bone transplantation may not share these immunologic vagaries with other organs. Therefore, interpretations of the test in terms of immunologic compatibility may not be valid. For this reason, a brief description of the tests performed might be appropriate. Hopefully, this will aid the reader in forming his/her own opinion regarding the significance of the immunologic laboratory results applied to bone transplantation.

Since the human population is composed of individuals produced by millennia of outbreeding, it is easy to understand why organs and many tissues are "rejected" when transferred from one such genetically unique individual to another. The recipient recognizes transplanted material as foreign. This, of course, means the cells of the immune system from the blood, lymph and probably interstitial fluid come into contact with foreign proteins (antigens) of the graft. Rapid re-establishment of circulation in transplanted organ or skin facilitates this contact. For grafted allograft skin, a classical model for study of the host response to transplantation, the rejection phenomenon is closely associated with graft revascularization (Guttman and Lindquist 1970). The ability of an individual to tolerate its own skin graft while recognizing and rejecting skin grafts from genetically different individuals is a fundamental aspect of transplantation immunology. Despite the enormous effort devoted to the study of this phenomenon our understanding of it is far from complete. The clinical successes with kidney transplantation and other organs are not based on the understanding of this phenomenon, but on the ability to suppress the immune response by selective poisoning of immunocytes and related cells.

The jargon of transplantation immunology is confusing; particularly since the terminology was changed after the use of such terms as "homograft" and "heterograft" had become well established. Now different prefixes are used to denote degrees of genetic relationship between the donor and the host. "Auto" (self) means the same individual is both the donor and the recipient of the graft. Thus, autograft denoted the transfer of tissue from one place to another in the same individual. The prefixes "iso" or "syn" are not commonly used clinically. These prefixes meaning "same" or "together" refer to different individuals that are genetically identical, i.e. identical twins or inbred animals. Isografts are not rejected when transferred between "isogeneic" individuals. "Allo", meaning other, is used to describe transplantation between genetically different individuals of the same species. The term "allograft" or "allogeneic graft" is now used instead of the old term "homograft". With the exception of identical twins, all humans are "allogeneic" in relation of one to another. The old term "heterograft" has been discarded in favor of a "xenograft". "Xeno", meaning "strange" or "foreign", denotes transplantation of tissues between different species, i.e. animal bone into humans, etc.

Laboratory Measurements of Immunologic Response

Since the rejection of allografts and xenografts is a process similar to other kinds of immune responses, it may be useful to review the nature of these responses and the means by which they are demonstrated in the laboratory.

The variety of foreign proteins against which the immune system reacts in a predictable manner is enormous. The foreign substances (antigens) are dealt with by the formation of antibodies (Bos et al 1981), synthesized within specialized cells of the lymphoid system. The main constituents of antibodies are serum globulins, usually having four polypeptide chains, two "heavy" and two "light". These gammaglobulins are referred to as immunoglobulins (Ig). Like any other large protein molecule, immunoglobulins may in turn act as antigens when injected in animals of other species. Therefore, they can be used to distinguish different antigenic and molecular types of Ig. The predominant component of Ig is immunoglobulin with a molecular weight of about 150,000 designated G or IgG. Besides IgG, in humans four other classes of immunoglobulins, IgM, IgA, IgD and IgE, are recognized. Each Ig molecule also contains some carbohydrate with the carbohydrate content of IgM being 12% and IgG being around 3%. IgM, the second most frequently encountered immunoglobulin, has a molecular weight of 900,000 and consists of five units each approximating the size of IgG.

IgM is the first antibody made in most immune responses. Its large molecular weight confines it to the bloodstream where it agglutinates foreign particles. Since IgG is smaller, it can diffuse into interstitial spaces and cross the placental barrier. Human plasma contains relatively high concentrations of IgA. IgD is present on the surfaces of blood lymphocytes where it probably acts in the control of lymphocyte activation and suppression.

There are several ways for detecting antibodies in a laboratory. These are: antigen binding, protection against antigen, precipitation, agglutination, hemolysis in a complement system and double-diffusion precipitation in gels. Each of these basic tests has a number of variations and modifications.

Antigens mixed with antibodies in solution can combine and form complexes which precipitate. The quantity of precipitate varies with the amounts of reagents used.

Most antibodies have two identical sites at which they combine with antigen, i.e. they are divalent. Antigens, on the other hand, can vary from small molecules with a single reactive site to substances that antibodies can attach to at numerous sites. If, antigen concentration is kept constant and antibody is added in increasing increments, a so-called zone of "antibody excess" will be produced at some point. In this zone of "antibody excess" all antigen molecules are surrounded by antibody. The addition of more antigen produces aggregates which precipitate. The ability of antibody and antigen to combine in variable proportions led to the lattice theory of Marrack. This simple scheme is often complicated by heterogeneity of antibodies, but it does illustrate the principle of the reaction (Marrack 1955).

Antigen and antibody placed in holes in agar gel diffuse toward each other and form a precipitate where there is optimum antibody-antigen mixture. The precipitate forms a visible opaque line. A relationship between two antigens can be studied by observing the precipitation reaction from adjacent holes. When the reagents are present in balanced proportions the lines formed are concave because of the slower diffusion of larger molecules present in the antigen or antibody. This method of gel diffusion is accurate and can be used for quantitative assessment of antigens. There are several modifications of this technique.

Immunoelectrophoresis is used to detect antigens according to their electrophoretic mobility. By this technique it is easy to show a number of precipitation arcs among proteins under study. Electrophoresis can be combined with immunoprecipitation. In this circumstance, the movement of an electric field drives the antigen toward the antibody. Several variations of the technique, including countercurrent electrophoresis and two-dimensional electrophoresis, are used.

Radioimmunoassay determines the antibody levels by measuring the amount of immunoglobulin binding to an insoluble antigen. If one takes into account the sensitivity of the test, the quantitative estimation of antibody content can be made. The test involves the addition of excess antigen labeled with radioisotopes to the antiserum, followed by the assessment of antigen that forms a complex with the antibody. The binding of radioactively labeled antigen to a fixed amount of antibody can be inhibited by the addition of unlabeled antigen. The extent of inhibition is used to measure the amount of material labeled

Immunofluorescence (immunohistochemistry) uses fluorescent dyes coupled to antibodies that will not alter the specificity of the latter. The fluorescent dyes coupled to antibody then combine with antigens in tissue sections. This allows for localization of antigens in the tissues and cells. This technique referred to as immunofluorescence or immunohistochemistry, has been widely used in transplantation immunology. The test is usually performed as a "direct", "indirect", or "sandwich test". In a direct test the antibody conjugated with fluorescent dye is applied directly to the tissue section. Fluorescence occurs at the sites of antigen-antibody interaction. In a direct test a double layer technique is employed. An unlabeled antibody is applied to the tissue substrate and then treated with a fluorescent-dye-conjugated anti-immunoglobulin serum. The fluorescence in an indirect test is brighter than with a direct test because several fluorescent immunoglobulins bind the antibody molecules. The sandwich test is a double layer procedure designed for visualization of specific antibodies; especially antibodies in low concentrations. The tissue under study is fixed with ethanol to prevent the antibody from being washed away during staining and then flooded with antigen. After washing, a fluorescently labeled antibody is added to localize the cells that bind the antigen.

Enzyme, labeled antibody methods couple antibodies to enzymes, which when stained can be visualized in a light or electron microscope. The enzyme most commonly used for this purpose is peroxidase. There are several drawbacks to this technique and, in general, it is not as accurate as immunofluorescence.

Agglutination

Agglutination of cells differs from precipitation because antibodies are allowed to clump together and form large aggregates. Since cells are electrically charged, a large number of antibody complexes are needed to overcome the repulsion. Therefore, agglutination of the cells cannot usually be achieved unless it is enhanced by special methods. Agglutination reactions are used to type red blood cells. They have also been applied to leukocytes. Because of its sensitivity and ease, the test has been used to identify antibodies to certain antigens that have been coated on various particles. Erythrocytes can be coated with proteins after their surfaces have been altered by chemicals such as chromium or tannic acid.

Antibodies directed against surfaces of particular cells can fix certain components of extracellular fluids collectively known as complement and with their aid lyse the target cells, hence the term cytotoxic reaction. This activity of complement depends on nine protein components of complement that consists of three major subfractions. When the first component binds to a cell of an antibody it activates the next components in the system. This produces a sequential effect. Complement like the blood coagulation cascade involves the activation of an enzymatic sequence of events. The residue of the complement reaction activated by an antigen-antibody complex binds to the cell membrane and eventually destroys the membrane resulting in cell lysis.

Cellular Participation in Immune Response

In addition to the antibody production, there is a second type of immune response, the so-called cellular response. Both of these responses are specific, i.e. the establishment of immunity to one antigen does not offer protection against another unrelated antigen. The cellular response is centered on the lymphocyte unit. Lymphoid tissue, from which these cells originate, is divided into central (thymus and bone marrow) and peripheral (lymph nodes, spleen and lymphoid accumulation in the GI tract) lymphoid organs. The distinction is important since, under normal circumstances in an adult, lymphocytes are produced only in the central organs.

Lymphocytes are relatively inactive cells. They cannot divide, but on stimulation small lymphocytes change (transform) into large cells. The function of lymphocytes was unknown until recently, although their morphology was extensively studied and described. During the last several decades lymphocytes were divided into two major categories, designated T (originating from the thymus) and B (originating from the bone marrow). They have very different functions. B lymphocytes produce antibodies, while T lymphocytes are responsible for "cell-mediated" immune reaction and for general regulation of the immune response. B and T lymphocytes appear identical when examined in a phase-contrast microscope or in stained preparations. There may be some differences in their enzyme content and their ultrastructure, although the differences are not striking. These cells differ in their surface markers, but practically T and B lymphocytes are distinguished from each other on the basis of their physiologic properties. *In vitro* T cells are not affected by anti-Ig antisera even in the presence of complement. T cells do not attach themselves to nylon wool under certain conditions while B lymphocytes do. These cells can also be fractionated by electrophoresis. "Antilymphocyte sera" can preferentially deplete animals of T cells. The vascular system of the body contains fluid which diffuses from capillary vessels into the surrounding tissues. On the way to the veins, this fluid passes through the lymph nodes. Lymphocytes pass from the blood into the lymph nodes, and then again return to the bloodstream via lymphatic ducts. Animal studies show that this process of recirculation is rapid. Both B and T lymphocytes recirculate, although T cells predominate in the thoracic duct and B cells tend to localize in the spleen. As mentioned earlier, B-lymphocytes produce immunoglobulins.

Immunofluoresecence was the first method of demonstrating immunoglobulins in lymphocytes. This technique involved combining a fluorescent dye with an antibody that was then allowed to react with the cells that were either segregated or contained in a histologic section. The antigens against which the antibody was directed were highlighted by bright fluorescence.

Micromanipulative techniques can demonstrate how many different antibodies are produced by one lymphocyte. The technique is based on segregation of individual lymphocytes from lymph nodes undergoing immune response and incubating them individually in micro-plates of suitable medium.

One of the most commonly used techniques in the study of immune reactions is the hemolytic plaque technique. In this test, lymphoid cells from an animal immunized with red blood cells of another species are mixed with complement and red blood cells for the type used for immunization. The mixture is then incubated between slides. Antibody released by active cells attaches to the surrounding red blood cells and lyses them, leaving a plaque that is approximately 0.5 cm in diameter. By this technique a single antibody-producing cell can be detected. Thus, this test is much more sensitive than the serum antibody titration method which measures only the cumulative effects of many lymphoid cells.

When lymphocytes from two different allogeneic populations are cultured together, a vigorous blast transformation occurs. This reaction is thought to be caused by one strain of lymphocytes reacting against the antigens on the surface of the others, and vice versa. Cultured lymphocytes respond only to living allogeneic lymphocytes, while the cells which proliferate in culture are genetically capable of responding to other cells. Oddly, allogeneic combinations provide a stronger reaction than do xenogeneic combinations.

Mixed lymphocyte culture (MLC) reaction was developed in an attempt to devise an *in vitro* measure of histocompatibility. The degree of blast formation as shown by radioactive thymidine uptake was indeed related to incompatibility for transplantation antigens. In unrelated individuals it was showed that less than 1 in 500 pairs of randomly selected individuals have a negative MLC reaction. The MLC provides no information about the cellular nature of the target in an allograft since the killer cells attack serologically determined antigens on all cells.

The concept of what is now known as "major histocompatibility gene complex or system" (MHC or MHS) was introduced to make a distinction between genes associated with the acute rejection of allogeneic tissue from those that control chronic rejection of allografts, usually kidneys. Genes of this complex control the expression of alloantigens (H-antigens) on the cell surface. In mice, histocompatibility or H antigens can be identified serologically (Granja et al 1996).

Histological assessment of allografts, particularly those that are non-vascularized, remains one of the most accurate means of determining the fate of transplanted musculoskeletal tissues. By these means cellular invasion of the graft can be easily visualized. However, with bone, the rejection phenomenon and the normal pattern of bone healing and regeneration can sometimes be confused by inexperienced observers.

In an immunologically stimulated individual, further contact with an antigen leads to reactivation, acceleration and intensification of the immune response. Under some circumstances this secondary reaction termed "hypersensitivity" may be so intense as to produce tissue damage.

The hypersensitivity lesions result from an aberration of a normal cell-mediated immune reaction, that requires T lymphocyte participation. Following earlier exposure, T cells recognize the antigen together with other complexes on the macrophages and are stimulated into blast formation. Some proliferating T cells release a number of factors into extracellular fluids that function to attract macrophages. Upon interaction with specifically sensitized lymphocytes, some lymphocytes form antibodies, others form cell products such as cytotoxins and migratory inhibition factor, while others serve as mediators of cellular immunity. Cytotoxic T-lymphocytes are normal components of the host's immune response to cell-associated antigens. The initiation of lysis by T cells appears simple. However, to date the process remains largely uncharacterized.

Transplantation Immunology and Bone Grafting

The rejection of a bone or soft tissue allograft is a process (Buxton 1951) that displays the same patterns as other organs and tissues. The reaction is specific. Bone allografts induce cell-mediated and antibody-mediated cytotoxicity specific for donor antigens similar to that seen after organ transplantation. However, this has been noted clinically only with some, not all, massive structural allografts (Burwell et al 1963). This immune response may be elicited by the so-called antigen presenting cells residing in the bone marrow (Czitrom et al 1985, 1986). Bone marrow is difficult if not impossible to remove from massive allografts. Although collagen and matrix can evoke an immune response, this response is relatively weak (Friedlander 1983). All of the studies with immunologic response in patients receiving musculoskeletal allografts have been performed with frozen osseous and osteochondral allografts. Most studies showed that massive frozen bone allografts are capable of inducing cytotoxic T lymphocytes with high affinity for donor-specific HLA antigens (Deijkers et al 1999). In additional studies, donor specific antibodies and cell mediated immunity has been found after transplantation of frozen bone allografts. Histologic changes suggestive of chronic rejection have also been described (Muscolo et al 1996). Since chronic rejection of bone allografts is presumed to be mediated by T cells, further study of these interactions might elucidate the relationship between these responses and incorporation of massive bone allografts in humans. However, to date, definitive information in this regard is insufficient due to the lack of interpretable retrieval data following transplantation of massive fresh allografts in human subjects.

Studies in rodents demonstrated frozen syngeneic or frozen allogeneic bone to be histologically inferior to fresh syngeneic bone. However, frozen allotransplants from related donors elicited barely detectable immune response. Although the relationship between the results of human bone transplantation and the demonstration of immune response in laboratory rats might be somewhat tenuous, the study demonstrated histocompatibility dependence on the in vivo immune response in rats to frozen corticocancellous grafts. Frozen grafts elicited an identical biologic response by the host with regard to remodeling and repair regardless of histocompatibility. Therefore, the role of immune response in the rat model remains puzzling. Does immune response play a critical role in the incorporation of the frozen graft? Immunosuppression was not detected (Bos 1983). Skin grafts survived the same amount of time in bone grafted animals as in ungrafted controls. This is at variance with observation on the immunology of freeze-dried allografts. In animals receiving freeze-dried bone allografts, prolonged skin survival was noted (Chalmers 1959, Burwell et al 1963). The difference might be attributed to specific differences between freeze-dried and frozen allografts. However, because of the significance of skin grafting in these experiments, it seems prudent to briefly review transplantation immunology of skin that provides a clear cut demonstration of immune responses to specific tissue.

In experimental animals, the skin graft has been the transplanted tissue from which many of the current concepts of transplantation immunology have evolved.

In rodents, skin autografts become vascularized within a few days of transplantation and remain permanently on the host. However, if the skin graft is an allograft, circulation stops, the skin becomes necrotic and is sloughed off in about 10 days. A second graft from the same donor is rejected much faster, usually in about 3–4 days. The "second set" response is specific. A third graft from a different donor will be rejected at the same rate as the first one. The ability to mount a "second set" response is transferable from one animal to another by injection of lymphocytes. The response to skin grafts initiated by secondary allograft rejection differs qualitatively as do primary and secondary antibody responses. The second set reaction is rapid with invasion by polymorphonuclear granulocytes, plasma cells and other lymphoid cells. Revascularization may be absent altogether.

After rejection of an allograft, humoral antibodies specific for the graft may be recognized in the serum of the recipients. In humans, lymphocytotoxins may appear (Goldberg et al 1982, 1984).

The specificity of antigens involved in graft rejection is well established. Genetically identical individuals have identical transplantation antigens. This allows for free exchange of grafts between such individuals. The Mendelian segregation of genes controlling these antigens has been demonstrated by experiments with mice. In fact, the entire concept of what is now known as major histocompatibility gene complex (MHC) came about as a result of genetic experiments on these animals. The shorthand used to describe the classical transplantation antigen, the so-called H-Z system, looks unnecessarily complex. In order to identify allelic genes within the H-Z complex in different strains of mice, one usually starts with certain pure inbred strains to provide the prototypes. The collection of genes in the entire H-Z complex is called the haplotype and the haplotype of each prototypic strain is allotted a superscript.

The discovery of leukocyte agglutinins led to the characterization of the Human Leukocyte Antigen (HLA) system. Transplantation of kidneys aided by tissue typing within the HLA system led to the intensive study and definition of this HLA system. The first antigen of the system (HLA-A2) was defined on the basis of the reaction pattern of leukoagglutinating antibodies in the sera of pregnant women. In the subsequent years, the HLA system was recognized as one autosomal region consisting of at least two separate loci (HLA-A and HLA-B). A third locus (HLA-C) was added in 1973. The serologically defined HLA antigens are coded by a system of three closely linked loci, HLA-A, B and C. The definition of HLA-A, B and C locus antigens is generally accomplished by a microcytotoxicity test using lymphocytes from peripheral blood.

Major histocompatibility complex antigen is divided broadly into Class I (HLA-A, HLA-B, and HLA-C) or Class II (HLA-D). Class I antigens are found virtually on all human nucleated cells. Class II antigens are expressed on the surface of B lymphocytes, activated T lymphocytes, macrophages, and other cells of macrophage lineage including osteoblasts (Horowitz and Friedlaender 1991, Goldberg et al 1984). In mice, one pattern of immunogenicity by cells native to bone was characterized (Horowitz and Friedlaender 1991). Bone activated allogeneic T-cells proliferated in response to stimulation by both Class I and Class II antigens. T-cells were of the killer/suppressor phenotype commonly associated with rejection of skin. However, the significance of activated T-cells with respect to bone allografts has not been determined to date. Therefore it is unknown whether or not the pattern of reactivity found in mice occurs in humans. The relationship between the systems responsible for bone-remodeling and the immune response remains ill-defined.

A, B, O and Rh Systems

The A and B antigens of red blood cells may be present on cells of other tissues including bone (Weinberg 1959). Antibodies to antigens can cause immune reactions, which preclude, not only success of blood transfusions by causing severe hemolytic reactions, but also participate in the rejection of transplanted tissues and organs (Sasama 1990, Rydberg 1987, Thorpe et al 1991). However, the significance of these antigens in the transplantation of osseous tissues is uncertain. Fresh cancellous bone shows weak A and B antigens, but these are probably derived from bone marrow residues (Weinberg et al 1959). A and B antigens could not be demonstrated in cortical bone (Ezra-Cohen and Cook 1961).

Bone allografts are usually non-viable and incorporation depends to a large extent on the functional glycoproteins of bone matrix (Mohan and Baylink 1991, Mundy 1995). A and B antigens are carbohydrate in nature and in addition to erythrocytes are confined mainly to epithelial tissues and vascular endothelium (Oriol 1987, Slurman 1962).

Fresh and frozen, unwashed bone allografts contain A and B antigens. Transplantation of such bone allografts can induce a red cell antibody response in the recipient (Curtis 1956). However, in practice such antibodies have not been detected in allograft recipients (Stassen et al 1993). Freeze-dried bone does not stimulate A and B antibody response.

In humans, bone marrow containing bone allografts can stimulate production of Rh and other red cell antibodies in the recipients (Cheek et al 1995, Jensen 1987). Red cell antibodies induced in women by bone allografts can cross the placenta barrier and cause hemolysis in the fetus if there is an Rh factor incompatibility. Several instances of Rh positive bone allografts causing formation of Rh antibodies in the Rh negative mothers have been reported. The amount of erythrocyte in a bone allograft that can elicit alloantibody formation may be quite small. However, bone marrow containing frozen allografts do not always induce antibody formation in Rh negative recipients (Stassen et al 1993). Presumably the risk of Rh antigen sensitization is low with well washed frozen or freeze-dried bone allografts. Nevertheless, it is inadvisable to transplant bone from an Rh positive donor into an Rh negative female recipient of child bearing age.

The transplantation of bone allografts in orthopedic surgery and other surgical specialties has increased exponentially in recent years. However, the degree to which immune responses effect the survival and acceptance of frozen and freeze-dried bone allografts remain ill-defined (van de Vord 2005). There appears to be a discrepancy between the laboratory findings in animals and the clinical observations in humans. In this regard the influence of MHC's in the biologic performance of allografts may not be as important as it may seem. Laboratory studies likewise provide different results. In a recent study involving frozen orthotopic bone allografts, the differences between the MHC's of the donors and the recipients did not play a decisive role in the healing and incorporation of these grafts. In addition no antibody production to the MHC-1 antigens was detected (Reikeras et al 2011).

Fresh (Untreated) Bone Allografts

The notion that fresh bone allografts invoke a severe immune response (Goldberg 2003) and provoke a vigorous inflammatory reaction has been echoed over the years by several investigators (Friedlaender 1983). This sentiment gained wide acceptance to the point that it became dogma (Burchard et al 1987, Stevenson 1987). However, the clinical evidence is not entirely consistent with these postulates.

Information regarding antigenicity of fresh bone allografts has been gained from many laboratory studies in a variety of animal models. There was a spectrum of results, some of which were disparate and frankly contradictory. Major questions which had not been answered involve a lack of definition as to what are the antigens in bone. Unquestionably, participation of major-histocompatibility-complex antigens have been implicated in the immune response to fresh allogeneic bone transplants, but it is still unknown whether these responses are harmful. The immune response to fresh bone allografts may not be elicited by cells as is the case with skin and vascularized organ allografts. In bone, immune response can be modified by removing bone marrow and soft tissue (Lewandowski et al 2001). Furthermore, characterization of immunoreactive proteins reveals that most antigenic targets are fragments of collagen molecules. There is also evidence that the collagen matrix, proteoglycans and proteins can evoke immune responses (Horowitz and Friedlaender 1991). Such being the case, it is easy to understand the similarity of immune responses to fresh as well as frozen allogeneic bone transplants.

In one of the early studies in rabbits, investigators at the US Navy Tissue Bank detected humoral and cell mediated immunity evoked by fresh and frozen, but not freeze-dried bone allografts (Friedlaender 1983). From these and other studies it was concluded that fresh bone allografts evoke significant host inflammatory response which inhibits new bone formation (Chalmers 1959, Elves 1974, Burwell 1963). However, it was also noted, data concerning antigenicity of bone allografts in animals cannot be used as the sole criterion for predicting the efficacy of graft material in humans. A frequently cited study, using a canine model, found fresh major mismatched grafts to be antigenic. Minor mismatched frozen allografts were least able to sensitize the recipient (Stevenson 1987). The conclusion was that graft incorporation was most successful when histocompatibility differences were minimized. On the other hand, in another canine study mismatched fresh cortical bone allografts elicited a very minor immune response, in only a few of the animals tested (Cox et al 1999). In this regard, an important observation was made by Schweiberer and co-workers. They observed in fresh bone transplants the initial destruction of proliferating osteoblasts by an antigen-antibody reaction elicited by T-antigens. After the initial response, the immunologic reaction subsided and osteogenesis was again initiated by the substances still present in the allografts (Schweiberer et al 1981). Proceeding up the evolutionary tree, in non-human primates, fresh osteochondral allografts failed to unite with host bone, whereas cryoperserved allografts did (Malinin et al 1994). In the same animal model, histologic evidence of an immune response was present at the host graft junction of fresh cancellous bone allografts.

However, this did not prevent the eventual healing of the graft (Malinin et al 2006).

Bone Allografts for Clinical Transplantation

Bone autografts were frequently inadequate and their harvesting met with complications (Arrington et al 1996). Therefore, bone allografts gained in popularity. There are varieties of grafting materials available. Freeze-dried and frozen allografts are the most common. Refrigerated allografts can also be used (Wilson 1951). When bone allografts, which retain bone morphogenic proteins (BMPs), are placed in contact with vascularized host bone, healing occurs when the calcified matrix is replaced by new bone. The individual peculiarities of the human skeleton are such that each bone has its own requirements for healing, immobilization and bone grafting. Therefore, there is no universal, all-purpose bone allograft, and there is no single way for preparing all bone allografts. To date, the most successful bone allografts have been aseptically excised and aseptically processed freeze-dried cortical, cancellous and corticocancellous grafts which have not been subjected to extensive manipulations, such as exposure to chemical agents, heating, irradiation, etc. (Malinin 1992).

Figures 5A and B (A) Freeze-dried particulate allograft of three different sizes. From left to right microparticulate, ground and crushed bone grafts; (B) Microparticulate bone allograft, reconstituted in saline, is used to pack an osseous defect

Bone allografts commonly used in reconstructive procedures of the skeletal system can be broadly divided into particulate and structural grafts. The former are used most frequently. Particulate grafts can be crushed cancellous or cortical bone (bone chips), ground bone, morselized bone or microparticulate grafts (**Figs 5A and B**). These are used for filling defects with largely intact walls.

The structural grafts are bone plates (bone struts), sections of mandibles, bone blocks, bone shaft segments, wedges, ribs, etc. Bone blocks of various sizes and shapes are used to support implants and to obliterate bone voids (**Figs 6A to C**).

Many types of preserved bone allografts have been studied experimentally and observed clinically. Since osteocytes in transplanted bone are non-viable, the grafts themselves do not contribute to the cellularity of the newly formed bone. The basis for the complex activity resulting in the new bone formation is the stimulation of the recipient's mesenchymal cells. These cells induce osseous bridging at the host-graft interface and gradually replace the graft itself. To be effective, bone allografts must possess osteoinductive properties that are maintained by some methods of preservation, reduced by some and destroyed by others.

Morphologic analysis of allogenic bone grafts removed from patients before complete healing and incorporation shows them to be largely acellular. The graft is surrounded by host mesenchymal tissue that undergoes metaplasia and ossification. This basically outlines the entire spectrum of bone allograft interaction with the host. Aside from temporal considerations, no quantitative differences have been noted between the autograft and freeze-dried allografts. To a lesser degree, the same patterns have been noted in frozen allograft incorporation as healing progresses through revascularization, osteoclastic resorption, new bone formation and remodeling.

The response to allograft implantation is modified by the ways in which the grafts are prepared. There are a number of methods that may include: irradiation, exposure to chemicals, etc., but aseptically excised and processed allografts serve as a standard (**Fig. 7**). Exposure to hydrogen peroxide, a commonly used component of many commercial chemical sterilants, diminishes or abolishes the osteogenic potential of the graft (Carpenter et al 2006).

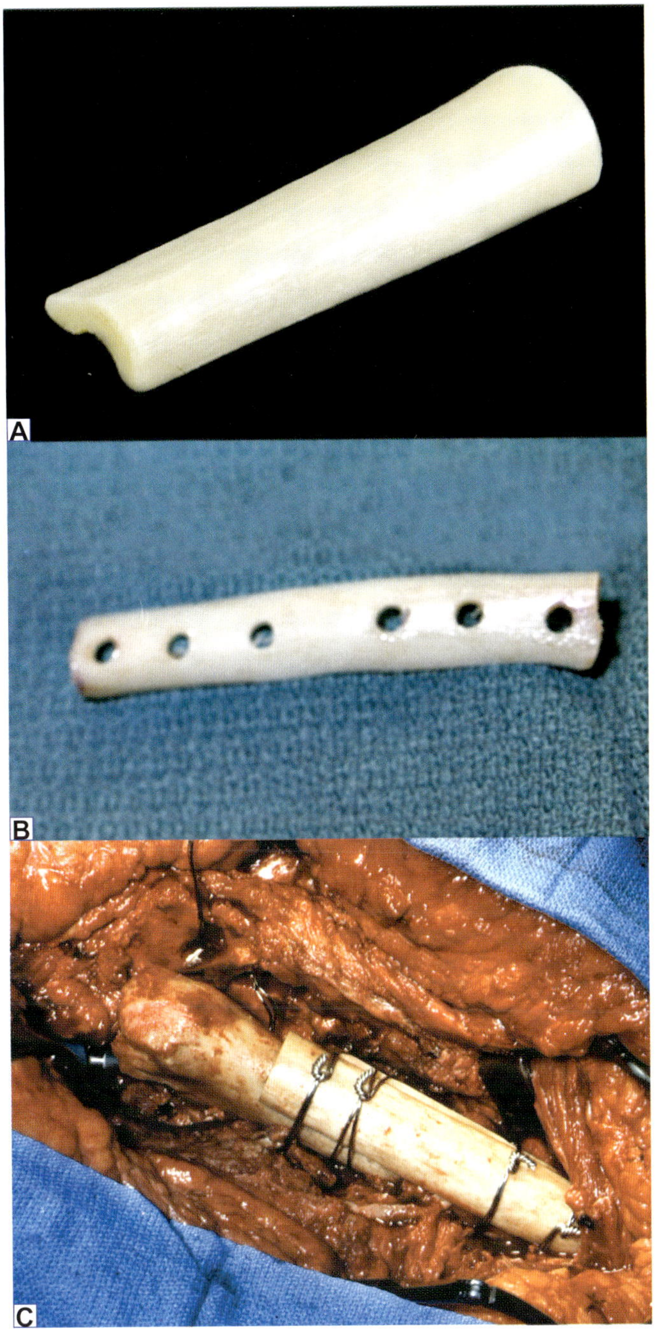

Figures 6A to C (A) Freeze-dried cortical plate allograft; (B) Thin bone plate with predrilled holes; (C) Onlay cortical plate allograft superimposed on a deficient femur. If a proximal or distal wire becomes loose during the procedure it should be retightened or repaired. It is important to ensure the circlage wires are tight

The clinical acceptance of freeze-dried bone allografts is based on the reduced immunogenicity of these preparations (Horowitz and Friedlaender 1987). The reason for the latter is most likely due to destruction of the antigen

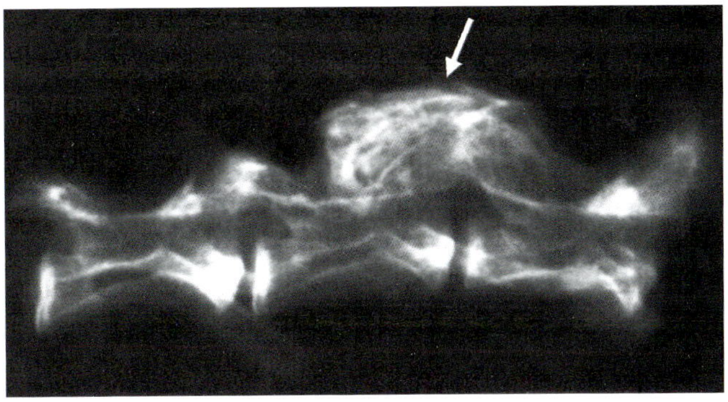

Figure 7 Spinal fusion in a rabbit with aseptically excised and process particulate bone allograft. The bone mass of the allograft is fused with the underlying bone of the spinal column (arrow)

presenting hematomyelopoietic cells that reside in the osseous trabeculae (Czitrom et al 1985) and the alterations in bone collagen.

Particulate Bone Allografts

Particulate bone allografts have been used to fill cavitary and periprosthetic defects and for spinal fusion for several decades with considerable clinical success (Henry and Geist 1933, Henry 1937, Spence et al 1972, 1976). Radiographic assessment of the grafts showed incorporation in over 90% of the patients. Complications were few.

Spence et al packed cystic defects in bone with cancellous bone croutons and crushed cortical bone. The results were comparable with packing, being more complete with cortical bone. Likewise freeze-dried allograft bone and autogenous bone provided nearly equivalent results. Radiographs of a patient with a unicameral bone cyst packed with crushed cortical bone are shown in **Figures 8A and B**.

Impaction grafting likewise with crushed bone allografts was shown by Sloof and associates to be an effective method in dealing with periprosthetic acetabular and femoral defects (Sloof et al 1996). Impaction grafting has gained popularity in dealing with large periprosthetic cavitary defects with or without cement (Dakes et al 2006). Impaction grafting with polished, tapered femoral prosthesis, as advocated by Gie and associates, is accompanied by a number of failures due to prosthetic subsidence (Jazrawi et al 1999). On the basis of these worrisome findings caution with impaction grafting and the polished femoral implant is advocated (Meding et al 1997, Dakes et al 2006). The relationship between particle size and bone incorporation is important. Some investigators feel the optimal size of particulate bone grafts should be in the 7–10 mm range (Gie et al 1993). Others advocate graft particles 2–5 mm in size (Schreurs et al 2001). However, despite clinical success with particulate allografts, ideal properties of these grafts were not well-understood until recently. In practice, tissue banks prepare particulate allografts in different sizes and forms and by a variety of methods. Recently, it became clear that one of the parameters which warrant attention is the size of the particle itself (**Figs 9A to C**).

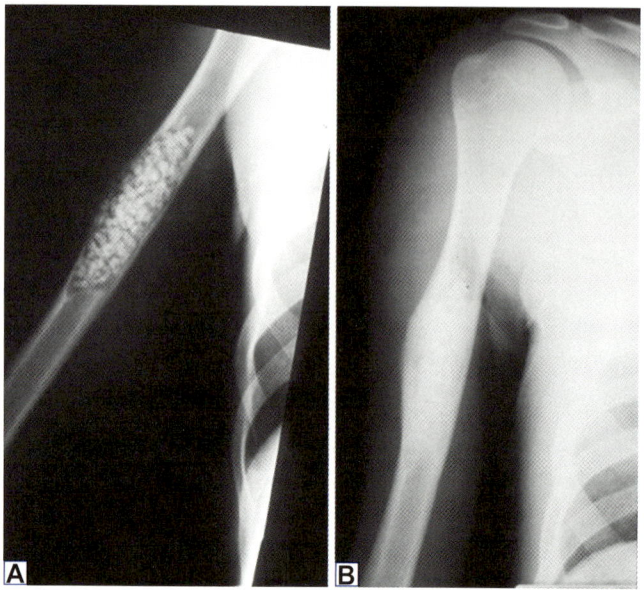

Figures 8A and B (A) Aneurysmal bone cyst packed with crushed cortical bone immediately post-packing; (B) One year later, the cavity is obliterated with newly formed bone and cortical remodeling

The size bears relation to osteoinduction and osteoconduction. Osteoinduction depends on the biological property of the graft reflected in its ability to stimulate ingrowth of neovasculature, mobilize the host mesenchymal cells and transform these mesenchymal precursors into osteoprogenitor cells. This process is mediated by the release of various growth factors, principal of which are bone morphogenetic proteins (BMPs). To be effective, BMPs must be present in physiologic quantities (Urist 1972). In addition, BMP depends on intraosseous lipids for transport, which facilitates BMP delivery to the site (Urist et al 1997). Ideally, particulate grafts should possess osteoconductive as well as osteoinductive properties. They must provide direct apposition between the graft and the host, and in addition to mechanical support allow for ingrowth of newly formed bone. Densely packed bone particles of appropriate sizes satisfy both requirements (Malinin et al 2007). Graft washing with removal of bone marrow and extraosseous fat allows for compacting of the graft material in the defect. Avoidance of

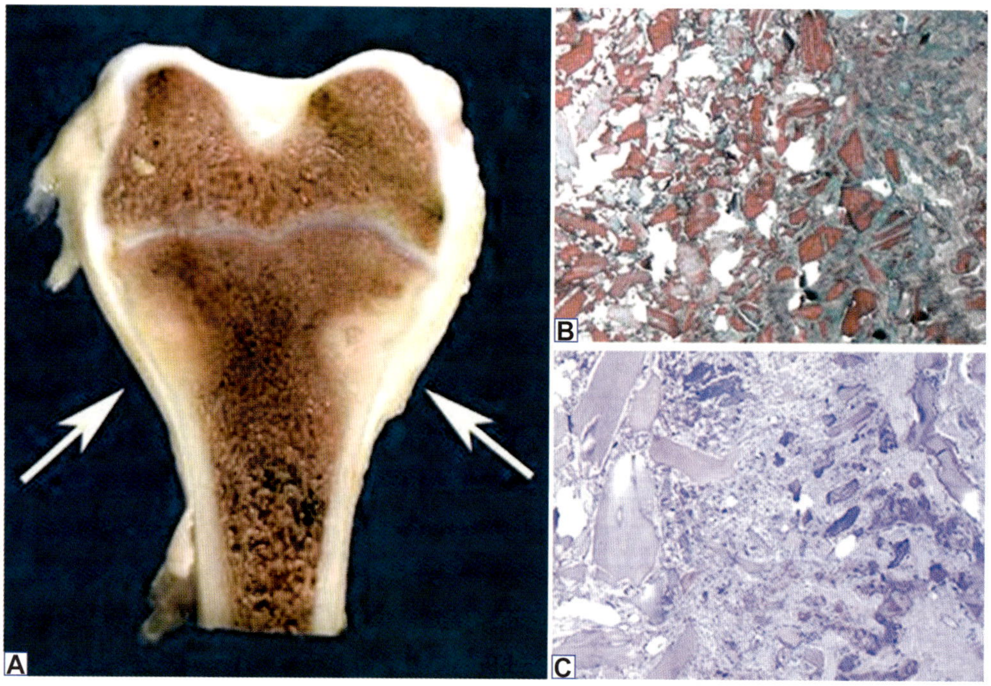

Figures 9A to C (A) Specimens of osseous defects in animals filled with mineralized microparticulate bone allograft, 3 weeks post-transplantation. There is complete incorporation of the allograft (arrows); (B) Allograft particles have been replaced with new osteoid (Masson's trichome stain × 25; (C) Intense osteoblastic and osteoclastic activity is present (PAS-hemotoxylin × 25)

alcohols and other solvents during processing used by many banks as decontaminants and chemical solvents dissolve the intraosseous lipids that are important to BMP transport.

In laboratory experiments bone allografts with particles of different sizes demonstrates statistically significant differences in the healing patterns and osteogenic properties. Particles in the range of the 300–90 microns produce rapid healing by direct ossification. Particles below 100 microns had a significantly reduced osteoinductive potential. Particles larger than 300 microns are much slower in healing and incorporation than 300–90 micron particles. Small-sized powdery bone particles below 75 microns induced little osteogenesis and were similar to small particles of hydroxyapatite that inhibit osteoclastic activity **(Fig. 10)** (Sun et al 1998). Because of these undesirable properties of truly powdered preparations, the term "bone powder" should not be applied across the board. Powder means a dry substance composed of minute dust-like particles, the composition which does not enhance osteogenesis. For this reason, and to delineate bone particle sizes most effective in inducing bone healing, the term "microparticulate bone allograft" is more descriptive and preferable. It clearly delineates particulate compositions between small granules and powders.

The process of bone incorporation using microparticulate grafts has been observed in human specimens. Mesenchymal cells are mobilized from the periphery at the allograft-host interface within days of transplantation. As time progresses, the particles are remodeled and later fuse with other particles to recapitulate the trabecular architecture of normal bone. The process is complete with the migration of hematomyelopoietic cells **(Fig. 11)**.

Frozen microparticulate allografts lag behind freeze-dried particles of similar size in inducing bone healing (Malinin and Temple 2007). Thus from practical considerations that include: easy handling, room temperature storage and superior osteoinduction, freeze-dried microparticulate allografts appear to be most suitable for filling bone defects. Microparticulate bone, when densely packed into a defect provides substantial compressive resistance. They can be added to a carrier either biologic (blood and marrow or platelet rich plasma), or non-biologic (glycerol), and injected into closed segment defects. They are radiodense and thus easy to ascertain adequate defect filling. Radiographically, in 6 weeks, the border between host and allograft bone becomes indistinct and the process continues from the periphery to the center of the defect. Graft material inadvertently left in the adjacent

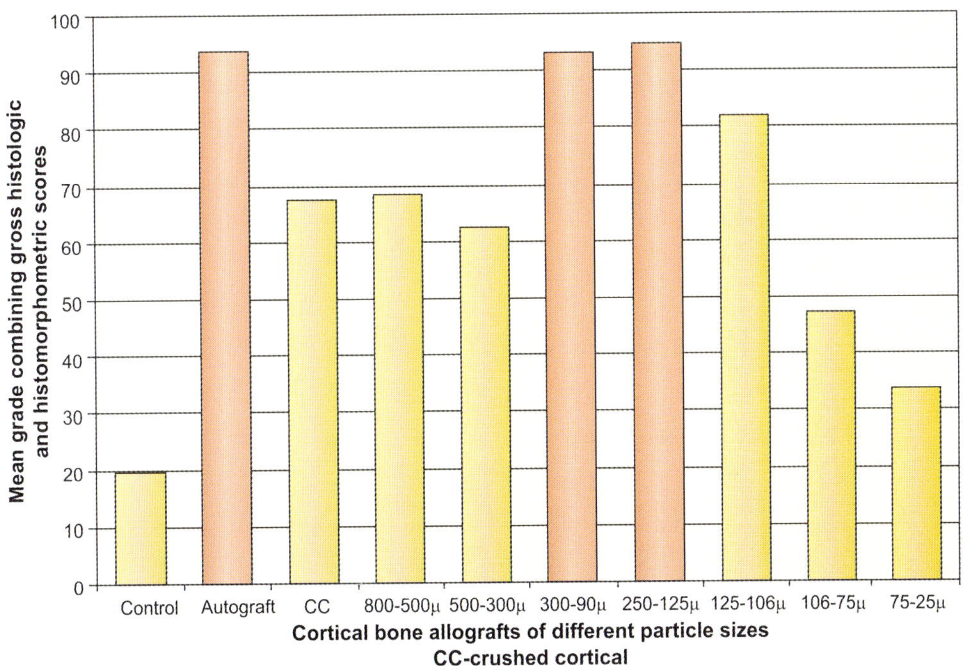

Figure 10 Graph comparing healing of intraosseous defects filled with particulate allografts of various particle sizes. Data from the gross, radiographic and histomorphometric scores is combined. Results were averaged for each experimental group

soft tissue during the grafting procedure is resorbed within 2–3 months (**Figs 12A and B**).

The most striking difference between cancellous allografts and autografts is the time sequence of the healing process. Otherwise, freeze-dried or frozen cancellous allografts follow the pattern of healing exhibited by autografts. Cortical bone when implanted as a solid structure, exhibits a different pattern of incorporation than cancellous grafts. However, in case of particulate allografts there is no difference between cortical and cancellous bone preparations since the basic structure of cancellous bone trabeculae and bone cortex is the same (Malinin et al 2007). Bone incorporation for microparticulate cancellous and cortical bone is the same based on clinical observations.

Demineralized Particulate Bone

There is a considerable confusion with regard to demineralized bone matrix (DBM) and demineralized, or more precisely, partially decalcified bone. The methods of preparation of these allografts are distinct, as are their biological properties. Demineralized bone matrix is prepared by simply demineralizing bone in an acid medium, typically hydrochloric acid, until the calcium content is reduced to less than 2.3%. Since DBM is prepared from particulate bone, the preparation frequently entails freeze-drying the bone, grinding it, demineralizing it, and repeating the freeze-drying process. Thus, in contradiction to other freeze-dried bone allografts, DBM is freeze-dried twice. This may account

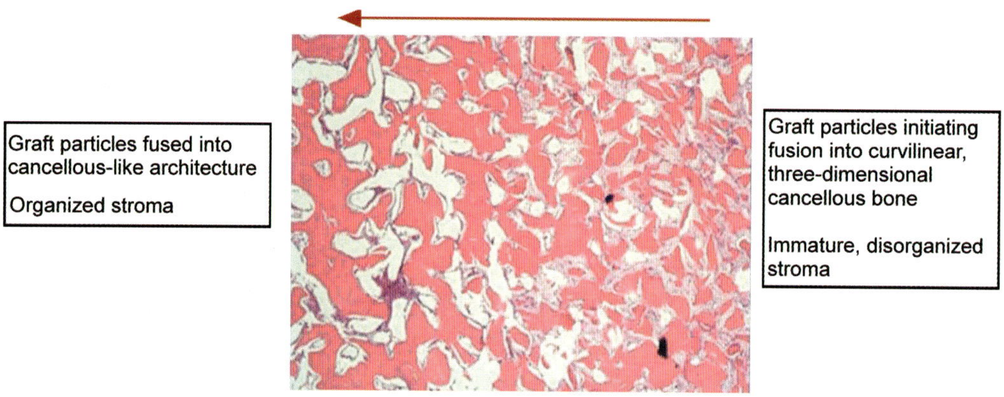

Figure 11 Specimen showing progressive conversion of isolated graft particles of various shapes and sizes into a well-organized, cancellous structure that serves as a scaffold for bone marrow development and osseous structure support (H&E × 100)

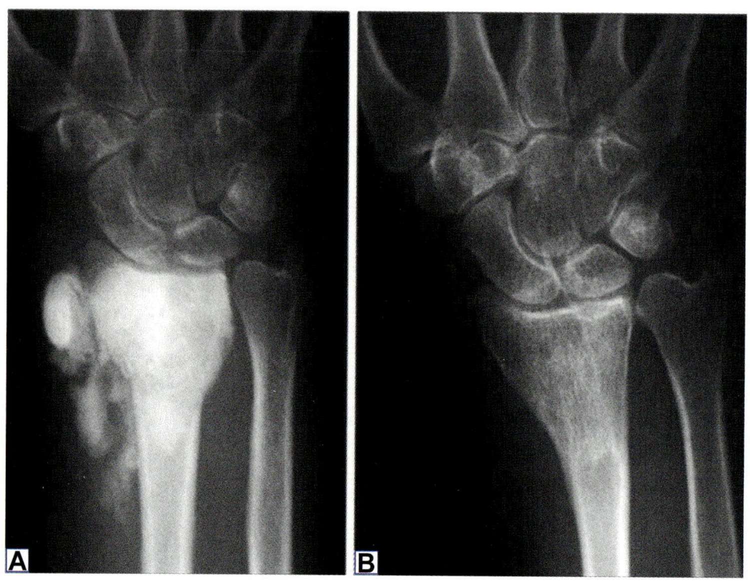

Figures 12A and B (A) Immediate postoperative AP radiograph of the wrist following extended curetting and microparticulate bone grafting. The lesion is filled with radiodense microparticulate bone. There is a moderate amount of material in the soft tissue; (B) One year postoperative AP radiographs in the same patient. There is complete incorporation of the graft and no residual bone in the soft tissue for its reduced immunogenicity. Demineralized bone matrix is grossly amorphous, soft and does not provide structural support.

Demineralization is said to increase the availability of bone matrix-associated bone morphogenetic proteins (BMPs) rendering these grafts osteoinductive. However, it must be noted, non-demineralized particulate grafts are also osteoinductive. There seems to be little difference in clinical results between demineralized and non-demineralized particulate allografts (Commack 2005). Thus, the extra effort and expense of using demineralized bone matrix might not be justified. Although demineralization releases BMP from bone, it also facilitates BMP elution and loss into the acid bath (Pietrzak et al 2009) or the surrounding tissue **(Fig. 13)**.

Partially demineralized, partially decalcified or surface decalcified bone is exemplified by Urist's chemosterilized, antigen-extracted, autolyzed allogeneic (AAA) bone (Urist 1983). Preparation of this allograft is complex and time-consuming, but its clinical efficacy has been documented **(Fig. 14)**.

The calcium content of AAA (Chemosterilized Antigen-Extracted Surface-Demineralized Autolyzed Allogeneic Bone of Urist) bone is in the neighborhood of 10–15%. The calcium content of surface demineralized bone allografts produced by tissue banks is usually somewhat higher, around 20% or more depending on the surface area and volume of bone. Both of these types of preparations maintain osteoconductive properties not present in the so-called demineralized bone matrix. Demineralized bone matrix seems to evoke a more intense inflammatory reaction than its mineralizaed counterpart. This was observed in experiments with non-human primates as shown in **Figure 13** where there is abundant sclerotic bone in the periphery and no apparent healing centrally. This should be compared with the specimen in **Figure 9A** where healing appears to be complete and uniform without peripheral sclerosis.

Structural Bone Allografts

Structural bone allografts are organized skeletal tissue grafts that can either support or replace anatomical skeletal structures. These grafts are osteoconductive. They serve to provide structural support for the skeleton and serve as a scaffold for new bone growth perpetuated by the recipient's bone. In this respect structural allograft is different from particulate allograft (Neufeld et al 2002).

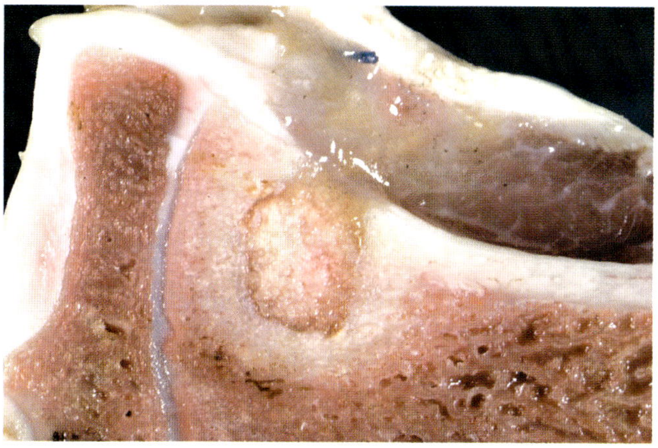

Figure 13 Intraosseous defect in an animal packed with demineralized bone matrix 3 weeks post-transplantation. Note the reaction around the defect

Figure 14 Chemosterilized antigen-extracted surface-demineralized autolyzed allogeneic bone of Urist. The acronym "AAA bone" stands for "chemosterilized, antigen extracted, and autolyzed allogeneic bone. The AAA bone prepared by Dr Urist was aseptically excised and processed. Bone allografts prepared in this manner incorporates more rapidly than their lyophilized or irradiated counterparts

Cancellous Bone Allografts

Cancellous bone grafts composed of trabecular bone serve to promote osteogenesis (Horowitz 1949). Cancellous bone allografts have greater capacity to induce new bone formation than do cortical bone allografts. An immobile trabecular bone graft allows the process of "creeping substitution" to proceed with some rapidity. In this process new bone is formed and deposited in the scaffold formed by dead trabeculae. The structure of cancellous bone allows for the formation of microvascular anastomoses and the diffusion of nutrient fluid.

Cancellous bone allografts are used primarily in filling smaller bone defects, in osteotomies, acetabular revision arthroplasty, foot fusions with bone loss and with non-unions (Horowitz and Richard 1947). The use of cancellous allografts in spine fusions is discussed separately in Chapter 11: Bone Grafts in Spine Stabilization.

During the first 2 weeks following transplantation, blood coagulum surrounds the cancellous allograft, thus preventing further bleeding (Burchard 1983). In animals, an inflammatory response occurs followed by vascular infiltration of the graft bed. By the second week granulation tissue becomes prominent. Osteoclasts make their appearance leading to new bone formation that replaces the autolyzed bone of the graft.

Freeze-dried cancellous bone allogafts, in most instances, approximate the behavior of cancellous bone autografts. Such is not the case with frozen allografts that take considerably longer to heal (Tomford et al 1990). Demineralized and decalcified cancellous allografts induce some osteoinduction, but not as much as their freeze-dried or frozen graft counterparts (Gazdaq et al 1995).

Surgeons contemplating the use of cancellous bone allografts should consider foremost the amount of bone loss and the biologic and biomechanical properties of the graft. All laboratory studies as well as clinical experience show the advantage of the biocompatibility of freeze-dried cancellous allografts with host bone. Of all available allografts, excluding the unique characteristics of freeze dried microparticulate bone; freeze-dried cancellous grafts come closest to autologous grafts. However, it must be remembered, freeze-dried allografts are not viable but they are significantly osteoinductive. Host bone tissues surrounding the graft intimates in progressive graft incorporation. During this process, a mild to moderate immunologic response may also exist, but with freeze-dried cancellous allografts it is usually transient. At times, the immune response may be difficult to distinguish from graft incorporation and new bone formation.

Freeze-dried cancellous bone structural allografts are among the most successful allografts. These can be used in combination with particulate bone allografts.

Long-term studies show small statistically significant differences in the morphology and repair between autograft and allograft (Neufeld et al 2002, Goldberg and Stevenson 1987). The use of cancellous bone allografts was found to have a particular advantage in the surgery of the foot and ankle. Since donor site morbidity is mitigated, the use of allografts made it possible to perform foot and ankle arthrodesis as an outpatient procedure. Thus, structural cancellous allografts offer an attractive alternative in reconstructive surgeries of foot and ankle, reduces hospital stay, potential donor morbidity, patient discomfort and give ultimately better overall outcomes. Many studies suggest similar or even better results when substituting cancellous allograft for autograft (McGarvey and Braly 1996, Neufeld et al 2002).

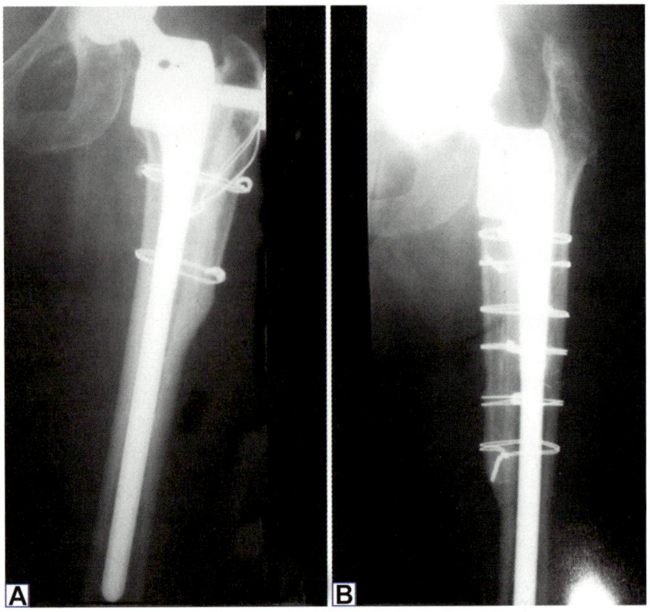

Figures 15A and B Radiographs of proximal femurs with superimposed cortical bone allografts. The edges of the grafts have rounded off, and the grafts have united with the underlying host bone. (A) Three years post-transplantation; (B) Five years post-transplantation

Cortical Bone Allografts

Cortical bone allografts provide immediate structural support after insertion. The belief that cortical bone grafts are not appreciably osteogenic and biologically weak is unfounded (Enneking et al 1975). In an animal model, healed bone plate allografts restored the mechanical strength of the entire underlying bone (Emerson et al 1990). Bone morphogenic proteins content measured in accordance with a technique described by Urist et al is high in these grafts (Urist et al 1970).

Freeze-drying and allograft processing methods may have been important in allowing for the preservation of BMP. These allografts can increase the thickness of the deficient wall by more than 1 cm. Therefore, satisfactory clinical results may be anticipated only with freeze-dried allografts prepared in an aseptic minimally manipulated fashion which preserves BMP activity (Head et al 1998). Other sterilization and preservation methods such as allograft sterilization by chemical disinfection, irradiation or ethylene oxide that are known to reduce BMP activity produce allografts which may not render comparable results. The events in cortical bone allograft healing became apparent when cortical bone plate allografts rigidly fixed to the host were examined sequentially (Malinin et al 1985).

Repair of cortical bone allografts is initiated along the periosteal and endosteal surfaces of the graft. There is mesenchymal cell recruitment and ingrowth into the Haversian canal system that is accompanied by osteoclastic resorption of the surrounding bone. This sequence of events is dictated by the bone architecture. Space must be made within the non-viable graft, before new bone formation can take place. During the early phases of repair, large channels are opened in the cortical graft, and there is direct vascular ingrowth into these channels from the adjacent host bone. Later, it is converted into a callus-like structure. At this juncture, cortical graft resembles cancellous bone. Highly vascularized mesenchymal tissue fills the space between the graft and the underlying recipient bone. Eventually, new bone formation exceeds resorption and with subsequent remodeling in accordance with Wolff's law, the bone mass is restored. This, of course, occurs only in allografts intimately attached to the recipient bone (**Figs 15A and B**).

As previously noted onlay cortical bone plate allografts in experimental animals unite with the host bone and undergo remodeling, provided there is rigid internal fixation and an adequate blood supply of the host bed. Biomechanical characteristics of the host bone-bone plate allograft construct approach those bones of contralateral control limbs weeks after transplantation (Malinin et al 1984).

In a dog, 2 weeks after transplantation, bone plate allografts are easy to separate from the underlying host bone and the fracture site and appear to be distinct from the underlying bone. By 4 weeks, callus is formed on the side of the fracture opposite to the plate. The plate could still be pried away from the underlying host bone. In this timeframe, when force was applied to the bone plate allograft constructs the radii refracture through the original fracture sites. In the 8-week specimens, intense osteogenic activity was noted at the edges of the bone plate. The edges of the graft were surrounded by dense mesenchymal tissue. Osteoclasts and Howship's lacunae were prominent. Mesenchymal tissue interposed between the bone plate and the host bone binds the two structures together. Ingrowth of mesenchymal tissue and neovascularization are present as are osteoblasts. Extramedullary callus was prominent and the fracture site healed. When refractured, the fractures now occurred through the screw holes rather than the original fracture sites. Twelve weeks after the fracture and bone plate transplantation, mesenchymal tissue between the plate and the host had been replaced with callus-like structures containing intense osteoblastic activity. The bone plate was now very firmly attached to the host, with the edges of the plate blending with the bone of the host. Grossly and microscopically, vascularization of the graft was marked as was new bone formation. Radiologically, the bone plate allograft was united with the underlying host bone.

In specimens examined 24 weeks after bone plate transplantation, the plate persisted and remained vascular. However, in longitudinal sections callus-like tissue was seen interposed between the host and the graft. Histologically, lamellar bridging was observed between the host bone and the allograft plate

As stated in Chapter 2: Form, Structure and Peculiarities of Bone, bone plate allografts, when superimposed on fractured bone produce a typical pattern of bone fracture healing. No delay in healing occurs. The bone plates themselves, participate in the repair process and incorporate into the host in the manner first described by Kreutz et al in 1951 (Kreutz et al 1951). The allograft bone plates are converted into callus-like structures, and undergo vascularization accompanied by intense osteoblastic activity. The vascularization and accelerated bone formation persist throughout the period of graft maturation. Intense biologic activity is manifested by canalization of the cortical bone in the host tissue directly opposed to the allograft. This can be explained by preservation of BMPs and possibly other growth factors within the graft.

Clinical Use of Freeze-Dried Bone Plate Allografts

The use of freeze-dried onlay cortical plate allografts, to compensate for bone loss, particularly in the proximal femur due to osteolysis, fracture, implant penetration and related causes has been an accepted procedure in revision joint replacements for decades (Chandler 1992, Emerson et al 1992, Head et al 1993). At times, these allografts have also been used in primary arthroplasties in patients with significantly compromised bone quality (Rock 1996).

Although, the use of onlay cortical plate allografts in hip revision arthroplasties has gained popularity, the cortical plate allografts themselves, were investigated as long as 6 decades ago by Kreutz and associates (Kreutz et al 1951) wherein they described the features of freeze-dried bone plate incorporation in an experimental setting. Subsequently, Gresham (1964) reported healing in 68 of 80 patients (85%) with non-unions of forearm fractures treated with onlay cortical plate allografts, and noted revascularization and replacement of these grafts with host bone. After many years, his clinical observations were confirmed (Carneiro et al 1989).

Much of the anticipated and subsequently realized success of cortical bone plates was based on the information gained from animal experiments (Malinin et al 1984, 1985). It became evident from these experiments, that within 8 weeks, cortical bone allografts united with the underlying host bone. By 12 weeks, the allografts had reconstituted the mechanical strength of the underlying radius. These studies, as well as the subsequent clinical observations, emphasize the need for close contact between the cortical allograft and underlying host bone, as well as preservation of the osteoinductive properties (BMP) in the allografts.

Since the introduction of bone plates or bone struts in revision surgery for failed hip arthroplasties, the use of these allografts has increased in popularity. Surgical techniques relating to graft application have been refined and reported in several papers (Head et al 1993, Emerson et al 1990, 1992, Chandler 1992, Robertson 1992). Observations in large numbers of patients with freeze-dried onlay cortical bone allografts emphasize several key technical points.

An important technical consideration, which results from acquired experience relates to the placement of grafts. Structurally, a compromised femur can indeed be salvaged by onlay bone plate allografts, but the graft cannot be used primarily for prosthetic support. It is intended only, for the augmentation of the deficient wall of the femur.

The rapid union and incorporation of the allografts into the wall of the host femur can be most likely explained by the osteogenic potential of the allograft itself and by the large area of contact between the graft and the host. Allografts used in our studies were excised and processed aseptically; they were not exposed to chemical agents or secondary sterilization. Applications of irradiated and ethylene oxide sterilized allografts in a small number of patients produced failures. Since presence of BMPs in minimally manipulated grafts has been confirmed by animal studies, it is reasonable to assume that the progressive changes observed along the host-graft junctions are modulated by BMPs. This is reinforced by findings suggesting that bone plate allografts do not undergo the usual pattern of maturation unless they are in direct contact and well fixed to the underlying host bone. Portions of bone plate allografts placed over bone cement or metal remain relatively unchanged for years. Radiologic studies on a large number of patients produced consistent clinical and radiographic outcomes (Malinin et al 1997, Head et al 1998). These observations led to the routine use of cortical bone plate allografts to augment and heal periprosthetic femoral fractures (Haddad et al 2002).

The radiologic appearance of bone plate allografts is consistent and predictable. Remodelling in the form of rounding off the ends of the allografts occurs usually by 7 months, but the length of the grafts remains virtually unchanged. Scalloping or radiologic evidence of the formation of Howships lacunae on the surface of the allograft plates becomes detectable about the same time. Time does not appreciably change the width of the graft, regardless of the period of observation.

Healing of the host-graft junctions, as judged by complete or partial bone bridging, occurs by 12 months. Radiographic union is considered complete when the junction lines become blurred by bridging bone trabeculae. This occurs in over 90% of the patients who undergo bone plate allograft repair. In our series (Head et al 1998) bridging was not apparent radiographically in two patients in whom the bone plates were placed anteriorly onto the femoral shaft, but it did take place in others. Measurements of the diameter of the femur shafts before and after application of plate allografts show a consistent increase of about 25% of the diameter of the femur (Emerson et al 1992).

Appearance of the trabecular patterns in the onlay allografts is present in bone plates for 2 years or longer. Radiographs of proximal femurs with bone plate allografts taken 3–5 years postoperatively are shown in **Figures 15A and B.**

Although subsidence is difficult to measure and quantitate we estimate that approximately 3% of the implants subsided following allograft plate augmentation of the proximal femur.

Studies of Retrieved Specimens

In all retrieved specimens, bone plate allografts, were firmly attached to the underlying host bone. The contouring and rounding off of the end of a bone plate allograft is demonstrated in a photograph of a specimen in **Figures 16A and B**.

The bone plate was excised 2 years post-transplantation. The surface of the graft shows a few shallow areas of erosion. The bone plate is firmly attached to the underlying femur. An increase in the diameter of the femur where the plates have been applied is also evident. A specimen sectioned longitudinally grossly showed ingrowth of the underlying host bone into the allograft plates and a firm union between the graft and the host bone. A histologic whole mount section of the plate shows canalization and remodeling by apparent enlargement of pre-existing Haversian channels and their transformation into confluent sinusoids (**Fig. 17**).

Cross section of distal femur with a bone plate allograft obtained 2 years post-transplantation shows doubling of the wall thickness at the site in which the allograft had been applied. The allograft is united with the underlying bone. In this specimen, the same process of cancellization involves both the allograft and the underlying host bone.

The earliest specimen available for examination was obtained 8 months post-transplantation. In this specimen, the allograft was already united to the host bone. The bone was revascularized through pre-existing channels. Viable osteocytes were present around the channels and early formation of new randomly scattered osteons was noted.

The gross and histological pattern of host-graft union, revascularization, cancellization of the allograft, as well as the formation of new osteons was uniform in all examined specimens.

Clinical observations in a large number of patients with freeze-dried tibial onlay allografts with a follow-up of over 15 years confirm our previously published results made on a smaller number of patients with short term follow-up (Malinin et al 1997).

The rapid union and incorporation of the allografts into the wall of the host femur is best explained by the osteogenic potential of the allograft itself and by the large area of contact between the graft and the host. As outlined in Chapter 13 Tissue Banking, we believe that satisfactory long-term results are dependent on the way in which the allografts were recov-

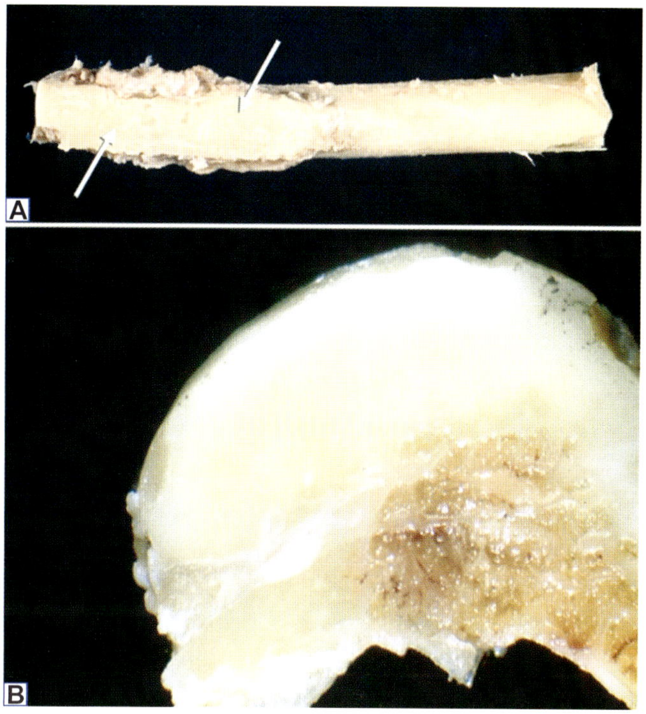

Figures 16A and B (A) Bone plate (arrows) attached to underlying femur; (B) Cross section of proximal femur with superimposed cortical plate allograft which has increased the diameter of deficient femur. Specimens obtained 2 years post-transplantation

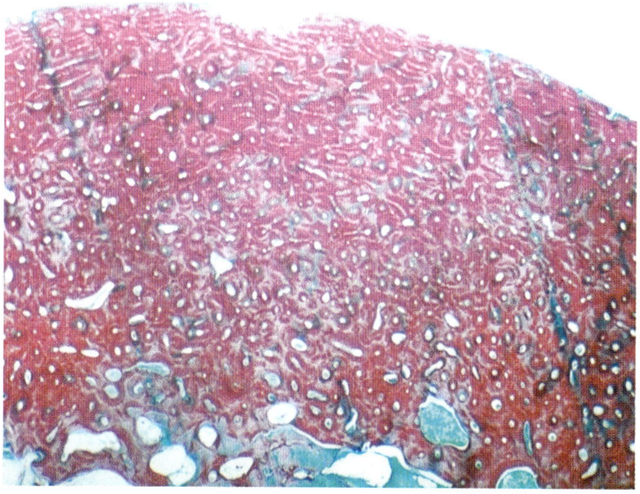

Figure 17 Bone plate allograft had undergone cancellization while increasing the thickness of the femur wall almost two-fold (Masson trichome stain × 35)

ered and processed. We likewise believe intimate contact with host bone is critical. Bone morphogenetic protein and related growth factors modulate the healing process via complex cell signaling. Finally, healing occurs predictably in a carefully choreographed and temporal sequence of biological events (Mohan and Baylink 1991).

Morphological observations on human biopsies and retrieved specimens demonstrate the same pattern of revascularization and bone replacement as that observed in experimental animals. However, in humans, as can be expected, the time sequence of bone remodeling is much longer than that in experimental animals. The histologic phases of bone incorporation are not clear cut or sharply delineated but blend one into another. The process appears to be remarkably uniform, and ongoing. Although, retrieved bone plates grossly appear as strong cortical structures firmly united to the underlying host femurs, histologially vascular canalization and new osteon formation are present in allografts as long as 6 years post-transplantation. Whether the speed of bone incorporation and remodeling is dependent on the recipient's age or other peculiar host factors are not known, but it certainly remains a possibility.

In humans, as in animals, the initial host response to the allograft is mobilization of mesenchymal cells that initiate osteogenesis. This is reflected in the X-ray findings of partial bridging. The radiological findings correlate well with morphologic data. The radiologic blurring of the cortex and the trabeculation of the allograft proceed concomitantly with active and ongoing repair of the host femoral cortex. The latter finding, although apparent in an animal model, was an unexpected finding in human subjects.

Whether or not the same pattern of host-graft interaction can be obtained with secondarily sterilized bone allografts is not known at this time. Therefore, in comparing results with allografting of the proximal femur the methods of allograft preparation must be taken into account.

The contact between the graft and the host is of paramount importance. Animal studies show that areas of allograft not in direct contact with underlying host bone will resorb, but in areas of direct contact, vascularized tissue will invade the graft and new bone formation will follow. This observation underscores the need for close approximation of the allograft to the underlying host femur. Most allografts are applied to the medial side of the femur and retained in place with secure cerclage wires, bands, cables or screws. Particulate autologous graft can also be placed between the plate and host bone to fill the areas where host-graft contact is deficient.

Clinical observations on patient with femoral revisions utilizing bone plate allografts have shown satisfactory end-results with an average Harris hip score of 80 and a 2.7% subsidence rate. However, it must be again noted that bone plate allografts by themselves do not provide enough stability to support an implant. Incorporation is necessary to achieve this goal. In the final analysis, implant stability can be achieved only by a combination of the prosthesis design, host response, graft quality and proper surgical technique. Since the onlay bone plate allograft by itself will not provide full support to the femoral component, these grafts will not be effective in the presence of complete loss of the proximal femur integrity such as occurs in cases of severe osteolysis and bone fragmentation. For patients with defects of this magnitude, replacement of the entire proximal femur is indicated.

Intercalary Allografts

Intercalary allografts show the same pattern of healing as bone plate allografts, except that healing takes place only from the distal and proximal ends of the grafts. Clinical experience and laboratory data gained over the last several decades indicate intercalary allografts secured in place by rigid internal fixation, show radiologic healing at host-graft junctions and replacement of the graft with host bone (**Fig. 18**). Without rigid internal fixation, intercalary allografts fail. This explains the dissolution of fibular allografts reported by Burchard et al in 1978 (Burchard et al 1978) where the grafts were not rigidly fixed to host bone. The interpretation of findings in this study created a great deal

Figure 18 Motion and the dissolution of freeze-dried intercalary allograft not subjected to rigid internal fixation. The allograft failed despite the limb being placed in a cast

of confusion among those who routinely use these grafts to reconstruct segmental defects. In comparing the healing pattern of freeze-dried autografts to that of fresh autografts, Burchard and colleagues noted little difference in healing between the two. However, when they compared freeze-dried allogeneic cortical grafts to fresh autografts, they stated the former repaired incompletely and failed to establish skeletal continuity.

In addition, these investigators failed to observe the decrease in antigenicity in freeze-dried allografts. These findings are at variance with the findings by other investigators and with clinical experience. Failure to reduce antigenicity can be attributed to inadequate freeze-drying (Malinin et al. 1983). A number of years later, investigators from the same institution showed healing and incorporation of frozen intercalary allografts, fixed with compression plates in a canine model **(Figs 19A and B)** (Beebe et al 2009).

In our experiments and clinical practice, freeze-dried intercalary allografts, rigidly fixed with compression plates, healed host-graft junctions and were gradually replaced with host bone (Malinin et al 1985). Non-stabilized grafts wedged into the defects were used as controls.

All animals in the non-stabilized group were found to have displaced grafts or non-unions. Non-unions developed at either one or both host-graft junctions. In animals transplanted with freeze-dried intercalary bone allografts, grafts were secured in place with stainless steel compression plates. Healing at the host-graft junctions in these animals was complete **(Figs 20A and B)**.

Healed allografts showed union of the host-graft junction with cancellous bone trabeculation in the end of the allografts. A uniformly clear area was present in the center of allografts during the first 5 months of observation **(Fig. 21)**. The clear space was obliterated in specimens examined 9 months post-transplantation.

The space (clear area) in the center of the medullary canal of the allografts was filled with gelatinous-like material that stained pink with phenol red added to the fixative. It was composed of a densely packed network of mesenchymal and connective tissue cells and was well vascularized. Islands of osteoid tissue were found throughout this gelatinous material.

Bone replacement, as evidenced by the uptake of tetracycline, was present in the cortical portion of the allograft and to a lesser extent in the gelatinous material in the center of the 4 and 5 month specimens. This was associated with vascular ingrowth into the cortical bone and osteoblastic activity.

The pattern of healing at the host-allograft junctions consisted of an external callus that enveloped the allograft and ingrowth of bone from the medullary canals of the host into the medullary canal of the graft. As the healing process progressed the callus remodeled. In human subjects similar patterns of healing was observed. An example of an intercalary allograft in a patient used to reconstruct a limb is given in **Figures 22A and B**.

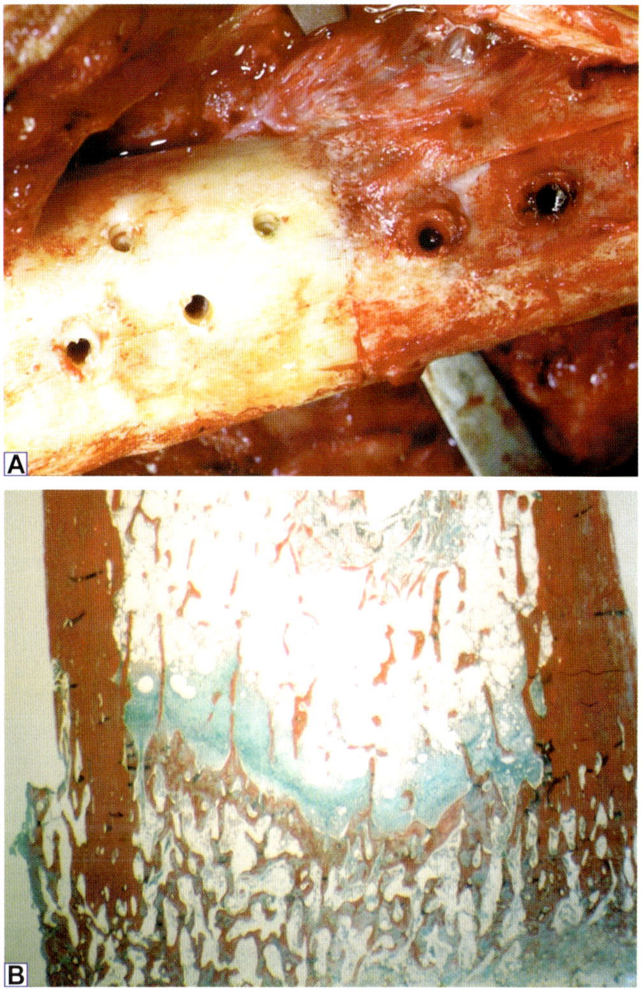

Figures 19A and B (A) Frozen intercalary allograft in a patient 3½ years post-transplantation. The host-graft junction has healed; (B) Histologic section of the host-graft junction showing bony union (Masson trichome stain × 10)

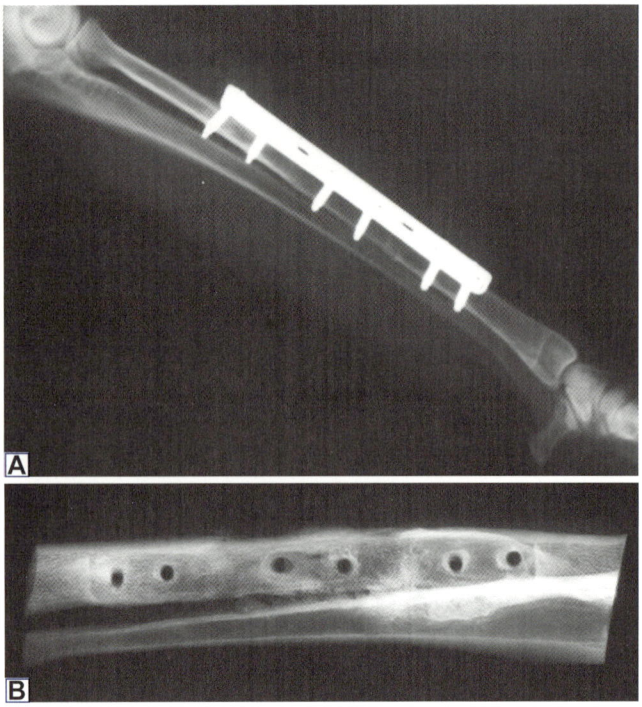

Figures 20A and B (A) Intercalary allograft secured in place with a dynamic compression plate 12 weeks post-transplantation; (B) Photograph of the same specimen with plate removed

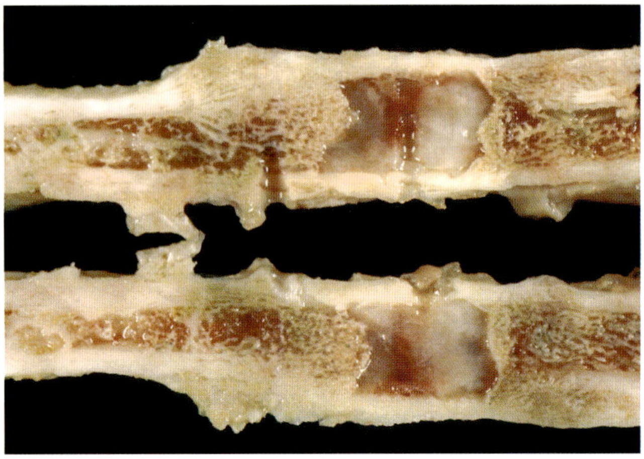

Figure 21 Gross specimen of a freeze-dried intercalary allograft 12 weeks post-transplantation showing the healing of the allograft host junctions and the ingrowth of the host bone into the allograft

Osteochondral Allografts

The long standing interest in the transplantation of osteochondral allografts and autografts (Buchmann 1908) has generated a large number of clinical and laboratory studies. All of these have several things in common. Clinically, osteochondral allografts represent satisfactory material for the

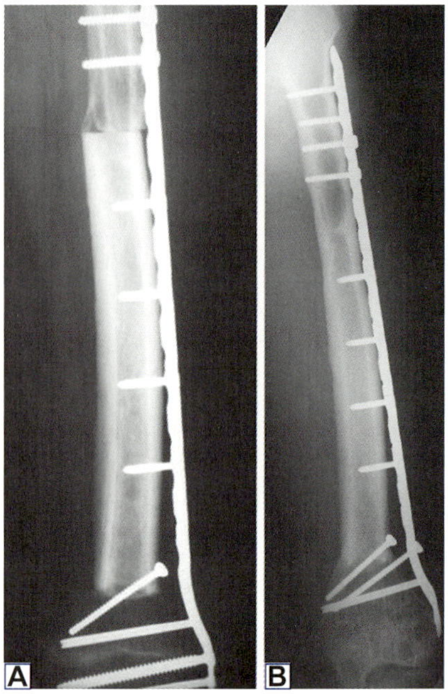

Figures 22A and B (A) Radiograph of a freeze-dried intercalary allograft in a 10-year old boy immediately after transplantation; (B) The same patient 6 years later, he now has practically normal function of that limb. At 8 months, when the distal allograft healed the screws across the epiphysis and those in the condyles were removed

reconstruction of skeletal defects resulting from the resection of tumors, trauma and degenerative disorders (Zatzepin 1984). Complication rates associated with osteochondral allograft transplantation are high (Mankin 1983, Mnaymneh and Malinin 1989, Tomford et al 1990). Despite the complexities of these procedures, use of osteoarticular allografts in the treatment of bone tumors has been rewarding (Maymneh et al 1994, Balsamo et al 2007). Immunologic responses to transplanted osteochondral allografts have been recorded, but their significance remains unknown (Friedlaender 1983).

Most of the laboratory studies on large animals were directed to immunology. The sequences of biologic events associated with incorporation of osteochondral allografts are similar to autologous grafts. On the other hand, quantitative and temporal aspects of repair vary depending on the preservation techniques employed and genetic disparity between the donor and the recipient. The original studies that were performed in order to determine the response of the host to osteochondral allografts still hold true (Herndon and Chase 1952, Curtis 1956, Heiple et al 1963). According to these studies, fresh allogeneic bone transferred across a strong histocompatibility barrier induced a vigorous inflammatory tissue reaction as well as a specific immunologic response.

Bone closely matched in terms of histocompatibility characteristic of tissues that had been subjected to deep-freezing usually exhibited the same pattern of graft incorporation observed with autografts except that it was delayed.

These studies also demonstrated that fresh osteochondral allografts were subjected to an unpredictable and unsatisfactory incorporation process (Herndon 1960). The results of these laboratory studies are not universally accepted in terms of applicability to human transplantation, since satisfactory clinical outcomes with fresh osteochondral allografts were recorded (Lexer 1925, Gross et al 1983, Rodrigo et al 1976). Some light on the subject is shed by the study of osteochondral allograft transplantation in non-human primates, baboons (Malinin et al 1985). The knee structure in baboons *(Papio hamadryas)* is anatomically closer to humans than those of other experimental animals. In addition, these animals make good subjects for long-term observations since, after healing of the allografts they can be returned to their original animal groups without restraining activity. The natural lifespan of baboons is around 20–30 years, so in human terms each year of observation of a baboon is equivalent to 3–5 human years. Distal femoral allografts transplanted into baboons were either fresh or control-velocity frozen after exposure to glycerol, a cryoprotective agent.

Distal femurs were excised under aseptic conditions. Those to be transplanted fresh were placed into tissue culture medium CMRL-1415, kept in a refrigerator at 4°C and transplanted within 72 hours of excision. Those to be frozen, were also placed into the same tissue culture medium and refrigerated. The allografts were then transferred to a laboratory, where at room temperature under strict aseptic precautions, they were cleaned with removal of bone marrow. The periosteum and soft tissue attachments other than the capsule, the cruciate ligaments and collateral ligaments were removed. The allografts were then placed into a solution of 15% glycerol in CMRL-1415. They remained in the solution for 30 minutes, and were then placed into double plastic bags that were heat sealed and wrapped in towels. The allografts were frozen at a freezing rate of 2°C per minute until a temperature of -60°C was reached. They were then placed into liquid nitrogen storage containers where they remained at -150°C until transplantation. On the day the allografts were to be transplanted, they were removed from liquid nitrogen (vapor phase), placed on dry-ice and transported to the operating room. The same procedure is used with human osteoarticular allograft transplantation (Buck and Malinin 1994).

The experimental animals have received either fresh or cryopreserved allografts. Seven cryopreserved allografts were rigidly fixed with stainless steel plates and screws. Four fresh allografts were also transplanted. The ligaments were reattached with non-absorbable sutures. The posterior capsule and the fascia lata were tightly closed. The skin was closed with subcuticular absorbable sutures. The posterior capsule and the fascia lata were tightly closed. On completion of the operation, the leg was wrapped with cast padding and placed in a fiberglass cast in moderate flexion to prevent weight bearing. Postoperatively, the animals were given antibiotics. The animals were placed in a restraining chair for 6 weeks. This prevented the animals from destroying the casts and injuring the operated legs, as was the case with the dogs. At the end of 6 weeks, the casts were removed, and the animals were returned to their original colonies. The animals received food or water at will. They also moved and jumped without restriction. Every 4–6 months, the animals were removed from the enclosures and examined. Roentgenograms of the operated limbs were taken. If limb impairment was noted (the animals were routinely observed daily), the animal was removed immediately and an X-ray of the limb was obtained.

There were no wound complications. During the second year post-transplantation, one animal dislocated the knee. The knee was immobilized with a pin, but after several weeks the pin broke. The animal was sacrificed 2 ½ years post-transplantation. Severe degenerative changes were present in the knee joint, both in the transplanted distal femur and in the tibia. In the remaining animals with cryopreserved transplants, radiologic evidence of healing of the host-graft junctions was evident 6 weeks post-transplantation. Rapid healing of host-graft junction did not occur in animals with fresh distal femur transplants. In the latter, host-graft junction healing was not apparent radiologically until 6 months after transplantation.

During the second year of observation, one animal fractured the allograft distal to the plate. The plate was changed and the fracture autografted. The fracture healed, but the joint fused. In another animal, a fracture occurred at the same site 3 years post-transplantation. In the second animal the joint was also fused and degenerated. The proximal ends of the allografts were solidly fused to the host bone in both animals. In this animal, the medullary canal of the allograft was filled

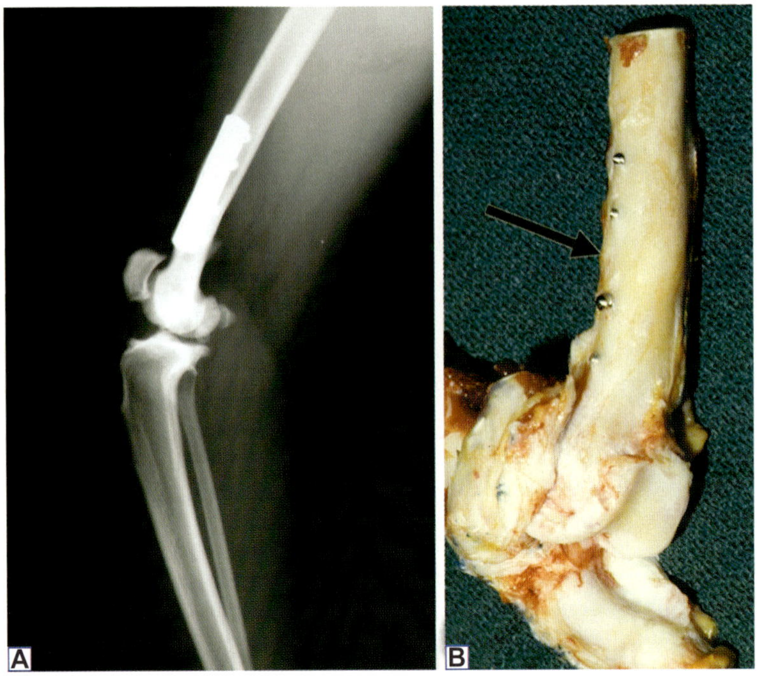

Figures 23A and B (A) Radiograph of a healed osteochondral graft in an experimental animal (baboon) 5 years post-transplantation; (B) Gross specimen from the same animal. Host-graft junction is indicated by an arrow. The results of the experiments with non-human primates parallel clinical experience. The host-graft junctions usually unite, but the cartilage in cryopreserved allografts undergoes gradual degeneration

with cancellous bone and bone marrow. Cancellous bone was present at the fracture sites. Histologically, formation of new lamellar bone was present in the transplanted distal femurs.

In the remaining animals (N-6), the knee function was normal 3 years post-transplantation. On roentgenograms, knee joint space was maintained and only minimal degenerative changes could be detected.

At 4 years, a third fracture occurred in an animal with a cryopreserved allograft. The animal was sacrificed. There was a solid union between the host and the graft, and the bone of the allograft was replaced. At the end of 5 years, complete union between the host and the graft was present in all animals. The host-graft junctions were no longer apparent (**Figs 23A and B**).

The medullary cavities were replaced with cancellous bone or fatty bone marrow to the subchondral bone. Vascularization and new bone formation was present in the cortical portions of the allografts. All animals had degenerative joint changes (**Fig. 24**).

However, all animals had a good range of motion and functional limbs. The cartilage on non-weight bearing surfaces was yellowish, but smooth. Microscopically, transition into fibrocartilage and the loss of chondrocytes and matrix were evident. On weight bearing surfaces, the cartilage was either eroded or pitted and irregular.

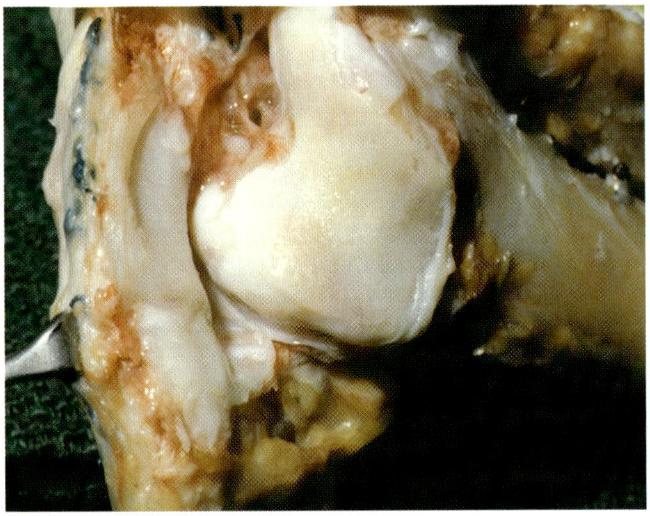

Figure 24 A close view of the articular surface of the osteochondral allograft. The articular surface is yellowish, but smooth

In animals with a fresh distal femur allograft, there was a replacement of trabecular bone with fibrous tissue in the metaphyseal portion of the graft. Sequestered bone and accumulation of round cells were also found. The junction was healed, but replacement of the bone marrow cavity was only partial. However, articular surfaces were smooth and glistening.

The reason for describing experiments with non-human primates in some detail is that the results parallel experience with human osteochondral allograft transplantation. In this connection, we must focus on the biologic behavior of osteochondral allografts in humans. To this end, Enneking and Mindell (1991) provided observations on retrieved osteochondral allografts. They documented a wide spectrum of changes in individual allografts, but noted unions at host graft junctions by the formation of an external callus derived from the host. Incomplete repair and avascularity were also present. Peripheral repair penetrated only a few millimeters. Our observations (unpublished) on over 30 retrieved allografts revealed extensive areas of vascularization and new bone formation. Fractures occurred through these sites.

Clinical Applications

Use of allografts in orthopedic oncology is described in a separate chapter, as is arthrodesis of the spine.

Revision Arthroplasties

Cavitary defects in the acetabulum are currently the most common clinical indication for bone allografting (Canale 2002). With the rapidly increasing number of hip and knee revision arthroplasties, more and more patients present with significant bone loss. In the hip, bone loss occurs both in the proximal femur and the acetabulum. Therefore replacement of the entire hip joint with an allograft is tempting. Unfortunately attempts to reconstruct failed hip arthroplasty with a joint transplant were not successful (Borja and Mnaymneh 1985).

A satisfactory result was experienced in only one out of five patients, who received whole hip joint allografts. The other four allografts failed within 2 years either because of articular cartilage degeneration and femoral head collapse, or because of non-union at the femoral host-graft junction.

Bone allografts commonly used in the reconstruction of the acetabulum are particulate, structural or combined. Particulate grafts can be crushed cancellous or cortical bone (bone chips), ground bone, or morselized bone and microparticulate bone. The latter types of bone grafts have been used for filling cavitary acetabular defects and femoral condyles. The structural grafts for acetabular reconstruction are heads of femurs, femoral condyles, and the acetabula with adjacent structures. Proximal femurs with femur heads and proximal tibias have also been used for acetabular reconstruction by a few surgeons. These offer little advantage over the previously listed grafts.

When major acetabular defects exist, decisions must be made as to whether the hip axis can be corrected without bone grafting or whether bone grafting is necessary. In recent years the availability of a wide variety of acetabular components allows for correction of many acetabular defects without bone grafts. However, with large defects, grafting may be necessary if the principles of acetabular reconstruction are to be met (Paprosky and Magnus 1994). When an allograft is used, rigid fixation and extensive graft-host bone contact is mandatory.

Acetabular bone grafting can restore the bone stock, thus recreating the normal hip center that is important to establish equal limb lengths and normal abductor function. It also allows for easier re-revision. The disadvantages of acetabular bone grafting are added operative time, and graft failure. The potential for disease transmission is negligible if the allografts are prepared from carefully selected and screened donors and processed maintaining strict aseptic technique and microbial surveillance. In general, acetabular allografting is indicated when implant stability cannot be achieved and/or sustained by the host bone. This occurs when there is a loss of the weight bearing portion of the acetabulum, and when there are major alterations in the hip mechanics (Chandler and Penenberg 1989).

Particulate bone allografts have been used to fill cavitary defects of the acetabulum for several decades with good outcomes. More recently microparticulate bone has been shown to provide excellent defect fill and incorporate quicker and more completely than larger sized particles. Particulate grafts have been used in combination with cemented and cementless acetabular components. Firm impaction of bone particles into acetabular defect with subsequent reaming on the construct leads to initial stability, providing the defect is contained. If the defect is not contained, a structural graft can be added (**Figs 25A to D**). Sloof was first to report over 10 years experience with these grafts (Sloof et al 1996). In his series, the most frequently encountered defects were medial cavitary (30%), superior medial cavitary (26%), and medial segmental cavitary (18%). Radiographic assessment of the

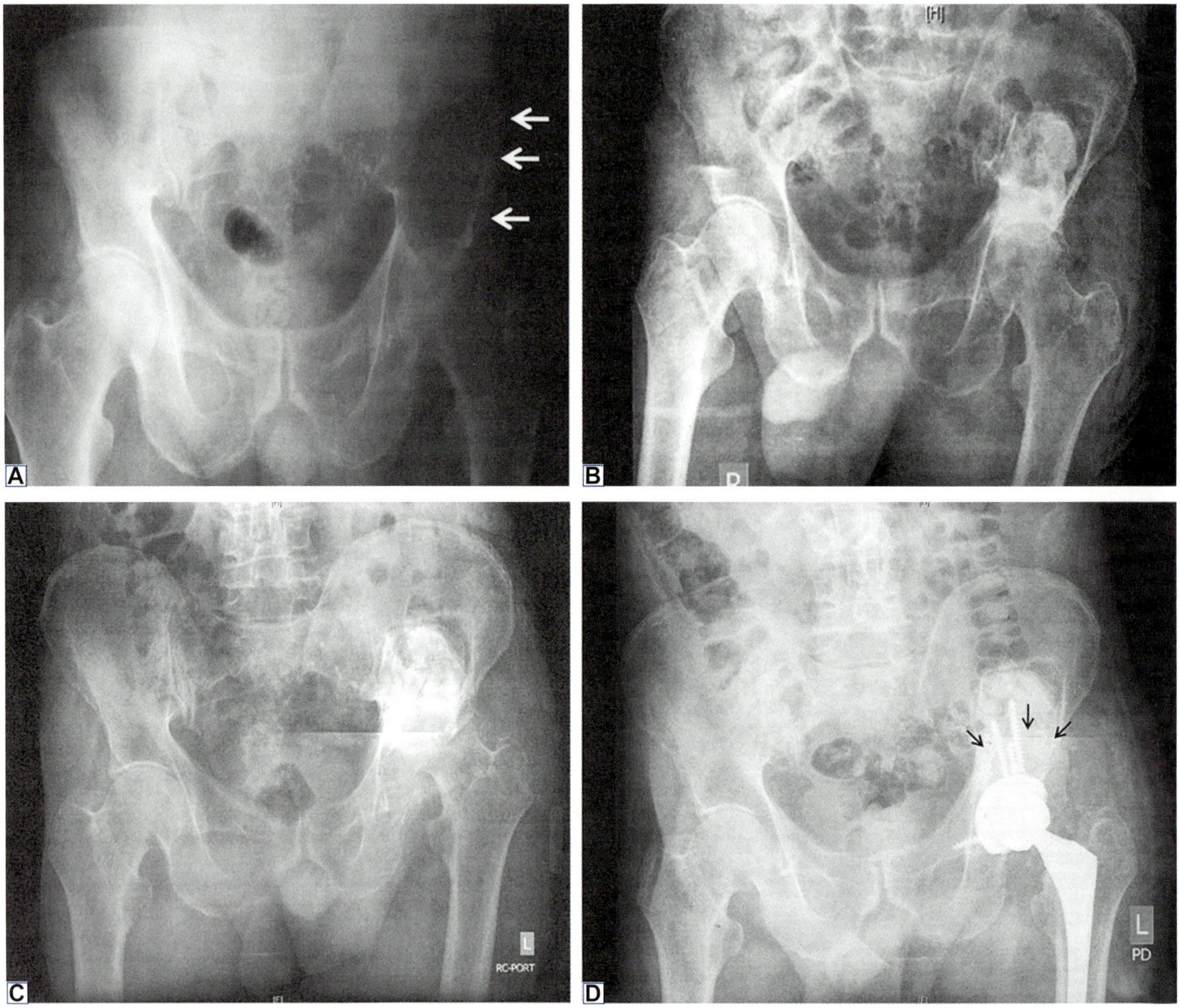

Figures 25A to D (A) 75-year-old male with plasmacytoma in periacetabulum (arrows); (B) AP radiograph of pelvis following curetting and packing acetabular defect with particulate bone; (C) Protrusio defect following fracture through subchondral bone; (D) Reconstruction of the acetabulum with freeze-dried femoral heads (arrows delineate graft). The hip is restored

grafts showed incorporation in over 90% of the patients. A Harris hip score of more than 80 was achieved in 81% of the patients. Two patients developed infections, and 8% showed radiographic loosening. These results were comparable to those obtained with autografts. Similar results were obtained by other authors.

Sculptured grafts to fit three-dimensional osseous defects can be fashioned from the heads of the femur or femur condyles to satisfy the defect. If the defect is large, the entire distal femur may be required for the obliteration of the defect. Stabilization of the graft within the host is essential to the success of this procedure. In extremely large defects with pelvic discontinuity, the entire acetabulum with adjacent bone may be transplanted. The technique of acetabular transplantation is similar to that used in oncologic surgery.

Proximal Femur Allograft

For most patients, including the elderly, allograft-prosthesis composites offer several advantages over custom made or totally metal devices. The allografts are immediately available in many sizes and shapes. Their fit can be further enhanced at the operating table. The mechanical properties of frozen or freeze-dried proximal femur allografts are comparable to those of normal bone. There are soft tissue attachments on the allograft to re-establish the abductor attachments to the femur, thus improving abductor function. Intramedullary support with long stem prosthesis diminishes future fractures.

A preferred technique for whole proximal femoral allografts includes cementing the graft and the implant. This assures stability and eliminates the possibility of fracturing the allograft in the process of press fitting the non-cemented prosthesis. The graft-cement-implant composite is then press fitted into the host femur. A unicortical plate across the junction will prevent rotation and nonunion.

A technical issue that occurs frequently is the discrepancy between the diameters of the medullary canals of the allograft and the recipient. Allografts are usually obtained from younger and healthier individuals. Consequently, the cortex of the allograft is thicker than that of the recipient and the canal narrower. Rather than reaming the allograft to accommodate a wider stem, thus weakening it, it may be advisable to use a narrower prosthetic stem, stabilizing the stem in the medullary canal of the recipient with bone plates and particulate bone.

Although satisfactory results have been obtained with proximal femur allografts, the operations are associated with high complication rates. In a series of patients originally described by Head et al half were younger than 50 years of age at the time of the revision. For these patients, alternatives such as a custom prosthesis or resection arthroplasty would be either functionally unsatisfactory or of limited durability (Head et al 1998). We have observed resorption of the greater trochanter and proximal lateral cortex of the graft. We presume that this may be due to rapid revascularization of the allograft through the soft tissue attachments, in this case, the gluteal tendon.

Rigid fixation of the femoral component deserves special mention as does the preservation of a portion of vascularized cortex of the host. Rigid fixation and osteosynthesis without a cement interphase will ensure host-graft healing and loading of the bone rather than the stem of the prosthesis, as is the case in primary arthroplasties. The type of fixation depends on the level of resection and the quality of the host bone. Supra and infra trochanteric levels of resection are the most difficult. The short allograft segment will not allow three point fixation of the prosthesis, and will often require cementing both distally and proximally. In summary, whole proximal femur allografts frozen or freeze-dried combined with femoral components, offer a satisfactory means of restoring hip function when a total hip replacement is associated with severe bone loss in the proximal femur. With lesser bone loss, bone plate allografts and lesser surgical procedures are clearly preferable.

Knee Joint

In failed total knee arthroplasties, as in hip arthroplasties, loss of bone stock is often encountered. Tibial condyles are frequently the site of massive bone loss, more so than femoral condyles. Traditionally, polymethacrylate has been used to replace bone deficits around the knee. However, cement is not the best choice due to loosening and further osteolysis over time. Custom metal wedges have also been used for smaller, more focal defects, in one or more areas of the proximal tibia. Conversely, bone grafts are optimal choices for filling large cavitary defects in bone. By and large, if the tibial deficit occupies more than 50% of the surface of the tibial condyle, or requires a cement column greater than 5 mm, a bone graft is recommended (Dorr and Ranawat 1984). Block grafts that provide structural support to the overlying prosthesis are preferred to morselized bone or cancellous chips. In primary arthroplasties, autogenous local bone, usually a resected distal or posterior femoral condyle can also be used as graft material. Unfortunately, in revision arthroplasties, bone in the femoral condyles is usually deficient as well and cannot provide an adequately sized graft. Similarly, in an occasional primary arthroplasty with deficient bone, more graft is needed than the femoral condyles could provide. In such cases allografts are used in conjunction with prostheses as graft-prosthesis composites.

In the tibia, usually only one condyle is deficient (medial tibial condyle in varus knees and lateral condyle in valgus knees). Iliac crest allograft to augment a deficient condyle had been used by many surgeons. In revision cases, all the cement and fibrous debris are removed. The underlying sclerotic bone is burred out or penetrated to expose bleeding bone.

In central deficiency, an attempt is made to preserve the peripheral cortical rim and sculpture the allograft to be press fitted in the defect. In peripheral deficiency, the recipient bone bed is squared off and then the proximal tibial horizontal cut is carried out in the usual manner using the tibial cutting jig, making sure that the cut is perpendicular to the long axis of the tibia. The residual condylar defect is then accurately measured and the iliac crest allograft is cut and sculptured to optimally match the defect. The cancellous surface of the iliac crest allograft is placed against the floor and sides of the osseous bed to achieve maximum bone contact. The graft is screwed to the host bone with vertical or horizontal screws. Particulate bone is used to fill in any residual crevices and prevent bone cement from seeping between the graft and the host bone. Sometimes, a thin layer of cement is used to seal off the surface of the residual crevices. The tibial component, usually with a proximal metal boss and long stem is used. A few small and shallow drill holes are made in the upper surface of the graft for cement fixation of the tibial component to the graft. Regretfully experience with total knee transplantation has not been a satisfactory one (Borja et al 1985, Deutschländer 1914).

Vascularized Bone Allografts

Although requirements for vascularized bone autografting have been reduced, by recent advances in surgical technique, vascularized autografts still remain a good source of bone replacement (Gabl et al 1999, Wilden et al 2006). Not so with vascularized bone allografts which still remain in the realm of scientific discovery and experimental design. Immunologic response to fresh vascularized allografts is severe, (Innis et al 1991) and requires host immunosuppression. In a cryopreserved allograft, the vasculature is compromised. Therefore it cannot be reanastamosed to the vasculature of the host. A compromise might be reached by transferring a vascular bundle of the host into a medullary cavity of the allograft. If revascularization of an allograft could be achieved, it would provide numerous advantages in bone transplantation. Experiments performed on dogs involved placement of a vascularized muscle flap into the medullary canals of freeze-dried allografts. This resulted in proliferation of tissue rich in mesenchymal cells that differentiate into osteoprogenitor cells with subsequent new bone formation (Carneiro and Malinin 1991).

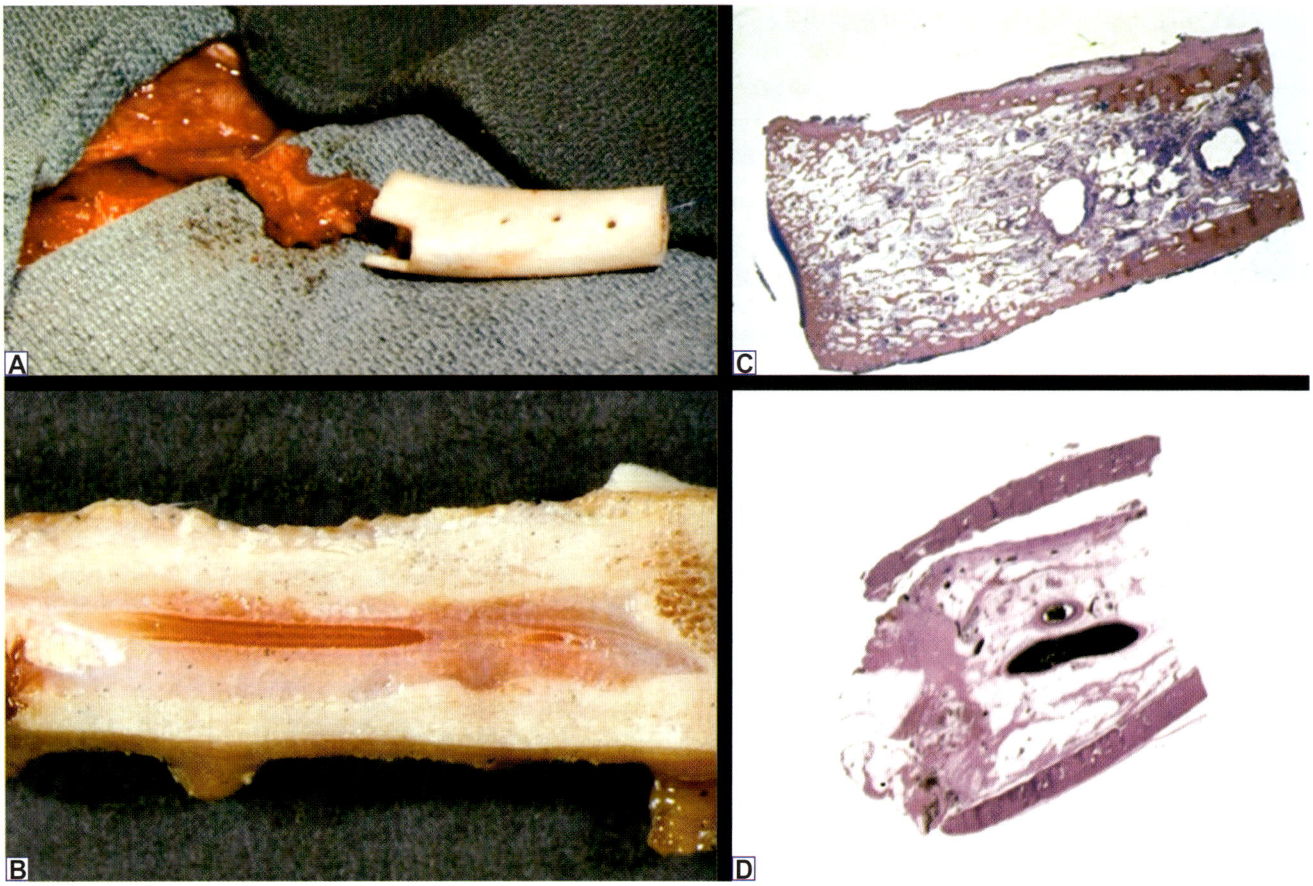

Figures 26A to D (A) Freeze-dried intercalary allograft with a vascularized bundle inserted into the medullary canal; (B) Gross section of a healed allograft 1 year post-transplantation; (C) Histological section viewing the filling of the medullary canal with newly formed cancellous bone; (D) Patent vascular channel injected with India ink

New bone formation was present both in the medullary canal and the cortical portion of allografts (**Figs 26A to D**).

If this procedure was used clinically the allograft could be implanted in an area where vascular pedicle could be obtained, providing vascular supply to a transplanted allograft. The allograft could then be transplanted directly into an area of bone loss or transferred directly into another location by means of microvascular anastomosis and transferred later. Positive effects of vascular bundle transplantation with allografts have been noted by other investigators (Ozyurek et al. 2008)

SUMMARY

Bone allografts are now used in many surgical procedures from grafting of minor defects to replacement of large bone segments. Various types of bone grafts are used and are available. Surgeons performing bone allograft transplantation will serve their patients well by knowing the capabilities and limitations of the large variety of grafts available. Unlike industrial products, bone allografts are not standardized; nor do they all have the same mechanical and biologic properties. To serve the patients' needs bone allografts must successfully interact with the recipient, and be selected according to its intended use.

BIBLIOGRAPHY

1. Adamkiewicz A. Über Knochentransplantation. Wiener Med Blätter, 1889;12 (No.23):355-6.
2. Albee FH. Evolution of bone graft surgery. American Journal of Surgery, 1944;63:421-36.
3. Albee FH. Fundamentals in bone transplantation: Experience with three thousand bone graft operations. Journal of the American Medical Association, 1923;81:1429-32.
4. Albrektsson T, Linder L. Intravital, long-term follow-up of autologous experimental bone grafts. Archives of Orthopaedic and Trauma Surgery, 1981;98(3):189-93.
5. Alldredge RH. Discussion of paper by MO Henry. Homgraft in Orthopaedic Surgery; Jounral of Bone and Joint Surgery, 1948;30:75.
6. Anderson KJ, LeCocq JF, AkesonWH, et al. Endopoint results of processing heterogeneous, autogenous and homogenous bone transplants in the human: a histologic study. Clinical Orthopaedics and Related Research, 1964;33:220-36.
7. Annersten S: Experimentelle Untersuchungen uber die Osteogenese und die Biochemie des Fracturcallas. Acta Chir Scand. 1940; 84. Supp. 160.
8. Arrington ED, Smith WJ, Chambers EG, et al. Complications of iliac crest bone graft harvesting. Clinical Orthopaedics and Related Research, 1996;329:300-9.
9. Axhausen G. Die histologischen und klinischen gesetze der freien osteoplastik auf Grund von Lierversuchen. Archiv für Klinische Chirurgie, 1908;88:23-145.
10. Axhausen G. Arbeiten aus dem Gebiet der Knochenpathologie und Knochenchirurgie. I. Kristische Bemerkugen un neue Beiträge zur freien Knochentransplantation. Archiv für Klinische Chirurgie, 1911;94:241-81.
11. Axhausen G. Die pathologisch-anatomishcen Grundlagen der Lehre von der freien Knochentransplantation beim Menschen and beim Tier. Med Klin, 1908;2:23-58.
12. Axhausen G. Histologische Untersuchungen über Knochentransplantation am Menschen Deutsch. Zeitschrift für Klinische Chirurgie 1907;91:338-428.
13. Axhausen G. Über den Vorgang partieller sequestrirung transplantierten Knochengewebes, nebst neuen histologischen Untersuchungen über Knochentransplantation am Menschen. Archiv für Klinische Chirurgie, 1909;89:281-302.
14. Balsamo LH, Malinin TI, Temple HT. Distal tibia osteoarticular allografts. Clinical Orthopaedics and Related Research, 2007;459:92-5.
15. Banwart JC, Asher MA, Hassamein RS. Iliac bone graft harvest donor site morbidity. Spine, 1995;20:1055-60.
16. Barth A. Die Enstehungund und das Wachsthum der freien Gelenkkorper. Eine histologischeklinische Studie. Archiv für Klinische Chirurgie, 1898;56:507.
17. Barth A. Histologische Untersuchungen über Knochenimplantationen. Beiträge zur Pathologische Anatomie und zur Allgemeinen Pathologie, 1895;17:65-78.
18. Barth A. Über histologische Befunde nach Knochenimplantationen. Archiv für Klinische Chirurgie, 1893;46:409-17.
19. Baschkirzew NJ, Petrow NN. Beitrage zum Knochenuberplflanzung. Deutsche Zeitschrift für Chirurgie, 1912;113:490-531.
20. Beebe KS, Benevenia JB, Tuy BE, et al. Effects of a new allograft processing procedure on graft healing in a canine model. Clinical Orthopaedics and Related Research, 2009;467:273-80.
21. Behrend, Moses. Transplantation of the head and shaft of the fibula to the humerus. Surgery, Gynaecology and Obstetrics, 1930;51:717.
22. Blocker TG, Weiss LR. Use of cancellous bone in the repair of defects about the jaw. Annals of Surgery, 1946;123:622.
23. Borja FJ, Mnaymneh W. Bone allografts in salvage of difficult hip arthroplasties. Clinical Orthopaedics and Related Research, 1985;197:123-30.
24. Bos GD, Goldberg VM, Powell A, et al. The immunogenicity of frozen bone grafts in inbred rats. Transactions of the Annual Meeting Orthopaedic Research Society, 1981;6:97-8.
25. Bos GD, Goldberg VM, Powell PE, et al. The effect of histocompatibility matching of canine frozen bone allografts. Journal of Bone and Joint Surgery (Am), 1983;65:89-96.
26. Bonfiglio M, Jeter WS, Smith CL. The immune concepts: its relation to bone transplantation. Annals of the New York Academy of Sciences, 1955;59:417-33.

27. Boyd HB. Pathology and natural history of congenital pseudarthrosis of the tibia. Clinical Orthopaedics and Related Research, 1982;166:5-13.
28. Brooks OB, Heiple KG, Herndon CH, et al. Immunological factors in homogenous bone transplantation. IV. The effects of various methods of preparation and irradiation on antigenicity. Journal of Bone and Joint Surgery (Am), 1963;45:1617-26.
29. Buchmann P. Behandlung knocherner Ellbogengelenksankylosen mittels Uberpflanzung von ganzen Gelenken. Zentralblatt für Chirurgie, 1908;35:582-6.
30. Buck BE, Malinin TI. Human bone and tissue allografts. Clinical Orthopaedics and Related Research, 1994;303:8-17.
31. Büdinger K. Über den Verschluss von Defecten am Schadel durch Knochenheteroplastik. Wiener Klinische Wochenschrift, 1900;13:1067.
32. Bull CR. Experimentelle Studien uber Knochentransplantation und Knochenregeneration. Norske Videnskaps-Akademie; Oslo; 1928;9:105.
33. Burchard H. The biology of bone graft repair. Clinical Orthopaedics and Related Research, 1983;174:28-42.
34. Burchard H, Jones H, Glowczewskie F, et al. Freeze-dried allogeneic segmental cortical grafts in dogs. Journal of Bone and Joint Surgery (Am), 1978;60:1082-90.
35. Burwell RG. Studies in the Transplantation of Bone. V. The Capacity of Fresh and Treated homografts of bone to evoke transplantation immunity. Journal of Bone and Joint Surgery (Br), 1963;45:386-401.
36. Burwell RG. Marrow in incorporation of bone graft. Clinical Orthopaedics and Related Research, 1985;200:125-41.
37. Burwell RG, Gowland G, Dexter F. Studies in the transplantation of bone. VI. Further observations concerning the antigencity of homologous cortical and cancellous bone. Journal of Bone and Joint Surgery (Br), 1963;45:597-608.
38. Buxton St JD. Bone grafting. Medical Illustration, 1951; 6:541-6.
39. Cammack GV II, Nevins M, Clem DS III, et al. Histologic evaluation of mineralized and demineralized freeze-dried bone allograft for ridge and sinus augmentations. International Journal of Periodontics and Restorative Dentistry, 2005;25:231-7.
40. Canale ST. Campbell's Operative Orthopaedics, 10th edition, St. Louis, Mosby; 2002.
41. Carnevale V. A proposito de injecto oseo conservado en parofenol (Communication previa). Sociedad Argentina De Orthopedia Traumatologia, 1951;16:121.
42. Carneiro R, Malinin TI. Vacularized bone allografts: an experimental study in dogs. Journal of Reconstructive Microsurgery, 1991;7(2):101-3.
43. Carneiro RB, Burkhalter W, Malinin TI, et al. Treatment of foreman fractures with freeze-dried cortical bone plates. American Society of the Surgery of the Hand, 1989;13-6.
44. Carpenter ET, Gendler E, Malinin TI, et al. Effect of hydrogen peroxide on osteoinduction by demineralized bone. American Journal of Orthopaedics, 2006;35(12):562-7.
45. Carrel A. The preservation of tissues and its application to surgery. Journal of American Medical Association, 1912;59:523-7
46. Carter WW. The ultimate fate of bone when transplanted into the nose. Archives of Otolaryngology, 1932;15:563.
47. Chalmers J. Transplantation Immunity in Bone Homografting. Journal of Bone and Joint Surgery, 1959;41:160-79
48. Chalmers J. Transplantation immunity of bone homografting. Journal of Bone and Joint Surgery, 1975;(32)57:36.
49. Chandler HP, Penenberg BL (Eds). Bone deficiency in total hip replacement. NJ, Slack: Thorofare; 1989.
50. Chandler HP. The use of strut allograft in the reconstruction of failed total hip replacements. Orthopaedics, 1992;15:1027.
51. Chase SW, Herndon CH. The fate of autologous and homologous bone grafts. Journal of Bone and Joint Surgery (Am), 1955;37:809-41.
52. Cheek RJ, Harmon JV, Stowell CP. Red cell alloimmunization after a bone allograft. Transfusion, 1995;35:507-9.
53. Christophe L. Recherses sur les greffes d'os fixe a l'alcohol et sur le mechanism de l'osteogenese. Archive Francois-Belgian Surgery, 1926;26:13-56.
54. Coventry MB, Tapper EM. Pelvic instability: a consequence of removing iliac bone for grafting. Journal of Bone and Joint Surgery (Am), 1972;54:83-101.
55. Cox E, DeCock H, Godderies BM, et al. Antigenicity of cortical bone allografts in dogs and effect of ethylene oxide sterilization. Veterinary Immunoly and Immunopathology, 1999;69:47-59.
56. Curtis PH, Herndon CH. Immunologic factors in homogenous-bone transplantation. Serologic studies. Journal of Bone and Joint Surgery, 1956;38:103-10.
57. Curtis BF. Case of bone implantation and transplantation for cysts of tibia, osteomyelitic cavities and ununited fracture. American Journal of Medical Science, 1892;106:30.
58. Curtis PH, Chase SW, Herndon CH. Immunologic factors in homogenous bone transplantation. II. Histological studies. Journal of Bone and Joint Surgery (Am), 1956;38:324-76.
59. Czitrom AA, Axelrod, Fernandes B. Antigen presenting cells in allo-transplantation. Clinical Orthopaedics and Related Research, 1985;197:27-31.
60. Czitrom AA, Langer F, McKee N, et al. Bone and cartilage allotransplantation. Clinical Orthopaedics and Related Research, 1986;208, 141-5.
61. Dakes DA, Cabanella ME. Impaction bone grafting for revision hip arthroplasty: Biology and clinical applications. Journal of American Academy of Orthopaedic Surgery, 2006;14:620-8.
62. Darkschewitzch LO, Widenhammer WW. Über den Ersatz von Trepanationstucke des Schadels. Zentralblatt für Chirurgie, 1892;19:835.
63. David M. Über die histologischen Befunde nach Replantation trepantirter Knochenstucke des Schadels. Archiv für Klinische Chirurgie, 1896;53:740.
64. David M. Zur Frage des Verhaltens austerepanirter und wieder eingeheilter Schadelstucke. Deutsche Medizinische Wochenschrift, 1897;23:751.

65. Deijkers RL, Bouma GJ, Van der Meer-Prins EM, et al. Human bone allografts can induce T cells with high affinity for donor antigens. Journal of Bone and Joint Surgery (Br), 1999;81:538-44.
66. Delangeniere H, Lewin P. A general method of repairing loss of bony substance and of reconstructing bones by osteoperiosteal grafts taken from the tibia. Surgery, Gynaecology and Obstetrics, 1920;30:441.
67. Dell PC, Burchard H, Glowszewskie FP. A roentgenographic; biomechanical and histologic evaluation of vascularized and non-vascularized segmental fibular canine autografts. Journal of Bone and Joint Surgery (Am), 1985;67:105-12.
68. De Palma AF, Tsaltas TT, Mauler GG. Viability of osteochondral grafts as determined by uptake of S35. Journal of Bone and Joint Surgery (Am), 1963;45:1565-78.
69. Deutschländer C. Beitrag zur Verpflanzung ganzer Kniegelenke. Deutsche Zeitschrift für Chirurgie, 1914;128:183-99.
70. Dorr LD, Ranawat CS. Bone grafts for tibial deficits in total knee arthroplasty. In: LD Dorr (Ed). Techniques in Orthopedics: revision of total hip and knee. Baltimore: University Park Press; 1984. p. 143-50.
71. Dorrance GM. Osteoperiosteal bone grafts. Annal Surgery, 1930;42:161.
72. Dyas FG. Treatment of fractures by autogenous bone transplants. Surgery, Gynaecology and Obstetrics, 1915;21:115.
73. Elliott H, Scott HJ. Bone bank in neurosurgery. British Journal of Surgery, 1951;39:31.
74. Elmer G, Schmincke A. Eine 15 ½ Jahre Alteshomoioplastisches Knochentransplantat beim Menschen. Zentralblatt für Chirurgie, 1925;52:562.
75. Elves, MW. Humoral Immune Response to Allografts of Bone. International Archives of Allergy, 1974;47:708-15.
76. Ely LW. Healing of fractures. Archives of Surgery, 1922;5:527-60.
77. Ely LW. An experimental study of buried bone. Journal of Orthopaedic Surgery, 1919;1:100-8.
78. Emerson RH, Head WC, Malinin TI, et al. Femoral reconstruction in hip arthroplasty, Part II. Surgery Rounds for Orthopaedics, 1990;4:31-8.
79. Emerson RH, Head WC, Malinin TI, et al. Allograft femoral reconstruction in hip arthroplasty. Part I. Surgery Rounds in Orthopaedics, 1990;4:15-22.
80. Emerson RH, Head WC, Malinin TI, et al. Allograft femoral reconstruction in revision hip arthroplasty. Part II. Surgery Rounds in Orthopaedics, 1990;4:31-40.
81. Emerson RH, Malinin TI, Cuellar AD. Cortical strut allografts in the reconstruction of the femur in revision total hip arthroplasty. A basic science and clinical study. Clinical Orthopaedics and Related Research, 1992;285:35-44.
82. Engstrom H. Über das Auftreten von Riesenzeller bei der Resorption intermuskular implantierter gekochter Knochensubstanz. Zeitschrift für Mikroscopische Anatomie, 1937;41:273.
83. Enneking WF, Burchard H, Pul JJ, et al. Physical and biological aspects of repair in dog cortical-bone transplants. Journal of Bone and Joint Surgery (Am), 1975;57(2):237-52.
84. Enneking WW, Mindell ER. Observation on massive retrieved human allografts. Journal of Bone and Joint Surgery (Am), 1991;73:1123-42
85. Ezra-Cohen HE, Cook SF. Blood typing of compact human bone tissue. Nature, 1961;191:1267-8.
86. Friedlaender GE. Immune responses to osteochondral allografts. Clinical Orthopaedics and Related Research, 1983;174:58.
87. Gabl M, Reinhart C, Lutz M, et al. Vascularized bone graft from the iliac crest for the treatment of nonunion of the proximal part of the scaphoid with avascular fragment. Journal of Bone and Joint Surgery (Am), 1999;81:1414-28.
88. Gallie WE. The repair of bone. British Journal of Surgery, 1919;7(26):211-59.
89. Gallie WE. Transplantation of bone. British Medical Journal, 1931;2:840-4.
90. Gazdaq AR, Lane JM, Glaser D. Alternatives to autogenous bone grafts: Efficacy and indications. Journal of American Academy of Orthopaedic Surgery, 1995;3:1-8.
91. Ghormley RK. Choice of graft methods in bone and joint surgery. Annals of Surgery, 1942;115:427-34.
92. Gie GA, Linder L, Ling RSM, et al. Cancellous allograft and cement for revision total hip arthroplasty. Journal of Bone and Joint Surgery (Br), 1993;75:14-21.
93. Goebell R. Ersztz von Fingergelenken durch Zehenglenke. Münchener Med Wochenschr, 1913;60:1598.
94. Goldberg VM, Stevenson S. Natural history of autografts and allografts. Clinical Orthopaedics and Related Research, 1987;225:7-169.
95. Goldberg VM, Heiple KG. Experimental hemijoint and whole joint transplantation. Clinical Orthopaedics and Related Research, 1983;174:43.
96. Goldberg VM, Lance EM. Revascularization and accretion in transplantation. Journal of Bone and Joint Surgery (Am), 1972;54:807-16.
97. Goldberg VM, Bos G, Heiple KG, et al. The effect of immunosupression on frozen bone allografts in histocompatibility mismatched dogs. Transactions of the Annual Meeting. Orthopaedic Research Society, 1982;7:173-4.
98. Goldberg VM, Powell AE, Heiple KG, et al. Improved acceptance of frozen bone allografts in genetically mismatched dogs by immunosupression. Journal of Bone and Joint Surgery (Am), 1984;66:937-50.
99. Goldberg VM. The biology of bone grafts. Orthopaedics, 2003;26(9):923-4.
100. Granja CB, Salazar M, Yunis EJ. Population genetics and human leukocyte antigen polymorphism. In: NL Tilney, TB Strom R, C Paul (Eds). Transplantation Biology. New York, Philadelphia: Lippincott-Raven; 1996. pp. 311-39.
101. Gresham RB. The freeze-dried cortical bone homograft: a roentgenologic and histologic evaluation. Clinical Orthopaedics and Related Research, 1964;37:194-201.
102. Gross AE, McKee NH, Pritzker KPH, et al. Reconstruction of skeletal deficits at the knee: a comprehensive osteochon-

dral transplant program. Clinical Orthopaedics and Related Research, 1983;174:96-106.
103. Groves EWH. Methods and results of transplantation of bone in the repair of defects caused by injury and disease. British Journal of Surgery, 1917;5:205-42.
104. Gruber R, Mayer C, Bobaczk K, et al. Effect of cartilage derived morphogenic protein and osteogenic protein-1 on osteochondrogenic differentiation of Periosteum-derived cells. Endocrinology, 2001;142:2087-94.
105. Guttman RD, Lindquist RR. The microvasculature during renal allograft rejection. In: TI Malinin, BS Linn, AB Callehan, et al. (Eds). Microcirculation, Perfusion and Transplantation of Organs. New York: Academic Press, 1970. pp. 207-18.
106. Haas SL. Transplantation of bone into joints. Archives of Surgery, 1926;12:887.
107. Haddad FS, Duncan CP, Berry DJ, et al. Periprosthetic femoral fractures around well fixed implants: use of cortical onlay allografts with or without a plate. Journal of Bone and Joint Surgery (Am), 2002;84:945-50.
108. Harbin M, Liber KE. The behavior of transplanted bone. Surgery Gynaecology and Obstetrics, 1949;59:149.
109. Head WC, Malinin TI, Mallory TH, et al. Onlay Cortical allografting fro the femur. Orthopaedic Clinics of North America, 1998;29:307-12.
110. Head WC, Wagner RA, Emerson RH, et al. Restoration of femoral bone stock in revision total hip arthroplasty. Orthopaedic Clinics of North America, 1993;24:697.
111. Heiple KG, Chase SW, Herndon CH. A comparative study of the healing process following different types of bone transplantation. Journal of Bone and Joint Surgery (Am), 1963;45:1593-616.
112. Heiple KG, Goldberg VM, Powell AE, et al. Biology of cancellous bone repair. In: GE Friedlaender, HJ Mankin, KW Sell (Eds). Osteochondral Allografts. Boston/Toronto: Little Brown and Co.; 1983. pp. 37-49.
113. Helsop BF, Zeiss IM, Nisbet MW. Studies on transference of bone: I. A comparison of autologous and homologous bone implants with reference to osteocyte survival, osteogenesis and host reaction. British Journal of Experimental Pathology, 1960;41:269.
114. Henderson MS. The massive bone graft in ununited fractures. Journal of American and Medical Association, 1936;107:1104.
115. Henry M. Homografts in orthopaedic surgery. Journal of Bone and Joint Surgery (Am), 1948;30:70-6.
116. Henry MO, Geist ES. Spinal fusion by simplified technique. Journal of Bone and Joint Surgery, 1933;14:622-5.
117. Henry MO. Chip grafts in orthopaedic surgery. Journal of Bone and Joint Surgery, 1937;19:1057.
118. Herndon CH, Chase SW. Experimental studies in the transplantation of whole joints. Journal of Bone and Joint Surgery (Am), 1952;34:564-78.
119. Herndon CH. Principles of bone graft surgery-different methods of operative procedure and indications for each. Instructional Course Lectures. American Academy of Orthopaedic Surgery. St. Louis: C.V. Mosby Co.; 1960. pp. 149-64.
120. Higgs SH. The use of cancellous chips in bone grafting. Journal of Bone and Joint Surgery (Am), 1946;28:15-8.
121. Horowitz MC, Friedlaender GE. Immunologic aspects of bone transplantation: a rationale and future studies. Clinical Orthopaedics and Related Research, 1987;18:227-3.
122. Horowitz MC, Friedlaender GE. Induction of specific T-cell responsiveness to allogeneic bone. Journal of Bone and Joint Surgery (Am), 1991;73:1157-68.
123. Horowitz MC, Friedlaender GE. Immune response to bone grafts. In: GE Friedlaender, V Goldberg (Eds). Bone and Cartilage allografts. American Academy of Orthopaedic Surgery. Park Ridge; 1991. pp. 85-101.
124. Horowitz TL, Richard G. Massive iliac bone grafts in the treatment of ununited fractures and large defects of long bones; the combined bone graft-metallic plate technique. Surgery, Gynaecology and Obstetrics, 1947;84:435.
125. Horowitz T. The behavior of bone grafts. Surgery, Gynaecology and Obstetrics, 1949;89:310.
126. Hutchison J. Observations of bone transplants in the anterior chamber of the eye. Glasgow Medical Journal, 1949;30:357.
127. Hutchison J. The fate of experimental bone autografts and homografts. British Journal of Surgery, 1952;39:552-61.
128. Innis PC, Randolph MA, Paskert JP, et al. Vascularized bone allograft: in vitro assessment of cell mediated and humoral responses. Plastic and Reconstruction Surgery, 1991;87(2):315-25
129. Jakimowitsch. Versuche über das Wiederanheilen vollkommen getrennter Knochensplitter. Deutsche Zeitschrift für Chirurgie, 1881;15:201-46.
130. Jazrawi LM, Della Valle CJ, Kummer FJ, et al. Catastrophic failure of a cemented, collarless, polished, tapered cobalt-chromium femoral stem used with impaction grafting. A report of two cases. Journal of Bone and Joint Surgery (Am), 1999;81:844-7.
131. Jensen TT. Rhesus immunization after bone allografting. Acta Orthopaedica Scandinavica, 1987;58:584.
132. Kandel RA, Gross AE, Ganel A, et al. Histopathology of failed osteoarticular shell allografts. Clinical Orthopaedics and Related Research, 1985;197:103-10.
133. Kassowitz M. Die normale Ossification. L Braumuller Wien; 1881. p. 37.
134. Kreutz FP, Hyatt GW, Turner TC, et al. The preservation and clinical use of freeze-dried bone. Journal of Bone and Joint Surgery (Am), 1951;33:863-72.
135. Küttner H, Die transplantation aus der Leiche. Beiträge für Klinische Chirurgie, 1911;75, 1-52.
136. Küttner H. Einige Dauerresuetete der Transplantation aus der Leiche und aus der Affen. Verhandlund Deutsche Gesellschaft für Chirurgie; 1913. pp. 353-61.
137. Lance EM, Fisher RL. Transplantation of rabbits patella. Journal of Bone and Joint Surgery (Am), 1970;51:145.
138. Lance EM. Some observations on bone graft technology. Clinical Orthopaedics and Related Research, 1985;200:114-24.

139. Langer F, Gross AE, West M, et al. The immunogencitiy of allograft knee joint transplants. Clinical Orthopaedics and Related Research, 1978;132:155.
140. Lecroix P. The organization of bones. London: J and A Churchill; 1951. p. 23, 190-4.
141. LeRiche R. Transplantation d'os most dans une pseudoarthrose. Lyon Chirurgie, 1922;19:219.
142. Levander G, Willstead H. Alcohol soluble osteogenetic substance from bone marrow. Nature, 1946;157:587.
143. Lewandowski KU, Rebmann V, Passler M, et al. Immune response to perforated and partially demineralized bone allografts. Journal of Orthopaedic Science, 2001;6(6):545-55.
144. Lewin JL. Case of successful grafting of ribs into skull for cranial defect. Medical Journal of South Africa, 1924;20:61.
145. Lexer E. Joint transplantation and arthroplasty. Surgery, Gynaecology and Obstetrics, 1925;40:782.
146. Lexer E. Substitution of whole or half joints from freshly amputated extremities by free plastic operation. Surgery, Gynaecology and Obstetrics, 1908;6:601.
147. Lexer E. The classic: joint transplantation. Clinical Orthopaedics and Related Research. 1985;197:4.
148. Lexer E. Freie Knochenplastic and Gelengimplantation. Med Klin, 1908;4:686.
149. Lexer E. Über freie Transplantationen. Archiv für Klinische Chirurgie, 1911;95:827.
150. Lexer E. Zwanzig Jahre Transplantationforschung in der Chirurgie. Archiv für Klinische Chirurgie, 1925;138:251.
151. Lexer E. Die Freien Transplantatione. Zweite Teil Verlag von Ferdinand Enke in Stuttgart, 1924.
152. Lexer E. Joint transplantation and arthroplasty. Surgery, Gynaecology and Obstetrics, 1925;40:782-809.
153. Macewen W. Observations concerning transplantation of bone. Illustrated by a case of interhuman osseous transplantation whereby over two-thirds of a shaft of a humerus was restored. Proceedings of Royal Society of London, 1881;32:232.
154. Macewen W. Intrahuman bone grafting and reimplantation of bone. Annals of Surgery, 1909;50:959.
155. Malinin TI, Carpenter EM, Temple HT. Particulate bone allograft incorporation in regeneration of osseous defects, importance of particle sizes. The Open Orthopaedic Journal, 2007;1:19-24.
156. Malinin TI, Head WC, Emerson RH, et al. Reconstructions of the deficient proximal femur with bone plate allograft in revision total hip arthroplasty. Orthopaedics Transaction, 1997;21:182.
157. Malinin TI, Mnaymneh W, LO HK, et al. Cryopreservation of articular cartilage. Clinical Orthopaedics and Related Research, 1994;303:18-32.
158. Malinin TI, Temple HT, Buck BE. Transplantation of osteochondral allografts after cold storage. Journal of Bone and Joint Surgery (Am), 2006;88:762-70.
159. Malinin TI, Temple HT, Garg A. Cancellous and cortical microparticulate allograft for dental implantation: an experimental study in non-human primates. Implant Dentistry, 2009;18:420-6.
160. Malinin TI, Temple HT. Comparison of frozen and freeze dried particulate bone allografts. Cyrobiology, 2007;55:167-70.
161. Malinin TI, Wu NM, Flores A. Freeze-drying of bone for allotransplantation. In: GE Friedlaender, HJ Mankin, KW Sell (Eds). Osteochondral Allografts. Boston/Toronto: Little Brown and Co; 1983. pp. 181-92.
162. Malinin TI, Wagner LL, Brown MD. Freeze-dried cortical plate allografts in fracture healing. Orthopaedic Transactions, 1985;9:284.
163. Malinin TI, Mnaymneh W, Lo HL, et al. Cryopreservation of articular cartilage. Clinical Orthopaedics and Related Research, 1994;303:18-32.
164. Malinin TI, Mnaymneh W, Wagner JL, et al. Healing of internally fixed intercalary canine allografts of freeze-dried bone. Orthopaedic Transactions, 1985;9:339.
165. Malinin TI, Mnaymneth W, Wagner JL, et al. Healing of internally fixed intercalary canine allografts of freeze-dried bone. Transaction of 31st Annual Meeting. Orthopaedic Research Society, Las Vegas, 1985;10:289.
166. Malinin TI. Acquisition and banking of bone allografts. In: MB Habal, AH Reddi (Eds). Bone Grafts and Bone Substitutes. Philadelphia: WB Saunders; 1992. pp. 206-25.
167. Malinin TI. University of Miami Tissue Bank: Collection of postmortem tissues for clinical use and laboratory investigation. Transplantation Proceedings, 1976;8(suppl 2):53-8.
168. Malinin TI, Latta LL, Wagner JL, et al. Healing of fractures with freeze-dried cortical bone plates. Clinical Orthopaedics and Related Research, 1984;190:281-6.
169. Malinin GI, Hornicek FJ, Banovac K, et al. Interaction of cultured rat osteoblasts with 8-methoxypsoral during irradiation with long-wave ultraviolet light. Biochemical and Biophysical Research Communications, 1995;207:877-81.
170. Mallory TH. Preparation of the proximal femur in cementless total hip revision. Clinical Orthopaedics and Related Research, 1988;235:67.
171. Mankin HJ, Doppelt SH, Tomford WW. Clinical experience with allograft implantation. Clinical Orthopaedics and Related Research, 1983;174:69-86
172. Mankin HJ, Friedlaender GE. Perspectives of bone allograft histology. In: GE Friedlaender, HJ Mankin, KW Sell (Eds). Osteochondral Allografts. Boston/Toronto: Little Brown and Co.; 1983. pp. 4-5.
173. Marrack JR. Structure of antigen-antibody aggregates and complement fixation. Annual Review of Microbiology, 1955;9:369-86.
174. Masse G, Gallinaro P, Garelli R, et al. Study sull'immunita da trapianto. IV. Induzione di risposte accelerate e iperimmuni dopo "primo trapianto" omologo allogenico di corticale ossea. Minerva Ortopedica, 1965;16:530.
175. Mayer L, Werner E. An experimental study of osteogenesis. American Journal of Orthopaedic Surgery, 1914;12:213.
176. Mnaymneh W, Malinin T. Massive allografts in surgery of bone tumors. Orthopaedic Clinics of North America, 1989;20: 455-67.

177. Mnaymneh W, Malinin TI, Lackman RD, et al. Massive distal femoral allografts following resection of bone tumors. Clinical Orthopaedics and Related Research, 1994;303:103-15.
178. Mnaymneh W, Malinin TI, Mackley JT, et al. Massive osteoarticular allografts in the reconstruction of extremities following resection of tumors not requiring chemotherapy and radiation. Clinical Orthopaedics and Related Research, 1985;187:76-87.
179. McCallum WG. Textbook of Pathology. Philadelphia: WB Saunders Co.; 1942.
180. McDermott AGP, Langer F, Pritzker KPH, et al. Fresh small-fragment osteochondral allografts: long-term follow-up study on the first 100 cases. Clinical Orthopaedics and Related Research, 1985;197:96.
181. McGarvey WC, Braly WG. Bone graft in hindfoot arthrodesis: allograft versus autograft. Orthopaedics, 1996;19:389.
182. McKeever F. Discussion of experimental studies in the transplantation of whole joints by CH Herndon and SW Chase. Journal of Bone and Joint Surgery (Am), 1952;34:578.
183. McWilliams C. A discussion of bone transplantation and the use of the rib as a graft. Annal Surgery, 1912;56:377.
184. Meding JB, Ritter MA, Keating EM, et al. Impaction bone-grafting below femoral stem with cement in revision total hip arthroplasty. A minimum two-year follow up study. Bone Joint Surgery (Am), 1997;79:1834-41.
185. Meyers MH. Resurfacing of the femoral head with fresh osteochondral allografts: long-term results. Clinical Orthopaedics and Related Research, 1985;197:111.
186. Meyers MH. The treatment of osteonecrosis of the hip with fresh osteochondral allografts and with the muscle pedicle graft technique. Clinical Orthopaedics and Related Research, 1978;130:202.
187. Meyers MH, Jones RE, Bucholz RW, et al. Fresh autogenous grafts and osteochondral allografts for the treatment of segmental collapse in osteonecrosis of the hip. Clinical Orthopaedics and Related Research, 1983;174:176-8.
188. Mohan S, Baylink D. Bone growth factors. Clinical Orthopaedics and Related Research, 1991;30-48.
189. Mowlem R. Bone and cartilage transplants: Their use and behavior. British Journal of Surgery, 1941;29:182.
190. Muller W. Demonstration zur Frage Osteoplastik. Munch Med Worchenschr, 1900;47:1472.
191. Mundy GR. Local control of bone formation. Clinical Orthopaedics and Related Research, 1995;313:19-26.
192. Murphy JB. Contributions to the surgery of bones, joints and tendons. Journal of Americal Medical Association, 1912;58:985-1004.
193. Muscolo DL, Ayerza MA, Calabrese ME, et al. Human leukocyte antigen matching, radiographic score and histologic findings in massive frozen bone allografts. Clinical Orthopaedics and Related Research, 1996;326:115-26.
194. Nageotte J. Formation de pieces squelettiques surnumeraires provoguees par la presence de griffons morts dan l'orleille du lapin adulte. CR Soc Bio, 1918;81:113.
195. Neufeld SK, Uribe J, Myerson MS. The use of structural allograft to compensate for bone loss in arthrodesis of the foot and ankle. Foot Ankle Clinic, 2002;7(1):1-17
196. Neuhof H. The transplantation of Tissues. D. New York: Apleton and Co.; 1923. pp. 185-9.
197. Ottolenghi CE. Massive osteoarticular bone grafts. Transplantation of the whole femur. Journal of Bone and Joint Surgery (Br), 1966;48:646-59.
198. Olier L. Traite experimental et clinique de la regeneration des os et de la production artificielle du tissu osseux. Paris: Victor Masson et Fils; 1867.
199. Oriol R. ABH and related tissue antigens. Biochemical Society Transactions, 1987;15:596-9.
200. Ozyurek A, Leblebicioglu G, Bilgiliti, et al. Effect of vascular bundle implantation on autograft, fresh allograft and xenograft incorporation in rabbit model. Orthopaedics, 2008;31(2):135.
201. Paacola CA, Xavier CA, Conclaves RP. Fresh immature articular cartilage allografts. Archives of Orthopaedic and Trauma Surgery, 1979;92:253.
202. Pap K, Krompecher S. Arthroplasty of the knee. Experimental and clinical experiences. Journal of Bone and Joint Surgery (Am), 1961;43:523-37.
203. Pappenheimer AM. Experimental studies upon lymphocytes. I The reaction of lymphocytes under various experimental conditions. Journal of Experimental Medicine, 1914;25:633-44.
204. Paprosky WG, Magnus RE. Principles of bone grafting in revision total hip arthroplasty. Clinical Orthopaedics and Related Research, 1994;298:147.
205. Peer LA. Transplantation of tissues. Vol. 1. Baltimore: The Williams and Wilkins Co.; 1955. p. 196
206. Petrow N. Ein 25 Jahre altes Knochentransplantat. Arch F Klin Chir, 1933;175:176.
207. Phemister DB. The fate of transplanted bone and regenerative power of its various constituents. Surgery, Gynaecology and Obstetrics, 1914;19:303-33.
208. Pietrzak WS, Ali SN, Chitturi D, et al. BMP depletion occurs during prolonged acid demineralization of bone: characterization and implication for graft preparation. Cell Tissue Bank. 2009; Published online, 29 Dec 2009.
209. Polettini B. Su neoformazioni cartilaginee ed ossee determinate da innesti di frammenti di cartilagine e d'osso fissati. Arch Ital Chir, 1922;6:179.
210. Pollock WE, McKenney PW, Blaisdell FE. The viability of transplanted bone. Archives of Surgery, 1929;18:607-23.
211. Poncet A. Greffe osseuse massive. Lyon Medicine, 1887; 54:437.
212. Poncet A. Transplantation osseuse interhumaine (greffe massive) dans un cas de pseudarthrose du tibia gauche chez un jeune homme de dix-neuf ans. Paris. Comp Rend Acad Sciences, 1887;104:929-32.
213. Reikeras D, Reinholt FP, Zinocker S, et al. Healing of long-term frozen orthotopic bone allografts is not affected by MHC differences between donor and recipient. Clinical Orthopaedics and Related Research, 2011;469:1479-86.

214. Reynolds F, Oliver OR, Ramsey R. Clinical evaluation of merthiolate bone bank and homologous bone grafts. Journal of Bone and Joint Surgery (Am), 1951;33:873-83.
215. Robertson JR. Proximal femoral bone loss after total hip arthroplasty. Clinical Orthopaedics and Related Research, 1992;23:291.
216. Rock MG. Revision Joint Replacement: Proximal Femoral Replacement. In: Orthopaedic Allograft Surgery. AA Czitrom and H. Winkler (Eds). New York: Springer-Wien; 1996. pp. 91-100.
217. Rodrigo JJ, Sakovich L, Travis C, et al. Osteocartilaginous allografts as compared with autografts in the treatment of knee joint osteocarilagenous defects in dogs. Clinical Orthopaedics and Related Research, 1989;134:132-349.
218. Rodrigo JJ, Block N, Thompson EC. Joint transplantation. Veterinary Clinics of North America, 1978;8:523-38.
219. Rodrigo JJ, Fuller TC, Mankin HJ. Cytotoxic HLA antibodies in patients with bone and cartilage allografts. Transactions of the Annual Meeting-Orthopaedic Research Society, 1976;1:131.
220. Rohde C. Beitrage zur Frage der Metaplasie des Bindegewebes in Knochen. Archiv für Klinische Chirurgie, 1924;128:302
221. Rohlich K. Uber die Beziehungen zwishchen der Knochensubstanz und der Bluthildung in Knochenmark. Zeitschrift für Microscopische Anatomie Forschung, 1941;49:425.
222. Rydberg S, Breimer M, Samnelson BE, et al. Blood Group ABO incompatible (A_2 to O) kidney transplantation in human subjects: a clinical, serologic and biochemical approach. Kindey, 1987;19:4528-37
223. Sabanas AO, Janes JM, Dahlin DC, et al. Comparison of homologous bone grafts preserved by acetone and formaldehyde and by acetone alone; experimental studies. Proceedings of the staff meetings. Mayo Clinic, 1955;30(19):432-6.
224. Saltykow S. Uber Replantation der Knochen. Beiträge zu Pathologische Anatomie und Allgemeine Pathologie, 1909;45:440.
225. Sasama K. Report of 355 transfusion associated deaths; 1976 through 1985. Transfusion, 1990;30:583-90.
226. Schreurs BW, Sloof TJ, Gardeniers JW, et al. Acetabular reconstruction with bone impaction grafting and a cemented cup: 20 years' experience. Clinical Orthopaedics and Related Research, 2001;393:202-15.
227. Schweiberer L, Brenneisen P, Dambe LT. Derzeitigen stand der auto-hetero und homoplastischen Knochentransplantation. In: Cotta, Hand Marti AK (Eds). Implante und Transplante in der Plastischen und Wiederherstellurgschirurgie. New York: Springer-Berlin-Heidelberg; 1981.
228. Sloof TJ, Buma P, Schreurs BW, et al. Acetabular and femoral reconstruction with impacted graft and cement. Clinical Orthopaedics and Related Research, 1996;324:108-15.
229. Sloof TJ, Schimmel JW, Buma P. Cemented fixation with bone grafts. Clinical Orthopaedics and Related Research, 1993;24:667-77.
230. Slurman AE. The histological distribution of the blood group substances in man as disclosed by immunofluorscence II. H Antigen and its relation to A and B antigen. Journal of Experimental Medicine, 1962;115:977-1005.
231. Smith AD. Use of homologous bone grafts in cases of osteogenesis imperfect. Archives of Surgery, 1937;34:687-96.
232. Spence KF, Bright RW, Fitzgerald SP, et al. Solitary unicameral bone cysts: Treatment with freeze-dried crushed cortical-bone-allografts. Journal of Bone and Joint Surgery (Am), 1976;58:636-41.
233. Spence KF, Sell KW, Brown RH. Solitary Bone Cysts: Treatment with Freeze-Dried Cancellous Bone Allograft. A Study of 177 Cases. Journal of Bone and Joint Surgery (Am), 1969;51:87-96.
234. Spence K, Sell KW, Harris B, et al. A Comparative Evaluation of Freeze-Dried Cancellous and Cortical Bone Allografts in Dogs. In Proceedings of the American Academy of Orthopaedic Surgeons. Journal of Bone and Joint Surgery (Am), 1972;54:1350.
235. Stassen JG, van Dijk BA, van Horn JR, et al. No irregular erythrocyte antibodies observed after bone allografts in 144 patients. Acta Orthopaedica Scandinavica, 1993;64:354-56.
236. Stevenson S. The immune response to osteochondral allografts in dogs. Journal of Bone and Joint Surgery (Am), 1987;69:573-82.
237. Stringa Gabriele. Studies of vascularization of bone grafts. Journal of Bone and Joint Surgery (Br), 1957;39:395.
238. Sun JS, Liu HC, Chang LH, et al. Influence of hydroxyapatite particles size on bone cell activities; an in vitro study. Journal of Biomedical Materials Research, 1998;39(3):390-7.
239. Tanaka H, Inoue M, Suzuki M, et al. A study on experimental homocartilage transplantation. Archives of Orthopaedic and Trauma Surgery, 1980;96:165.
240. Tauber AI. Metchnikoff and the phagocytosis theory. Nature Reviews Molecular Cell Biology, 2003;4:897-901.
241. Thorpe SJ, Hunt B, Yacoub M. Expression of ABH blood group antigens in human heart transplantation. Transplantation, 1991;51:1290-5.
242. Tietze A. Ersatz des resecierten unteren Radiusendes durch eine Grosszehenphalange. Verhandlung das Deutsche Geselschaft für Chirurgie, 1902;1:77.
243. Tomford WW, Thomgphasuk J, Mankin HJ. Frozen musculoskeletal allografts. Journal of Bone and Joint Surgery (Am), 1990;72, 657-63.
244. Tuffier T. Sur les greffes osteo-articulares. Apropos d'une observation de M. Pierre Duval. Bull.et Meme. Societe de Chirurgie de Paris, 1913;39, 1078.
245. Urist MR. Chemosterilized antigen-extracted surface demineralized allogeneic bone for arthodesis. In: G Friedlander, H Mankin, KW Sell (Eds). Osteochondral Allografts. Boston: Little Brolin; 1983. pp. 193-201.
246. Urist MR. Fundamental and clinical bone physiology. J.B. Lippincott Philadelphia: 1980.
247. Urist MR, Benham K, Kerendi F, et al. Lipids closely associated with bone morphogenetic protein (BMP) and induced heterotropic bone formation. With preliminary

observations of deficiencies in lipid and osteoinduction in lathyrism in rats. Connective Tissue Research, 1997; 36:9-20.

248. Urist MR. Osteoinduction undermineralized bone implants modified by chemical inhibitors of endogenous matrix enzymes. Clinical Orthopaedics and Related Research, 1972;87:132-9.
249. Urist MR, Jurist JM, Duboc Fl, et al. Quantitation of new bone formation in intramuscular implants of bone matrix in rabbits. Clinical Orthopaedics and Related Research, 1970;68:279-93.
250. van de Vord PJ, Nasser S, Wooley PH. Immunologic responses to bone soluble proteins in recipients of bone allografts. Journal of Orthopaedic Research, 2005;23:1059-64
251. von Eiselberg. A Zur Heilung grosserer Defecte der tibia durch gestielte Haat-Periost-Knochenlappen. Archiv für Klinische Chirurgie, 1897;55:443.
252. Wade, H. Report of a patient six years after implantation of a hompastic bone graft. Edinburgh Medical Journal, 1920;24:37.
253. Weinberg H, Makin M, Nelken D, et al. A and B antigens in human bone tissue. Journal of Bone and Joint Surgery (Br), 1959;41:151-3
254. Wereschinski OA. Vergleichende Untersuchungen uber Explanation und Transplantation von Knochen, Periost und Endost. Archiv für Pathologie, Anatomie und Physiologie, 1924;251:268.
255. Wilden JA, Moran SL, Dekutoski MB, et al. Results of vascularized autogenous fibular grafts. Journal of Bone and Joint Surgery (Am), 2006;832-9.
256. Wilson PD. Follow-up study of the use of refrigerated homogenous bone transplants in orthopaedic operations. Journal of Bone and Joint Surgery (Am), 1951;33:307-23.
257. Zatzepin ST. Limb sparing operations in bone tumors. Moscow: Medizina; 1984.

Chapter 7

The most frequent source of tissues for xenografts were cows and pigs.

(*Reference:* Peer LA. Transplantation of Tissues. Vol 1. Baltimore; Williams and Wilkins, 1955.)

Bone Xenografts

INTRODUCTION

Bone xenografts, earlier known as bone heterografts, have a long and checkered history. Xenogeneic bone transplants were performed by a number of surgeons, generally with little or no success. The notion of interchanging parts between different species goes back many millennia. Centaurs, mermaids and other mythical creatures come to mind when recollecting various classical legends. Sphinx, chimera and lamassu are other examples of creatures existing with various parts of different animals. (Cooper et al 2007).

Many anecdotal accounts of surgical xenotransplantations have endured through the ages. The most commonly quoted account dates back to 1668 AD and relates the story of a Russian physician repairing the skull of a wounded nobleman with a dog bone (Meekeren 1668). The story does not have a happy ending. The recipient was barred from entering the church because dogs were not allowed in church. He finally had his xenograft removed.

Ready access to animal tissues makes the idea of transplanting bone xenografts appealing and commercially profitable. Consequently, xenogeneic skeletal transplants which include bovine bone, cartilage, dog bone, cow horn, ivory from several sources and bone from other animals and membranes have been tried (Baer 1918). Many surgeons who pioneered the use of bone autografts and allografts have, at one time or another, investigated xenografts. In the past, many types of xenografts have been described. At present only a few bone xenografts are promoted for some clinical applications.

Although experience accumulated with bone xenografts spans over a century, critical analysis of xenogeneic bone transplantation is difficult if not impossible (Burwell 1985). Beginning with the end of the 19th century, numerous papers dealing with the transplantation of animal bone grafts into humans have been published. Nevertheless, long-term and systematic studies of the behavior of animal bone transplants in humans are lacking. The matter is further complicated by periodic surges of clinical reports describing and advocating particular types of xenografts. Many of these are usually favorable accounts of experiences with xenografts published in a relatively short span of time which were, in subsequent years, followed by silence on the subject.

Descriptions of failures and complications with several types of xenografts have largely discouraged their use in the United States (McMurray 1982, Antonutti and Labbadia 1968, Espersen et al 1984). Several surveys among orthopedic, neurologic and plastic surgeons indicate the majority of the respondents have not used, nor were they contemplating using xenogeneic bone grafts (Peer 1955, Ray 1972, Albee 1944, Burwell 1969). The few reports on clinical use of xenografts seem to represent the experiences of a small number of surgeons outside the United States (Salama and Weissman 1978, Goran and Murthy 1978, Taheri and Gueramy 1972). Thus, the sentiment expressed by Albee in 1944, namely that "one important principle was proved in the very beginning, and that was untrustworthiness of heterologous bone graft" is still relevant in today's practice for American surgeons (Albee 1944). However, in other countries, xenografts seem to have enjoyed wider popularity and a greater degree of acceptance (Hallen 1966, Sarang et al 1978, Arbes 1981, Loktev 1976).

It is generally accepted that untreated xenogeneic grafts are unsuitable for transplantation because they evoke a strong immune response (Katthagen 1987). The inferior properties of untreated xenogeneic transplants have been documented as far back as 1867 (Oilier 1867), but they have not been subjected to intensive study. Subsequent studies have demonstrated the appearance of cytotoxic antibodies whose production was caused by xenogeneic bone transplants (Elves and Salama 1974) as well as the presence of chronic inflammatory cells in and around the grafts (Malinin et al 1941). These studies unequivocally documented a strong immune response to xenogeneic iliac bone. Because of this observed immunologic response to untreated xenogeneic grafts, a number of regimens aimed at abolishing or modifying this response were devised. Likewise, a variety of xenogeneic tissues were produced and at some time or other transplanted into patients.

IVORY AND COW HORN

Ivory screws, plates, pegs and nails have been used in the past to a considerable extent. The desirable properties of ivory which led to its employment in surgery are several. It expands on immersion in water or saline, and thus fills a void when implanted in bone. Ivory pegs did not produce necrosis of surrounding bone and underwent time-dependent resorption.

Cow horn is tough and firm, but at the same time flexible and non-brittle. For this reason it was used as an intermedullary peg or an onlay splint.

Since it is obviously impossible to obtain aseptically either cow horns or ivory, these had to be sterilized before implantation. Both the ivory and the cow horn xenografts were sterilized by autoclaving.

Ivory plates and pegs were introduced by König in 1913 with reported success. Encouraged by his initial results, he began to replace portions of long bones of the extremities and of mandibles with implants made of ivory. In several patients, the implants failed and had to be removed (König 1913). Magnuson inserted ivory strips into the medullary canals to stabilize fractures and noted a lack of softening of the recipient's bone adjacent to the ivory strips. However, ivory pegs gradually became absorbed. Magnuson found the resorptive properties to be an advantage, and subsequently made ivory screws which he used to fix oblique fractures. He reported no foreign body reaction to ivory implants (Magnuson 1913). Groves described a patient in whom ivory from a walrus tusk was used to fill a large cavity in a femur. This ivory proved inert, and 16 years later, appeared the same as when it was placed into the patient (Groves 1939). Although it was said that Groves advocated the use of ivory pegs in reducing the fractures of the femoral neck, this is not entirely accurate. What he actually said was that "there still remains a difference of opinion whether the peg should be made of a living graft taken from the patient's own skeleton or a piece of animal bone or ivory". Initially, for pegging fractures of the femoral neck, Groves used autologous bone exclusively, but then switched to "dead bone", not ivory. However, he fashioned an ivory nail to create a new femoral head to treat a patient with an ankylosed hip (Groves 1917). Wolf used ivory for the correction of saddle nose deformity but without great success (Wolf 1938).

The use of ivory as xenografts has not persisted; ivory implants on an occasional case is rarely observed **(Fig. 1)**.

Cow horn and deer antlers were likewise used as xenografts. Lever, Dahl Iverson (Dahl Iverson 1930), and Boliarsky used cow horn xenografts in a few patients. Fowler (Fowler 1934) believed it stimulated fractured bone to form excessive callus. He further noted that cow horn was bioabsorbable making it a desirable surgical implant. Groves fashioned a peg from a stag horn and used it for pinning a fractured hip in one patient. In radiographs taken 3 years postoperatively, the peg appeared to be completely replaced by host bone (Groves 1939). Carrell used cow horn for internal fixation in 40 patients. He stated it provided adequate mechanical support, stimulated callus formation and did not evoke a foreign body reaction (Carrell 1936). Ghormley may have used screws made of walrus ivory as well as of stag's antlers (Ghormley 1942). However, laboratory studies cast doubts on the clinical efficacy of cow horn xenografts. Fowler showed that plates made of cow horn, when placed over fractured ulnas in dogs were almost completely absorbed in

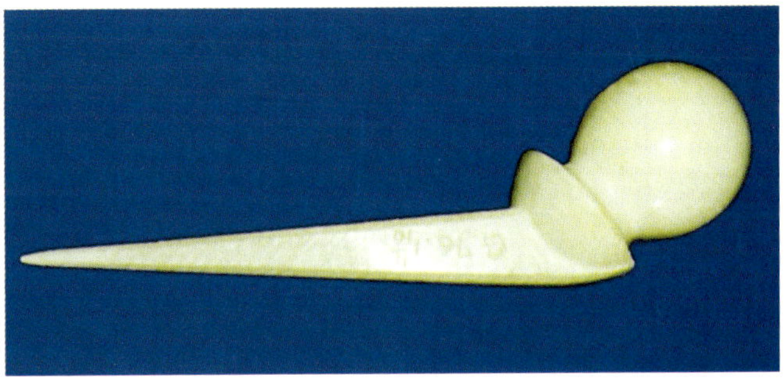

Figure 1 Proximal femur replacement component made from elephant ivory

3 months (Fowler 1934). In a histological study by Hughes, neither cow horn nor ivory pegs implanted into dogs or rabbits produced union between the host bone and the graft (Hughes 1950). Hughes suggested beef bone was a much better substitute for autogenous bone than ivory or cow horn.

Fresh Bone Xenografts

Fresh calf bone xenografts were shown from the outset to be unsatisfactory and the general consensus is that there is no place for these grafts in orthopedic surgery (Peer 1955). Gallie and Robertson in 1918 performed experiments with xenografts, transplanting fresh bone from cats into dogs. By the end of 8 weeks these transplants were surrounded by fibrous tissue with only an occasional osteoclast in the lacunae on the surface of the grafts. There was no new bone formation. Therefore they considered xenogeneic bone "useless for transplantation" (Gallie and Robertson 1919). As early as 1867, Oilier noted that fresh xenogeneic periosteal transplants formed only a fibrous membrane and not bone (Oilier 1867). On the other hand, xenografts stored in Ringer's solution in a refrigerator produced good results in experimental animals (Gaudioso 1924). Frozen xenografts fared even better. In experimental studies on rabbits, bovine heterografts preserved at -25–30° C were used as intermedullary grafts to provide fixation in fractured femurs. The fractures healed by periosteal callus, but heterografts became absorbed. This led to the conclusion that frozen bovine heterografts might be suitable for fracture stabilization in humans (Gongalsky 1974), the presumption being that the freezing process diminished the antigenicity of the graft, making it more suitable for transplantation in humans.

Decalcified Bovine Bone

Attempts to use bovine xenografts processed by various methods have been long lasting. In 1889 Senn reported the healing of bone cavities filled with decalcified chips of ox bone in 4 patients (Senn 1889). Miller likewise used decalcified pieces of ox rib mixed with autologous bone chips to fill a cavity in the proximal tibia, substituting carbolic acid for hydrochloric acid to decalcify the bone. Introduction of this material into the bone cavity of a patient was followed by postoperative drainage which stopped only after the extrusion of bone xenograft chips (Miller 1890). In a more recent experimental study, decalcified calf and sheep bone were found to be unsatisfactory (Salama 1978).

Chemically Preserved Bone

Calf cortical bone preserved by phenol alcohol was transplanted into a phalanx of a patient with osteomyelitis by Kronacher (Kronacher 1897). Years later the entire implant was outlined on a radiograph. It was surrounded by reactive bone formation. Alcohol-preserved beef bone splints were advocated by Brenizer for the repair of fractures of long bones (Brenizer 1920). On the other hand, Tavernier expressed the prevailing opinion that alcohol-preserved beef bone is a surgical material without much value (Tavernier 1930). His clinical experience indicated that even under the most favorable conditions the alcohol-preserved xenografts did not unite with host bone with which it was in contact and ultimately resorbed. Years later Lacroix showed that alcohol preservation inhibited or even abolished the osteogenic potential in all bone grafts (Lacroix 1951).

Ether-preserved cancellous calf bone, used to produce arthrodesis of the spine in patients with tuberculosis or degenerative spine disease, supposedly met with great success (Calvé 1935). Ether-preserved xenografts were washed in serum before they were implanted. These xenografts were well tolerated by 15 recipients and the clinical results were said to be good. However, no further reports on the subject could be found.

Bone Xenografts of Other than Bovine Origin

In addition to bovine bone, bone from several other animal species were sporadically transplanted into humans, mostly in the first part of the twentieth century. The accounts of these transplants are anecdotal. Küttner (1913) reported on transplanting aseptically an excised fibula from a Java monkey (Macaca nigra) into a child with congenital absence of the fibula. Grafted bone was visible on radiographs 4 years later (Küttner 1917). Neuhoff interposed femur from a rabbit between fragments of a fractured femur in a human. The clinical result confirmed by radiography 6 months postgrafting was excellent (Neuhoff 1923). Bird bone was also used for fractures, pseudarthroses, osteosynthesis and other operations. Judet, who accumulated considerable experience with xenografts, in addition to bovine bone, transplanted bone from pigs and young horses. These were preserved by freezing. Judet considered bone from colts to be the best xenografts for transplantation into humans (Judet 1953). Stagnara et al used sheep bone for filling bone cavities, pseudarthroses, infections and arthrodeses (Stagnara et al 1950). These grafts were removed aseptically under ultraviolet light and were stored in a freezer at -70°C.

STERILIZATION AND FREEZING

Since it is difficult to obtain animal bone aseptically, it is usually subjected to sterilization. In the past it was most often either boiled or autoclaved. Thus the reports on transplantation of frozen xenografts also have to take into consideration the sterilizing procedure. Frozen calf bone was at one time rather popular. However, Kingna, who transplanted it in clinical trials using calf bone stored at -40°C, concluded the ultimate results were disappointing and he abandoned the practice (Kingna 1960).

Early on Gallie and Robertson showed that, in contrast to fresh bovine bone, boiled beef bone placed in close contact with the living bone of the recipient stimulated new bone formation (Gallie and Robertson 1919). Henderson observed that beef bone screws were absorbable and well tolerated by the host. These were used widely (Henderson 1920). Sterilized and frozen beef bone grafts were used for fusions in scoliosis. Kleinberg in 1929 reported on 58 patients who underwent successful fusion with preserved beef bone xenografts (Kleinberg 1929). Danis used ox bone in several patients with pseudarthrosis and was favorably impressed with this grafting material. However, late results on his patients were not available (Danis 1936).

OS PURUM

Os purum is a bone that has been soaked in warm potassium hydroxide, extracted with acetone and washed in salt solution. It was introduced by Orell in 1937 (Orell 1937). Both bovine and human bones were treated by this method. Orell devised this treatment of bone grafts to remove connective tissue, fat and proteins; however, os purum did contain residual collagen. Os purum was sterilized by boiling.

Os novum was another graft devised by Orell. It combined xenogeneic and autologous bone. The graft was produced by implanting a narrow strip of os purum subperiosteally over the anteromedial surface of the tibia. With time the os purum splint replaced with new bone and the new bone, os novum, could be excised and transplanted. Orell employed these grafts with considerable success (Orell 1938, 1953). Several other surgeons subsequently undertook studies to determine the efficacy of this graft. Goff used os purum in 80 operations and encountered complications attributable to the graft in at least six (Goff 1944). Os purum was used with considerable success in tuberculous spondylitis and for non-union (Esnaurrizar 1940). Van der Noff was intrigued by the elasticity of os purum and used it for fracture stabilization. However, he noted later that advantages of os purum did not withstand further scrutiny. Untreated beef bone seemed to do as well (Van der Noff 1944). On the other hand, Nyst held that in many cases os purum could replace autogenous bone grafts. He described in detail the insertion of os purum xenografts into fractured bone with delayed callus formation (Nyst 1941). Os purum is not used in modern practice.

DEPROTEINIZED BONE

As noted above, os purum was not truly deproteinized bone. Consequently, procedures for "deproteinizing" bone were devised and investigated (Schweiberer 1970). This was done on the assumption that removal of organic matrix from the bone would result in the removal of most, if not all, immunogenic material while retaining the biomechanical and osteogenic properties of the graft. "Boplant", deproteinized bone produced by Squibb Co., was one of the most notable examples of deproteinized xenografts. This freeze-dried, ethylenediamine-extracted bovine bone prepared by the Squibb Institute for Medical Research **(Fig. 2)** became in the mid-sixties the subject of intense laboratory and clinical investigations.

Ethylenediamine extraction of bone was first described by Williams and Irvine (Williams and Irvine 1954). Morphologically, inorganic crystals in ethylenediamine-extracted bone were positioned in the same way as in the collagen-containing bone. Thus, ethylenediamine extraction was thought to be different from extraction by other chemical solvents because it removed only the organic component of bone with little or no change in the cation or anion composition of its inorganic component. Losee and Hurley stated that ethylenediamine-extracted bone could serve as a foundation for new osseous in-growth in long bone defects

Figure 2 Ethylenediamine-extracted bone (Boplant) prepared and distributed by the Squibb Company in the 1960s

(Losee and Hurley 1956). Studies of the same material as mandibular grafts in monkeys produced data which led Losee and Boyne to conclude that "there was biological host acceptance of heterogeneous "anorganic" bone placed into mandibular and maxillary osseous defects in the oral cavity" (Losee and Boyne 1957). Similar results were reported by other investigators (Heiple et al 1967). Hancox, Owen and Singleton tested ethylenediamine-extracted bone in sheep and concluded that since it did not produce an unfavorable reaction in animals it would be difficult to imagine that the human body would behave differently towards such graft material (Hancox et al 1961).

This bone preparation was said to be non-immunogenic and was reported to incorporate in the same manner as autologous bone (Anderson et al 1961, 1963). Some authors went so far as to state "after 6 months the heterografts were more thoroughly remodeled than the autografts as judged by gross, radiographic, histologic and microradiographic criteria (Harmon 1968). Bassett and Creighton compared the host response to cortical "Boplant" xenografts with that to cortical autografts and noted little difference (Bassett and Creighton 1962). Hurley et al advocated use of these xenografts in spinal surgery, but noted cortical onlay grafts persisted as dense shadows in roentgenograms throughout as many as 4 years of observation (Hurley et al 1960). Scopp et al investigated "Boplant" implants for infraosseous oral lesions and reported favorable results with this graft compared to autogenous bone grafts (Scopp et al 1966). However, later experimental studies showed ethylenediamine-extracted freeze-dried bone to be the least satisfactory of many types of preserved bone studied (Heiple 1967). Finally the initial enthusiasm was tempered by long term clinical results which showed graft failures in over one-third of all patients. Postoperative pyrexia was also recorded in a large number of patients (Pieron et al 1968). The use of "Boplant" xenografts has been discontinued as a result of these findings and further investigation likewise ceased.

Bovine cancellous bone extracted by hydrogen peroxide and ethylenediamine (Kershaw 1963) was used clinically by Roaf who concluded that it was effective only when placed into a well vascularized bed. Under these circumstances it produced desired results when used in filling bone cavities, in posterior spinal fusions and as a supplement to autogenous grafts (Roaf 1979, Burwell 1969). The grafts prepared in this fashion did not evoke an abnormal tissue response in animals (Hancox 1961). However, this xenograft preparation was not produced commercially and thus was never subjected to extensive clinical evaluation. In general bovine xenografts, no matter how they were prepared, produced poor results. The general observation was that these grafts remained inert for many years (**Figs 3A and B**).

KIEL BONE

"Kiel" Bone is the only xenograft which has been available commercially for a long time and has been used clinically with some consistency. Either calves or young pig bones were used to prepare "Kiel" bone. The bone was cut, washed in distilled water for 24 hours, macerated and extracted by 20% hydrogen peroxide at 37°C for 48 hours with fat solvents, washed and dried in ether or acetone. The bone was then sterilized either by ethylene oxide or by gamma irradiation. This procedure did not deproteinize bone completely.

"Kiel" bone was developed by Maatz and Baumeister (Maatz and Baumeister 1961) concomitantly with ethylenediamine extracted bone. Baumeister claimed incorporation of Kiel bone on the basis of the "spongiosa test" (Maatz et al 1954), in which small plugs of bone to be tested were placed side by side in the cancellous bone of an experimental animal in a canine model. "Kiel" bone was reported to be superior

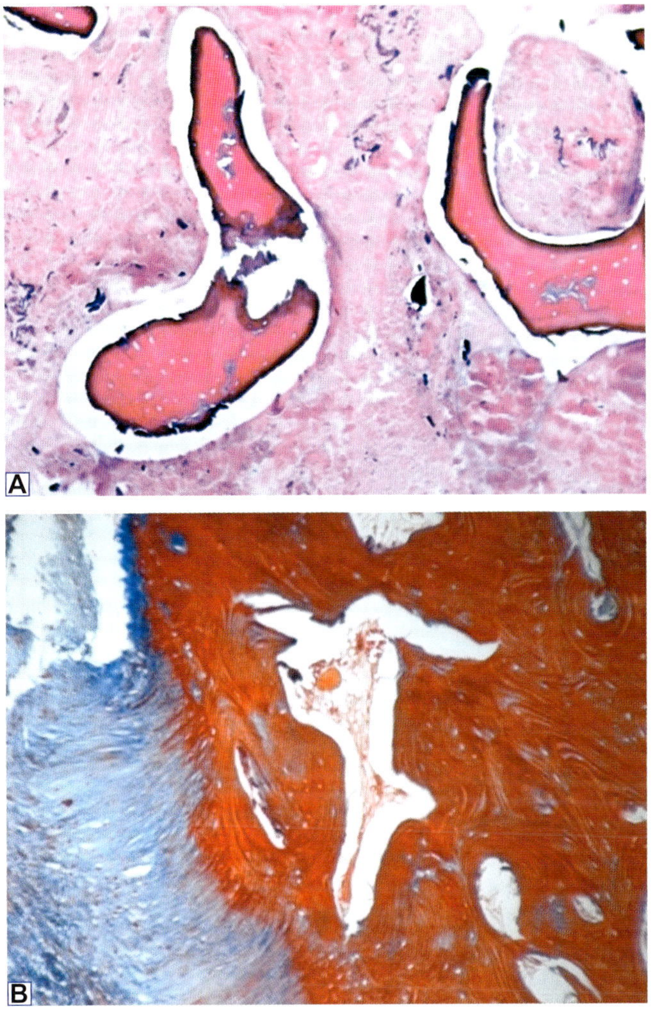

Figures 3A and B (A) Biopsy of the spinal fusion with deproteinized bovine xenograft 14 years after implantation. The bone remained unchanged. It is surrounded by fibrous connective tissue. (B) Edge of the graft showing attachment of connective tissue fibers but no osteogenesis (Masson trichrome × 100)

to autologous bone with respect to osteogenic activity (Baumeister 1958). The treatment with hydrogen peroxide was thought to reduce the antigenic properties of proteins by oxidizing the amino acid side chains (Hallen 1966). However, a rise in cytotoxic antibodies was demonstrated in animals receiving "Kiel" bone (Elves 1974). Immunogenic properties of preserved xenogeneic bone were demonstrated in rabbits (Tato 1979) Immunologic studies of human recipients suggested "Kiel" bone was weakly antigenic in the human host (Salama and Weissman 1978).

Experimentally, Hallen (Hallen 1966) showed that in dogs, "Kiel" bone did not differ appreciably from autologous or allogeneic transplants. However, the experiments were conducted in such a way as to provide optimal conditions for healing. On the other hand, when "Kiel" bone was placed into subchondral defects, osseous integration did not take place. Such was not the case when autologous bone was placed in an identical location (Mommsen et al 1979). The healing of the lateral wall of the maxilla was studied in the pig by Rosenquist et al (Rosenquist et al 1982). In this study, the effect on healing of "Kiel" bone and "AAX" bone (autolysed antigen extracted allogeneic bone) (Urist 1975) was compared with that of autografts (Urist 1983). It was found that both autograft and "AAX" bone implants improved the rate of healing, but "Kiel" bone had no effect. "Kiel" bone transplanted into the rabbit ilium was resorbed unless the xenografts were coated with a fibrin adhesive (Bösch et al 1980). "Kiel" bone was not osteoinductive when implanted into muscles of rodents (Schilling and Nockeman 1962).

Clinically, "Kiel" bone has been used primarily in Europe with varying degrees of success. Hopf recorded failures in 44% of 142 patients in whom "Kiel" bone xenografts were implanted. About half of these failed because of infections (Hopf and Cit Haasch 1963). However, generally lower failure rates, around 18%, have been reported by other investigators (Hallen 1966).

Successful use of "Kiel" bone xenografts for anterior cervical fusions has been described in a rather large number of patients (Taheri 1972, Sarang 1978). On the other hand, Ramani et al after a careful study of "Kiel" bone in the same operation concluded it had no place in orthopedic surgery (Ramani et al 1975) **(Fig. 4)**.

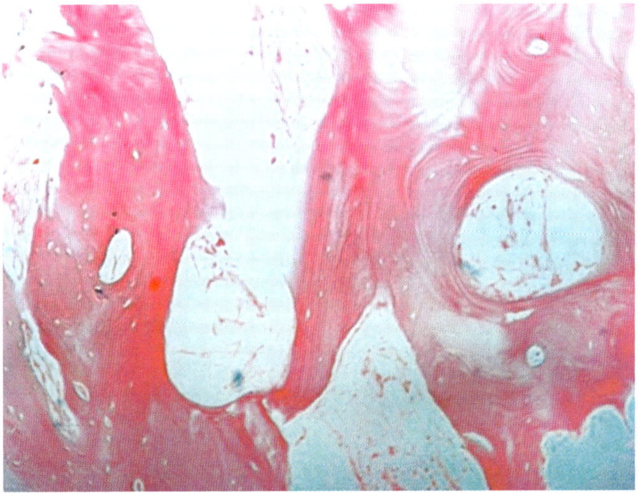

Figure 4 Biopsy of the fusion site with "Kiel" bone. The exact time after fusion is not known, but the patient stated it was "many years". The bone xenograft remains unchanged. There is no new bone formation (H&E × 100)

In other orthopedic procedures satisfactory results were described by some authors (Hallen 1966, Koch and Dahlmen 1962), but from these studies it became evident "Kiel" bone xenografts would tolerate only moderate compressive stresses and were unsuitable for bridging bone defects. Resorption of "Kiel" bone xenografts has also been reported (Antonutti 1968). Although "Kiel" bone was reported to yield good results in calcaneal osteotomy (Lemperg and Smith 1976) in the hand, tumor-like granulation was reported after transplanting "Kiel" bone for navicular pseudarthrosis (Bongert 1970).

After 10 years of observation Schweiberer concluded "Kiel" bone served no useful purpose as a graft material. In fact, it was felt to inhibit rather than promote new bone formation (Schweiberer 1970, Schwelberon et al 1981).

OSWESTRY BONE

Oswestry bone is similar to "Kiel" bone. Either sheep or calf bone is used as starting material. The bone is processed by the method described by Kenshaw (1963) using ethylenediamine and hydrogen peroxide which was supposed to deproteinize bone completely. When transplanted experimentally, oswestry bone grafts appeared as amorphous material containing plasma cells. Very little bone formation was elicited by these grafts (Salama et al 1973). Oswestry bone grafts elicited an immune response in sheep, indicating the grafts were not completely non-antigenic (Elves 1974). Some residual organic material remains in the "oswestry" bone despite its complete deproteinization (Wheeler and Hyatt 1960).

XENOGRAFTS COMBINED WITH AUTOLOGOUS BONE MARROW

When autologous red marrow is transplanted to a heterotopic site in rodents, it forms bone (Burwell 1961). This property of the bone marrow has been utilized for some time by combining it with allogeneic bone grafts, thus accelerating allograft incorporation and remodeling (Kline and Rimer 1983). Since results with transplantation of xenografts have been discouraging, Salama and Weissman thought that supplementing xenografts with autologous bone marrow would enhance their osteogenic potential. Preliminary clinical experience in 28 patients showed these investigators to be correct. Subsequently, Salama described his experience with 98 patients who received "Kiel" bone and bone marrow composite grafts (Salama et al 1973). The results were very satisfactory except in tibial and femoral pseudarthrosis. Similar results have been obtained by Graham (Graham 1982) insofar as large amounts of autologous bone marrow have been obtained by implanting xenogeneic bone into defects in the ilium and leaving them there for 10 weeks before grafting (Sarvary 1981).

If healing from os novum can be expected, the use of these composite grafts may well be justified. Otherwise, a direct autogenous bone graft would seem preferable.

It is of interest to note that Gelfoam did not provide a suitable framework for osteogenesis by marrow cells while xenogeneic bone did (Salama and Weissman 1978, Salama et al 1983). This would suggest a bone structure is needed to facilitate osteogenesis by autologous bone marrow cells. In this context, deproteinized xenogeneic bone may serve as a useful bone framework provided it is impregnated with living autologous marrow. However, in the same context we must not overlook the antigenicity of xenograft bone.

Ghormley, who espoused the use of xenogeneic bone screws, believed circumstances must be extraordinary to warrant the transplantation of bone xenografts (Ghormley 1942). Peer stated, "Considering the availability of frozen and preserved homogenous bone grafts when autogenous bone may not be used, there is little if any reason to use animal grafts in the human." Past experiences with xenogeneic bone grafts tend to substantiate this contention.

There are other problems with xenotransplantation, not the least of which is the choice of an animal as a source of bone grafts. In the past, bovine bone was used primarily for bone xenografts. However, cow may not be the best suited source of bone xenografts.

The most likely source of animal tissue to be transplanted into humans is the pig (Cooper et al 2007) due to the immunologic similarities between porcine and human tissues. Not only is the pig used due to antigenicity of bovine bone, but also because of the possibility of disease transmission from calf. Bovine spongioform encephalopathy is a case in point.

Pigs free from designated pathogens have been raised because of similarities or tolerance to porcine tissue. Despite controlled conditions, the improved understanding of the human immune system and years of research and observation of xenograft transplants in humans, xenogeneic bone grafts still present a problem posed by crossing the species barrier. Pigs have been approved by the FDA as a source of livers for a temporary life support until a human liver becomes available, but this requires immunosuppressive therapy, something neither patients nor surgeons would readily accept for bone grafting.

If the use of bone xenografts were to increase, it would raise additional concerns. Much attention has been directed to animal pathogens, particularly bovine spongioform encephalopathy (BSE). Past experience had shown that the causative agent of BSE had not been inactivated by irradiation. Thus, if present in a xenograft, it could infect the recipient. For this reason, in addition to other possible diseases, current American Association of Tissue Banks (AATB) standards prohibit the recovery of tissue for the purpose of transplantation if the donor had a history of prior xenograft transplantation. Should xenograft use become widespread, this would seriously impact tissue and blood donation. Endogenous retroviruses present in virtually all animal species would pose another threat (Cooper 2007).

Despite limitations of xenografts treated in various ways demonstrated during the last 50 years, there is, from time to time, a resurgence in investigation (Liu 2007, Luchetti 2007) and commercial promotion of these grafts. Collagen is still present in deproteinized bone (Liu 2007) as is nitrogen (Wheeler and Hyatt 1960) meaning "deproteinized" bone is not entirely inorganic. In patients receiving dental implants, deproteinized xenografts remain unchanged for a long time. Therefore the basic problem remains and unless something entirely new appears on the horizon, bone xenografts must continue to be approached with great caution.

From time to time xenogeneic grafts have been produced commercially with little information on their preparation and properties furnished to the surgeon **(Fig. 5)**.

Few human studies with xenografts, with the exception of spinal and maxilofacial applications, have been reported (Block and Poser 1993). Despite a recent concerned effort to improve the biologic utility of xenografts; immunologic rejection, prolonged integration and biomechanical incompatability still remain (Hancock 1997, Cook and Cook 2009, Bansal et al 2009). Significant potential risks also include possible pathogen transmission and delayed rejection with sterile fistula formation.

Despite the above listed problems and risks, xenografts are still used in the treatment of osseous defects, although

Figure 5 Strips of fascia lata prepared and distributed by the Ethicon Company in 1946. The source of fascia lata is not specified

in smaller numbers. Such is the case because of their ample supply and initial osteoconductive characteristics (Schweiberer 1971).

Xenograft dissolution is common in patients undergoing chemotherapy. Likewise, surprisingly, in elective procedures the infection rate is also higher than that with allografts.

Charalambides et al reported 12% infections with xenografts, while Levai and co-workers recorded 16% infection rate (Charalambides et al 2005, Levai et al 2003).

Failure of xenografts to incorporate after two and a half years was noted in over 10% of the patients. These authors felt that xenografts have unpredictable outcomes as well as a high incidence of complications. In view of this, transplantation of xenografts in humans is difficult to justify.

BIBLIOGRAPHY

1. Albee, Fred H. Evolution of bone graft surgery. American Journal of Surgery, 1944 ;63:421-36.
2. Anderson KJ, Dingwall JA, Schmidt J, et al. The effect of particle size of heterologous bone transplant and host tissue: histological study. Journal of Bone and Joint Surgery (Am), 1961;43:996-1006.
3. Anderson KJ, Le Cocq, Mooney JG. Clinical evaluation of processed heterogeneous bone transplants. Clinical Orthopaedics and Related Research, 1963;29:248-63.
4. Antonutti A, Labbadia S. On a case of resorption of the Kiel bone graft. Friuli Medicine, 1968;23:367-74.

5. Arbes H, Bösch P, Lintner F, et al. First clinical experience with heterologous cancellous bone grafting combined with the fibrin adhesive system (FAS). Archives of Orthopaedic and Trauma Surgery, 1981;98:183-8.
6. Baer WS. Arthroplasty with the aid of animal membrane. Journal of Bone and Joint Surgery, 1918;16:171-99.
7. Bansal MR, Bhagat SR, Shukla DD. Bovine cancellous xenograft in the treatment of tibial plateau fractures in elderly patients. International Orthopaedics, 2009;33:779-84.
8. Bassett CA, Creighton OKJ. A comparison of host response to cortical autografts and processed calf heterografts. Journal of Bone and Joint Surgery (Am), 1962;44:842-54.
9. Baumeister A. Experimentelle Grundlagen für den Aufbau einer neuen Knochenbank. Berlin: Springer-Verlag; 1958.
10. Baumeister A. Die Behaudlung von Zysten, Tumoren und enzandlichen Prozessen des Knochens mit dem "Kieler Knochenspan". Bruns Beiträge für Klinische Chirurgie, 1961;203:287-310.
11. Block JE, Poser J. Does xenogeneic demineralized bone matrix have clinical utility as a bone graft substitute? Medical Hypotheses, 1995;45:27-32.
12. Bongert H. Tumorartige Granalationsbildung nach Bolzuug einer Navikularpseudarthrose mittels Kieler Knochenspan. Handchirurgie, 1970;2:214-5.
13. Bösch P, Lintner F, Arbes H, et al. Experimental investigation of the effect of the fibrin adhesive on the Kiel heterologous bone graft. Archives of Orthopaedic and Trauma Surgery, 1980;96:177-85.
14. Brenizer AG. The use of intramedullary and extracortical beef bone splints in the repair of fractures of long bone. Surgery, Gynaecology and Obstetrics, 1920;30:209-10.
15. Burwell RG. A study of homologous cancellous bone combined with autologous red marrow after transplantation to a muscular site. Journal of Anatomy, 1961;95:513-24.
16. Burwell RG. The fate of bone grafts. In: Recent Advances of Orthopaedics. In: AG Apley (Ed). London: J and A Churchill; 1969. pp. 115-207.
17. Burwell RG. The function of bone marrow in the incorporation of a bone graft. Clinical Orthopaedics and Related Research, 1985;200:123-41.
18. Calvé J. De l'emploi du tissue spongieux hétérogène en chirurgie osseuse. Bulletin Societe National d'Chirurgie, 1935;61:1170-4.
19. Carrell WB. Cow horn fixation in bone surgery: its use in 40 cases. Surgery, Gynaecology and Obstetrics, 1936;63:636-9.
20. Charalambides C, Beer M, Cobb AG. Poor results after augmenting autografts with xenograft (Surgibone) in hip revision surgery: a report of 27 cases. Acta Orthopaedica, 2005;76:544-9.
21. Cook EA, Cook JJ. Bone graft substitutes and allografts for reconstruction of the foot and ankle. Clinical Podogeriatrics of Medical Surgery, 2009;26:589-605.
22. Cooper DKC, Ezzelarab M, Hara H. Xenotransplantation Encyclopedia of Life Sciences; 2007. pp. 1-6.
23. Dahl-Inversen E. Osteosynthesis mit resorbierbarem Material. Archiv für Klinische Chirurgie, 1930;680.
24. Danis MR. Les greffes auto-et heteroplatiques dans l'osteosynthese. Bull Acad Roy. Medicine Belgian, 1936;1:88.
25. Elves MW, Salama R. A study of the development of cytotoxic antibodies produced in recipients of xenografts (heterografts) of iliac bone. Journal of Bone and Joint Surgery, 1974;56(B):331-9.
26. Espersen JO, Buhl M, Eriksen EF, et al. Treatment of cervical disc disease using Cloward's technique. Acta Neurochirurgica, 1984;70:97-114.
27. Esnaurrizar ML. Heterogenous bone grafts. Journal of International College of Surgeons, 1940;3:151-6.
28. Fowler EB. Cow's horn for fixation of fractures: Its stimulating effect on callus formation and simplified technic. Illinois Medical Journal, 1934;231.
29. Gallie WE, Robertson DE. The repair of bone. British Journal of Surgery, 1919;7:211-61.
30. Gaudioso ET. Recerche experimentali suglinnesti ossee immediatie a distanza. Policlinico, 1924;31:735.
31. Ghormley RK. Choice of bone graft-methods in bone and joint surgery. American Journal of Surgery, 1942;115:427-34.
32. Goff CW. The os purum implant. A substitute for the autogenous implant. Journal of Bone and Joint Surgery, 1944;20:758-67.
33. Gongalsky VI. Some theoretical premises for use of heterobones in dyaphysary fractures. Journal of Orthopaedics and Traumatology Prosthetics, 1974;35:12-4.
34. Goran A, Murthy KK. Fracture dislocation of the cervical spine. Value of anterior approach with bovine bone interbody fusion. Spine, 1978;3:95-102.
35. Graham CE. Further experience with bone grafting of fractures using xenografts mixed with autologous red bone marrow. In proceedings of the Australian Orthopaedic Association. Journal of Bone and Joint Surgery (Br), 1982;64:123.
36. Groves EWH.Methods and results of transplantation of bone in the repair of defects caused by injury and disease. British Journal of Surgery, 1917;5:185-242.
37. Groves EWH. New bones for old. Lancet, 1939;1:69-72.
38. Hallen LG. Heterologous transplantation with Kiel bone. Acta Orthopaedica Scandinavica, 1966;37:1-19.
39. Hancock WW. Delayed xenograft rejection. World Journal of Surgery, 1997;21:917-23.
40. Hancox NM, Owen R, Singleton A. Cross-species grafts of deproteinized bone. Journal of Bone and Joint Surgery (Br), 1961;43:152-61.
41. Harmon P. Processed heterogenous bone implants (Boplant, Squibb) as grafts in spinal surgery. Acta Orthopaedica Scandinavica, 1964;35:98-116.
42. Heiple KG, Kendrick RE, Herndon CH, et al. A critical evaluation of processed calf bone. Journal of Bone and Joint Surgery (Am), 1967;49:1119-27.
43. Henderson MS. The use of beef bone screws in fractures and bone transplantation. Journal of the American Medical Association, 1920;74:715-7.

44. Hopf A, Cit Haasch K. Klinische Erfahrungen mit dem Kieler Span. Der Chirurg, 1963;34:21.
45. Hughes CH. Rate of absorption and callus stimulating properties of cow horn, ivory, beef bone and autogenous bone. Surgery, Gynaecology and Obstetrics, 1950;91:665-71.
46. Hurley LA, Zeier FG, Stinchfield FE. Anorganic bone grafting. American Journal of Surgery, 1960;100:12-21.
47. Judet J, Judet R. Greffes osseous animals en chirurgie humaine. Acta Orth Belg, 1953;19:135-9.
48. Katthagen BD. Bone regeneration with bone substitutes. Berlin-Heidelberg: Springer-Verlag; 1987.
49. Kershaw R. Preparation of anorganic bone grafting material. The Pharmaceutical Journal, 1963;8:537.
50. Kingma MJ. Results of transplantations with preserved calf bone. Archivum Chirurgicum Neerlandicum, 1960;12:221-35.
51. Kleinberg S. Spine fusion for scoliosis. Journal of Bone and Joint Surgery, 1929;11:66-80.
52. Kline SN, Rimer SR. Reconstruction of osseous defects with freeze-dried allogeneic and autogenous bone. Clinical and histologic assessment. American Journal of Surgery, 1983;146:471-3.
53. Koch W, Dahlmen G. Experimentelle und Kinkische Erfahrungen mit dem heterologen Knochenbanspan. Zentralblatt für Orthopaedie Orthopaedie, 1962;96:348.
54. König F. Uber die implantation von Elfenbein zumtion des os et de la production artificielle du tissue osseux Paris: Victor Masson et fils. 1913;1867. pp. 412-57.
55. Kronacher N. Casusistisches zur Heteroplastik Einheilung eines Kalbsknochensfukes in einin Defekt der 1 Phalanx des rechten Zeigefingers. Münchener Medizinische Wochenschrift, 1897;44:416.
56. Küttner D. Dauerresultete der Transplantation aus der Leiche und aus dem Affen. Verhandlung Deutsches Gesselschaft der Chirurgie, 1913;11:353-61.
57. Küttner D. Die Transplantation aus dem Affen un ihre Dauereerfolge. Münchener Medizinische Wochenschrift, 1917;64:1449-52.
58. Lacroix P. The Organization of Bone. London: J A Churchill; 1951.
59. Lemperg RK, Smith TW. The calcaneal osteotomy of Dwyer: an indication for Kiel bone. Acta Orthopaedica Scandinavica, 1976;47:125-8.
60. Levai JP, Bringer O, Descamps S, et al. Xenograft-related complications after filling valgus open wedge tibial osteotomy defects. Revue de Chirurgie Orthopedique et Reparatrice de l'appareil Moteur, 2003;89:707-11.
61. Loktev NI. Use of Kiel bone in plastic surgery of jaw defects. Vestnik Khirurgii, 1976;116:92-94.
62. Losee FL, Hurley LA. Bone treated with ethylenediamine as successful foundation material in cross-species bone graft. Nature, 1956;177:1032-3.
63. Losee FL, Boyne PJ. Response of oral tissues to grafts of ethylene diamine-treated heterogeneous bone. Nature, 1957;179:818.
64. Maatz R, Lentz W, Graf R. Spongiosa test of bone grafts for transplantation. Journal of Bone and Joint Surgery (Am), 1954;36:721-33.
65. Maatz R, Baumeister A. A method of bone maceration-results of animal experiments. Journal of Bone and Joint Surgery (Am), 1957;39:153-66.
66. Maatz R, Baumeister A. 36 Klinische Erfahrungen mid dem Kiel Span. Langenbecks Arch Chir, 1961;298:239-44.
67. Magnuson PB. Holding fractures with absorbable material: Ivory plates and screws. Journal of the American Medical Association, 1913;61:1514-6.
68. Malinin IM. Histological changes in transplants of bovine bone. In: Sokolv's Bovine Transplants in Plastic Surgery. Monographs of Kuban Medical Institute, Krasnodar. 1941.
69. McMurray GN. The evaluation of Kiel bone in spinal fusions. Journal of Bone and Joint Surgery (Br), 1982;64:101-4.
70. Meekeren van J. Heel en geneeskundige aanmerkingen. Commelijn; 1668.
71. Miller AG. A case of bone graft with decalcified chips. Remarks. Lancet, 1890;2:618-9.
72. Mommsen U, Scheer H, Jungbluth KH, et al. Reaktion des Gelenkknorpels auf le subchondrale Defektauffullung mit autologer spondiosa, Kieler Knochenspan und Knochenzement. Chir Forum Exp Klin Forsch, 1979;199-202.
73. Neuhof H. The transplantation of tissues. New York: Appleton and Co; 1923. p. 184.
74. Nyst PME. Purified beef bone. Nederl Tijdschr Geneesk, 1941;85:1428.
75. Oilier L. Traite experimental et clinique de la regeneration des os et de la production artificielle du tissue osseux. Paris: Victor Masson et fils; 1867.
76. Orell S. Surgical bone grafting with os purum, os novum and boiled bone. Journal of Bone and Joint Surgery, 1937;19:873-81.
77. Orell S. The use of os purum in bone implantation with special reference to its use in tuberculous bone and joint lesions. Surgery Gynaecology and Obstetrics, 1938;66:23.
78. Orell S. Implantation of autoplastic, homoplastic, and heteroplastic bone. Journal of International Surgery, 1953; 13:328-40.
79. Peer LA. Transplantation of Tissues. Vol I. Baltimore: Williams and Wilkins; 1955.
80. Pieron AP, Bigelow O, Hamonic M. Bone grafting with Boplant. Journal of Bone and Joint Surgery (Br), 1968;50:364.
81. Ramani PS, Kalbag RM, Sengupta RP. Cervical spine interbody fusion with Kiel bone. British Journal of Surgery, 1975;62:147-50.
82. Ray RD. Bone grafts and implants. Editorial Comment. Clinical Orthopaedics and Related Research, 1972;87:2.
83. Roaf R, Hancox N. Fate of heterogenous deproteinized bone implants. Journal of Bone and Joint Surgery, 1917;45(Br):617.
84. Rosenquist JB, Rosenquist K, Sund G. Effects of bone grafting on maxillary bone healing in the growing pig. Journal of Oral and Maxillofacial Surgery, 1982;40:566-9.
85. Salama R, Burwell RG, Dickson IP. Recombined grafts of bone and marrow. Journal of Bone and Joint Surgery (Br), 1973;55:402-17.

86. Salama R, Weissman SL. The clinical use of combined xenografts of bone and autologous bone marrow. Journal of Bone and Joint Surgery (Br), 1978;60:111-5.
87. Sarang I, Birosz B, Salamon A. Our experiences with the Kiel type heterlog conserved bond graft. Magyar Traumatologia, Orthopaedia es Helyreallito Sebeszet, 1978;21:89-94.
88. Sarvary A. A new way of producing autogenous cancellous bone graft. Magyar Traumatologia, Orthopaedia es Helyreallito Sebeszet, 1981;24:65-71.
89. Schilling H, Nockeman PF. Histologische Betrachtungen uber tier experimentelle und Klinische Erfahrungen mit einem heteroplastischen Knochenmaterial. Umfallheilkunde, 1962;65:227-40.
90. Schweiberer L. Experimentelle Untersuchungen von Knochentransplantaten mit unveranderter und mit denaturierter Knochengrundsubstanz. Hefte Unfallheilkunde: Springer-Heidelberg; 1970. p. 103.
91. Schweiberer L. New data on bone regeneration and their clinical significance. Langenbecks Archiv für Chirurgie, 1971;329:886-96.
92. Schweiberer L, Brenneisen R, Dambe LT, et al. Derzeitiger Stand der auto-hetero und homoplastischen Knochentransplantation. In: H Cotta, AK Marti (Eds). Implante und Transplante in der Plastischen und Wiederherstellungschirurgie. Springer-New York-Berlin-Heidelberg; 1986.
93. Scopp IW, Morgan FH, Dooner JJ, et al. Bovine bone (Boplant) implants for infrabony oral lesions. Journal of Periodontology, 1966;37:400-7.
94. Senn N. Healing of aseptic bone cavities with decalcified chips. American Journal of Medical Science, 1889;98:219-47.
95. Stagnara P, Dubost-Perret T. Greffes osseus: Transplants homogenes et heterogenes. Orthopaedic review journal, 1950;36:404-16.
96. Taheri ZE, Gueramy M. Experience with calf bone in cervical interbody spinal fusion. Journal of Neurosurgery, 1972;36:67-71.
97. Tato FB. Immunizing properties of a preserved xenogeneic bone graft in the rabbit. Chirurgia degli Organi di Movimento, 1979;65:707-10.
97. Tavernier L. A proposs des greffes d'os tue. Lyon Chirurgie, 1930;27:233.
98. Urist HR. Chemosterilized antigen-extracted surface-demineraliized autolysed allogeneic (AAA) bone for arthrodesis In: GE Friedlaender, HJ Mankin, KW Sell (Eds). Osteochondral allografts. Boston/Toronto, Little: Brown and Co; 1983. pp. 193-201.
100. Urist HR, Mikulski AJ, Boyd SD. A chemosterilized antigen extracted bone morphogenic alloimplant. Archives of Surgery, 1975;110:416-28.
101. Van der Noff HLM. Os purum from compact substance and preparation thereof. Acta Chirurgica Scandinavica, 1944;90:352.
102. Wheeler TE, Hyatt GW. A study of Residual Nitrogen in Bone Following Extraction with Ethylenediamine. Journal of Bone and Joint Surgery (Am), 1960;42:1435-41.
103. Williams JB, Irvine JW. Preparation of the inorganic matrix of bone. Science, 1954;119:771-2.
104. Wolf MM. Technic for ivory implant for correction of saddle nose. Archives of Surgery, 1938;37:800.

Chapter 8

Mats Brittberg MD (1953), Professor, Cartilage Research Unit, University of Gothenburg, Sweden. Dr Brittberg and his associates are pioneers in the transplantation of autologous chondrocytes for the healing of articular defects.

(*Reference:* Brittberg M, Lindahl A, Nillsön A, et al. Treatment of deep cartilage defects in the knee with autologous chondrocyte transplantation. New England Journal of Medicine. 1994;331:889-95.)

Transplantation of Autologous Cartilage and Chondrocytes

INTRODUCTION

According to many reports in the literature, autologous chondrocyte transplantation is a safe and efficient procedure for the treatment of isolated chondral lesions of the knee (Marlovits 2008). According to others the procedure produces a high rate of complications (Gomoll et al 2009, Niemeyer et al 2008), and not altogether satisfactory results (Minas 1998). The truth must be somewhere in between. In a group of patients enrolled in an international multicenter study, improvement was recorded in 73.6%, deterioration in 8.8% and unchanged status in 17.6% (Erggelet et al 2000). However, in evaluating autologous chondrocyte transplantation results, several factors must be taken into consideration. Age is one of these. Younger patients are more likely to obtain better results than their older counterparts. Location and size of the lesion are other factors influencing the final results (De Windt et al 2009).

Work with autologus chondrocytes dates to the 1980s. Initial experience with chondrocyte transplantation was performed in Gothenburg, Sweden by a group of surgeons which included Mats Brittberg and Lars Peterson (Brittberg et al 2003).

The initial goal of the experiments performed in Gothenburg was to increase the number of chondrocytes to have enough cells to fill a local defect in articular cartilage. This was necessary because of a small amount of cartilage which could be obtained arthroscopically. Taking larger cartilage pieces would have compromised the patient's knee, making the treatment worse than the disease. To this end, removed cartilage was dispersed and segregated with chondrocytes placed in cell culture to expand cell population and provide an adequate number of cells to fill the defect. Once a sufficient number of cells was obtained in a monolayer culture in which the chondrocytes were suspended in the culture medium, they were implanted into an articular defect. The defect was then covered with a periosteal patch. For convenience, the periosteal patch was sutured in place first with chondrocytes injected under it. A photograph of expanded chondroctyes in culture is provided in **Figure 1**.

One drawback of the autologous chondrocyte transplantation (ACT) is the time required to expand chondrocytes in culture. Once a lesion is judged suitable for ACT, a biopsy of healthy cartilage is taken. The most common sites for such biopsies are the intercondylar notch and the medial edge of the trochlea. Pieces of cartilage so obtained usually measure about 0.5–1 cm and contain approximately 300,000 chondrocytes. These must be expanded to some 12 million cells per culture flask and then suspended in 0.4 ml of culture

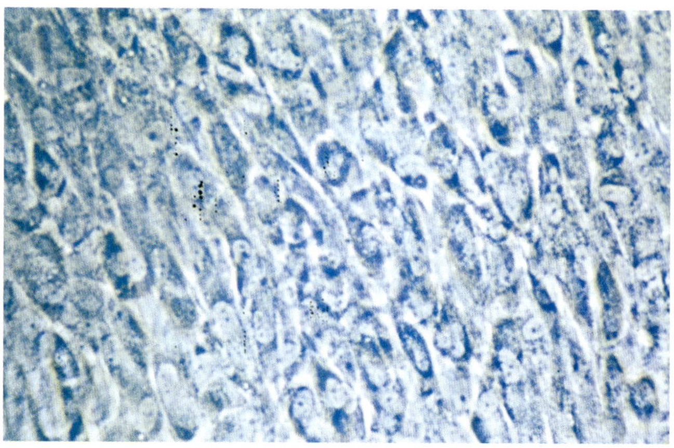

Figure 1 Expanded chondrocytes in culture (Phase contrast × 100)

media in an implantation vial (Minas and Peterson 1999). To achieve this goal requires 6 weeks or more. This means the patient must undergo two operative arthroscopic procedures, 6 weeks apart. Slight as it might be, each surgical procedure and anesthesia carry a certain risk.

The second problem lies with the cultured chondrocytes themselves. Enzymatic digestion of tissues is not innocuous. Even if everything goes well, it kills and injures a certain percentage of cells. In addition there is a natural attrition of cells in culture, as well as their alteration. Therefore, the biological characteristics and potency of expanded chondrocytes assume special significance (Marlovits 2008).

The loss of specific morphologic and physiologic attributes change with passages in cell culture. Modifications and changing of growth media have to date been able to modify these changes, but not abolish them. The expansion of chondrocytes results in dedifferentiation, a process which, according to Brittberg and Peterson (2003), reverts cultured chondrocytes to the fetal stage. Implantation of these cells into defects initiates cell condensation and cartilage formation in a manner similar to that occurring during formation of limbs. However, according to other authors, defects into which autologous chondrocytes have been implanted become filled with fibrocartilage and not hyaline cartilage (Horas et al 2003).

Clinical transplantation of autologous chondrocytes is a relatively recent development, but the interest in chondrocyte transplantation spans several decades. Andrey Smith isolated chondrocytes in 1965 and succeeded in freezing the cells (Smith 1965). Postnatal human chondrocytes were maintained and studied in tissue culture by Manning and Bonner in 1967 (Manning et al 1967). Shortly thereafter isolated chondrocytes were transplanted into chondral defects in rabbits. The latter healed (Chesterman and Smith 1968). These experiments provided impetus for further studies with chondrocyte transplantation and the development of human autologous chondrocyte transplantation.

Several other reports documenting the feasibility of chondrocyte transplantation in animals were published subsequently. Of these studies, those by Green (1977) were particularly noteworthy. He showed that under appropriate conditions rabbit chondrocytes maintained their chondrocytic differentiation in tissue culture. These produced chondroid tissue which superimposed on decalcified bone and repaired a large articular defect in 10 days! But this was not the transplantation of isolated chondrocytes, just one of the earlier examples of tissue engineering.

Results with the experiments by the Gothenburg group were published in 1984 (Peterson et al 1984) and 1989 (Grande et al 1989). These provided the technique for clinical transplantation of chondrocytes in humans.

Human autologous chondrocytes from primary monolayer cultures were first implanted into patients in 1987 (Brittberg). This led to a wide interest in this therapeutic modality which continues to be investigated to this day. The entire procedure depends on the expansion (culture) of chondrocytes. The technique of chondrocyte isolation and culture has been described in detail by Brittberg et al (1994) as follows.

Cartilage slices weighing 300–500 mg are obtained from a minor load bearing area of the upper medial condyle of the damaged knee. Cartilage samples are placed into cold saline and the isolation procedure is started within 5 hours. Cartilage specimens are minced and washed × 3 in Ham's F-12 medium with 10 mmol per liter of HEPES buffer. About 50 μg of gentamicin, 2 μg/ml of amphotericin B and 50 μg/ml of ascorbic acid were also added to the medium. Minced cartilage was digested in a spinner bottle for 16 hours in culture medium containing clostridial collagenase (1 mg/ml) and deoxyribonuclease (0.1 mg/ml). The suspension was then filtered through nylon mesh with a pore diameter of a micron. Cells were then washed three times, and counted. Total count ranged between 180,000 cells and 455,000 cells. Cells were then resuspended in culture medium supplemented with 15% of autologous serum, and seeded at a cell density of 5,000–10,000 cells per cm^3 in 25 cm^2 or 75 cm^2 culture flasks. The culture medium is changed twice weekly.

Cultured cells are transplanted 14–21 days after initial surgery. Chondrocytes are detached from the monolayer with trypsin, pelleted in a centrifuge and washed × 3 with medium containing 20% autologous serum. The final volume of cell suspension is 50–100 microliters containing 2.5–5 million cells. At the time of the initial description of the procedure, chondrogenic phenotype was determined by the evaluation of clonal growth. Now, some 15 years later, the chondrocytic phenotype is determined by the use of more precise techniques. These include determination of type II collagen, aggrecan, cartilage matrix protein, chondrocyte expressed protein-68, osteonectin, Sox 9 and others. By employing these techniques the preservation of the chondrocyte phenotype or its loss can be determined.

Initially, chondrocyte transplantation as pioneered by Brittberg and colleagues was combined with transplantation of a periosteal patch which covered the defect. However, subsequently it was shown that the rate of complication is higher with periosteum than without it. As an alternative, a synthetic patch can be used (Niemeyer 2008). The technique for chondrocyte culture had been also modified with some surgeons transplanting autologous chondrocytes as late as 6 weeks after initiation of cultures. Whether or not delay with preparation of autologous chondrocytes, accompanied by the likelihood of dedifferentiation, has a bearing on outcomes is not known. The surgical technique for the autologous chondrocyte implantation has been described in a number of publications (Peterson 1984, Minas 1998, Marlovits 2008).

The procedure, as currently practiced, includes arthrotomy, debridement of the defect, excision of a periosteal patch, usually from the tibia, periosteum suturing over the defect, fibrin sealant, implantation of chondrocytes and wound closure.

In conclusion, when 15 years experience with autologous chondrocyte transplantation is viewed in toto, the procedure appears to offer an efficient and safe option in the treatment of circumscribed, isolated articular defects. The best results have been reported in post-traumatic lesions and osteochondritis dissecans.

Surgical expertise, the patient selection and the ability to expand chondrocytes *in vitro* contribute directly to the success of autologous chondrocyte transplantation.

Osteochondral Autografts for Transplantation into Full Thickness Articular Defects Extending into Subchondral Bone

Two forms of osteochondral autografting are employed: mosaicplasty and a single osteochondral cylinder, otherwise known as osteochondral autograft transplantation systems (OATSR). Both techniques rely on the transfer of cylinders of articular cartilage from non-load-bearing portions of the knee to the weight-bearing portions of the knee joint. There are no human studies which would demonstrate whether or not cartilage with subchondral bone derived from non-weight bearing surfaces of the knee can withstand loads to which load-bearing knee cartilage is subjected (Redman et al 2005). The extent of morbidity associated with osteochondral autograft excisions is not known.

In the laboratory, clinically, transplantation of autologous osteochondral plugs is successful and technically feasible. Both short-term and long-term animal studies show survival of osteochondral autografts transplanted into weight bearing areas **(Figs 3A and B)** (Malinin et al 1994, Malinin and Ouellette 2000).

Autologous grafts have been limited by the small size of available donor sites. Mosaicplasty can be performed either arthroscopically or by a mini-arthrotomy. The most advantageous site for autograft excision is the periphery of the medial femoral condyle above the sulcus terminalis. Placement of the tubular harvester perpendicular to the surface is mandatory. To prevent dislodgement, the minimal length of the graft should be at least double its diameter. After the grafts are cut, these are ejected from the round cutting chisel with a tamp. The grafts are then inserted into the preprepared defect. A photograph of mosaicplasty autografts is shown in **Figures 2A and B**.

After the tunnels in the recipient site are created, the harvested grafts are press-fitted individually into each tunnel. Donor sites are usually left open. They eventually fill with cancellous bone and fibrocartilage. Advocates of mosaicplasty believe the procedure reduces donor site morbidity and results in an even articular surface (Scapinelli et al 2002).

The osteochondral autograft transfer system procedure is similar to mosaicplasty, except only one cylinder, rather than multiple plugs is used to fill the prepared defect. It is performed, often in combination with anterior cruciate ligament (ACL) replacement.

Clinical studies with autologous osteochondral transplants are relatively few, although several authors state that joint function and symptoms as a consequence of osteochondral autografting review of scientific literature lacks controlled clinical trials. Therefore, it is difficult to recommend osteochondral autografting as the treatment of choice for isolated full thickness articular defects. On the other hand, Horas et al (2003) noted that improvement provided by the autologous chondrocyte implantation lagged behind that offered by osteochondral cylinder transplantation. They noted that the defects filled with cultured autologous chondrocytes were primarily filled by fibrocartilage.

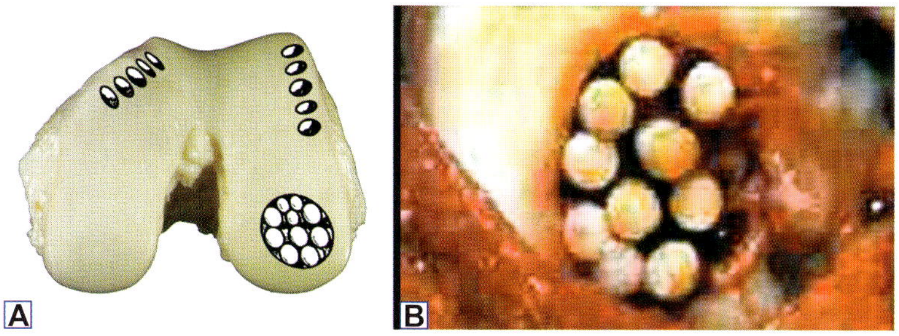

Figures 2A and B (A) Schematic representation of mosaicplasty procedure; (B) Intraoperative photograph of mosaicplasty with autologous transplants

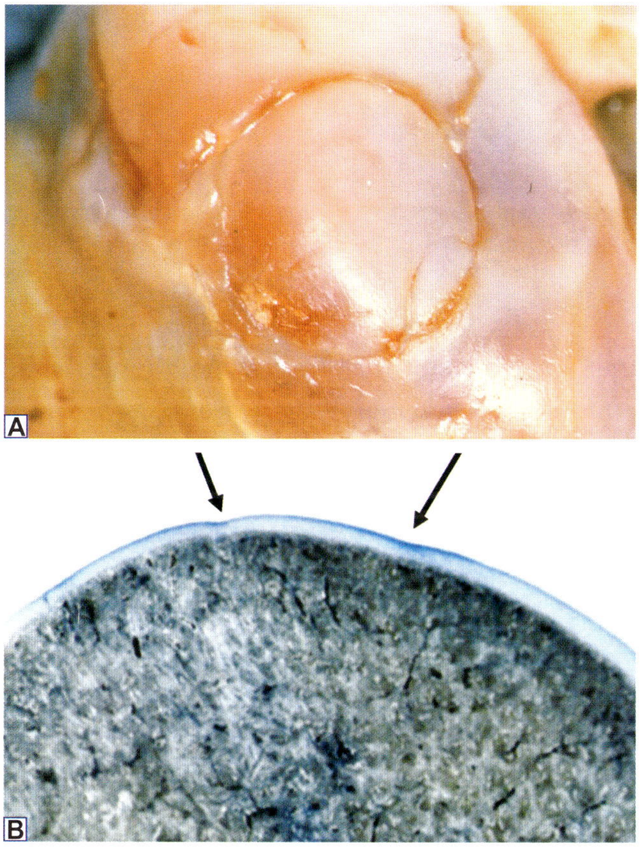

Figures 3A and B (A) Autologous osteochondral plug placed into an experimental animal (dog) 6 months post-transplantation. Gross appearance of the cartilage is normal; (B) Section of osteochondral transplant in an animal (baboon) 5 years post-transplant. The specimen is injected with India ink. The osseous portion of the autograft is completely revascularized. The cartilage plate appears normal, but the host-graft junctions are identifiable

Figure 4 Histologic section of autologous transplant showing normal histologic architecture (Romanowski-Giemsa stain × 250)

Osteochondral transplants retained their hyaline cartilage character, but the interface between the cartilage of the host site and the implant was filled with fibrocartilage.

Evaluation of patients with mosaicplasty with data gathered by one investigator revealed 92% good to excellent results in femoral condylar implantation cases, 88% of tibial resurfacements and 81% of patellar or trochlear implants (Hangody et al 2002).

Fibrocartilage at the interface between the original cartilage and the cartilage of the transplant was also observed in animals as long as 5 years post-transplantation (Malinin and Ouellete 2000). Despite the presence of fibrocartilage at the cartilage junction, the cartilage in the transplant remained viable during the entire period of 5 years observation (**Fig. 4**). Subchondral bone was revascularized and completely regenerated. These, as well as previous experimental studies (Malinin et al 1985), support the concept of clinical transplantation of autologous osteochondral grafts and can be offered as an addition to limited clinical results with these grafts. However, it must be recognized that graft-host junctions become united by fibrocartilage and not regenerated hyaline cartilage (**Figs 5A and B**).

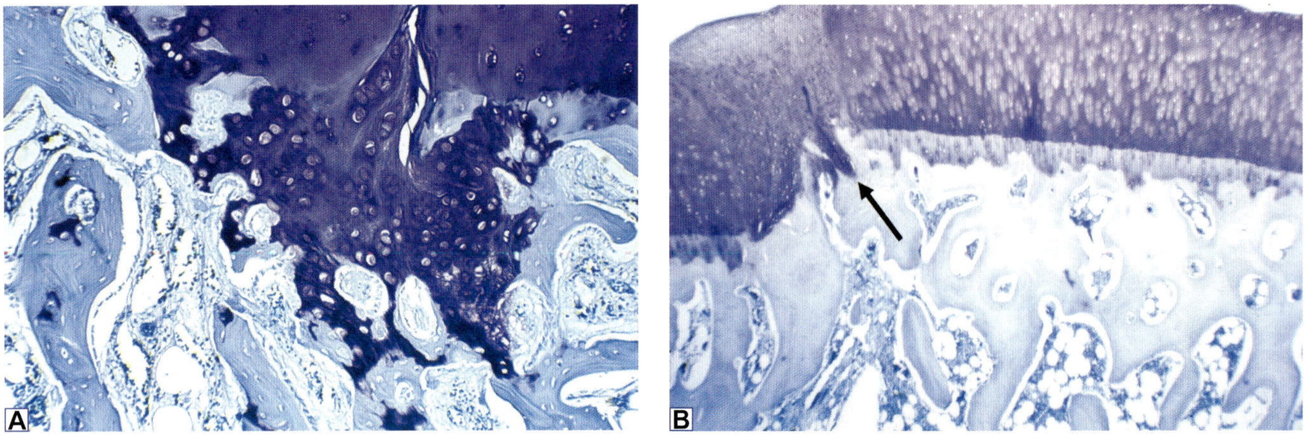

Figures 5A and B (A) The formation of hyaline cartilage in the subchondral bone in response to injury, (Romanowski-Giemsa stain × 250), (B) Healed host-graft junctions composed of fibrocartilage (Romanowski-Giemsa × 100)

BIBLIOGRAPHY

1. Brittberg M, Lindahl A, Nilsson A, et al. Treatment of deep cartilage defects in the knee with autologous chondrocyte transplantation. New England Journal of Medicine, 1994;331:889-95.
2. Brittberg M, Peterson L, Sjogren-Jansson E, et al. Articular cartilage engineering with autologous chondrocyte transplantation. Journal of Bone and Joint Surgery (Am), 2003;85(supp 13):109-23.
3. Chesterman PJ, Smith AU. Homotransplantation of articular cartilage and isolated chondrocytes. An experimental study in rabbits. Journal of Bone and Joint Surgery (Br), 1968;50:184-97.
4. De Windt TS, Bekkers JEJ, Creemers LB, et al. Patient profiling in cartilage regeneration. American Journal of Sports Medicine, 2009;37(supl I):585-625.
5. Erggelet C, Brown JE, Fu F, et al. Autologous chondrocyte transplantation for treatment of cartilage defects of the knee joint. Clinical results. Zentralblatt für Chirurgie, 2000;125:516-22.
6. Gomoll AH, Probst C, Farr J, et al. Use of a type I/III bilayer collagen membrane decreases reoperation rates for symptomatic hypertrophy after autolgous chondrocyte implantation. American Journal of Sports Medicine, 2009;37 (supp I):215-35.
7. Grande D, Pittman MI, Peterson L, et al. The repair of experimentally produced defects in rabbit articular cartilage by autologous chondrocyte transplantation. Journal of Orthopaedic Research, 1989;7:208-18.
8. Green WT. Articular cartilage repair, Behavior of rabbit chondrocytes during tissue culture and subsequent allografting. Clinical Orthopaedics and Related Research, 1977;124:237-50.
9. Hangody L, Duska Z, Karpati Z. Autologous osteochondral mosaicplasty. Techniques in Knee Surgery, 2000;1:13-22.
10. Horas U, Pelinkovic D, Herr G, et al. Autolgous chondrocyte implantation and osteochondral cylinder transplantation in cartilage repair of the knee joint. Journal of Bone and Joint Surgery (Am), 2003;85:186-92.
11. Malinin T, Ouellette EA. Articular cartilage nutrition is mediated by subchondral bone: a long term allograft study in baboons. Osteoarthritis and Cartilage Journal, 2000;8:483-91.
12. Malinin TI, Mnaymneh W, Lo HK, et al. Cryopreservation of articular cartilage. Clinical Orthopaedics and Related Research, 1994;303:18-32.
13. Malinin TI, Wagner JL, Pita JC, et al. Hypothermic storage and cryopreservation of cartilage. Clinical Orthopaedics and Related Research, 1985;197:15-26.
14. Manning WK, Bonner WM. Isolation and culture of chondrocytes from human articular cartilage. Arthritis and Rheumatism, 1967;10:235-39.
15. Marlovits S. Autologous chondrocyte cartilage repair. In: WS Pietrzak (Ed). Musculoskeletal Tissue Regeneration: Human Press; 2008. pp. 369-94.
16. Minas T, Peterson L. Advance techniques in autologous transplantation. Clinics of Sports Medicine, 1999;18:13-44.
17. Minas T. Chondrocyte implantation in the repair of chondral lesions of the knee: economics and quality of life. American Journal of Orthopaedics, 1998;27:739-94.
18. Niemeyer P, Pestka JM, Kreuz PC, et al. Characteristic complications after autologous chondrocyte implantation for cartilage defects of the knee joint. American Journal of Sports Medicine, 2008;36:2091-9.
19. Peterson L, Menche D, Grande D, et al. Chondrocyte transplantation- an experimental model in the rabbit. Transactions of Orthopaedic Research Society, 1984;9:218.
20. Redman SN, Oldfiel SF, Archer CW. Articular cartilage repair. European Cells & Materials Journal—Open and Free Access. Swiss Society for Biomat. The Official Journal of AO Trauma, 2005;9:23-32.
21. Scapinelli R, Aglietti P, Baldovin M, et al. Biologic resurfacing of the patella: current status. Clinical Sports Medicine Journal, 2002;21:547-73.
22. Smith AU. Survival of frozen chondrocytes isolated from cartilage of adult mammals. Nature, 1965;205:782-4.

Chapter 9

Allan E Gross MD (1938), Professor of Orthopaedics, University of Toronto; Mt. Sinai Hospital, Toronto, Canada is a pioneer in transplantation of allogeneic articular cartilage. His contribution and that of his associates paved the way for general acceptance of clinical transplantation of cartilage allografts.

(*Reference:* Gross AE, Beaver RJ, Zukor, et al. The use of fresh osteochondral allografts to replace traumatic joint defects. In: Orthopeadic Allograft Surgery; AA Czitrom and H Winkler (eds). Springer, Wien, New York. 1996. pp. 281-92.)

Transplantation of Allogeneic Cartilage

INTRODUCTION

Transplantation of articular cartilage allografts has an extensive clinical history. It is not new. Despite this, the procedure has achieved only qualified successes. After Lexer's experience with massive fresh osteochondral allograft transplantation, the procedure remained dormant until it was adopted for the treatment of articular defects by Gross, Meyers and associates (Lexer 1908, Gross et al 1975, 1983, Meyers 1985, Pritzker et al 1977).

Transplantation of massive osteochondral allografts in oncologic surgery is dealt with in Chapter 12. In the interim, the interest in chondral transplantation has not been lost as attested by the publication of the results of a number of experimental studies (Pap and Krompecher 1961, De Palma et al 1963, Campbell et al 1963, Lance and Fisher 1970). The inherent issue with transplantation of fresh osteoarticular allografts has been tissue acquisition and storage. The feasibility of storing viable articular cartilage in tissue culture media at hypothermic temperatures has been demonstrated by Shadle and Brighton in 1977 (Shadle and Brighton 1977). Some programs and full-service tissue banks began providing osteochondral allografts stored in tissue culture media at hypothermia since the mid eighties (Malinin et al 1985), but on a limited basis. These grafts are available in limited quantity because there are, among other things, few qualifiable donors. Chondrocyte viability of allografts stored at hypothermia in cell culture media is time dependent. Twenty one days appears to be the upper storage time limit (Malinin et al 1985). The need to exclude possible microbial contamination further limits the time window in which these grafts can be distributed for transplantation. Careful donor selection, aseptic recovery and effective microbial surveillance mitigate safety concerns in graft recipients.

Historically, logistical challenges presented by the acquisition and preparation of fresh osteochondral allografts have limited their availability to a few programs and tissue banks with dedicated highly trained personnel and resources committed to the recovery and processing nuances of these tissues. The lack of specialized training and scientific inquiry in some cases has resulted in the distribution of suboptimal fresh grafts for clinical use. These grafts were so rendered due to unreasonably extensive lengths of storage, which diminish articular cartilage viability to a degree making them suboptimal for transplantation. These smaller osteochondral allografts designed to fill a single defect were usually cylindrical in shape. Larger grafts with a thin subchondral bone shelf were transplanted fresh. This subchondral bone shelf served to reduce the number of immunogenic cells present in the bone and bone marrow. Limiting the age of the donor and visually ascertaining the quality of the cartilage were additional factors contributing to the quality of cartilage transplants. Visual storage time related changes, however, generally occur late as we have shown in *in vivo* and *in vitro* studies corroborating optimal storage times for chondrocyte viability under 21 days. The availability of good quality allografts is essential to the success of cartilage transplantation (Ahmed and Hincke 2010).

GENERAL CONSIDERATIONS

Lack of vascularity, lymphatics, nerves and viable cells capable of effecting tissue repair limits the capacity of articular cartilage to regenerate in an adult. Response to articular cartilage injury has been described in detail (Buckwalter 2002). Tears or cuts limited to the cartilage layer alone do not heal. Tears, fractures and cuts penetrating into subchondral bone heal by forming fibrocartilage composed primarily of collagen type I fibers. Normal hyaline cartilage surface is not restored.

Osteochondral allografts are transplanted to restore smooth articular surfaces. The transplants derive their nutrition through the diffusion of the synovial fluid and by re-establishment of subchondral blood vessel networks (Malinin et al 1985). The subchondral bone of the allograft offers both advantages and disadvantages.

Articular cartilage transplants contain variable amounts of subchondral bone. Transplanted bone can be either non-viable or viable. Non-viable subchondral bone must be replaced by the host as is the case with all bone transplants (Rodrigo et al 1987). Viable bone, which includes bone marrow, may provide antigenic stimulation with a subsequent immune response.

There are several theoretical and practical advantages of osteochondral allografts over other methods of articular cartilage restoration, such as osteochondral autografts, autologous chondrocyte implantation and bone marrow stimulation (Latterman and Romine 2009). The principal advantage is transplantation of a normal biologic structure with viable cells capable of maintaining extracellular matrix (Czitrom et al 1990, Convery et al 1996).

At this time, no other method of biologic repair can achieve this. The clinical advantages of osteochondral allografts include availability, preparation of the grafts in a variety of sizes, avoidance of donor site morbidity, and decrease in operative time and blood loss. Disadvantages include potential immunological reaction, at times limited availability and a theoretical risk of disease transmission.

Transplantation of fresh osteochondral allografts is a unique and frequently effective treatment for articular cartilage defects. The success of the procedure depends, in a large measure, on the availability and quality of the transplant. To utilize the technique to its full advantage, the surgeon must understand the process of tissue excision, graft preparation and storage.

The reliance on fresh (unfrozen) osteoarticular allografts was precipitated by disappointing experimental and clinical results with frozen allografts independent of freezing methods. Although isolated chondrocytes can be successfully frozen, chondrocytes embedded in normal matrix of articular cartilage cannot. Cryopreserved osteochondral allografts, under the best of circumstances, undergo latent injury and in time, hyaline cartilage transitions into fibrocartilage. Fibrocartilage does not have the viscoelastic properties of normal hyaline cartilage and eventually results in fraying and deterioration of the articular surface, ultimately resulting in joint arthrosis.

The technique known as microfracture involves removing the damaged articulate cartilage down to the subchondral bone and drilling small holes into it. This results in gradual filling of the defect with fibrocartilage. The addition of expanded autologous chondrocytes (Carticel™) into a cartilage defect and sealing it with a membrane is expensive and devoid of extensive controlled clinical data.

Thus, the fresh cartilage allograft is the only viable option, pending the development of newer, perhaps better techniques that combine cells, matrices and growth factors to achieve a durable hyaline cartilage regenerative substrate. Before implantation, allografts must be matched to the recipient (Görtz and Bugbee 2007, Chu et al 1999, Convery et al 1996). This is done on the basis of size and ideally the radius of curvature. For the knee, the most convenient way to achieve this is by obtaining anteroposterior radiographs of the knee with a magnification marker for analog radiography or more commonly today, utilizing digital radiography. The medial-lateral dimension of the tibia or femoral condyle is measured in a standard fashion. The use of thin slice CT and 3-D reconstruction formatting may be more accurate in assessing size and radius of curvature of the defect and selecting the closest graft match. Condyles can also be measured directly at the time of recovery or, more appropriately at, processing (Highgenboten and Jackson 1994).

Osteoarticular allografts are uniquely suited for the treatment of complex articular cartilage defects (Farr et al 2011). Fresh osteochondral allograft transplantation is indicated for the treatment of lesions for which other forms of treatment may be inadequate (Görtz and Bugbee 2007). Conditions believed to be most amenable to treatment by transplantation of osteochondral allografts include post-traumatic defects (Shasha et al 2003), osteochondritis dissecans (Garret 1994) osteonecrosis, (Emmerson et al 2006), and patellofemoral arthrosis (Jamali et al 2005). Osteochondral allografts have also been used for the salvage of knees in which other cartilage resurfacing procedures have failed.

Allografts have been used successfully in the ankle and the hip joints. In the hip, osteochondral allografts were first pioneered by Meyers in the treatment of osteonecrosis (Meyers 1985). The results were mixed in this anatomic site where surgical technique is demanding and postoperative management is long and arduous. Furthermore, the success depends not only on the healing of the osteochondral shell allograft, but also on the healing of the drilled out core which typically is packed with autologous bone graft. Current indications include symptomatic lesions of relatively small size that have not been amenable to other forms of treatment. The operation is performed largely in younger people in an attempt to avoid hip arthroplasty. Careful preoperative imaging and, more recently, hip arthroscopy may aid in determining the size of the defect, noting reciprocal changes in the acetabulum and assessing associated labral pathology. Depending on location and size of the defect, the procedure may be combined with a femoral osteotomy to unload the repaired articular surface and to restore the hip biomechanics.

In the ankle joint, osteochondral allografts have been used for resurfacing of tibiotalar joint arthrosis (Gross et al 2001, Meehan et al 2005), osteonecrosis of the talus, and for osteochondritis dissecans (Tasto et al 2003).

Osteochondral allograft reconstruction has been performed in the shoulder and other joints, but there is scant clinical data on the subject.

With regard to the transplantation of osteochondral allografts, it is becoming obvious that despite the advent of several operative procedures, new products and cell therapies designed to restore articular cartilage, osteochondral allografting is currently the only procedure that has been clinically proven to restore articular cartilage in an adult. The success of the procedure depends on the long-term survival of chondrocytes within the allograft; this has been demonstrated by several investigators (Convery et al 1996, Jamali et al 2007, McGovern et al 2002). This, of course, depends on the presence of an adequate number of viable chondrocytes in the transplant to begin with. Therefore, the method of the preparation and storage of osteochondral allografts assumes particular importance.

Finally, before contemplating transplanting osteochondral allografts, one must consider indications and contraindications for the procedure. Osteochondral allografting is indicated primarily for the treatment of full thickness defects of 1 cm or larger in younger individuals. These defects usually occur secondary to trauma. Other conditions, innumerated earlier, such as osteochondritis dissecans, etc. has been also successfully treated with osteochondral allografts. Although allografting has been performed with partial success in unicompartmental arthritis, there is at this time no place for this procedure in the treatment of patients with multicompartmental osteoarthritis (Görtz and Bugbee 2007).

Shell Allografts

Shell allografts are carefully sculpted to match the articular surface of the observed and carefully measured cartilage defect. Each case is different. Once the defect is defined, the allograft is shaped using a freehand technique to fit the defect while minimizing damage to the normal adjacent cartilage. Subchondral bone thickness is maintained at a minimum. Initially the graft is slightly oversized. Excess bone and cartilage are then removed, and the graft is fashioned to the right size through multiple trimmings and fittings. Once the graft is shaped, it is placed in the defect flush with the articular surfaces of the host. Fixation is based on the stability of the new construct. When fixation is required, bioabsorbable pins or screws are used.

Obviously, the difficulty with this technique is the demand it places on the artistic skill and the manual dexterity of the surgeon. For this reason, many surgeons do not choose to perform operations involving transplantation of shell allografts.

The success with fresh osteochondral shell allografts must be attributed primarily to the efforts of Gross and his associates (Gross et al 2005), who in 1972 transplanted their first fresh osteochondral allograft for a post-traumatic lesion of the tibial plateau. Their long-term prospective study confirmed the value of these allografts in the reconstruction of articular defects of the knee in the young active patients. In the latest study, these authors reported 95% graft survival at 5 years, 80% at 10 years and 65% at 15 years. They have also demonstrated that in older patients, bipolar transplants, improper loading of the graft and grafts for osteoarthritis and steroid induced avascular necrosis have less predictable and, in general, poorer clinical outcomes (McDermott et al 1985, Ghazavi et al 1997).

Dowel Allografts

The success of dowel allografts depends on instruments used for their insertion. Several proprietary instrumentation systems are available for the implantation of press-fit allografts up to 3.5 cm in diameter.

Dowel allografts, the most frequently transplanted allograft, are inserted either in single or multiple cylindrical grafts. One surface of the graft consists of full-thickness cartilage, while the remainder of the cylinder is made up of compact cancellous bone.

Unlike shell allografts, dowel allografts are retained in place by press-fitting. They do not require additional fixation (Farr and DeNovo 2010).

Results with fresh dowel osteochondral allografts in the knee have been encouraging (Bugbee and Convery 1999, Williams et al 2007, Garret 1994). However, success rates reported by different authors range from 10 to 90% (Henning and Abate 2007).

Although results with osteochondral allografts have been promising, the conclusions must be guarded because of the newness of the procedure and the relatively small number of patients included in the studies (Farr and Yao 2010). Ghawazi et al (1997) reported on 123 patients with a mean age of 35. In these patients, allografts were used to treat post-traumatic defects. The authors recorded 85% success rate. Davidson et al (2007) in a study of 67 patients with full thickness condylar lesions found improvements in the International Knee Documentation Committee and SF-36 scores. The International Cartilage Repair Society scores (ICRS) were almost normal. Similar improvements were reported by the other authors. However, despite the positive reports the success of the procedure must be included in the "qualified" category. Therefore, until rigorous clinical trials become available, decisions regarding the transplantation of fresh osteochondral allografts must be guided by available data and individual patient needs (Farr and Yao 2010).

Allograft failures have occurred both early and late. Early failures may be related to chondrocyte viability in the transplanted cartilage, improper indications and errors in techniques (inadequate fixation and size mismatching). Late failures have been associated with the fracture of the graft, incomplete healing at the host-graft junction and graft resorption (Görtz and Bugbee 2007). Some of the failures can be attributed to poor selection of patients early on in the history of osteochondral dowel transplantation.

Most of these failures occurred in the knees of patients with diffuse osteoarthritis, inflammatory disease and uncorrected instability and misalignment. Accumulated clinical experience would now exclude these patients. Currently osteochondral allografts are performed primarily in younger patients with focal articular cartilage defects in one or more surfaces of the knee (Bugbee 1999).

The use of cryopreserved or frozen osteochondral allografts would greatly simplify the procedure because one would not have to wait for the availability of fresh grafts. However, attempts in this direction have met with consistent failures. Studies have shown that the number of viable chondrocytes is dramatically reduced in cryopreserved material (Malinin et al 1985, Rodrigo et al 1987). The injury inflicted on hyaline cartilage takes place at the time of freezing. It is neither produced nor enhanced by storage at cryogenic temperatures (Malinin et al 1994).

Excision and Preparation of Fresh Osteochondral Allografts

The excision, preparation, storage and distribution of fresh osteochondral allografts must be carefully monitored. The process must guard the safety and integrity of the grafts. Transmission of infections with allografts used in knee reconstruction must be avoided at all costs (CDC 2001). This can be accomplished by careful microbiologic surveillance. If the samples are inoculated into blood culture media, the definitive results are available in 7 days. Therefore, there is little reason to delay the release of fresh allografts much beyond this time frame.

Dowel allografts are the most commonly transplanted and clinically successful osteochondral allografts **(Fig. 1)**. A single precisely shaped dowel can be placed at an angle to fit the contour of the recipient's condyle. The press-fit transplant has an advantage of not requiring fixation and leaving no space for the ingrowth of fibrocartilage.

Dowel allografts are usually 8–15 mm in diameter and are about 15 mm in length. Unlike shell allografts, dowel allografts do not require additional fixation, but must be inserted tightly. This limits the anatomical implantation to sites of the posterior femoral condyles, trochlea, tibia and patella. Sometimes, normal cartilage and subchondral bone must be sacrificed to obtain a perfect fit **(Figs 2 to 4)**.

Shell allografts are shaped free-hand. Grafts must be matched to the existing defect, minimizing the loss of normal articular cartilage. A condyle, an entire distal femur, a half tibial plateau, or the entire distal tibia or talus is provided to a surgeon by a tissue bank to construct a shell allograft. The graft must be meticulously sculpted to provide an exact fit. Bioabsorbable pins or low profile interfragmentary screws may have to be used to provide stable fixation.

Large allograft fragments can be furnished to surgeons for coring dowel allografts, but the procedure is wasteful and time consuming. A preferable technique is the preparation of dowel grafts outside the operating room.

Not only does this technique save time and effort, but it also allows for thorough depletion of marrow in the cancellous bone. Furthermore, the amount of osseous tissue transplanted can be minimized by retaining only a stalk of bone with a narrow rim of subchondral bone and a cartilage cap. Bone dowels can be taken from various anatomic sites to match the size, shape and contour of the recipient

Bone dowel allografts are press fitted into the recipient bed. No fixation is necessary.

Storage of osteochondral grafts at hypothermia in cell culture media with or without serum, results in a time-dependent loss of chondrocytes. Therefore, the length of preimplantation storage of these grafts is of paramount importance. The shorter the storage, the better the results. The accepted time of storage varies between 14 days and 21 days (Malinin et al 2006, Latterman and Romine 2009). Most investigators have relied on vital stains to support longer storage times. The contention that osteochondral allografts can be stored up to 42 days prior to transplantation is not supported by the data provided by recent studies

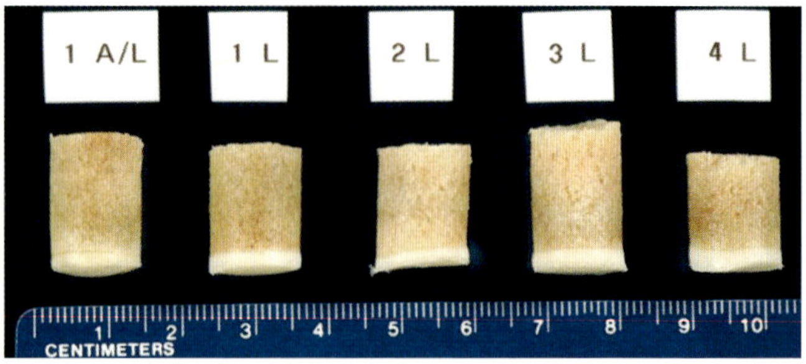

Figure 2 Osteochondral dowel allografts prepared by a tissue bank. Labels indicate the anatomical position of each graft

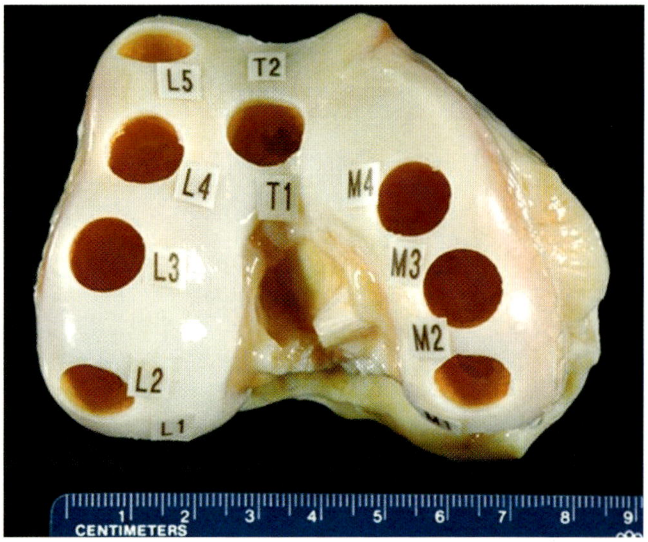

Figure 3 Osteochondral dowels can be prepared from various anatomic sites of the femoral condyle. (L) lateral condyle, (M) medial condyle, (T) trochlea. The number designates the location of the graft in the respective anatomic locations

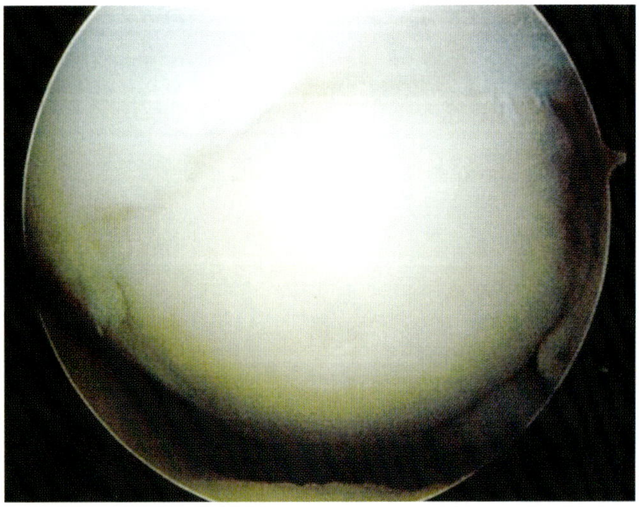

Figure 1 Hypothermia stored osteochondral allograft used to repair an articular cartilage defect caused by osteochondritis dissecans, 16 months post-transplantation

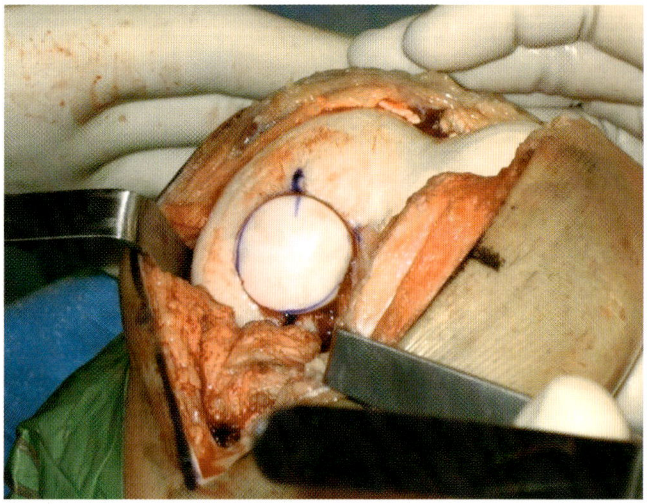

Figure 4 Large bone dowels with cartilage caps are press-fitted into the recipient's defect. No fixation is necessary

(Williams et al 2007). Explant data and *in vitro* studies in non-human primates have shown that beyond 21 days, chondrocyte viability is impaired. Thus, we do not advocate fresh cartilage transplantation in grafts that exceed 21 days of cold storage.

Particulate Cartilage Allografts

Regeneration of full thickness articular cartilage defects by implantation of particulate cartilage have been demonstrated experimentally (Malinin et al 2009) and clinically (Farr and Yao 2010). The procedure would offer several advantages since it avoids problems associated with the osseous portion of osteochondral allografts (Ahmet & Hincke 2010).

Poor healing of host graft junctions with osteochondral allografts is a well established phenomenon. The most likely cause of graft failure is host immunologic responses to the osseous and marrow components of the graft. According to Bugbee, there is a correlation between the amount of transplanted fresh osseous tissue and graft failure (Bugbee 2002).

The possibility of viable or non-viable cartilage being transplanted successfully without an osseous component would not have been thought to be possible in the recent past. However, during the last few years both experimental and clinical studies suggest that cartilage defects can be filled with hyaline cartilage utilizing fragments or particles of fresh allogeneic cartilage (Albrecht et al 1983).

As happens not infrequently, the original publication of results with particulate cartilage transplantation went largely unnoticed (Albrecht et al 1983). In this report, the investigators described cartilage repair induced by fresh cartilage allograft particles. Subsequently, fresh-stored juvenile cartilage allografts have been investigated (Farr and Yao 2010). In the laboratory, the technique results in the production of hyaline-like cartilage.

Presumably, chondrocytes escaped from the cartilage fragments, migrated, multiplied and formed new cartilage matrix. The escape of chondrocytes from cartilage fragments and their subsequent growth *in vitro* is predictable and has been amply demonstrated **(Fig. 5)**.

This phenomenon leads to a potential clinical application. Adult cartilage yields chondrocytes from cartilage fragments in a fashion that is similar to juvenile cartilage. Therefore, cartilage fragments can be collected in large volumes, tested for the production of chondrocyte outgrowth in culture, and transplanted into chondral defects with assurances of viability. Alternatively, cartilage fragments with documented viability or explant cells can be cryopreserved and transplanted at will at a later date, thus avoiding the time limitations and technical challenges inherent in fresh osteochondral grafts.

To expand the horizon of cartilage fragment transplantation, non-viable cartilage particles have also been investigated for their ability to induce cartilage regeneration of the host rather than repopulate the defect with new chondrocytes. Preparations of particulate cartilage either freeze-dried or dehydrated under hypothermic conditions, when placed in full thickness cartilage defects in experimental animals, induce regeneration of cartilage and healing of the defect **(Fig. 6)**.

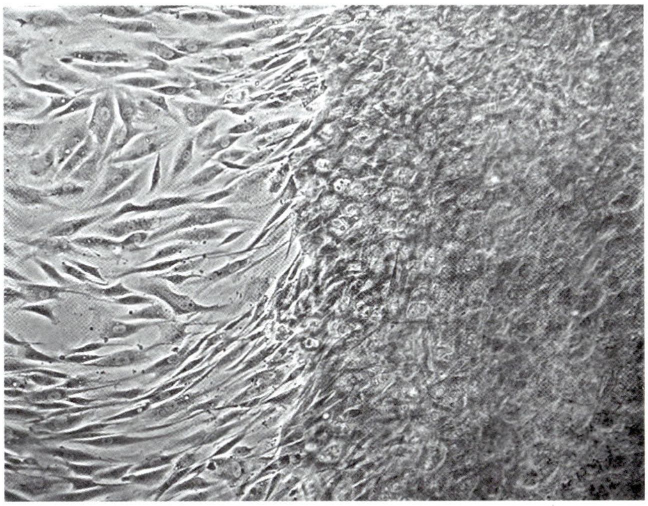

Figure 5 Viability of fresh cartilage fragments is demonstrated by cellular outgrowth *in vitro*. (Phase-contrast × 50)

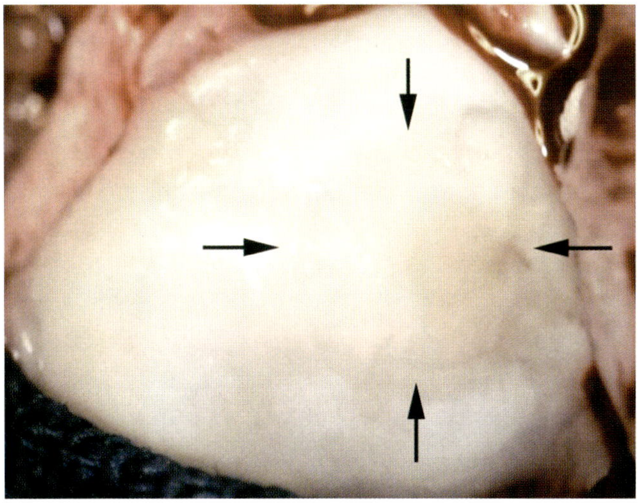

Figure 6 Full thickness chondro-osseous defect packed with freeze-dried microparticulate cartilage, 28 weeks post packing

When the same material is implanted intraosseously, it does not induce new bone formation (Malinin et al 2009). The preparation remains inert in the muscle of athymic rat. This suggests that non-viable particulate cartilage mediates cell signaling and induces chondrogenesis in experimental animals. Interestingly, these factors are present in and released from dried, but not frozen cartilage. This has very significant implications for cartilage healing in humans and will almost certainly be the subject of intense experimental and clinical investigations.

Allogeneic fresh cartilage grafts have been used clinically with considerable success. Costal cartilage is a particularly good source of this material.

Diced cartilage grafts were originally prepared by cutting rib cartilage into strips and then further dividing them into cubes. Later, small segments were used in the form of flat shavings. These grafts are particularly intriguing in the construction of new ribs. Brodkin and Peer (1954) introduced diced cadaver cartilage in a tantalum mesh, fastening the open ends of the mesh with parse-string wire sutures. The mesh with its cartilage content was then sutured to the sternum and cut of the rib. Mesenchymal tissue grew through the mesh, binding cartilage particles into a solid structure. Diced cartilage was also used with success in the skull, the ear, and in infarcts of the spine. Peer felt with justification that particulate cartilage grafts may have a much wider field of application than originally supposed (Peer 1955).

Meniscal Allografts

Transplantation of allogeneic menisci has become an acceptable treatment modality for patients who have injured their knees and required total meniscectomy. This development has taken place over the last 20–25 years. Menisci perform a vital function in the knee. Therefore, replacement of the menisci serves to prevent further degenerative articular changes which take place after the removal of the whole meniscus (Cummings et al 1997). In experimental animals, early as well as late, meniscal transplantation prevents or decreases degenerative changes which follow meniscectomies. Over time, knee supervenes in most patients, especially those who undergo lateral meniscectomy (**Fig. 7**).

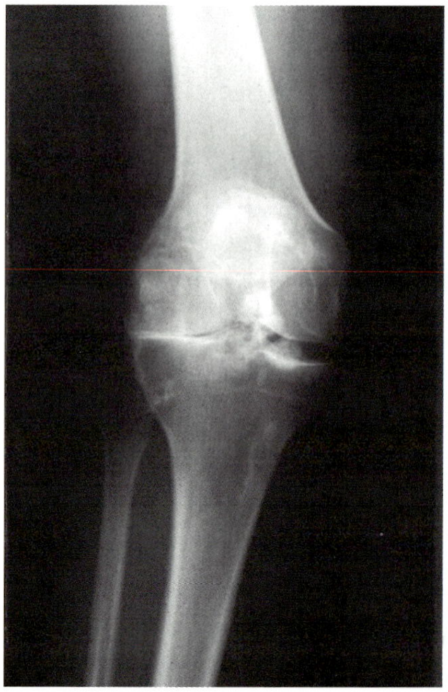

Figure 7 A standing AP radiograph of the left knee of a 53-year-old male who underwent total meniscectomy at the age of 18. There is significant joint narrowing, valgus deformity and squaring of the lateral femoral condyle

A patient with symptoms and diagnostic findings indicative of a meniscus deficient knee joint is the most common candidate for a meniscal transplant. However, the results are generally poor in patients with advanced joint degeneration (Packer and Rodeo 2009). Therefore, most authors advocate limiting meniscal allograft transplantation to patients with normal cartilage or early articular cartilage degeneration (Rodeo 2001, Carter 1999).

The meniscus is a crescent shaped disc of fibrocartilage attached to the superior articular surface of the tibia by dense connective tissue fibers called the meniscotibial ligament. In the knee, there are two menisci, the lateral meniscus and the medial meniscus. The lateral meniscus (meniscus lateralis) articulation is genus lateral semilunar cartilage, and external semilunar cartilage. These are nearly circular in form. In flexion and extension both menisci move with the tibia, but in rotation (pronation and supination) they move with the femur. The medial meniscus (meniscus medialis articulation is a genus, medial semilunar cartilage, and internal semilunar cartilage) is ovoid in shape and is more prone to injury than its lateral counterpart, because the attachment to the capsule of the lateral meniscus is not as robust as the attachment of the medial meniscus to the internal capsular ligament. A photograph of normal menisci of a human knee joint is provided in **Figure 8**.

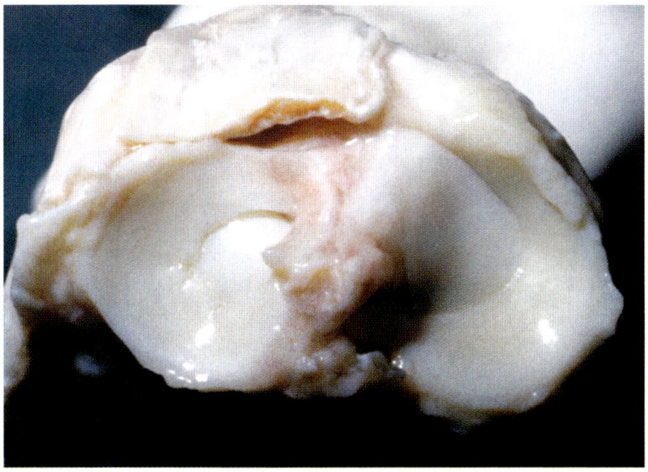

Figure 8 Normal human knee (Right) from a young adult. The photograph illustrates anatomical differences between the medial and the lateral menisci

The fibrocartilage of the meniscus has virtually no capacity for regeneration except in the vascularized periphery. Small fissures and tears might be sealed by cells from the overlying synovial membrane, but large tears usually persist without healing. For this reason, in the past, meniscectomy was a common method of treatment for torn menisci. This operation is the most frequent cause for meniscal transplantation. Understanding the importance of the menisci in the knee has resulted in fewer complete meniscectomies, more frequent repairs for unstable peripheral tears and the use of meniscal transplants in those in whom injury is extensive and repair, impossible.

Regeneration of the meniscus does occur after excision (Collins 1949). However, the new disc is formed by fibrous tissue rather than by fibrocartilage. It takes many months for the new disc to form. Hyalinization of the infolded synovial membrane may occur in the later stages of repair. However, newly formed structures do not cover as much of the tibia as the normal meniscus.

The hard-learned lesson is to preserve the menisci in whole or part whenever possible. A meniscal allograft is a salvage procedure for the young patient with persistent pain after complete meniscectomy with no instability and minimal degenerative articular changes.

Like articular cartilage, menisci are considered by many to be immunoprivileged because meniscal cells are imbedded deep in a dense matrix that is relatively avascular and thus inaccessible to immunoreactive cells. This notion is supported by most experimental and clinical studies which show that meniscal transplants do not elicit local or systemic immune responses. However, there are some notable exceptions. In the case of so-called fresh meniscal transplants there has been a confusion of terms. When most authors refer to "fresh" meniscal allografts they actually mean allografts stored in balanced salt solutions or tissue culture media at hypothermia or at normothermia prior to transplantation. The record of these transplants has been indeed very good both in clinical and experimental settings. On the other hand, in our experience in the study of goats, a large percentage of menisci were transplanted fresh, without storage in solutions, eliciting a strong immune response. This suggests immunogenicity of fresh menisci. The exact mechanism of this phenomenon is unclear.

Graft—Recipient Matching

The optimal biomechanical function of a transplanted meniscus allograft depends on the close fit of the allograft to the recipient knee. Allograft matching can be performed with either radiographic or gross matching of the meniscus and/or the tibial plateau. Sizing of the allograft and the tibial plateau through direct measurement is necessary and accurate. Matching the allograft size to the recipient is more difficult. Accurate radiographic measurement of the recipient's tibial plateau is a preferred method of measuring the recipient's size, since measurement of the meniscus itself might be difficult due to meniscal pathology. Several studies demonstrate a direct relationship between the size of the meniscus and osseous landmarks (Donahue et al 2006, Urban et al 1999, Shaffer et al 2000).

There is variability in imaging methods in determining the size of the meniscus. Some studies report measurement error of around 8% (Pollard et al 1995).

In addition to radiographs, magnetic resonance imaging (MRI) and computed tomography (CT) scan modalities have been used, but it is not clear which technique is the most accurate (Shaffer et al 2000, Kuhn and Wojtys 1996, Johnson et al 1994, Prodromos et al 2007). It seems the elaborate scanning techniques add little to standard roentgenogram measurements. In view of the cost and accuracy, the authors prefer digital radiography using standard osseous landmarks. It seems that MRI and CT imaging consistently underestimate the meniscal size (Garret 1992, Johnson et al 1994).

Accurate allograft size matching contributes to the success of the operation (Rijk 2004). Although estimates have been made that allograft size should be within 5% of the recipient's meniscus, the knee's tolerance for size mismatching has not been determined with precision (Wilcox et al 1996).

Meniscal allografts can be prepared by tissue banks in several configurations. The whole tibial plateau could be brought to the operation with the attached menisci, then detached and implanted. Either lateral or medial meniscus can be left attached to a section of bone of the intercondyloid fossa. Alternatively, anterior and posterior horns of the menisci can be left attached to blocks of bone (**Figs 9A and B**).

Meniscus Allograft Preparation

Many studies with meniscal allografts have reported satisfactory clinical outcomes. Studies also showed that irradiated meniscal allografts do not do well. This may be due to the amount of radiation that beyond a certain dose affects the structural integrity of the graft. Therefore, we will deal only with aseptically excised and processed transplants.

Fresh (stored at hypothermia), deep-frozen and cryopreserved allografts are used most frequently. Glutaraldehyde preserved menisci, and menisci stored in tissue culture media at normothermia have also been described. Freeze-dried meniscus allografts are now used rarely, if at all.

In one study, 25 meniscal allografts stored in tissue culture medium at hypothermia were transplanted into dogs. Eighteen healed completely, 3 healed incompletely and 4 failed (Mikic et al 1993). Similar results with fresh hypothermia stored meniscal allografts were obtained in rabbits (Cummings et al 1997, Tachibana 1990). The transplanted menisci were repopulated with cells from the synovium and the perimeniscal connective tissue.

In a seminal study using goats, Jackson and co-workers found no donor DNA within cells of the meniscal allografts 4 weeks after transplantation (Jackson et al 1993) this indicated the graft was repopulated with host's cells. Cryopreserved meniscal allografts transplanted into dogs maintain normal gross appearance and undergo cell repopulation (Arnoczky et al 1990). Although degenerative changes have been noted with time in animals receiving cryopreserved meniscal allografts, grafts prepared in this fashion still offer a viable option for clinical meniscal transplantation (Mikic 1993, Jackson et al 1992, Szomor et al 2000). Allograft stored at hypothermia media tissue culture would serve as a gold standard in meniscal transplantation if one could avoid an immune response. Interpretation of the results obtained with cryopreserved allografts is complicated by variations in

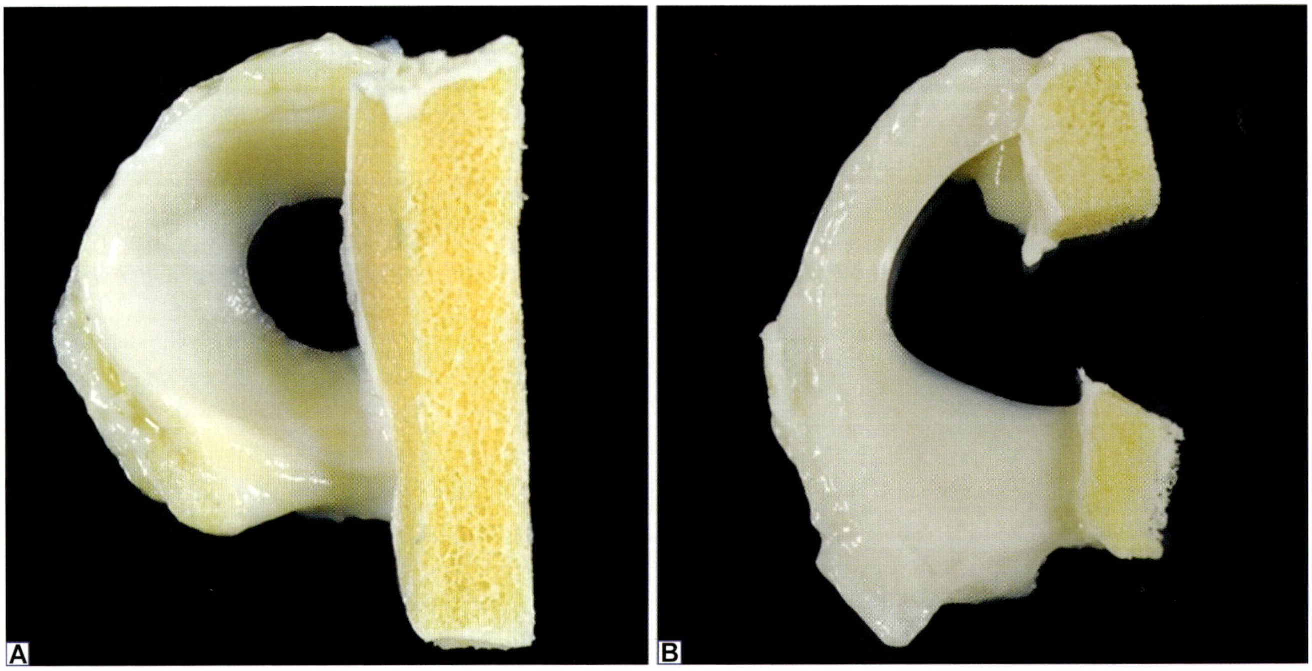

Figures 9A and B (A) Meniscus attached to a strip of bone from the intercondyloid region. This retains the exact size of the meniscus allograft; (B) Lateral or medial menisci with bone cubes at the attachment sites are probably the most versatile meniscal allograft constructs, as these allow surgeons to compensate for inherent inaccuracies in measurements cryopreservation techniques. Different investigators might use different cryoprotective agents, concentrations and time of exposure and freezing regimens.

In a study performed on goats, Keating et al reported excellent preservation and function of cryopreserved meniscal allografts 7 months post-transplantation (Keating et al 1988). At 3 months, these grafts showed complete incorporation by vascular synovial tissue. At 7 months, transplanted menisci had smooth and glistening surfaces and appeared normal in the necropsy specimens. There was neither arthritis nor subchondral erosions at the observed time intervals. Microscopically, cryopreserved menisci maintained normal architecture and viable cells (**Figs 10A and B**).

Ultrastructural studies revealed normal appearing chondrocyte nuclei. The cytoplasm contained an active golgi apparatus, numerous vesicles of rough endoplasmic reticulum, lysosomes and mitochondria. Exocytosis of proteoglycans was also present. The cells were embedded in a matrix of dense predominantly type I collagen and glycoproteins, an arrangement present in normal menisci (Ghadially et al 1983) (**Fig. 11**).

In these specimens, matrix stained positive with PAS, Safranin O and green with Masson trichome stain that suggests normal matrix including cells with nuclei containing DNA as revealed by FOTA technique (Malinin 1961).

These findings are supported by other investigators who also showed that cryopreserved human meniscal tissue maintains normal ultrastructure (Gelber et al 2009).

There is a general agreement on poor quality of menisci preserved either by rapid freezing or freeze-drying (Rodeo et al 2000), although others noted little difference between frozen and cryopreserved allografts 4 months post-transplantation. In another study, shrinkage was present in a significant percentage of frozen (20%) and freeze-dried (90%) allografts (Milachowski et al 1989). Our experimental results with freeze-dried meniscal allografts were dismal (Keating et al 1988).

A number of attempts have been made to use gluteraldehyde to preserve meniscal allografts. However, these have not met with success. A study of gluteraldehyde-treated meniscal allograft transplanted in a canine model resulted in a joint effusion and poor healing compared to autografts and allografts stored at hypothermia in tissue culture

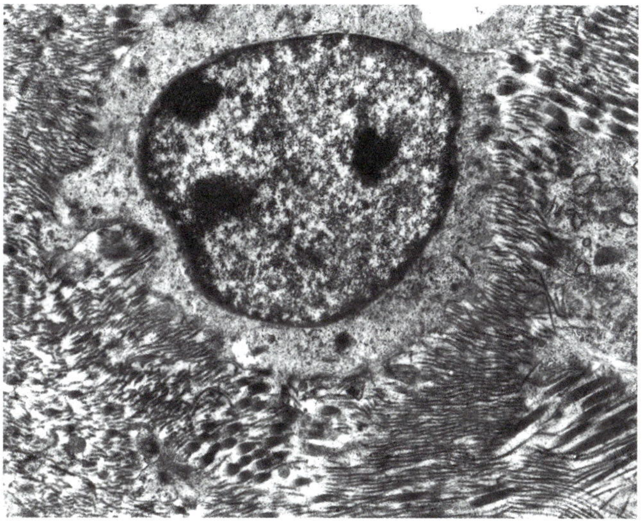

Figure 11 Electrophotomicrograph of a transplanted cryopreserved goat meniscus 7 months after transplantation. Collagen fibers maintain uniform orientation and normal appearance. The nucleus is surrounded by a clear rim. The nuclear membrane is intact, original magnification ×30,000

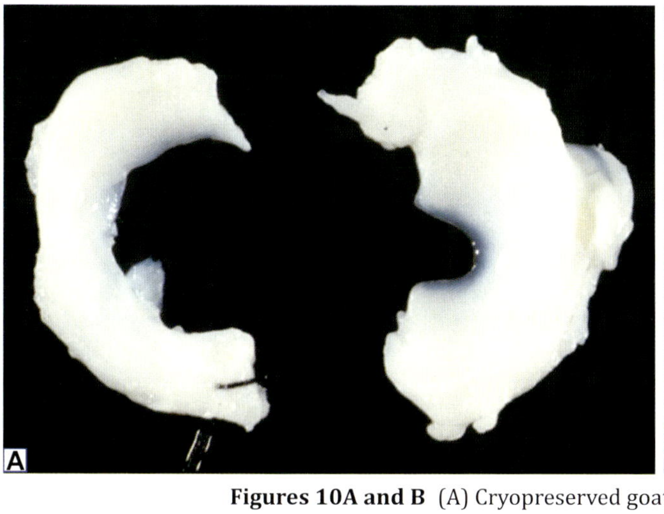

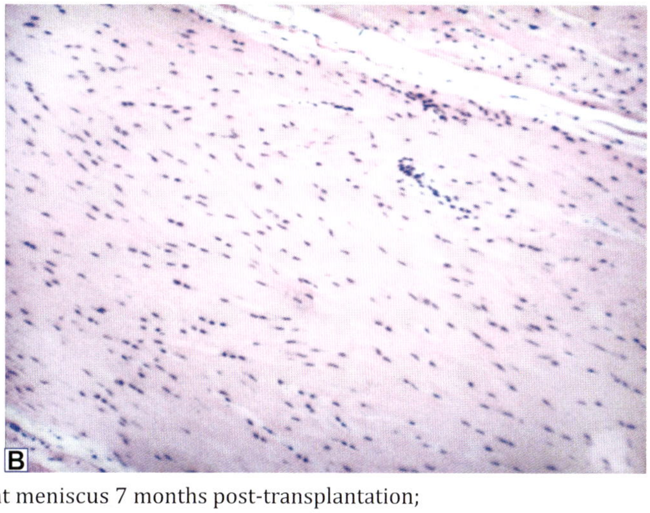

Figures 10A and B (A) Cryopreserved goat meniscus 7 months post-transplantation; (B) Transplanted meniscus demonstrates normal cellular architecture (H&E ×100)

media (Canham and Stanish 1986). Similar observations were reported by other investigators who have observed gluteraldehyde-treated allograft shrinkage and articular degeneration in a dog model (Powers et al 1988). Another disadvantage of gluteraldehyde preservation is the chronic synovitis produced by the leaching out of the compound from the graft (Jackson and Simon 1992). For these reasons, transplantation of gluteraldehyde-fixed meniscal allografts is no longer used.

In summary, cumulative experience with meniscal allograft preservation points to the advantages of the cryopreservation technique. Therefore, the additional costs and labor associated with the preparation of cryopreserved meniscal allografts appear to be justified. The performance of fresh (stored in tissue culture media) allografts has been largely satisfactory in practice. Complications have been relatively few, but the reports of chronic synovitis and a reported case of an acute rejection of an allograft should dampen our enthusiasm for processing by this technique (Milachowski et al 1989, Hamlet et al 1997). Our observation of an acute rejection of fresh (not stored at hypothermia) meniscal allograft in a goat model points to the antigenicity of the fresh tissue and its modification by hypothermic storage in cell culture media. Evidence of a synovial tissue proliferation surrounding meniscal allografts in experimental animals should be also of concern (Rijk and Van Noorden 2002, Ochi et al 1995).

Much information on meniscus allograft transplantation has been obtained from experimental animal studies. Most of these report healing and function of allografts prepared and preserved by various methods.

Fresh (unfrozen) allografts are considered by many to be ideal substitutes for missing native menisci, because fresh tissue contains viable cells. Viable cells are also present in cryopreserved menisci. Although, like articular cartilage, menisci are considered to be immunoprivileged, some experimental studies have shown immune responses to transplanted meniscal allografts. The nature and incidence of immune responses in humans following fresh meniscal transplantation are unknown.

A case of an acute rejection of a cryopreserved meniscal allograft has been reported by Hamlet and co-workers (Hamlet 1997). Fresh meniscal allografts in rats produce histologic changes compatible with rejection (Wada 1993). Two groups of investigators noted proliferation of synovial cells surrounding meniscal allografts in rabbits and in mice (Rijk and Van Noorden 2002, Ochi et al 1995).

Khoury et al showed Class I and II antigens in the cells of meniscal allografts, suggesting a possibility of an immune response (Khoury et al 1994). Others have shown sensitization to class I and class II in recipients of cryopreserved meniscal transplants (van Arkel et al 1997).

Immunologically active B lymphocytes and T-cytotoxic cells were also found in most transplanted allografts in humans. The presence of these cells suggests a mild immune reaction (Rodeo et al 2000).

Since bone allografts are more immunogenic than meniscal allografts, a suggestion has been made that meniscal allografts transplanted with bone plugs would pose a higher risk of an immune response than menisci alone. However, such was not the case (Zukor et al 1990). Therefore, meniscal allografts are frequently prepared and used with attached bone.

Many experimental studies with meniscal allografts have been published, but the results with regard to the best method of preparation of meniscal allografts are unclear. Currently, the most popular and widely accepted modes of meniscal allograft preparation are hypothermic storage and cryopreservation.

BIBLIOGRAPHY

1. Ahmed TA, Hincke MT. Strategies for articular cartilage lesion repair and functional restoration. Tissue Engineering Part B Reviews, 2010;16:305-29.
2. Albrecht F, Roesswer A, Zimmerman E. Closure of osteochondral lesions using chondral fragments and fibrin adhesive. Archives of Orthopaedic and Trauma Surgery, 1983;101:213-7.
3. Arnoczky SP, Warren RF, Mc Devitt CA. Meniscal replacement using a cryopreserved allograft. Clinical Orthopaedics and Related Research, 1990;252:121-8.
4. Buckwalter AE. Articular cartilage injury. Clinical Orthopaedics and Related Research, 2002;402:21-37.
5. Bugbee WD, Convery RF. Osteochondral allograft transplantation. Clinical Journal of Sports Medicine, 1999;18:67-75.
6. Bugbee WD. Fresh osteochondral allografts. Journal of Knee Surgery, 2002;15(3):191-5.
7. Campbell CJ, Ishida H, Takahashi H, et al. The transplantation of articular cartilage. An experimental study in dogs. Journal of Bone and Joint Surgery (Am), 1963;45:1579-92.
8. Canham W, Stanish W. A study of the biological behavior of a meniscus as a transplant in the medial compartment of the dog's knee. American Journal of Sports Medicine, 1986;14:376-9.
9. Carter T. Meniscal allograft transplantation. Sports Medicine and Arthroscopy Review, 1999;7:51-62.
10. CDC. Public health dispatch. Update: Unexplained death following knee surgery, Minnesota. Morbidity and Mortality Weekly Report, 2001;50(48):1080.
11. CDC. Septic arthritis following anterior cruciate ligament reconstruction using tendon allografts. Florida and

Louisiana 2000. Morbidity and Mortality Weekly Report, 2001;50(48):1081-3.
12. Chu CR, Convery RF, Akeson WH, et al. Articular cartilage transplantation: clinical results in the knee. Clinical Orthopaedics and Related Research, 1999;360:159-68.
13. Convery RF, Akeson WH, Amiel D, et al. Long-term survival of chondrocytes in an osteochondral articular cartilage allograft. Journal of Bone and Joint Surgery (Am), 1996;78:1082-7.
14. Cummings JF, Mansour JN, Howe Z, et al. Meniscal transplantation and degenerative articular change: an experimental study in the rabbit. Arthroscopy, 1997;13:485-91.
15. Czitrom AA, Keating S, Gross AE. The viability of articular cartilage in fresh osteochondral allografts after clinical transplantation. Journal of Bone and Joint Surgery (Am), 1990;72:574-81.
16. Davidson PA, Rivenburgh DW, Dawson PE, et al. Clinical, histologic and radiographic outcomes of distal femoral resurfacing with hypothermically stored osteochondral allografts. American Journal of Sports Medicine, 2007;35:1082-90.
17. DePalma AF, Tsaltas TT, Mauler GG. Viability of osteochondral grafts as determined by uptake of S35. Journal of Bone and Joint Surgery (Am), 1963;45:1565-73.
18. Donahue TL, Hull ML, Howell SM. New algorithm for selecting meniscal allografts that best matches the size and shape of the damaged meniscus. Journal of Orthopaedic Research, 2006;24:1535-43.
19. Emmerson BC, Görtz S, Jamali AA, Chung C, Amiel D, Bugbee WD. Fresh osteochondral allografting in the treatment of osteochondritis dissecans of the femoral condyle. Read at the 2006 Symposium of the International Cartilage Repair Society. San Diego CA, 2006.
20. Farr J, Cole B, Dwahan A, et al. Clinical cartilage restoration. Clinical Orthopaedics and Related Research, 2011;469:2695-705.
21. Farr J, Yao JQ. Chondral defect repair with particulated juvenile cartilage allograft. Cartilage, 2010;2:346-53.
22. Farr J, De Novo NT. Natural tissue graft. International Cartilage Repair Society, 8th World Congress; 2010. p. 242.
23. Garret JC. Fresh osteochondral allografts for the treatment of articular defects in osteochondritis dissecans of the lateral femoral condyle in adults. Clinical Orthopaedics and Related Research, 1994;303:33-7.
24. Garrett JC. Meniscal transplantation. In: Aichroth PC, Canon WD, Patel DV (Eds). Knee Surgery: Current Practice. New York: Raven; 1992. pp. 95-103.
25. Gelber P, Gonzales G, Torres R, et al. Cryopreservation does not alter ultrastructure of the meniscus. Knee Surgery, Sports Traumatology, Arthroscopy, 2009;17(6):639-44.
26. Ghadially FN, Lalonde JM, Wedge JH. Ultrastructure of normal and torn menisci of human knee joint. Journal of Anatomy, 1983;136:773-91.
27. Ghazavi MT, Prtizker KP, Davis AM, et al. Fresh osteochondral allografts for post-traumatic osteochondral defects of the knee. Journal of Bone and Joint Surgery (Br), 1997;79:1008-13.
28. Görtz S, Bugbee WD. Allografts in articular cartilage repair. AAOS Instructional Course Lectures, 2007;56:469-80.
29. Gross AE, Agnidis Z, Hutchison CR. Osteochondral defects of the talus treated with fresh osteochondral allograft transplantation. Foot and Ankle International, 2001;22:385-91.
30. Gross AE, McKee N, Pritzker KPH, et al. Reconstruction of skeletal defects at the knee: a comprehensive osteochondral transplant program. Clinical Orthopaedics and Related Research, 1983;174:96-106.
31. Gross AE, Shasha N, Aubin P. Long-term follow up of the use of fresh osteochondral allografts for post-traumatic knee defects. Clinical Orthopaedic and Related Research, 2005;435:79-87.
32. Gross AE, Silverstein EA, Falk J, et al. The allotransplantation of partial joints in the treatment of osteoarthritis of the knee. Clinical Orthopaedics and Related Research, 1975;108:7-14.
33. Hamlet W, Liu SH, Yang R. Destruction of cryopreserved meniscal allograft: a case for acute rejection. Arthroscopy, 1997;13:517-21.
34. Henning A, Abate J. Osteochondral allografts in the treatment of articular cartilage injuries of the knee. Sports Medicine and Arthroscopy, 2007;15(3):126-132.
35. Highgenboten CL, Jackson A, Trudelle-Jackson E, et al. Cross-validation of height and gender estimations of femoral condyle width in osteochondral allografts. Clinical Orthopaedics and Related Research, 1994;298:246-9.
36. Jackson DW, Mc Devitt CA, Simon TH, et al. Meniscal transplantation using fresh and cryopreserved allografts. An experimental study in goats. American Journal of Sports Medicine, 1992;20:644-56.
37. Jackson DW, Whelan J, Simon TM. Cell survival after transplantation of fresh meniscal allografts. DNA probe analysis in a goat model. American Journal of Sports Medicine, 1993;21:540-50.
38. Jamali AA, Emmerson BC, Chung C, et al. Fresh osteochondral allografts. Clinical Orthopaedics and Related Research, 2005;437:176-85.
39. Jamali AA, Hatcher SL, You Z. Donor cell survival in a fresh osteochondral allograft at twenty-nine years: a case report. Journal of Bone and Joint Surgery (Am), 2007;89(1):166-9.
40. Johnson D, Swenson T, Harner C. Meniscal reconstruction using allograft tissue: an arthroscopic technique. Operative Techniques in Sports Medicine, 1994;2:223-31.
41. Keating EM, Malinin TI, Belchick G. Meniscal transplantation in goats: an experimental study. Transactions 34th Annual Meeting Orthopaedic Research Society. Atlanta; GA. 1988. p. 147.
42. Khoury MA, Goldberg VM, Stevenson S. Demonstration of HLA and ABH antigens in fresh and frozen human menisci by immunocytochemistry. Journal of Orthpaedics Research, 1994;12:751-7.
43. Kuhn JE, Wojtys EM. Allograft meniscus transplantation. Clinical Journal of Sports Medicine, 1996;15:536-7.
44. Lance EM, Fisher RL. Transplantation of rabbit's patella. Journal of Bone and Joint Surgery (Am), 1970;52:145.
45. Latterman C, Romine SE. Osteochondral allografts: state of the art. Clinical Journal of Sports Medicine, 2009;28:285-301.
46. Lexer E. Joint transplantation. (Translated from German from Med Klin. 1908) Clinical Orthopaedics and Related Research, 1985;197:4-10.

47. Malinin T, Manymneh W, Lo H, et al. Cryopreservation of articular cartilage. Clinical Orthopaedics and Related Research, 1994;303:18-32.
48. Malinin T, Temple HT, Buck BE. Transplantation of osteochondral allografts after cold storage. Journal of Bone and Joint Surgery (Am), 2006;88:762-70.
49. Malinin TI, Martinez OV, Brown MD. Banking of massive osteoarticular and intercalary bone allografts: 12 years experience. Clinical Orthopaedics and Related Research, 1985;197:44-57.
50. Malinin TI, Temple HT, Carpenter EM. Induction of regeneration of articular cartilage defects by freeze-dried particulate cartilage allografts. 8th World Congress of the International Cartilage Repair Society. Miami, FL. 2009. p. 197.
51. Malinin TI, Wagner JL, Pita JC, et al. Hypothermic storage and cryopreservation of cartilage. Clinical Orthopaedics and Related Research, 1985;197:15-26.
52. Malinin TI. Feulgen oxidized tannin-azo (FOTA) technique for concomitant staining of DNA and protein. Stain Technology, 1961;36:198-200.
53. McDermott AG, Langer F, Prtizker KP, et al. Fresh small-fragment osteochondral allografts. Long-term follow-up study on first 100 cases. Clinical Orthopaedics and Related Research, 1985;197:96-102.
54. McGoveran BM, Pritzker KP, Shasha N, et al. Long-term chondrocytes viability in a fresh osteochondral allograft. Journal of Knee Surgery, 2002;15:97-100.
55. Meehan R, McFarlin S, Bugbee W, et al. Fresh ankle osteochondral allograft transplantation for tibiotalar joint arthritis. Foot and ankle International, 2005;26:793-802.
56. Meyers MH. Resurfacing of the femoral head with fresh osteochondral allografts: long-term results. Clinical Orthopaedics and Related Research, 1985;197:111-4.
57. Mikic ZD, Brankov MZ, Tubic MV, et al. Allograft meniscus transplantation in a dog. Acta Orthopaedica Scandinavica, 1993;64:329-32.
58. Milachowski KA, Weismeier K, Wirth CJ. Homologous meniscus transplantation: Experimental and clinical results. International Orthopaedics, 1989;13:1-11.
59. Ochi M, Ishida O, Daisaku H. Immune response to fresh meniscal allografts in mice. Journal of Surgical Research, 1995;58:478-84.
60. Ochi M, Ishide O, Daisaku H, et al. Immune response to fresh meniscal allografts in mice. Journal of Surgical Research, 1995;58:478-84.
61. Packer JD, Rodeo SA. Meniscal allograft transplantation. Clinical Journal of Sports Medicine, 2009;28:259-83.
62. Pap K, Krompecher S. Arthroplasty of the knee. Experimental and clinical experiences. Journal of Bone and Joint Surgery (Am), 1961;43:523-37.
63. Pollard ME, Kang Q, Berg EE. Radiographic sizing for meniscal transplantation. Arthroscopy, 1995;11:684-7.
64. Powers DL, Davenport ME, Wisnewski PJ. Gluteraldehyde-cross linked meniscal allografts: Clinical gross and histological results. Journal of Investigative Surgery, 1988;1:249-57.
65. Pritzker KPH, Gross AE, Langer F, et al. Articular cartilage transplantation. Human Pathology, 1977;8:635-51.
66. Prodromos CC, Joyce BT, Keller BL, et al. Magnetic resonance imaging measurement of the contralateral normal meniscus is a more accurate method of determining meniscal allograft size than radiographic measurement of the recipient tibial plateau. Arthroscopy, 2007;23:1174-9.
67. Rijk PC, Van Noorden CJF. Structure analysis of meniscal allografts after immediate and delayed transplantation in rabbits. Arthroscopy, 2002;18:995-1001.
68. Rijk PC. Meniscal allograft transplantation-part I. Arthroscopy, 2004;20:728-43.
69. Rodeo SA, Seneviratne A, Suzuki K, et al. Histological analysis of human meniscal allografts. Journal of Bone and Joint Surgery (Am), 2000;82:1071-82.
70. Rodeo SA. Meniscal allografts-where do we stand? American Journal of Sports Medicine, 2001;29:246-61.
71. Rodrigo JJ, Thompson E, Travis C. Deep-freezing versus 4 degrees preservation of avascular osteocartilaginous shell allografts in rats. Clinical Orthopaedics and Related Research, 1987;218:268-75.
72. Shadle C, Brighton CT. Preservation of articular cartilage. Proceedings 1977 meeting of the American Association of Tissue Banks KW Sell, VP Perry, M Vincent (Eds). p.109.
73. Shaffer B, Kennedy S, Klimkiewicz, et al. Preoperative sizing of meniscal allografts in meniscal transplantation. American Journal of Sports Medicine, 2000;28:524-33.
74. Shasha N, Krywulak S, Backstein D, et al. Long-term follow-up of fresh tibial osteochondral allografts for failed tibial plateau fractures. Journal of Bone and Joint Surgery (Am), 2003;85(suppl 2):33-9.
75. Szomor ZL, Martin TE, Bonar F, et al. The protective effect of meniscal transplantation on cartilage. An experimental study in sheep. Journal of Bone and Joint Surgery (Am), 2000;82:80-8.
76. Tachibana Y. Meniscus transplantation using a cryopreserved graft: an experimental study in a rabbit. Journal of Japanese Orthopaedic Association, 1990;64:583-92.
77. Tasto JP, Ostrander R, Bugbee W, et al. The diagnosis and management of osteochondral lesions of the talus: osteochondral allograft update. Arthroscopy, 2003;19(suppl 1):138-41.
78. Urban WP, Nyland J, Caborn DN, et al. The radiographic position of medial and lateral meniscal horns as a basis for meniscal reconstruction. Arthroscopy, 1999;15:147-54.
79. van Arkel ER, van den Berg-Loonen EM, van Wersch JW, et al. Human leukocyte antigen sensitization after cryopreserved human meniscal transplantations. Transplantation, 1977;64:531-3.
80. Wada Y. Meniscal transplantation using fresh and cryopreserved allografts. An experimental study in the genetically defined rat. Journal of Japanese Orthopaedic Association, 1993;67:677-83.
81. Wilcox TR, Goble EM, Doucette SA. Goble technique for meniscal transplantation. American Journal of Knee Surgery, 1996;9:37-42.
82. Williams RJ, Ranawat AS, Potter HG, et al. Fresh stored allografts for the treatment of osteochondral defects of the knee. Journal of Bone and Joint Surgery (Am), 2007;89(4):718-26.
83. Zukor DJ, Cameron JC, Brooks PJ. The fate of human meniscal allografts. In: JC Ewing (Ed). Articular Cartilage and Knee Joint Function. New York: Raven Press; 1990. pp. 47-152

Chapter 10

Douglas Wayne Jackson MD (1940), Long Beach Memorial Medical Center, Long Beach, California. Past President American Academy of Orthopaedic Surgeons, has made major contributions to both scientific investigation of ACL replacement allografts and their clinical transplantation.

(*Reference:* Jackson DW, Simon TM, Kurzweil PR, et al. Survival of cells after inter-articular transplantation of fresh allografts of the patellar and anterior cruciate ligaments. Journal of Bone & Joint Surgery (Am). 1994;74;112-118.)

Tendons, Ligaments and Membranes

INTRODUCTION

Next to bone, tendons and ligaments are the most frequently transplanted musculoskeletal tissue allografts. This can be explained by the popularity of using allografts for the replacement of anterior and posterior cruciate ligaments (Malinin 1993).

Interest in tendon transplantation has been longstanding. Both Pirogoff and Paget addressed the issue and demonstrated that regeneration of a segment of tendon can occur without an intermediate healing of the transplant (Bier 1917, Peer 1955).

Paget (1853) and Pirogoff (1840) described the healing pattern of Achilles tendon in rabbits, noting that the intermediate space of a divided tendon becomes filled with fluid or a semi-liquid substance that quickly transforms into lymph and cells. These in turn are transformed into cartilaginous germ tissue and later into fibrous tissue.

It is now recognized that adult tendons have a good capacity to heal. The body repairs most of the mechanically damaged tissue with scar tissue that is made up of collagen. Fibers constituting tendons are also made up of collagen, but there is a morphological difference between scar tissue and tendons. In undamaged tendons, collagen fibers are arranged in regular alignment parallel to each other. This allows the tendon to withstand the forces it is placed under. In scar tissue, collagen fibers are arranged randomly in many directions. This means that for the same width of tissue, tendon can withstand a much greater vertical pulling force than scar tissue.

Tendon has relatively few cells and abundant intercellular collagenous material. It is sparsely supplied with blood vessels since only a few are needed to support the cells. These enter at both ends, somewhat like nutrient vessels in bone. There are some exceptions. The flexor tendons of the fingers receive an additional blood supply in the middle of their course by means of "mesenteries," reflexions of the mesothelial lining of the tendon sheath, known as vincula vasculosa (Peer 1955). The structure is also referred to by various authors as constituting the mesotenon, paratenon and epitenon. The terms denote loose gliding connective tissue covering the tendon.

In the study of the reparative process of the guinea pig Achilles tendon, the importance of cellular participation in the reparative process became apparent (Enderlen 1893). Viering made a valuable contribution to this concept with his experimentation in rabbits (Viering 1891). He excised the Achilles tendons longitudinally and made observations on its healing up to 50 days. He believed connective tissue surrounding the tendon was essential for tendon healing. Mitosis of tendon cells in the injured areas was noted to take place as early as 4 days post injury. Under tension, the cell nuclei and the tendon fibers arranged themselves in rows parallel to the direction of tension. In subsequent experiments, Hoffa (1901) investigated tendons sutured to each other. He concluded that the peritendinous connective tissue of the tendon participated in tendon repair.

In 1903 Seggel performed the first animal experiments consisting of free autologous tendon transplantation. These experiments showed that both surrounding connective tissue and the tendon itself (Seggel 1903) participated in joining the tendon. Thus, the principles of autologous tendon transplantation were established.

Rehn first investigated allotransplantation of tendons in 1910 (Rehn 1910). Fresh autologous tendons transplanted by Lewis and Davis (1911) showed that tendon transplants are 2 to 3 times thicker than normal tendons. Kirschner also noted changes in transplanted tendon size (1909). In describing a tendon examined 33 days after transplantation, Kirschner noted that the transplanted tendon appeared as a rounded cord, containing twice the thickness of a tendon in the contralateral leg. Rehn stated, in describing one of the specimens removed after 7 weeks, that transplanted segments are three times as thick as the tendon on the opposite side.

As indicated by the references to the work performed at the end of the 19th century and the beginning of the 20th century, the process of tendon repair has been studied extensively through history. However, these usually did not include the cruciate ligaments of the knee.

Autologous or allogeneic tendon grafts in contact with the tendons of the recipient are reminiscent of bone grafts. In considering the matter, one might suppose that the cells in autologous tendon grafts might survive, as do cells in the autologous bone grafts. One may also consider that tendons are joined together by fibrous union as bones are by bony union. Allogeneic tendon grafts, on the other hand, may be replaced by creeping substitution from the recipient's tendon or the surrounding connective tissue in a manner analogous to the incorporation of bone grafts.

There is still some uncertainty about the exact mechanism of repair and replacement of autologous tendon grafts. There are two views on the subject. One holds that tendon graft repair is accomplished by tissue outgrowth from the cut ends of the tendon. This view is supported by experiments showing that the cruor between cut ends of the tendon is invaded and replaced by fibrous tissue outgrowths from the internal and external tendon sheaths (Peer 1955). This disorganized fibrous connective tissue is, in the end, replaced by specific, organized tissue from the tendon itself. Thus, the union is first achieved by cells proliferating from the tendon sheath, followed by formation of a new specific tissue from the cells of the tendon itself. This concept of tendon repair was established as far back as 1941 (Masson and Allen 1941) by investigators who showed that the tendon cells in free autologous grafts survive transplantation as viable cells and participate directly in the repair process.

Another opinion holds that tendons regenerate mainly from the sheaths that surround the tendon stump and that the tendon tissue itself does not participate in the process. A middle ground would suggest that peritenon is not vital to the repair of the tendon, but its presence is essential for normal function. The cells in a free autologous tendon graft degenerate after transplantation and the graft is replaced by infiltrating cells from the surrounding tissue of the host. With time, the new tissue is transformed into an exact replica of a normal tendon.

Knowledge concerning fresh or preserved tendon allografts is not as complete. Good clinical results reported by many investigators contain evidence of the allograft being replaced by the recipient's tendon or the surrounding connective tissue. According to some observers, the replacement structure does not reproduce the morphology of a normal tendon. According to others, the replacement structure is an exact replica of normal tendon.

The difference in opinions might be explained by the time the observations were made. The longer the transplant survives, the more likely it will replicate the normal structure.

Transplantation of xenograft tendons have been of interest to investigators for a long time, as the possibility of an unlimited supply of tendon grafts of various shapes and length is an attractive one.

As far back as 1885, Fargin and Assay transplanted tendon segments from rabbit into dog, from dog and ram into rabbit, from bird into rabbit and vice versa (Fargin and Assay 1885). They reported good results. However, after these reports tendon xengograft transplantations were not attempted until over 30 years later. Xenografts were then used for other than tendon substitutes. Horsley (1931) wrapped kangaroo tendon about the pylorus, the small intestine and the abdominal wall of the dog. The kangaroo tendon was quickly encapsulated and absorbed. On the other hand, chromatized kangaroo tendon fared better, showing resistance to absorption. From a small number of experiments performed in the first half of the 20th century and reported in the literature, one would be led to believe the xenografts did not do as well as allografts. However, Bahr in 1952 indicated that collagen fibers in xenograft tendons were absorbed and replaced with new fibers (Bahr 1952). Later studies cast doubt on the suitability of xenogeneic tendon for human transplantation (Allen et al 1987). Glutaraldehyde-treated tendon xenografts were reported to function well in chickens (Smith et al 1986).

Currently, the indications for the use of xenografts for human tendon repair are still ill-defined (Laurencin and El-Amin 2008). Laurencin and El-Amin, in an extensive treatise on the subject, provided an overview of the industry production and clinical use of xenotransplantation. Much of the material is based on animal experimentation, but is clearly not sufficient to provide clear-cut indications for tendon xenotransplantation in humans.

Xenotransplantation has a long and checkered history that continues to be surrounded by controversies. Safety issues, particularly zoonosis are still of concern. Theoretically, disease transmission is considered possible even by decellularized xenografts.

Immune response is another issue. Immune response to tendon xenografts has been studied in rabbits (Veillette et al 1998, Tauro et al 1991). Bovine fibroblast extracellular proteoglycans and glycoproteins have elicited immune response similar to that elicited by collagen matrix. Cellular elements within xenografts elicit the greatest degree of immune stimulation.

Several commercially produced xenografts approved by the FDA are offered to orthopedic surgeons. Most of these include material for rotator cuff repair. Presently, it has not been determined whether or not xenografts will be used as tendon replacement grafts. Most of the commercially promoted soft tissue xenografts are marketed on the basis of information provided by the manufacturers and not on the basis of unbiased scientific studies.

Cruciate Ligaments of the Knee

The normal function of the knee is dependent in large part on cruciate ligaments. Normal knee kinematics (geometry of the motion of bodies) depend on the anatomy of articulating surfaces, the biomechanical properties of the ligaments, ligament tension and the site of ligament insertion into the bone.

The cruciate ligaments perform the dual function on one hand, permitting motion of the articular surfaces of the knee and, on the other hand, restraining their motion (Huskies et al 1985, Hudson 1974). The anterior cruciate ligament (ACL) is of paramount importance in performing these tasks. Therefore most of the studies and observations dealing with the motion of the knee joint relate to this ligament. The function of ACL is determined by the anatomic structure of the ligament and its attachments **(Fig. 1)**.

Anterior cruciate ligament is a three-dimensional collagenous structure (van Dijk et al 1979). The ligament exerts its function partially by elongation. The force produced by this elongation has been measured in the laboratory by several investigators (Noyes and Grood 1979).

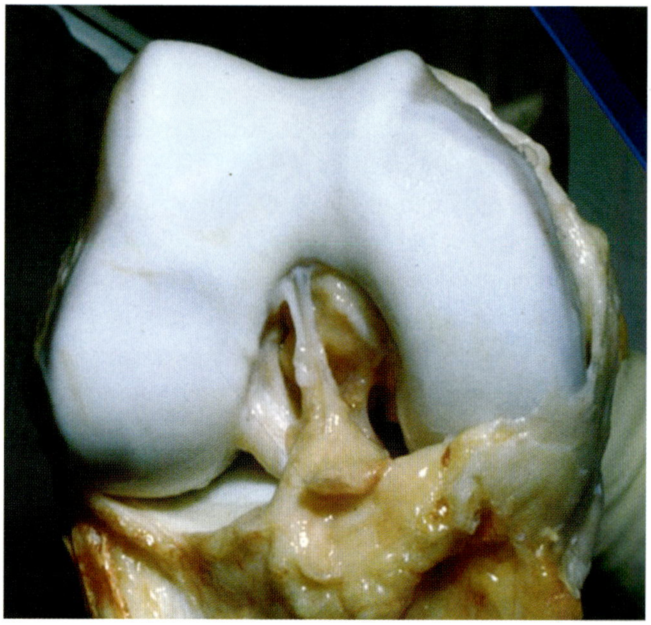

Figure 1 Normal knee of a young adult showing collateral ligaments. The ligaments cross each other in the center and are behind the meniscofemoral ligament. Anterior cruciate ligament inserts into the lateral condyle, while posterior cruciate ligament inserts into the medial condyle. In performing biomechanical testing, all ligaments and capsule other than the ligament being tested are severed

The biomechanical strength of ACL is cited in the literature to be around 1725 N +/- 269. This figure produced by Noyes and associates (1984) is taken to represent 100% of ACL strength and in fact serves as the "gold standard" against which biomechanical properties of all allografts are measured. However, the figure was obtained from younger subjects and does not necessarily represent biomechanical properties of ACLs in a cross section of population. There is considerable difference in biomechanical properties of ACL between age groups **(Table 1)**.

The biomechanical strength of ACLs in individuals 40 years and older is almost one-half of that in subjects 30 years old or younger. The difference is quite significant and should be borne in mind when selecting an ACL replacement allograft (Malinin 2009).

Histologic structures of normal cruciate ligaments reveal their collagenous nature. Sharpey's fibers at the distal and proximal ends of cruciate ligaments pass into subarticular bone up to the tide line (van Rens et al 1986). Collagen

Table 1: Strengths of anterior cruciate ligaments according to the age groups

Biomechanical strength (in Newtons) of 40 isolated anterior cruciate ligaments according to the age groups					
Age group and #	***Force to failure-mean***	***Standard deviation***	***Standard error***	***95% confidence mean interval***	
				Lower bound	***Upper bound***
<30 (4)	1575.00	270.74	135.37	1144.18	2005.81
31–40 (12)	1183.83	349.81	100.98	961.57	1406.09
41–50 (8)	864.87	143.68	50.79	744.75	984.99
51–60 (12)	856.75	376.50	108.68	617.53	1095.96
61+ (4)	966.75	145.35	72.67	735.45	1198.04

P > Anova*

	Sum of squares	Degrees of freedom	Mean squares	Significance
Between groups	2062917.2	4	515729.30	P = .002
within groups	3333167.5	35	95233.35	
Total	5396084.7	39		

*Analysis of one way variance

organization, elastic fibers, vascularity and an intimate relation with the synovium constitute a hallmark of the cruciate ligaments of the knee.

Anterior cruciate ligament is a band of dense connective tissue that runs from femur to tibia. It originates from within the depth of the notch of the distal femur and inserts in front of the intercondyloid eminence of tibia. There are two bundles of the ACL, anteromedial and posterolateral, named after the place of their insertion into the tibia (Strocchi et al 1992). The two-bundle nature of ACL becomes apparent when ACL fails under tension as shown in **Figures 2A to C**.

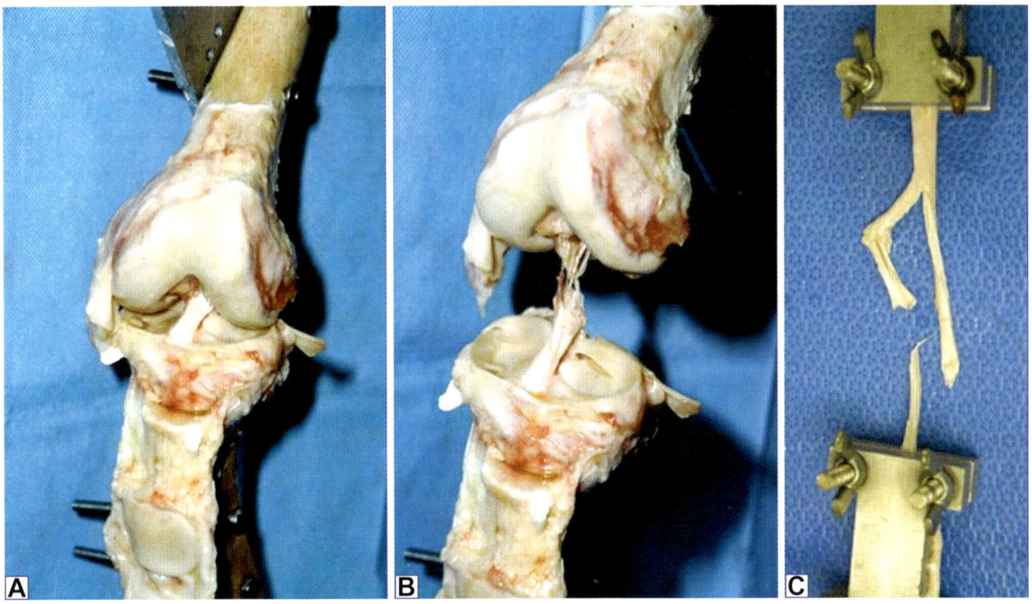

Figures 2A to C Failure mode of an anterior cruciate ligament. (A) Intact ligament before the force is applied; (B) Mechanical failure in response to distracting force; (C) In an isolated ACL, the two-bundle nature of the ligament is apparent

In turn, ACL bundles are made up of multidirectional collagen fibers, the majority of which are oriented along the axis of the ligament. Fascicles making up the bundles are characterized by their undulating appearance that becomes prominent in polarized light. Within the bundles, two types of collagen fibrils are present, small with an average diameter of 45 nm and large containing non-uniform thickness and diameters varying between 35 nm and 75 nm. Numerous fibroblasts with elongated nuclei are interspersed with these fibrils. The elongated fibroblasts have branches as well as short cytoplasmic processes. Elastic fibers are also present within collagen bundles. The microstructure of collagen bundles is of various types, but the collagen itself is made up primarily of collagen type I.

The varied orientation of the bundles, the complex ultrastructural organization and the abundant elastic tissue make the cruciate ligaments of the knee distinct and different from other ligaments and tendons (Strocchi et al 1992). Cruciate ligaments are composed of heterogeneous rod shaped and ovoid cell populations, intercellular matrix and parallel collagen bundles.

Anterior cruciate ligament is innervated by posterior articulate branches of the tibial nerve and is vascularized by branches of the middle genicular artery (Dutton et al 2006).

Replacement of Anterior Cruciate Ligament

When ligaments rupture, the load distribution across the knee joint is altered, resulting in abnormal knee kinematics. Because of this deficiency, in order to prevent further deterioration, it is advisable to replace torn or ruptured cruciate ligaments (France et al 1988). Repair of torn or ruptured cruciate ligaments is usually not effective. The clinical result of the nonoperative treatment of ruptured ACL is also shown to lead to functionally unacceptable results (Fu et al 1999). The poor quality of ACL healing observed clinically and experimentally has led orthopedic surgeons to perform reconstructions rather than repairs.

Anterior cruciate ligament and posterior cruciate ligaments (PCLs) can be repaired either with autologous or allogeneic grafts (Fu et al 1999). Prosthetic cruciate ligament replacement has, so far, proved unsatisfactory (Klein and Jensen 1992, Macnicol et al 1991). The most successful replacement for ACLs has been biologic tissue grafts.

Autografts have been used first. Autografts have virtually no risk of eliciting an immune response or disease transmission. The down side is donor site morbidity and limitation in size.

The most common autografts used for ACL replacement are bone-patella tendon-bone and semi tendinosis/gracilis tendon constructs. The bone-patella tendon-bone graft, taken from the center, is usually 8–11 mm wide and includes bone blocks from the patella and tibial tubercle.

The bone-patella tendon-bone construct is in vogue because of its biomechanical properties. It can tolerate a tensile load of over 2,000 N. The possibility of rigid fixation through bone blocks at either end of the graft is also attractive.

Hamstring tendon grafts are associated with less donor site morbidity than bone-patella tendon-bone grafts. Therefore, many surgeons prefer this graft for ACL reconstruction (Simonian et al 1997). Initially, hamstring tendon graft was used as a single semitendinosus tendon. It then evolved into a quadruple strand semitendinosus/gracilis graft by folding and combining the tendons (Butler et al 1986, van Rens et al 1986, Meijer et al 1987). The ligaments restrain mobility and produce kinematic couplings via articular surfaces. In this process the ligament does not function as a unit, but as a complex of collagen fibers that are recruited as required. Histologic structure of normal cruciate ligament reveals its collagenous nature.

Biomechanical properties of patella tendons are used as a standard against which other grafts are measured. Literature cites the figures of around 2,300 N for the ultimate tensile load for this structure (Schatzman et al 1998). Literature reports and our own experience are mostly in agreement regarding biomechanical properties of bone-patella tendon-bone constructs.

However, the values for the intact ACL vary from 2,160 N (Fu 1999) to 839 N* with an intermediate value of 1,725 (Noyes et al 1984). Single strand semitendinosus tendon tensile strength varies from 1,216 N to 675 N*.

The technique of replacement of torn or ruptured cruciate ligaments, particularly the ACL, came into prominence in the early eighties of the last century. The patellar tendon

*Average figure obtained in our laboratory from 18 randomly selected donors.

(bone-patella tendon-bone, BPTB) autograft emerged as a "gold standard" for ACL reconstruction. It has been used consistently by many orthopedic surgeons and still remains the graft of choice for patients active in sports with high demand on the knee function. The BPTB autograft produced excellent surgical outcomes with a 90–95% success rate in terms of returning to pre-injury level of activity in sports (Fadale and Johnson 2010).

The patellar tendon autograft is usually obtained through a vertical incision, some 7–10 cm long, medial to the patellar tendon. However, some surgeons used two horizontal, shorter incisions made over proximal and distal ends of the tendon.

The middle third of the tendon 1–1.2 cm wide is excised longitudinally with 2–2.5 cm bone blocks from the patella and the tibial tubercle attached to the tendon. This produces a composite bone-patellar tendon bone-autograft with desirable biomechanical properties. Tensile strength of such grafts has been reported to be around 2,900 N to failure (Noyes et al 1984).

The principal advantage of this construct is that its fixation is easy. To this end, a headless screw (interference screw) is inserted next to the bone ends of the graft in the bone tunnel, thus locking the construct in place. The insertion of the patella tendon into bone is strong. The autograft bone heals in about 8 weeks, resulting in a biomechanically stable construct providing the bone tunnels have been placed correctly. In addition to metallic screws, interference screws made of a bioresorbable material are now available and sometimes used. These are said to resorb in 2–3 years, but exact data on the resorbtion time is not available.

The disadvantages of the bone-patella tendon-bone autografts concentrate on the donor site. A considerable amount of donor site pain, compared to that encountered with other procedures, has been noted. In some patients, this may result in quadriceps muscle atrophy requiring prolonged physical therapy and rehabilitation.

Another problem is the incision scar that may present a cosmetically unacceptable appearance. In addition, almost all patients encounter an area of sensation loss 5–6 cm wide lateral to the incision. Patellar tendon rupture as well as fracture of the patella has also been encountered. Some studies report increased incidence of anterior knee pain with resumption of athletic activities (Re et al 1993, Freedman et al 2003).

Because of the difficulties with the "gold standard" bone-patella tendon-bone autografts including the technical considerations of the tunnel angle and graft length (McGuire and Hendricks 2009), hamstring tendon autografts began to be used as alternatives (Brown et al 1993). The excision and the after effects were said to be less painful compared with BPTBs. However, anterior knee pain remained a problem (Shelbourne and Trumper 1997).

Excision of hamstring grafts is technically not difficult, but the procedure requires surgical experience. Transection of tendon, injury to the nerves and other tendons can occur and must be guarded against.

Hamstring tendons are viable tissues taken from the back of the injured patient's knee. Because their excision is relatively easy, these tendons became popular as grafts. The tendons themselves are not strong and they have no bone attached to them. Therefore, healing of these grafts must occur entirely by tendon healing to the bone. This is an uncertain process that may take as long as a year. If during this process stretching occurs, laxity of the knee may be the end result. Another major problem with hamstring autografts is that excision of these grafts ablates two important muscle-tendon units that complement ACL by holding the tibia backwards. Such being the case, the burden of maintaining knee stability must be borne by the ACL replacement graft alone. This might well be another reason for failure.

Anterior knee pain with BPTB and hamstring autografts has led to quadriceps tendon being still another alternative for autologous tissue replacement for ACLs. Its proponents cite fewer complications associated with graft excision and lower incidence of pain (Chen et al 1999). Some surgeons use quadriceps tendon without bone block. However, at this time, information regarding quadriceps tendon grafts is too limited to form a definite opinion on their utility.

Semi-tendinosus tendons can be used either singly or doubly, i.e., folded over. They can also be combined with gracilis tendons. These provide biomechanical strength. These constructs are sometimes referred to as "quad" grafts because, folded over, the tendons provide a construct with four individual strands. Initially placing large sutures in the ends of the grafts and tying these to screws placed outside the bone tunnels accomplished fixation of semi-tendinosus-gracilis grafts. This method depended on the integrity of the sutures and healing of the tendons to bone inside bone tunnels, a notoriously slow and unpredictable process. Subsequently, different hamstring tendon fixation techniques were developed. The principal method was the fixation of the grafts with interference screws. These screws contained specially designed blunt threads that would not cut the tendon. The results approximated those obtained with BPTB constructs. Bioresorbable interference screws are now also available, but bioresorbable implants have their own complications. These include cyst formation, screw breakage and infections (Lubowitz 2010).

A study by Brown et al (1999) reported on the strength of a patented "knotless rope" loop. A pull-out strength of this device averaged 1,345 N versus 710 N for BPTB construct.

Hamstring graft excision does not violate patellar tendon, nor does it involve quadriceps extensor mechanism. Therefore patients with hamstring tendon autografts experience less pain and are rehabilitated quicker than their counterparts with BPTB autografts. Hamstring autografts also have some disadvantages. The principle of these, as was mentioned earlier, is reliance on tendon-to-bone healing. The process is unpredictable and bears a relationship to bone tunnel geometry and length. Tunnel widening is apparently more frequent with hamstring grafts versus BPTB grafts.

There have been several studies comparing the results of hamstring grafts versus patella tendon grafts in ACL reconstructions. Both types of grafts provide very satisfactory overall results. The difference in prospective and retrospective study results lack clear-cut differences, although some authors report tighter instrument measured values with BPTB autografts (Cory et al 1999, Otero and Hutcheson 1993). However, others have reported similar laxity values and functional scores between the two types of grafts (Marder et al 1991, Aglietti et al 1994, 1997, Ericksson et al 2001, Beard et al 2001).

Allografts

Since clinical results with autografts are quite satisfactory, why use allografts at all? The answer is simple. Allografts offer several advantages over autografts. They also eliminate the litany of complaints regarding complications inherent in the use of autografts (McGuire and Hendricks 2009).

Allograft tissues offer a wide variety of structures that can be used for ACL replacements. In addition to patella tendons with bone blocks and hamstring tendons, Achilles tendons, tibialis tendons and fascia lata are available as allografts. Although ACL reconstructions can be performed with any soft tissue grafts, the above listed structures are most convenient and reliable because of their biomechanical properties. Biomechanical data on ACL replacement allografts is summarized in **Table 2**.

Allografts for ACL replacement are usually preserved either by freezing or freeze-drying. These techniques have little effect on the biomechanical properties of the graft (Woo 1986, Jackson et al 1991). Many surgeons prefer frozen allografts because these do not have to be reconstituted several hours before surgery (Indelicato et al 1990, 1992).

The advantages of allograft use for ACL reconstruction are significant. The number of reconstructions continues to increase. Surgeons contemplating transplanting allografts must be, of course, cognizant of patient safety. Potential for disease transmission, biomechanical alterations due to sterilization procedures, as well as immunogenicity are common concerns. On the other hand, using autogenous grafts is not an altogether benign option (McGuire and Hendricks 2009).

Proponents of autografts argue that harvest site complications are less important than potential for disease transmission with allografts. Although these concerns are valid in actuality, they are highly exaggerated. Highly publicized reports of HIV transmission with allograft tendon and bacterial infection also transmitted with tendon allografts (Caspari et al 1995) were isolated instances that could and should have been avoided and prevented (CDC 2002). To

Table 2: Comparative strength of ligament allografts

Anatomic structure	*N*	*Load to failure (in Newtons)*	*SD*
Achilles tendon	8	2204.0	287
Bone-patella tendon-bone	7	2521.6	1175.4
Tibialis posterior tendon	10	900.0	182.9
Tibialis anterior tendon	9	840.5	235.2
Peroneus longus tendon	12	876.4	428.6
Semitendinosus-gracilis tendons	10	697.5	271.1
Anterior cruciate ligaments	18	839.4	315.7

date, there has not been a reported case of viral or bacterial disease transmission with freeze-dried allografts. However, this might be coincidental as freeze-drying does not sterilize tissues. There have been reports of incidents of disease with frozen allografts, but none in the last 10 years.

Methods of ACL replacement allograft excision and preparation have been described in several publications (Malinin 1993, Mc Allister et al 2007).

Allografts for cruciate ligament reconstruction are most commonly prepared from patella tendon with bone attached at each end. Achilles tendon is also used with attached bone on one side, as is quadriceps tendon. Semitendinosus, gracilis, tibialis posterior and anterior tendons and peroneus longus tendon lack bony attachments.

For ACL replacement, the allograft must be 9–11 cm in diameter and long enough to allow for femoral and tibial attachments. For fascia lata, to achieve desired dimensions, it must be divided superiorly at the tensor fascia lata, anteriorly along the medial intermuscular septum, laterally along the border of the biceps and inferiorly just above the knee joint. Fascia lata folded into an 8.4 mm tube is quite strong, with ultimate load to failure in 18 specimens of 3,266 N, SD 987 and stiffness in N/mm of 414, SD 151.

Patellar tendon is excised whole with a bone block from tibial tuberosity and whole patella or its inferior portion. Bone-patella tendon-bone constructs can be prepared from the central portion or from hemisections of the originally excised structure. A photograph of a hemisection is shown in **Figure 3**.

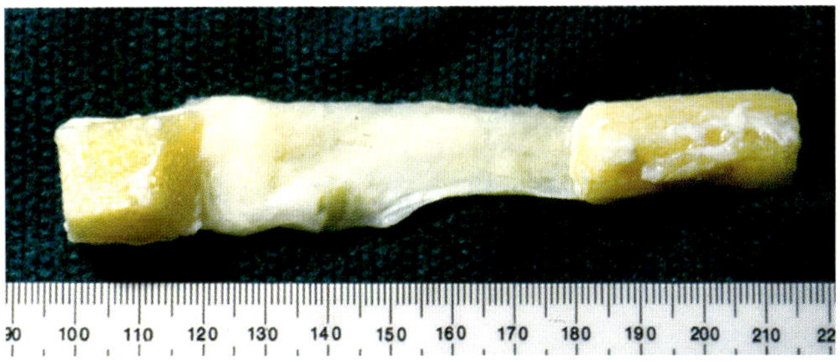

Figure 3 Fresh bone-patella tendon-bone allograft. The graft is usually divided in two. Holes are drilled in each block and the graft is tensioned by 24-gauge flex wires or by #5 woven sutures. The bone blocks are shaped to fit into 11 mm diameter tunnels. The tendon itself is trimmed as needed

In either case the tendinous section is incised longitudinally to obtain a graft 10–12 mm in width. Bone blocks at the end of the tendon can either be left intact or they can be shaped to the same width. Biomechanical properties of sectioned BPTB allografts are given in **Table 3**.

Table 3: Biomechanical properties of bone-patella tendon-bone constructs

	N	*Load to failure (in Newtons)*	*SD*
Whole	7	2310.5	1051.3
1/2 Medial and Lateral	10	1506.0	572.7
1/3 Central	7	1504.7	490.7

All tested specimens were frozen.

Compared to whole BPTB grafts, they are still very adequate. Patellar and tibial bone plugs are cut to about 2.5–3.5 cm in length. Bone plugs must be hard enough to allow for drilling and passing of the sutures through their ends. Osteopenic or osteoporotic bone is not suitable for BPTB constructs.

Like ACLs, the biomechanical properties of BPTB constructs vary with the donor's age. The differences are considerable. While BPTB allografts from donors under 30 years of age are very strong on the average (2,248 N to failure), those from donors over 60 years of age are much weaker. Comparative values for biomechanical properties of BPTB constructs from various age groups are given in **Table 4**. In view of the data presented in **Table 4**, claims by some tissue banks that BPTB allografts from various donors of various ages are comparable are incorrect. These claims, based on a paper published by Blevins et al, are not supported by our data (Blevins et al 1994). Therefore it might be prudent to select allografts on the basis of their biomechanical similarity with biomechanical properties of ACLs of the recipients.

As was mentioned earlier, there are two generally accepted methods of allograft preservation. These are freezing and freeze-drying. Detailed description of the techniques was provided in the chapter on Tissue Banking.

There are proponents of each technique for the preservation of BPTB allografts.

Melberg and Indelicato (1993) studied results with 116 frozen BPTB allografts. Eighty-seven percent of patients achieved good or excellent results when evaluated with Lysholm Knee Score (Lysholm and Gillquist 1982). These results are comparable to other reports in the literature

Table 4: Biomechanical strength (in Newtons) of 92 half patella ligaments according to age groups

Age group and #	Force to failure mean	Standard deviation	Standard error	95% Confidence interval	
				Lower bound	Upper bound
<30 (6)	2248.83	858.53	350.49	1347.85	3149.80
31–40 (25)	1531.98	608.91	121.78	1280.63	1783.32
41–50 (22)	1276.31	537.35	114.56	1038.06	1514.56
51–60 (28)	1279.17	572.78	108.24	1057.06	1501.27
61 + (11)	762.00	226.23	68.21	610.01	913.98

ANOVA

	Sum of squares	Degrees of freedom	Mean squares	Significance
Between groups	9738232.8	4	2434558.194	P = .000
Within groups	28018092	87	322047.039	
Total	37756325	91		

(Noyes and Barber 1991). The experience of these authors with fresh-frozen bone-patella tendon-bone allografts has been favorable (Barber et al 2010).

Freeze-dried tendon allografts have their proponents; (Weiner et al 1988), other investigators have used both frozen and freeze dried allografts (Levitt and Malinin 1991). The results between two types of allografts appear similar, but there are some differences in their biomechanical properties. However, these are not striking. Comparison between biomechanical properties of frozen versus freeze-dried BPTB allografts is provided in **Table 5**.

Table 5: Biomechanical properties of frozen versus freeze-dried (FD) bone-patella tendon-bone constructs

	N	Load to failure (in Newtons)	SD
Whole patella Tendon-frozen	7	2310.5	1051.3
Whole patella Tendon-FD	3	2241.0	1175.5
1/2 Patella Tendon-frozen	10	1506.0	572.7
1/2 Patella Tendon-FD	11	1417.2	493.5
1/3 Patella Tendon-frozen	7	1504.7	470.9
1/3 Patella Tendon-FD	6	1872.5	521.1

The BPTB allografts for ACL replacement did not reach a peak level of acceptance until the 1980s (Cooper and Beck 1993). Initially, reports with long-term follow-ups were relatively few. Later these were extended and valid comparisons between BPTB autografts and allografts were made. Shino and associates were the first to describe a series of patients receiving allogeneic tendon grafts for ACL substitution (Shino et al 1986). Since then they have published several studies on the subject, including one with histological analysis of ACL allograft biopsies (Shino et al 1988). Their arthroscopic observations indicated the grafts were "viable". Early surface hypervascularity subsided with time. Their observations suggested the grafts reached maturity at 18 months. Subsequent long-term observations (57 months) reaffirmed the original findings (Shino et al 1990).

Many other investigators have reported the results of clinical studies with allogeneic BPTB allografts used for ACL replacement (Indelicato et al 1992, Zoltan et al 1988, Weiner 1988, Silvagio and Fu 1990). The initial phase of reports dealing with the replacement of ACLs with allografts culminated with the publication of data on 300 patients with a follow-up time of up to 6 years (Levitt and Malinin 1991). In this study, both bone-patella tendon-bone allografts as well as Achilles tendon allografts were used.

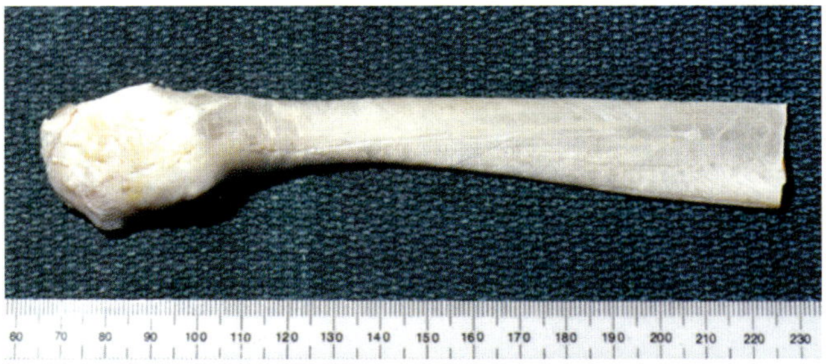

Figure 4 Achilles tendon allograft. The proximal portion of the graft is removed to produce the graft of a desired length

Table 6: Biomechanical properties of Achilles tendons

	N	*Load to failure (in Newtons)*	*SD*
Whole Achilles tendon	8	2204.0	287
Achilles tendon 1/2	2	1328.5	133.6

Achilles tendons are very similar in their biomechanical properties to the BPTB constructs. The principal difference lies in the fact that Achilles tendons have bone block only on one side. Achilles tendons are usually excised and furnished to the surgeon as in **Figure 4**, but they can also be divided longitudinally in half. If the Achilles tendon is long enough, it can be used for both ACL and PCL reconstructions (McGuire and Hendricks 2009).

Biomechanical properties of Achilles tendons are summarized in **Table 6**.

Clinical results with Achilles tendon allografts are comparable to those with BPTB allografts with recorded failure rates in the range of 2% (Rappe et al 2007, Malinin et al 2002). Siebold et al reported better results with Achilles tendon versus BPTB allografts (Siebold et al 2003).

Age of Patellar and Achilles Tendon Allograft Donors

Donor and recipient age discrepancy has been of concern from the start of the use of ACL replacement allografts. Originally it was thought that older donors would provide substandard allografts for younger athletic recipients (McGuire and Hendricks 2009). This concept is absolutely correct, despite a published study that claims there are no age differences on the biomechanical properties of BPTB allografts (Blevins et al 1994). Our findings do not support the conclusions concerning the lack of donor age on the biomechanical differences between BPTB constructs. In fact, as can be seen from **Table 4**, the difference in the biomechanical properties of BPTB constructs obtained from individuals of various ages are considerable and statistically significant. Allografts from donors 41–60 years of age are almost half as strong as those under 30 years of age. On the other hand, the differences are not as well expressed with Achilles tendon grafts. Achilles tendons are the strongest in individuals 41–50 years old and, oddly enough, weakest in individuals younger than 30. The remaining age groups are comparable (**Table 7**). In view of these findings, the current practice of transplanting ACL replacement allografts from older donors into younger individuals is justified in the case of Achilles tendon allografts. However, transplanting BPTB allografts from older donors into young, active, athletic individuals might not be such a good idea. The justification that tissue from older donors is being used on increasingly younger recipients with apparent success might be an illusory one. In the case of BPTB, no data exists to support this sentiment. It is true that most graft failures can be attributed to faulty surgical technique, rehabilitation errors, early re-injury and unidentified pathology. However, deficiency of the allografts themselves has been in many instances responsible for allograft failures (Gorschewsky et al 2005, Rappe et al 2007, Fideler 1995).

Muscle Tendon Allografts

Throughout the history of ACL reconstruction, failures of synthetic ligament replacements became apparent (Klein and Jensen 1992, Macnicol et al 1991, Fu et al 1999). Therefore, interest has shifted to biological replacement. Because of complications with BPTB autograft harvest site morbidity, alternative autografts were sought. Among these, interest became focused on hamstring autografts with their relatively low donor site morbidity (Simonian et al 1997). Because of their inherent weakness, hamstring grafts have evolved from a single strand semitendinosus graft to a quadrupled semitendinosus/gracilis graft with both tendons combined and folded in half (Aglietti et al 1994, Maeda et al 1996, Wallace et al 1997). However, the limitation for this procedure, aside

Table 7: Biomechanical strength (in Newtons) of 25 isolated Achilles tendons according to age groups

Age group and #	*Mean*	*Standard deviation*	*Standard error*	*95% Confidence interval for mean*	
				Lower bound	*Upper bound*
<30 (3)	1520.33	509.26	294.02	255.26	2785.40
31–40 (6)	2058.55	904.72	369.35	1109.10	3007.99
41–50 (7)	3338.42	1148.48	434.08	2276.26	4400.59
51–60 (4)	2060.00	384.69	192.34	1447.86	2672.13
61+ (5)	2059.20	562.93	251.75	1360.22	2758.17

ANOVA

	Sum of squares	Degrees of freedom	Mean squares	Significance
Between groups	10172678	4	2543169.566	P = .024
Within groups	14236898	20	711844.923	
Total	24409576	24		

from the still present but lesser donor site morbidity (Simonian et al 1997), is the length of the hamstring allograft itself.

In addition to hamstring tendons, allografts can also be prepared from tibialis anterior tendons, tibialis posterior tendons and peroneus longus tendons. Biomechanical properties of these anatomic structures are given in **Table 8**. Semitendinosus and gracilis tendons are the weakest, with one strand being only some 70% of the strength of ACL (Noyes et al 1984). However, semitendinosus and gracilis composite grafts are equal approximately to an 11 mm patellar tendon graft.

Tendon allografts perform satisfactorily clinically, comparable to autografts. Initial concerns with stretching over time have not been substantiated. Studies dealing with the subject provide data showing that non-irradiated hamstring allograft constructs do not stretch with time and do not demonstrate laxity (Edgar et al 2008). This data can probably be extrapolated to apply to other non-irradiated freeze-preserved tendon allografts.

It is generally believed that allograft tendons incorporate slower than do comparable autografts. However, data on autograft incorporation is scarce. One retrieved autograft tendon, at 2 years, demonstrated large acellular areas (Malinin et al 2002).

In view of time dependent healing and incorporation of ACL replacement allografts, the initial strength of the graft is of importance. In this report there is considerable difference between individual allografts, with Achilles tendon and BPTB constructs being the strongest and semitendinosus/gracilis tendons being the weakest.

Table 8: Strengths of allografts from randomly selected donors of both sexes from a donor pool of the University of Miami Tissue Bank

Comparative Strength of Tendon Allografts			
Anatomic structure	*N*	*Load to failure (in Newtons)*	*SD*
Tibialis posterior tendon	20	1867	806
Tibialis anterior tendon	22	1348	671
Peroneus longus tendon	16	1061	406
Semitendinosus-gracilis tendons	10	697.5	271.1

There was no statistically significant difference comparing strength (N) and donor age.

Fixation of tendon allografts in the tunnel presents a problem. Techniques include staples and endobutton, interference screws, cross-pins and others. Both the endobutton and tying of sutures over periosteal button constitute a weak link in the biomechanical properties of ACL replacement constructs (**Table 9**). Until healing of the graft-to-host bone takes place, this, rather than the integrity of the grafts themselves, constitutes the limitation of the ACL replacement techniques (Jarvinen et al 1995).

Nevertheless, the initial strength of the allograft is important. Until healing and replacement of the allograft takes place, the function of the knee will depend on it.

Table 9: Biomechanical properties of intra-articular ACL replacement allografts

Fixation in the Tunnels			
	N	*Load to failure (in Newtons)*	*SD*
Achilles tendon tib bone plug-staples	2	755.5	90.5
BPTB Fem 3.5 inter. Screw, tib-staple	2	438.5	133.6
Fem 7 inter. Screw tib-staple	2	363	56.6
Fem 7 mm tib-9 mm Kur. inter. screw	2	607	452
Fem 5.5 tib 9-mm Kur. inter. screw	2	366	108

Source: Malinin et al. 2001

Clinical comparison between autografts and allografts of bone-patella tendon-bone constructs reveal good results in both groups. There were no statistically significant differences of patients evaluated at 3, 6, 12 and 24 months postoperatively for the range of motion. However, early on there was a trend for more allografts to glide on the pivot shift as compared to autografts (20% versus 7%) (Shelton et al 1997).

Fascia Lata

Fascia lata offers very versatile autografts as well as allografts. Both fascia lata autografts and allografts have been transplanted into various anatomic sites with various degrees of success.

Kirschner was the first to suggest that fascia lata might be used to repair defects in tendons (Kirschner 1910). Fascia grafts have been used by Dean Lewis to transplant a muscle to a new insertion and in the form of a tube to repair defects in tendons. Rothschild corrected a functional disturbance of paralyzed deltoid muscle with a fascia lata graft (Lewis and Davis 1911, Rothschild 1910). Payer, mentioned by Kirschner, used pieces of fascia lata to correct a congenital ptosis in a 12-year-old boy. Thus, by the beginning of the last century, fascia lata became established as a multipurpose graft with over 100 years of history.

Fascia lata is a collagenous fibrous tissue structure located between the muscle and the subcutaneous tissue. Fascia lata is composed of two layers, a superficial layer and a deep layer. Superficial fascia is attached to the subcutaneous tissue by fat. Deep fascia underlies the superficial layer. It is strong and composed of densely packed collagenous fibers. Deep fascia invests individual muscles or muscle groups. Outgrowths of deep fascia form intermuscular septa that can also terminate in muscle attachments. Fascia lata acts on a spring; when muscles contract, it shifts in position. By doing so, it compresses or relaxes the vasculature, facilitating flow of blood towards the heart. Because of the sliding properties of fascia lata, it is relatively easily removed intact as shown in **Figure 5**. Fascia lata is rich in blood vessels and nerves.

An important function of deep fascia is to act as a restraining envelope for contact against thick and resistant subcutaneous tissue, muscle, bone and the thin-walled veins and lymphatic vessels by enhancing the flow through unidirectional valves. Therefore, removal of large portions of fascia lata results not only in an unsightly muscle herniation but also in pathophysiological alteration of the thigh. For this reason whenever an allograft of the fascia lata is available and provides a function comparable with a fascia lata autograft, the latter might be preferable. Yasuma observed in 1962 that there were no clinical differences between the frontalis suspension procedure when either autologous or allogeneic fascia was used (Yasuna 1962).

In the early part of the last century, surgeons used fascia lata for a number of applications. There were some instances in which the usage of fascia lata was unsuitable for its function. With autografts, herniation of the muscle at the donor site occurs with relative frequency, but usually disappears within 8–10 months postoperatively (Peer 1955). This can be explained by scarification and partial regeneration of the fascia.

Fascia grafts have been and continue to be used with success in several anatomic locations. Fascia grafts have functioned well to replace gaps in the extensor but not in the flexor tendons of the hand (Thatcher 1949). Fascia grafts

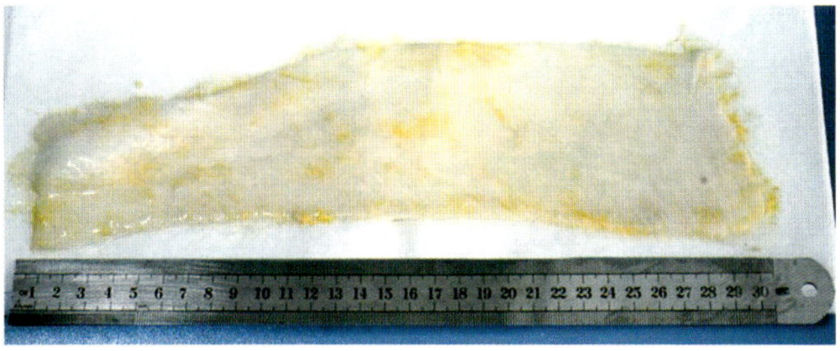

Figure 5 Excised facia lata provides adequate material for closure of large defects and for construction of ACL replacement allografts

have also been used for hernia repair and used extensively in ophthalmic surgery. It is from these endeavors that we have gained knowledge of the long-term fate of fascia lata autografts. A seminal paper on this topic describes a fascia lata autograft used to correct ptosis retrieved 42 years later (Orlando et al 1985). On a microscopic examination, the tissue of the graft was viable, but demonstrated increased vascularization, fibroblastic infiltration and integration with surrounding structures.

Autologous and allogeneic fascia lata have been used to correct many problems in ophthalmic surgery. A paper by Flanagan and Campbell describes their use to correct lid retraction secondary to thyroid ophthalmopathy, cicatricial entropion and extruding orbital implants. Ophthalmic surgeons of all subspecialties have used these procedures with good results for many years (Flanagan and Campbell 1981).

In other surgical applications, fascia lata autografts have yielded some interesting data dealing with their biology. Specimens of regenerated fascia lata used in hernia repair were removed at 8 months, 2 years and 15 years with the patient's permission during subsequent revisional operations. Regenerated fascia lata contained two layers, an inner and a transverse layer. The prominence and thickness of the inner layer strongly suggested the influence of the underlying muscle on fascia regeneration (Foshee 1947). This supports Gallie and Le Mesurier's statement to the effect that "experimental studies on animals and clinical experience with humans demonstrate that fascia lata, as an autogenous graft, will continue to live unchanged" (Gallie and Le Mesurier 1948). To further investigate the fate of autogenous fascia lata autografts, Peer buried these autografts in the abdominal fat of humans and removed the autografts at 24-hour intervals for 1–6 days. Grafts were also removed at 15 days and 18 days as well as at 4 months and 18 months. Patients volunteered as experimental subjects.

As would be expected, grafts transplanted between 1 day and 2 days show accumulations of RBCs around the graft and a beginning collection of cellular exudate, associated with blood vessel hyperemia fibroblast proliferation. Blood vessel channels in the graft were empty of blood. In specimens buried for 2 days, there were many polymorphonuclear granulocytes (PMNs) and lymphocytes present. There was no invasion of the graft by fibroblast. The grafts buried for 3 days showed an intense cellular reaction. Blood vessels within the grafts were distended and filled with erythrocytes and exudate cells. Endothelial lining appeared normal. Apparently circulation was re-established by end-to-end anastomosis between the grafts and the host blood vessels. At this point there was no invasion of the graft by fibroblasts. The re-establishment of circulation varied between grafts. In some grafts, circulation did not take place until the 4th and 5th day.

In the grafts buried for 5 days, there were dense collections of PMNs and lymphocytes. Plasma cells and eosinophils were also present. Fibroblasts contained dark staining nuclei. The number of fibroblasts increased or compared with controls. In fascia grafts transplanted for 6 days, the cells appeared to be viable. Edema had subsided.

After 15–18 days of transplantation, edema was absent and there was only a moderate infiltration of inflammatory cells within and around the graft. A fibrous capsule formed around the transplant. At 4 months, the fascia autograft had a normal appearance except it contained an increased number of fibroblasts compared to the controls. The graft transplanted for one and half years appeared exactly as did the graft after 4 months.

These observations were combined with those made on a retrieved allograft 41 years post-transplantation. The results clearly demonstrate that fascia lata autografts do not resorb; they initially evoke inflammatory response and revascularize in an identical fashion with skin autografts, i.e., by direct vascular anastomosis. The increased number of fibroblasts most likely indicates the replacement of the original cells in the autograft with new cells from the host.

Fascia lata autografts have proven to be clinically useful, but they have to be used correctly. Fascia is a tissue that is not generally found on the body surfaces or in lining hollow viscera. Therefore it should be transplanted as a completely buried graft and not with one surface exposed. Fascia grafts can be used as substitutes for injured portions of ligaments or as their reinforcements. Gallie and Le Mesurier reported using fascia lata autografts with success for chronic shoulder dislocations, for visceroptosis and for ventral, inguinal and femoral hernias. Plastic surgeons employ strips of fascia to support the paralyzed side of a patient's face. Ophthalmologists use it extensively in ophthalmic reconstructive procedures. In short, fascia lata autografts have many valuable clinical applications. However, as with tendons, limitations are the complication and morbidity at the donor site. For this reason, in many instances, surgeons prefer to use preserved fascia lata allografts.

The biology of preserved fascia lata allografts is rather different from that of autografts. Vascularization of the allografts proceeds by direct vascular ingrowth, not by end-to-end anastomosis. The initial inflammatory reaction to fascia allografts is stronger and of longer duration than that with autografts. Cells in fascia allografts appear to be replaced eventually with fibroblasts from the host. Nevertheless, fascia lata allografts have found a significant niche in reconstructive surgery, particularly in the replacement of cruciate ligaments of the knee.

Bright and Green (1981) studied 47 patients in whom freeze-dried fascia lata was used to effect ligament and shoulder repair and hip and ankle stabilization. Results were excellent in 72, good in 20 and poor in 8% of the patients. Allograft fascia did not stretch, leading the authors to the conclusion that freeze-dried fascia can provide satisfactory clinical results.

Fascia lata allografts have also been used as pubovaginal slings for treatment of intrinsic sphincter deficiency. There was no significant difference in the clinical outcomes between the allograft and autograft groups of patients. The success of the operation was high and no complications related to the allografts were encountered (Wright et al 1998). On the other hand, experience with irradiated and freeze-dried fascia lata was not a favorable one. High material failure rate with irradiated fascia lata allografts suggest that such tissue should not be used in urogynecologic procedures (Fitzgerald et al 1999).

Substitution of injured ACLs with fascia lata allografts has been reported in many publications. Both frozen and freeze-dried allografts have been used. Experimental and clinical data on fascia lata allografts used intra-articularly as ACL substitutes are abundant.

Curtis and co-workers studied ACL reconstruction in dogs. In these experiments, freeze-dried fascia lata was reconstituted and folded into tubes placed through tunnels and fixed with staples. At 24 weeks postoperatively, grafts provided knee stability with only 2 mm increases in anterior laxity. At 12 weeks, the grafts were revascularized. At 24 weeks, the grafts grossly resembled normal ACLs. However, histologically there was necrosis and disorganization in the central portion. Biomechanically, allografts achieved 67% tensile strength of contralateral ACLs (Curtis et al 1985).

In 1990, Noyes and co-workers compared the results of fascia lata ACL substitute grafts in 22 patients with results of 22 BPTB recipients. The majority of fascia lata allografts were sterilized with ethylene oxide and freeze-dried. Fascia lata grafts were rolled into tubes 10–11 mm in diameter and fixed to the tibia and femur with staples. In the entire group of 47 patients, 42 had good or excellent results. There was no statistically significant difference in the results between the two types of grafts (Noyes et al 1990). Defrere and Francart also used freeze-dried, but aseptically processed fascia rolled into a tube in 70 patients requiring ACL replacements. In these patients, followed for 4.5 years, there were 95% excellent or good objective results. Biopsies of six grafts revealed superficial and deep revascularization as early as 4 months postoperatively. In the later specimens, the grafts were covered with synovium (Defrere and Frankart 1994).

Goble and associates described 100 patients who received freeze-dried fascia lata allografts for ACL deficiency. They tightly rolled the tissue around a 7/12-inch diameter hex driver and sutured the leading edge of the graft. The resulting tube, 11–12 mm in diameter, was passed through the tunnels and secured to a threaded plastic stud. The results were excellent in 80% of the patients, good in 15% and fair in 5% (Goble et al 1993). Pritchard and associates recorded similar results in 67 patients. In this series, 87% of the patients had normal or nearly normal knee function and an average Lysholm score of 92.3. These results obtained after 10 years indicated the remarkable durability of fascia lata allografts that also demonstrated the longitudinal orientation of fibers and normal cellularity (Pritchard et al 1995). Other studies involving freeze-dried fascia lata allografts recorded good to excellent results in the 80% range (Myers 1991).

The cited clinical experiences as well as experimental studies, too numerous to cite, bring forth biological and biomechanical properties of fascia lata which might well qualify it as one of the best biologic materials for ACL substitution allografts. Specifically, freeze-dried fascia lata allografts might have several advantages over other grafts. Once reconstituted it can be rolled or folded to any desired diameter. It maintains its original strength after reconstitution (Thomas and Gresham 1964). Its biomechanical properties are comparable to other ACL replacement grafts. It undergoes ligamentization and maintains its structure for a very long time. Fascia lata can be fashioned into various constructs, including those obviating the need for metallic, plastic or suture fixation (**Fig. 6**).

Preservation of ACL Replacement Allografts

The majority, if not virtually all ACL replacement allografts, are preserved either by freezing or freeze-drying. Each method has its own advocates and proponents.

The storage temperature of frozen allografts varies, but temperatures above 100°C are generally unsuitable for the long-term preservation of tissues (Malinin 1966). Cryopreservation techniques, although used by some tissue banks, might not be desirable for ACL replacement graft preservation. Cryopreservation, i.e., controlled velocity cooling in the presence of cryoprotective agents is designed to maintain viable tissues or cells. This is precisely what one does not want in a biologic scaffolding graft that will be replaced by host cells in the process of ligamentization. It is well known that viable allogeneic cells and tissues are much more antigenic than frozen (non-viable) or freeze-dried tissues and cells. Transplantation of highly antigenic allografts may cause marked rejection, something that must be avoided if at all possible.

Frozen allografts are popular and convenient. Once thawed, they are pliable and easy to handle. Since frozen allografts are usually not thawed until the surgeon is sure he/she will transplant the graft, grafts can be returned if not used.

Clinical results with frozen ACL replacement allografts are good and extensive. According to several authors the results are comparable to those obtained with autografts (Melberg and Indelicato 1993, Harner et al 1996).

Freeze-dried allografts have their own proponents. The principal inconvenience with freeze-dried allografts is the time needed for their reconstruction. If freeze-dried tendon allografts are manipulated or cross clamped before fully reconstituting, the stiff fibers will fracture. How long the rehydration will take depends on packaging. Grafts stored in peel pouches or similar packages easily take an hour or longer, depending on the temperature. Freeze-dried allografts in vacuum-sealed glass containers (**Figs 7A and B**) can be reconstituted in about 15 minutes providing the vacuum is not completely released from the container.

Preservation of tendon allografts by solvent dehydration is not employed frequently. Clinical results with this technique are unfavorable (Gorschewsky et al 2005, Defrere and Frankart 1994).

Preservation in Cialit solution (Ethylmercury salt) is not generally used for ACL replacement grafts. Permeation

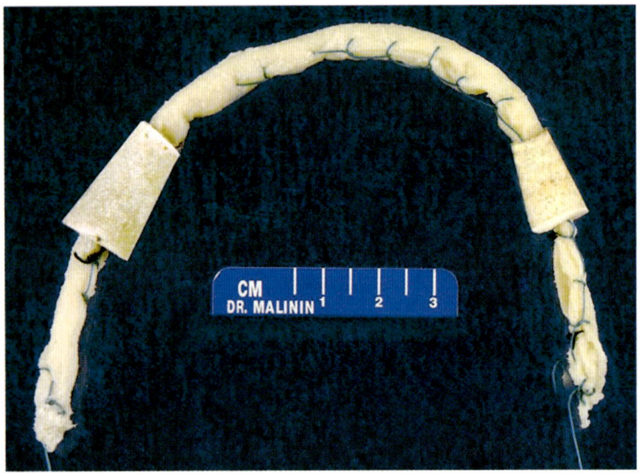

Figure 6 Rolled fascia lata allograft threaded through two conical bone plugs at each end. Conical bone plugs are inserted in the tibial and femoral tunnels drilled to accommodate and retain them. The soft tissue graft is retained in the plugs by tying a knot or folding the same. Tibial plug is inserted first. Desired tension is achieved by pulling the tendon graft through the femoral plug and securing it at a desired tension. For additional security, bone plugs can be supplemented with interference or transfixion screws

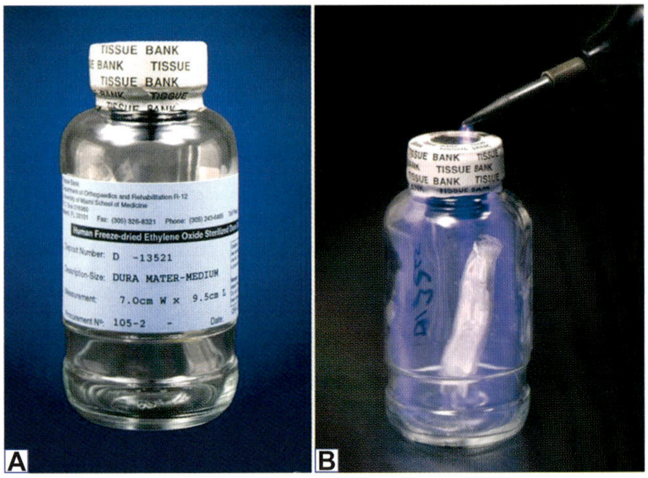

Figures 7A and B (A) Allograft stored under vacuum in a glass container. Reconstitution solution is injected through the rubber stopper, thus allowing the allograft to be rehydrated under vacuum; (B) Presence of vacuum can be ascertained by a "spark test"

enhanced extraction technique and chloroform solvent extraction technique did not produce biomechanically suitable grafts, at least in animals (Zimmerman et al 1994).

Secondary Sterilization

Ionizing irradiation is a secondary sterilization technique used by many tissue banks. However, in the case of ACL replacement allografts, it is considered an unsuitable method that produces poor results. The main reason one irradiates bone allografts is to prevent the transmission of infections, including those caused by HIV. Irradiation is ineffective against prions and, thus, cannot prevent transmission of Creutzfeldt-Jakob disease. Relatively high doses of irradiation are needed to inactivate HIV in bone, but the actual dose estimates vary. Although a 15–25 kGy dose is commonly used, 15 kGy would not reliably inactivate HIV in bone allografts. Therefore, doubling the dose to 30 kGy may be necessary. The same would apply to soft tissue allografts (Conway et al 1990).

Transplantation of allografts for ACL reconstruction has increased dramatically in the past 20 years. An extensive amount of practical information has been gained from this experience. Multiple studies support transplantation of non-irradiated tissue allografts in cruciate ligaments of the knee reconstruction (Rappe et al 2007, Bach et al 2005, Harner et al 1996, Kleipool et al 1998, Noyes and Barber-Westin 1996, Peterson et al 2001, Shelton et al 1997, Wainer et al 1988).

The reason tendon allografts are subjected to irradiation is the fear of transmission of infection with cadaver tissue. In fact such a possibility is rather remote in the present circumstances. Presently, the possibility of HIV carrier donors not being recognized is in the neighborhood of 1:1.5 million (Buck et al 1988, Barber et al 2003). This does not mean actual transmission, but the chance of the donor being a carrier. A case of HIV infection has actually occurred in the so-called "Life Net Incident" (Simmonds et al 1992).

In retrospect, the incident would have been avoided, had the medico-social history of the donor which came to light after the infection of transmission occurred, were known at the time. There were no subsequent transmissions of HIV with musculoskeletal tissue allografts. Transmission of bacterial infections had also occurred. The highlighted case was that of an ACL replacement allograft recipient who acquired *Clostridium septicum* infection, through a tendon allograft, from which he died (CDC 2002). A few other bacterial allograft associated infections have been reported in the same time framework (CDC 2001, 2003). These too could have been avoided by the use of careful and complete microbiologic monitoring. However, the CDC made an ill-advised recommendation that all allografts be secondarily sterilized. In the case of ACL replacement grafts, this led to a large number of allograft failures with accompanying clinical complications including reoperations.

Many studies comparing non-irradiated to irradiated ACL allografts demonstrate the inadequacy of the latter. Rapp and co-workers state that they have discontinued using irradiated allografts (Rappe et al 2007). It was discovered that a decrease in irradiation dose does not ameliorate the ill effect of radiation. Fiedler et al showed that the initial biomechanical strength of tendon allografts was reduced by 15% by exposure to 2 Mrads of irradiation (Fiedler et al 1995)*. He also showed that there was a dose-dependent significant decrease of the biomechanical integrity of grafts receiving 3–4 Mrads (Fiedler et al 1995). After exposure to 2 Mrads of irradiation, significant decrease in the tensile strength and modulus of tendon allografts was likewise shown (De Dayne and Haut 1991). Furthermore, increase in irradiation dose resulted in microscopic alterations in tissues. In animals, Gibbons et al, as well as Schwartz and colleagues, proved that irradiation at 2 Mrads adversely affects patella tendons by reducing stress and strain values.

High doses of irradiation (4 Mrad) adversely affect the viscoelastic and structural properties of ACL allografts (Gibbons et al 1991, Schwartz et al 2006). The failure rate with irradiated allografts exceeds 30%, a figure much too high to be acceptable (Sun et al 2009).

Ethylene oxide (ETO) gas penetrates materials such as cloth, cellophane and tissues. It has bactericidal and virucidal properties (Kerulek et al 1970, Klarenbeek and van Dongen 1954, Sidwell et al 1969, Prolo 1980) and has been used for tissue sterilization with success for many years (Prolo 1980). When tissue is sterilized with ethylene oxide, residue as well as its secondary products, ethylene glycol (EG) and ethylene chlorohydrin (ECH), remain in it.

*Irradiation doses are given as they appeared in cited publications

Therefore, before transplanting ETO sterilized allografts, the residues as well as the residual gas must be removed. These residues, in high enough concentrations, cause hemolysis and inflammation. For this reason, the FDA has published a limit on the quantity of these compounds on the implantable devices (Gardner 1978). This requires post-sterilization aeration and washing.

Unfortunately, ethylene oxide sterilization received bad press because of intra-articular reactions following transplantation of ETO sterilized BPTB allografts. The same response could not be elicited in animals (Jackson et al 1990). This discrepancy in results could probably be explained by quantitative differences in the ETO residuals instilled in the joints.

No subsequent incidence of synovitis induced by ETO sterilized tendon grafts has been reported. Thus, ETO sterilization of ACL replacement allografts remains a reasonable sterilization option.

Incorporation of ACL Replacement Allografts

The aim of ACL replacement with allografts is the reproduction of ACL biology. Anterior cruciate ligament is incorporated through junction with subchondral bone by means of Sharpey's fibers. Information concerning autografts is relatively scarce, and comes mainly from the specimens obtained during revision arthroplasties (Ishibashi et al 2001). Since time and sequence of healing in animals is vastly different from that of human studies, usage of animal models are of little help. In clinical studies, the maturation process has been evaluated mainly in biopsy specimens. These, of course, present for study only small and superficial portions of the grafts. On the basis of light and electron microscopic studies Abe et al concluded that autografts remain immature for at least 1 year (Abe et al 1993). Malinin et al described a complete autograft obtained 2 years post-transplantation in a patient with a periosteal osteosarcoma of the distal femur. Although portions of the autograft were vascularized and repopulated with cells, the central portion remained largely acellular (Malinin et al 2002). This finding supports the findings of Abe and colleagues. Rogoff et al's contention that autograft patella tendons undergo a prolonged process of ligamentization that may take up to 3 years also appears to be correct and similar to changes observed in allografts (Rougraff et al 1993).

Material available for the study of human allografts is much more voluminous than that for human autografts. However, the number of whole retrieved allografts available for study is small. Previously reported biopsy studies correlate well with the changes occurring on the surface of tendon allografts (Levitt et al 1994, Alm 1973, Horstman et al 1993).

Based on biopsy and retrieved allograft specimens, allografts undergo several phases of healing and incorporation. In the early phase, necrosis and inflammatory reaction are prominent. This is followed by vascularization, cell recruitment and cellular proliferation. Initially there is collagen replacement and allograft remodeling followed by completion of ligamentization. The completely remodeled tendon allograft is surrounded by synovium and biomechanically intact.

Intra-articular portions of the graft are composed of the doubly refractile collagen bundles. However, in the center there is still necrotic tissue (**Fig. 8**).

Allograft tendon appears similar in both the tibial and the femoral tunnels. At completion of the maturation period, the tendons were tightly fixed to the cancellous bone to the point that the allograft tendon edges blended with the walls of the tunnel (**Fig. 9**).

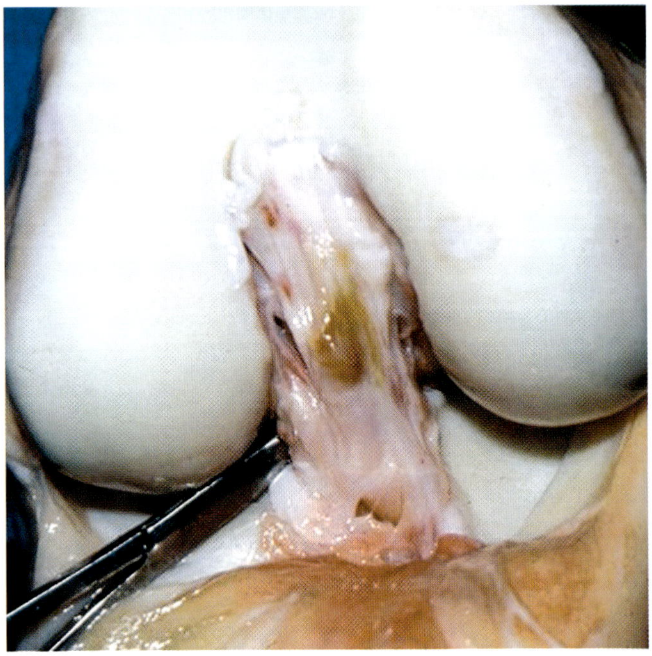

Figure 8 Frozen BPTB allograft 2 years post-transplantation. The graft is covered with a smooth and glistening synovium. The allograft appears to have completed its maturation cycle

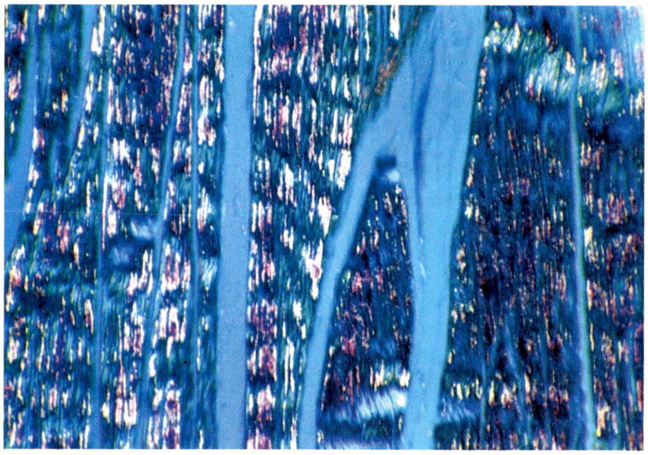

Figure 9 Freeze-dried Achilles tendon allograft, 3.5 years post-transplantation. Doubly refractive collagen fibers are visible in polarized light × 100

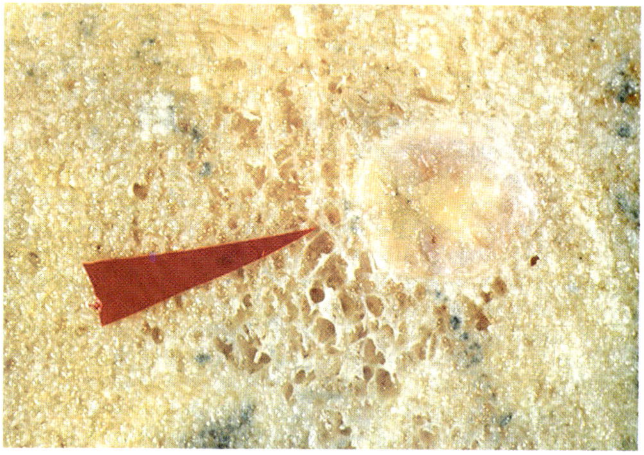

Figure 11 Frozen bone-patella tendon-bone in a femoral tunnel, 2 years post-transplantation. The margins of the tendon allograft blend into the walls of the bone tunnel. In the center of the graft, there is a yellow area indicating the necrotic area of the graft

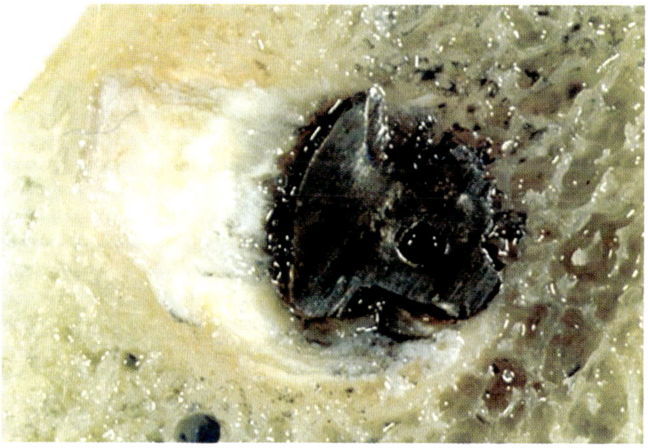

Figure 10 Cut section of an interference screw retaining the graft in the tunnel. The metal screw had undergone osteointegration

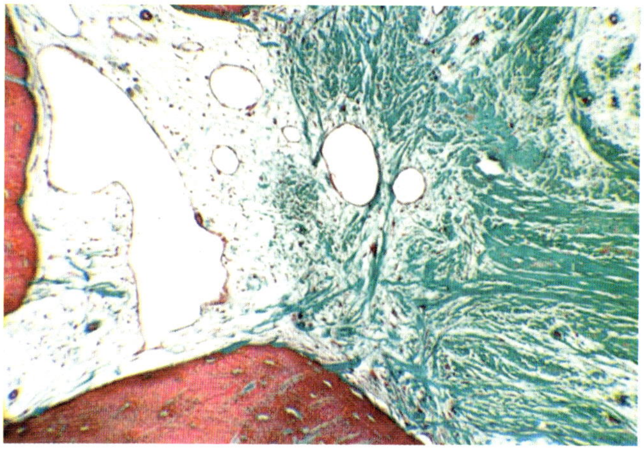

Figure 12 Freeze-dried Achilles tendon 3.5 years post-transplantation. Specimen is taken near the entrance into the bone tunnel. Vascularization and synovial tissue are present (Masson trichome stain × 100)

The metal interference screws went through the osteointegration process and were firmly secured to the tendon in the tunnel (**Fig. 10**).

As the tendon maturation progressed in three and a half years, the allografts near the tunnel entrance were vascularized and then came in intimate contact with synovial tissue (**Fig. 11**).

In the center of the tunnels, Sharpey's fibers connected dense collagenous tendon to the surrounding bone. Finally in 3.5–7 year specimens, complete ligamentization took place (**Figs 12 and 13**).

As maturation and ligamentization became complete, the architecture of the transplanted tendon changed to the appropriate normal structure (**Fig. 14**).

COMPLICATIONS AND FAILURES

The major and most serious complications of allograft tendon transplantation in the knee are, of course, infections. These incidents have been described in the preceding pages. It must be borne in mind, however, that the incidents of allograft-associated infections have been rare. None have

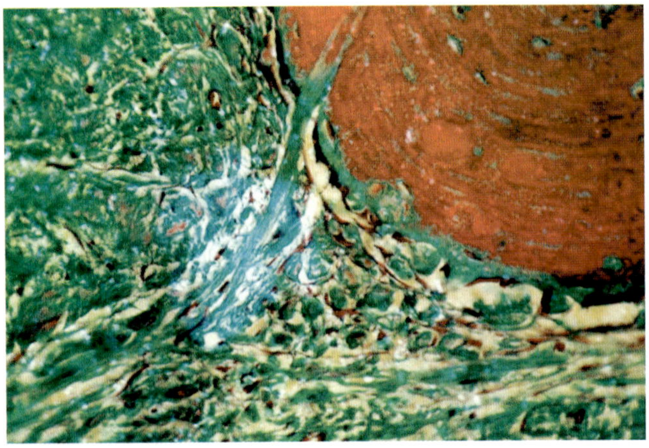

Figure 13 Same specimen as in Figure 12 but section is taken from the center of the tunnel. It shows Sharpey's fiber connecting the tendon to the surrounding bone. (Masson trichome stain × 50)

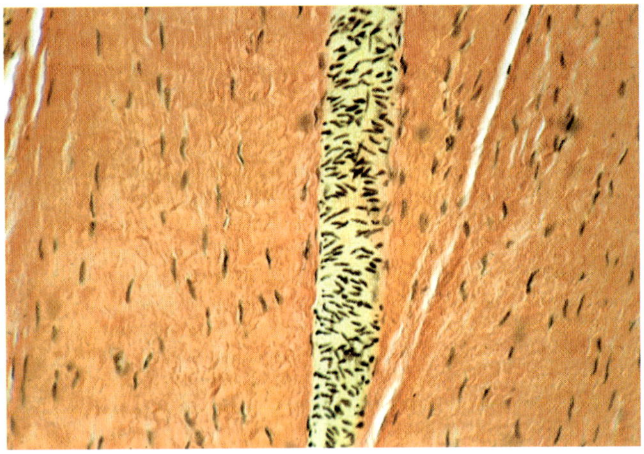

Figure 14 Ligamentized Achilles tendon allograft 3.5 years post-transplantation with a blood vessel in the center. Collagen fibers are arranged in rows with nuclei oriented in the same direction. (Van Gieson stain ×250)

been reported recently. Furthermore, they can be avoided by exercising due diligence in donor selection and tissue processing.

Mechanical and anatomic inadequacies can cause problems and have been reported (Tarandy et al 1995). These can be avoided by careful recipient-donor matching and by a visual inspection of the allograft prior to transplantation. Radiographs of the graft might also be helpful.

Several authors report allograft failures. These are difficult to evaluate as frequently the authors do not state how the allografts were prepared. Sterling et al, for example, attributed higher than expected failure rates in cruciate ligament allograft reconstruction to the time interval between freezing and freeze-drying. However, neither the freeze-drying regimen nor the temperature and length of frozen storage were provided.

On the other hand, the high failure rate of Tutoplast allografts is clearly attributable to the allograft properties (Gorschewsky et al 2005). Other failure rates such as that of fascia lata allografts in urogynecologic procedures can be attributed to the irradiation of these grafts (Fitzgerald 1999).

There have been several reports stating there is a high failure rate with ACL allografts, particularly in young athletes. However, as in other studies, details are difficult to uncover. Tibialis anterior allografts have been particularly singled out as the graft producing high failure rates in younger, active patients (Singhal et al 2007). Other studies, on the other hand, report similar outcomes for patients aged 40 and those younger than 40 with BPTB reconstructions (Barber et al 2003).

As with any surgical procedure, complications and failures occur with ACL replacement grafts. The prevailing opinion tends to favor autografts in young, active athletes. This seems justified in view of the faster healing of autografts and the desire to resume sporting activities. However, one must be mindful of other considerations. There is a learning curve for the operation and technical pitfalls for surgeons with a limited experience with the procedures. One complication that leads to graft failure is the inability to adequately tension the graft. In the future as techniques, instrumentation and allograft fixation improve, failures due to technical omission will undoubtedly decrease.

Pericardium

Both allogeneic and xenogeneic pericardium including pericardial patches and pericardial slings, are used clinically. However, information regarding their biology and clinical efficacy is generally not available in the scientific literature. What is available is a plethora of commercial promotional literature and advertisements. This does not provide an adequate basis for discussion.

Dura Mater Allografts

Dura mater provides versatile grafting material. Although initially intended for the closure of pachymeningeal defects (Rosomoff and Malinin 1976, Abbott and Dupree 1971), it found a place in many other surgical applications. Dura mater allografts have been used in oral and maxillofacial surgery (Kniepkamp 1994), gynecologic surgery (Sevin and Malinin 1990, Angioli et al 1996, Jerrel et al 1987), hernia repair, repair of the abdominal wall (Quilici et al 1985), reinforcement of colon anastomosis (Takahashi et al 1994), covering of femoral blood vessels (Fiorica et al 1991) and others.

Unfortunately, dura mater allografts received bad press because of the transmission of Creutzfeldt-Jakob disease (CJD), with the grafts produced in Europe by a single manufacturer. The highest incidence of CJD transmission occurred in Japan (CDC 1997).

These incidents were associated with a single brand of dura mater allograft, Lyodura. The other brand also distributed in Japan, Tutoplast, was not reported to be associated with CJD transmission. Sporadic cases of CJD transmission associated with Lyodura transplantation were also reported from Spain, Italy, USA, New Zealand and Great Britain.

It must be noted, however, that the number of CJD transmissions was not very high considering that, in the period covered by the surveys, some 20,000 dura mater allografts were transplanted in Japan alone. However, had there been only one incident of transmission, this would have been one of the many incidences. Therefore, dura mater allografts were put under a microscope, and various measures were implemented to avoid the repetition of similar occurrences.

In Japan, transmissions of CJD with dura mater allografts were investigated by a committee appointed by the Ministry of Health and Welfare (Tateishi 1997). All cases identified in Japan were related to a commercial preparation, Lyodura, first imported in Japan in 1973. Most of the CJD transmissions were recorded between 1984 and 1987. Between 1979 and 1996, 43 patients became infected with CJD through dura mater allografts. Seventy-seven percent of these received their grafts during 1983–1987.

The infection has occurred at a rate of approximately 1 per 3,000 Lyodura recipients. In Japan, some 60% of dura mater allograft recipients received Lyodura. The others received Tutoplast. Only one possible, but not definitive, transmission was recorded with dura mater allografts other than Lyodura. One case was reported in the US, six in Great Britain, four in Spain, two in Italy and a few in other countries.

Transmission of CJD with dura mater allografts suddenly decreased when new decontamination procedures with NaOH were introduced and individual processing of dura mater was insisted upon. Tutoplast and the University of Miami Tissue Bank dura mater allografts had been processed individually to begin with (Malinin, Oster 1997). Since 1996, there has been no incident of CJD transmission with a dura mater allograft.

The current practice involves rigorous donor selection, with elimination from the donor pool of all potential donors with suspected or present degenerative CNS disease, and histological examination of the cerebral cortex.

Before the publicity associated with CJD transmission by dura mater allografts, commercially prepared dura mater allografts were widely promoted and advertised throughout the world, including the US (**Figs 15 and 16**). The practice has been curtailed since the revelations of CJD transmission.

Dura mater allografts in the US are now produced in accordance with FDA guidelines (1997). No incidence of CJD transmission with dura mater allograft has been reported in the last 11 years.

Dura mater is a unique and complex anatomic structure which comprises one of the three meningeal layers covering the brain. The three layers combined (dura mater, pia-arachnoid, and the pia mater) are often referred to as pachymeninges. Dura mater is impervious to water and is tough. It is particularly adept to contain cerebrospinal fluid.

Dura mater is made up of two layers. The superficial layer, also called the endocranium, serves as the inner periosteum of the skull. The deep layer, which is the actual dura mater, is in contact with the brain.

The falx cerebri and tentorium cerebri, infoldings of the dura mater, are generally not used for transplantation.

The cerebral surface of the dura mater is smooth. Some areas, for example, excised dura mater specimens, contain numerous vascular channels filled with blood. Blood from these specimens is difficult to remove by washing. In other areas, blood vessels are empty (**Figs 17A and B**).

The success of dura mater allografts is due, primarily, to its mechanical properties and strength. These facilitate the ability of dura mater allografts to form water tight barriers and to withstand mechanical loads. This can be explained by the results of studies using small angle light scattering (SALS) technique which demonstrates that dura mater is, in fact, an anisotropic rather than an isotropic structure (Sacks et al 1998), as previously believed by several investigators (Van Noort et al 1981, Mc Garvey et al 1984, Wolfinbarger et al 1994).

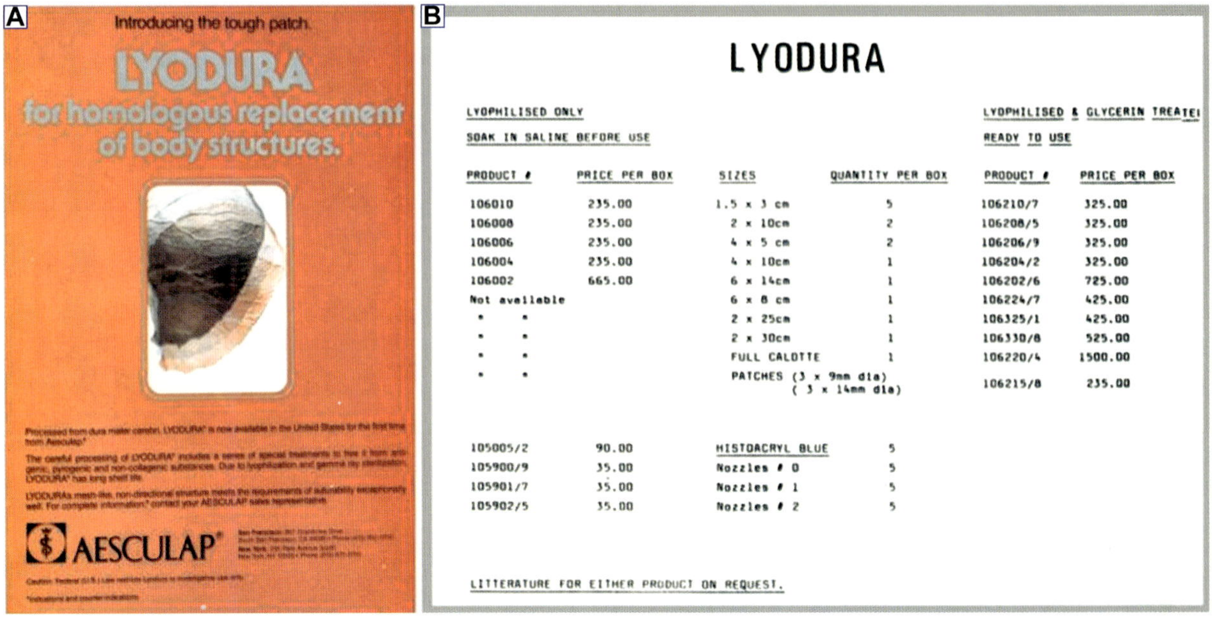

LYODURA

LYOPHILISED ONLY
SOAK IN SALINE BEFORE USE

LYOPHILISED & GLYCERIN TREATE[D]
READY TO USE

PRODUCT #	PRICE PER BOX	SIZES	QUANTITY PER BOX	PRODUCT #	PRICE PER BOX
106010	235.00	1.5 x 3 cm	5	106210/7	325.00
106008	235.00	2 x 10cm	2	106208/5	325.00
106006	235.00	4 x 5 cm	2	106206/9	325.00
106004	235.00	4 x 10cm	1	106204/2	325.00
106002	665.00	6 x 14cm	1	106202/6	725.00
Not available		6 x 8 cm	1	106224/7	425.00
" "		2 x 25cm	1	106325/1	425.00
" "		2 x 30cm	1	106330/8	525.00
" "		FULL CALOTTE	1	106220/4	1500.00
" "		PATCHES (3 x 9mm dia) (3 x 14mm dia)		106215/8	235.00
105005/2	90.00	HISTOACRYL BLUE	5		
105900/9	35.00	Nozzles # 0	5		
105901/7	35.00	Nozzles # 1	5		
105902/5	35.00	Nozzles # 2	5		

LITTERATURE FOR EITHER PRODUCT ON REQUEST.

Figures 15A and B (A) Advertisement in an American neurosurgery journal promoting Lyodura; (B) Promotional sheet mailed to neurosurgeons in the USA

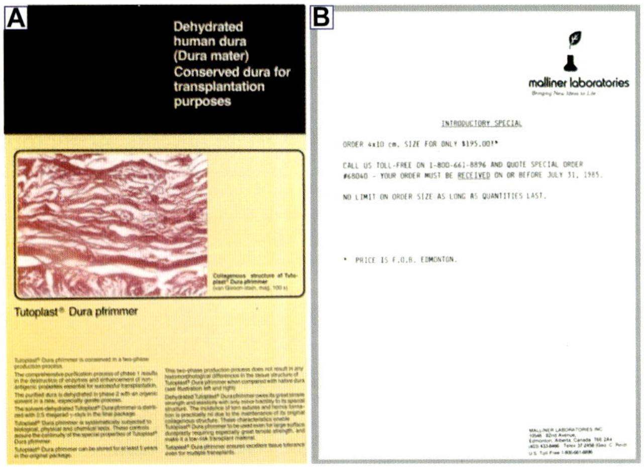

Figure 16 Advertisement promoting Tutoplast dura mater allografts

Isotropy, of course, means uniformity of all fiber orientations and identical properties in all directions, while anisotropy means being directionally dependent.

Further, SALS studies emphasized unique fiber architecture of dura mater (Hamann et al 1998) as distinguished from other connective tissue membranous structures. Collagen fibers over most calvaric dura mater have multidirectional orientation. Cranial dura mater is made up primarily of sparse fibroblasts and an extensive extracellular collagen network with interspersed elastic fibers **(Fig. 18)**.

Collagen is interwoven in lamina and oriented in various directions. The collagen fibers are relatively uniform in diameter, measuring about 20–55 nm (Vandenabeele et al 1996). Individual collagen fibers, in longitudinal sections, are cross striated with a periodicity of 67 nm along their length **(Figs 19A and B)**. There is a close relationship between the microstructural anatomy and the biomechanical properties of the dura mater (Runza et al 1999). Like other connective tissues, dura mater has an extracellular matrix composed largely of collagen fibers which have the strongest structural components of soft tissue proteins (Fung 1981).

The biomechanical properties of dura mater depend on the multilayered distribution of collagen thickness, which has regional distribution. However, the initial tensile strength of multiple dura mater samples reveals the strong mechanical properties of this structure. The average force to failure of dura mater strips on average measure 37 +/- 0.98 N (Wolfinbarger et al 1994).

Biological compatibility of freeze-dried rehydrated dura mater is good. In experimental animals, no adhesions were formed between the cerebral portion of the dura mater allografts and the underlying cerebrum **(Figs 20A and B)**.

Dura mater allografts can be preserved by several methods: freezing, freeze-drying, chemical dehydration and immersion in glycerol. All are equally effective.

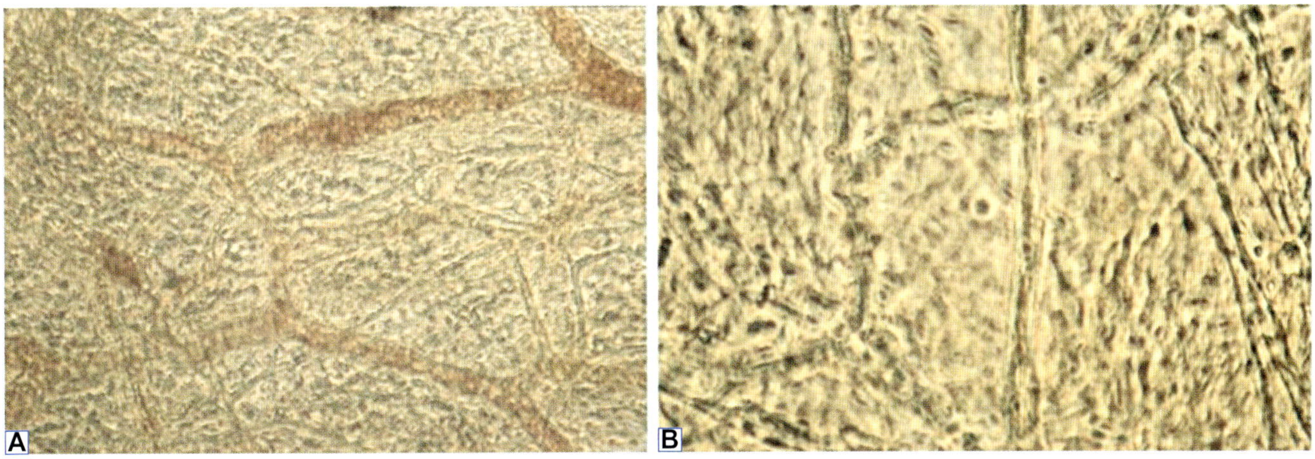

Figures 17A and B (A) An area with blood vessels filled with blood; (B) Cerebral surface of dura mater empty of blood

Figure 18 Cranial dura mater injected with radiopaque dye. The photograph taken in polarized light illustrates the distribution of collagen fibers around the blood vessels. Collagen fibers are not preferentially aligned around the meningeal vasculature

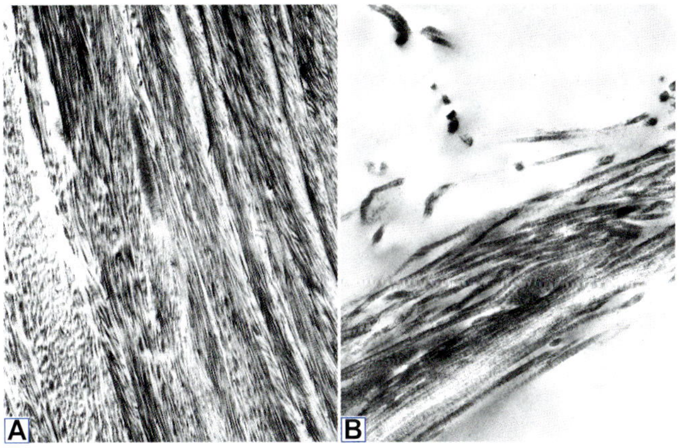

Figures 19A and B Electronphotomicrograph of dura mater showing (A) closely packed fibers with different orientation (original magnification 30,000); and (B) cross striations of collagen fibers (original magnification 60,000)

Although it is possible to excise dura mater aseptically, this is not usually done because of convenience. Instead, dura mater allografts are excised "clean" without aseptic precautions and are then sterilized secondarily. Two principal methods of sterilization are exposure to ethylene oxide and irradiation. Ethylene oxide requires removal of ethylene oxide residuals (ethylene glycol and ethylene chlorohydrin) following sterilization.

Irradiation, although effective for microbial and viral inactivation, is not effective for inactivating the CJD prion.

Dura mater allografts are usually prepared in standard sizes of 8–30 cm^2, 31–50 cm^2, 51–99 cm^2, and those greater than 100 cm^2.

Clinically, dura mater allografts found wide application in the closure of pachymeningeal defects. Biopsies of human allografts of dura mater showed these do not evoke immune response (**Fig. 21**).

Technique of closure of a large defect is illustrated in **Figures 22A to C**. Evaluation of immune response to freeze-dried dura mater allografts in animals showed the immune response to dura mater to be almost identical to that evoked by autologous fascia lata (Kadioglu et al 2002). In 170 neurosurgical patients, it was shown to be a safe and effective material for dural closure. In this study, low complication rates were encountered (Abbott and Dupree 1971).

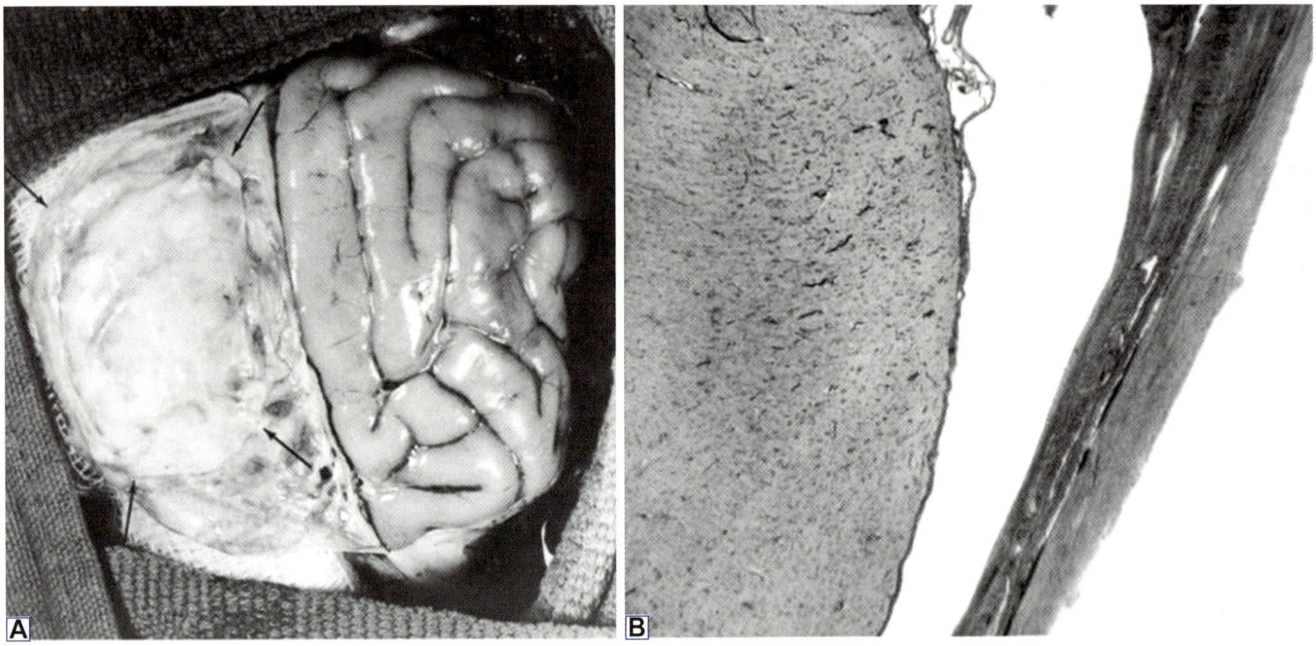

Figures 20A and B (A) Freeze-dried dura mater allograft in a canine model 6 weeks post-transplantation. Dura mater with transplanted allograft has been reflected over the contralateral hemisphere to reveal smooth and glistening cerebral cortex; (B) Histologic section showing lack of adhesion between the dura mater allograft and the underlying cerebral cortex

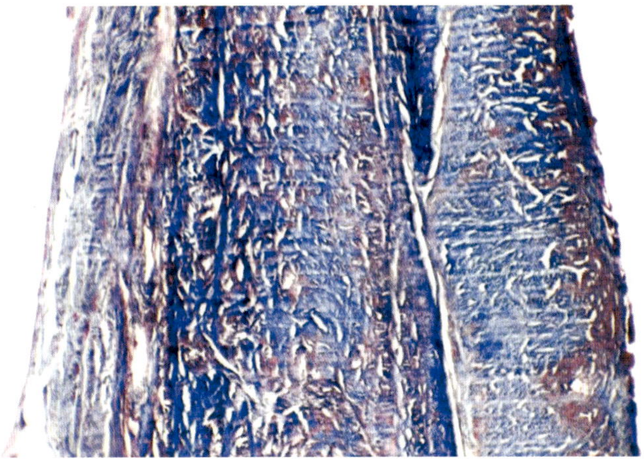

Figure 21 Biopsy of a dura mater allograft 3 years post-transplantation. (Mallory connective tissue stain ×50)

Dura mater allografts have also been used in the repair of abdominal hernias (Upshur 1958, Arango 1977), and as mentioned earlier, in gynecological surgery. Repair of large abdominal wall hernias with good results was described by Klein et al (1990).

A most interesting result has been obtained in Brazil, with heart valves constructed of dura mater (Puig et al 2003). These valves were used in Brazil for mitral or tricuspid valve replacement between 1971 and 1972, before being replaced with glutaraldehyde fixed porcine valves. Seventy patients with prosthetic valves of dura mater were followed for 30 years. The results were favorable. Freedom from reoperation in this series of 70 patients was reported to be 18%. Dura mater prostheses were associated with a lower incidence of calcification as compared with pericardial and porcine prostheses treated with glutaraldehyde.

In summary, dura mater allografts offer several alternatives for repair of various defects and structures dealt with by several surgical specialties. Unfortunately, many investigative activities have been discontinued because of unfortunate incidences with CJD transmission. However, once the fear is overcome, dura mater allografts should again become prominent in surgical endeavors.

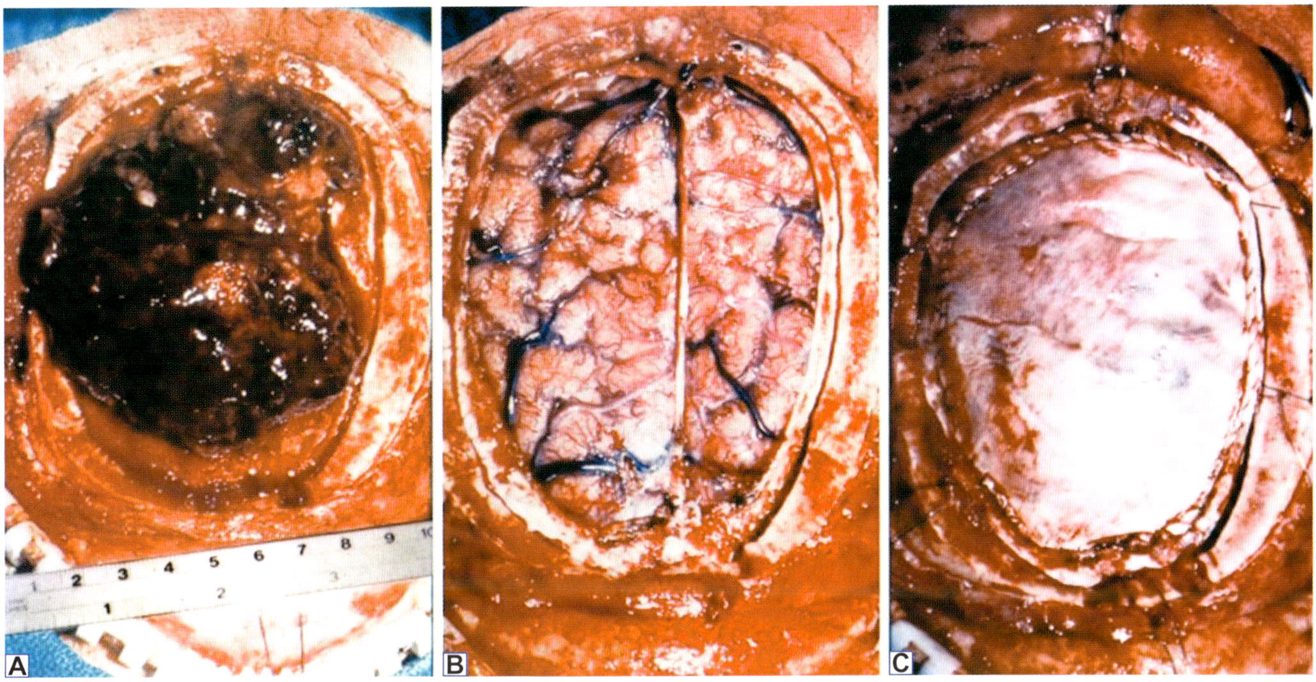

Figures 22A to C Techniques of closure of large pachymeningeal defect with dura mater allograft. (A) Bifrontal dura metastatic Ewing's sarcoma; (B) Exposed bifrontal lobes postdural and sagittal sinus resection; (C) Defect closed with a 10 × 7 cm dura mater allograft
Courtesy: Raymond Sawaya, MD

BIBLIOGRAPHY

1. Abe S, et al. Light and electron microscopic study of remodeling and maturation process in autogenous graft for anterior cruciate ligament reconstruction. Anthroscopy, 1993;9:394-405.
2. Abbott WH, Dupree EL. Clinical results of lyophilized human cadaver dura transplantation. Journal of Neurosurgery, 1971;34:770-3.
3. Aglietti P, Buzzi R, Zaccherotti G, et al. Patellar tendon versus doubled semitendinosus and gracilis tendons for anterior cruciate ligament reconstruction. American Journal of Sports Medicine, 1994;22:211-8.
4. Aglietti P, Zaccherotti G, Buzzi R, et al. A comparison between patellar tendon and doubled semitendionsus/gracilis tendon for anterior cruciate ligament reconstruction. A minimum five-year follow-up. Journal of Sports Traumatology and Related Research, 1997;19:57-68.
5. Allen PR, Amis AA, Jones MM, et al. Evaluation of preserved bovine tendon xenografts: a histological, biomechanical and clinical study. Biomaterials, 1987;8(2):146-52.
6. Alm A. Survival of part of patellar tendon transposed for reconstruction of anterior cruciate ligament. Acta Chirurgica Scandinavica, 1973;139:443-7.
7. Angioli R, Sevin B-U, Penalvar M, et al. Pelvic floor reconstruction with dura mater allograft after pelvic exenteration. Journal of Pelvic Surgery, 1996;2:58-62.
8. Arango A. Repair of large abdominal-wall defects with dura mater allografts. Proc meeting Am Assoc Tissue Banks. Rockville, MD. 1977. p. 158.
9. Bach BR, Aadalen KJ, Dennis MG, et al. Primary anterior cruciate ligament reconstruction using fresh-frozen, nonirradiated patellar tendon allograft: minimum 2-year follow-up. American Journal of Sports Medicine, 2005;33:284-92.
10. Bahr GF. Untersuchungen an implantierten Kollagen-material. Archiv für Dermatologie und Syphilis, 1952;195:99-115.
11. Barber FA, Aziz-Jacobo, Oro FB. Anterior cruciate ligament reconstruction using patella tendon allograft: An age dependent outcome evaluation. Arthroscopy, 2010;26:448-93.
12. Barber FA, McGuire DA, Johnson DH. Should allografts be used for routine anterior cruciate ligament reconstructions? Arthroscopy, 2003;19:421-5.
13. Beard DJ, Anderson JL, Davies S, et al. Hamstring vs patellar tendon for anterior cruciate ligament reconstruction: a randomized controlled trial. Knee, 2001;8:45-50.
14. Bier A. Beobachtung über Regeneration beim Menschen. Deutsche Medizinische Wochenschrift, 1917;43:134-40.
15. Blevins FT, Hecker AT, Bigler GT, et al. The effects of donor age and strain rate on the biomechanical properties of bone-patellar tendon-bone allografts. American Journal of Sports Medicine, 1994;22:328-33.
16. Bright RW, Green WT. Freeze-dried fascia lata allografts: A review of 47 cases. Journal of Pediatric Orthopaedics, 1981;1:13-22.

17. Brown C, Steiner M, Carson E. The use of hamstring tendons for anterior cruciate ligament reconstruction. Clinical Journal of Sports Medicine, 1993;12:723-8.
18. Brown CH, Wilson DR. Hecker AT, et al. Comparison of hamstring and patellar tendon femoral fixation: cyclic load. American Journal of Sports Medicine, 1999;27:35.
19. Buck B, Malinin T, Brown M. Bone transplantation and human immunodeficiency virus: an estimate of risk acquired immunodeficiency syndrome (AIDS). Clinical Orthopaedics and Related Research, 1989;240:129-36.
20. Butler DL, Kay MD, Stouffer DC. Comparison of material properties in fascicle-bone units from human patellar tendon and knee ligaments. Journal of Biomechanics, 1986;19:425-32.
21. Caspari RB, Bottenfield S, Hurwitz RL, et al. HIV transmission with allograft organs and tissues. Sports Medicine and Arthroscopy Review, 1993;1:42-6.
22. CDC Update. Allograft associated bacterial infections. United States 2002. Morbidity and Mortality Weekly Report, 2002;51:207-10.
23. CDC. Creutzfeld-Jacob disease associated with cadaveric dura mater grafts-Japan, January 1979-1996. Morbidity and Mortality Weekly Report, 1997;45/46:1066-9.
24. Centers for Disease Control. Invasive *Streptococcus pyogenes* after allograft implantation-Colorado. Morbidity and Mortality Weekly Report, 2003;52:273-4.
25. Centers for Disease Control. Septic arthritis following anterior cruciate ligament reconstruction using tendon allografts-Florida and Lousiana. Morbidity and Mortality Weekly Report, 2001;50:1081-3.
26. Centers for Disease Control and Prevention. Update. Allograft-associated bacterial infections-United States. Morbidity and Mortality Weekly Report, 2002;51(10):207-10.
27. Chen CH, Chen WJ, Shih CH. Anthroscopic anterior cruciate ligament reconstruction with quadriceps tendon-patella bone autografts. Journal of Trauma, 1999;46:678-82.
28. Conway B, Tomford WW, Hirsch MS, et al. Effects of gamma radiation on HIV-1 in a bone allograft model. Transactions of the Annual Meeting-Orthopaedic Research Society, 1990;15:225.
29. Cooper JL, Beck CL. History of soft tissue allografts in orthopaedics. Sports Medicine and Arthroscopy Review, 1993;1:2-16.
30. Cory IS, Webb JM, Cingefeffer AJ, et al. Anthroscopic reconstruction of the anterior cruciate ligament. A comparison of patellar tendon autograft and four-strand hamstring tendon autograft. American Journal of Sports Medicine, 1999;27:444-54.
31. Curtis RJ, De Lee JC, Drez DJ. Reconstruction of the anterior cruciate ligament with freeze-dried fascia lata allografts in dogs. A preliminary report. American Journal of Sports Medicine, 1985;13:408-14.
32. De Dayne P, Haut R. Some effects of gamma irradiation on patellar tendon allografts. Connective Tissue Research, 1991;27:51-62.
33. Defrere J, Frankart A. Freeze-dried fascia lata allografts in the reconstruction of anterior cruciate ligament defects. Clinical Orthopaedics and Related Research, 1994;303:56-66.
34. Dutton VB, Barea C, Abrassard S, et al. Anatomy of the anterior cruciate ligament. Knee Surgery, Sports Traumatology, Arthroscopy, 2006;14:204-13.
35. Edgar CM, Zimmer S, Kakar S, et al. Prospective comparison of auto and allograft hamstring tendon constructs for ACL reconstruction. Clinical Orthopaedics and related research, 2008;466:2238-46.
36. Enderlen W. Über Sehnenregeneration. Archiv für Klinische Chirurgie, 1893;46:563-74.
37. Eriksson K, Anderberg P, Hamberg P, et al. A comparison of quadruple semitendinosus and patellar tendon grafts in reconstruction of the anterior cruciate ligament. Journal of Bone and Joint Surgery (Br), 2001;83:348-54.
38. Fadale PD, Johnson DH. Controversies in ACL reconstruction-Part II. Sports Medicine Arthroscopy Review, 2010;18:1-25.
39. Fargin, Assay. Recherches experimentales sur la greffe tendineuse et sur la regeneration des tendons. Comptes Rendus Societe Biologique de Paris, 1885;2:624.
40. FDA. Transmissible spongioform encephalopathies advisory committee meeting. Bethesda, MD; 1997.
41. Fideler BM, Vangness CT, Lu B, et al. Gamma irradiation: effects on biomechanical properties of human bone-patella tendon-bone allografts. American Journal of Sports Medicine, 1995;23:643-6.
42. Fiorica JV, Roberts WS, La Polla JP, et al. Femoral vessel coverage with dura mater after inguinofemoral lymphadenectomy. Gynecologic Oncology, 1991;42(3):217-22.
43. Fitzgerald MP, Mollenhouer J, Bittman P, et al. Functional failure of fascia lata allografts. American Journal of Obstetrics and Gynecology, 1999;181:1339-44.
44. Flanagan JC, Campbell CB. The use of autogenous fascia lata to correct lid and orbital deformities. Transactions of the American Ophthalmological Society, 1981;79:227-39.
45. France EP, Paulos LE, Rosenberg TD, et al. The biomechanics of anterior cruciate allografts. In Prosthetic Ligament Reconstruction of the Knee (MJ Friedman and RD Ferkel eds.) WB Saunders, Philadelphia, 1988. pp.180-5.
46. Freedman KB, D'Amato MJ, Nedeff DD, et al. Arthroscopic anterior cruciate ligament reconstruction: a meta-analysis comparing patella tendon and hamstring allografts. American Journal of Sports Medicine, 2003;31:2-11.
47. Fu FH, Bennett CH, Lattermann C, et al. Current trends in anterior cruciate ligament reconstruction. Part 1: Biology and biomechanics of reconstruction. American Journal of Sports Medicine, 1999;27(6):821-30.
48. Fung YC. Biomechanics: mechanical properties of living tissues. New York: Springer-Verlag, 1981. pp. 196-260.
49. Gallie WE, Le Mesurier AB. Recurring dislocation of shoulder. Journal of Bone and Joint Surgery (Br), 1948;30:9-18.
50. Gardner S. Ethylene oxide, ethylene chlorohydrins and ethylene glycol. Proposed maximum residue limits and maximum levels of exposure. Federal Register 1978;43:27474-83.
51. Gibbons M, Butler D, Grood E, et al. Effect of gamma irradiation on the initial mechanical and material properties

of goat bone-patellar tendon-bone allografts. Journal of Orthopaedic Research, 1991;9:209-18.
52. Goble EM, Downey DJ, Walton CN. Freeze-dried non-irradiated experience: Fascia lata ACL reconstruction. Sports Medicine and Arthroscopy Review, 1993;1:71-6.
53. Gorschewsky O, Klakow A, Riechert K, et al. Clinical comparison of the tutoplast allograft and autologous patellar tendon (bone-patellar tendon-bone) for the reconstruction of the anterior cruciate ligament: 2 and 6 year results. American Journal of Sports Medicine, 2005;33:1202-9.
54. Gresham RB. Freeze-drying of tissues for clinical use. Cryobiology, 1964;1:150-6.
55. Hamann MC, Sacks M, Malinin TI. Quantification of the collagen fiber architecture of human cranial dura mater, Journal of Anatomy, 1998;192:99-106.
56. Harner CD, Olson E, Irrgang JJ, et al. Allograft versus autograft in anterior cruciate ligament reconstruction. Clinical Orthopaedics, 1996;324:134-44.
57. Hoffa A. Die experimentale Begrundung der Sehnenplastik. Münch med. Wochenschrif, 1901;48:2036.
58. Horsley GW. The behavior of alcohol preserved fascia lata of the ox, autogenous fascia and chromicized kangaroo tendon in dog and man. Annals of Surgery, 1931;94:410-21.
59. Horstman JK, Ahmadu-Suka F, Norrdin RW. Anterior cruciate ligament fascia lata allograft reconstruction: progressive histologic changes toward maturity. Arthroscopy, 1993;9:509.
60. Huiskes R, van Dijk R, de Lange A, Woltrink HJ, Selvik G, van Rens T. Kinematics of the human knee joint. In: N Berke et al. (Eds). Biomechanics of Normal and Pathological Human Articulating Joints. Dordrecht: Nijhoff, 1985. pp. 165-88.
61. Hudson A. Biomechanische Probleme des Kniegelenks. Orthopädie, 1974;3:119-26.
62. Indelicato PA, Bittar ES, Prevo TJ, et al. Clinical comparison of freeze-dried and fresh patella tendon allografts for anterior cruciate ligament reconstruction of the knee. American Journal of Sports Medicine, 1990;18:335-42.
63. Indelicato PA, Linton RC, Huegel M. The results of fresh-frozen patellar tendon allografts for chronic anterior cruciate ligament deficiency of the knee. American Journal of Sports Medicine, 1992;20:118-21.
64. Ishibashi Y, Toh S, Okamura Y, et al. Graft incorporation within the tibial bone tunnel after anterior cruciate ligament reconstruction with bone-patellar tendon-bone autografts. American Journal of Sports Medicine, 2001;29:473-9.
65. Jackson DW, Grood ES, Cohn BT, et al. The effects of in situ freezing on the anterior cruciate ligament. Journal of Bone and Joint Surgery (Am), 1991;73(2):201-13.
66. Jackson DW, Windler GE, Simon TM. Interarticular reaction associated with theuse of freeze-dried, ethylene-oxide sterilized bone-patellar tendon-bone allograft in the reconstruction of anterior cruciate ligament. American Journal of Sports Medicine, 1990;18:1-11.
67. Järvinen M, Jozsa L, Johnson RJ, et al. Effects of anterior cruciate ligament reconstruction with patellar tendon or prosthetic ligament on the morphology of other ligaments of the knee joint. An experimental study in dogs. Clinical Orthopaedics and Related Research, 1995;311:176-82.
68. Jerrel MA, Malinin T, Averette HE, et al. Human dura mater allografts in repair of pelvic floor and abdominal wall defects. Obstetrics and Gynecology, 1987;70:280-5.
69. Kadioglu HH, Tacki E, Arik M, et al. Immune response to dehydrated human dura mater: evaluation in a rabbit model. Neurology India, 2002;50:256.
70. Kerulek K, Gammon RA, Lloyd RS. Microbiological aspects of ethylene oxide sterilization. II. Microbial resistance to ethylene oxide. Applied Microbiology, 1970;19:152-6.
71. Kirschner M. Über freie Sehnen und Fascientransplantation. Beitr Z Klin Chir, 1909;65:472.
72. Kirschner. Die praktischen Ergebnise der freien Fascien transplantation. Arch F Klin Chirurgie, 1910;42:888.
73. Klarenbeek A, van Dongen HAE. Virucidal action of ethylene oxide. Journal of Hygiene, 1954;52, 525-8.
74. Klein W, Jensen KU. Synovitis and artificial ligaments. Arthroscopy, 1992;8:116-24.
75. Klein L, Mericka P, Pinter L. Use of dura mater allografts in operations for abdominal wall hernias. Beitrage zur Orthopadie und Traumatologie, 1990;37:499-503.
76. Kleipool AEB, Ziji JAC, Willems WJ. Arthroscopic anterior cruciate ligament reconstruction with bone-patellar tendon-bone allograft for autograft: a prospective study with an average follow-up of 4 years. Knee Surgery, Sports Traumatology, Arthroscopy, 1998;6:224-30.
77. Kniepkamp HE. Safety of lyodura. Journal of Oral and Maxillofacial Surgery, 1994;52:896-7.
78. Laurencin C, El-Amin S. Xenotransplantation in orthopaedic surgery. Journal of American Academy of Orthopaedic Surgeons, 2008;16:4-8.
79. Levitt RL, Malinin T, Posada A, et al. Reconstruction of anterior cruciate ligament with bone-patella tendon-bone and Achilles tendon allografts. Clinical Orthopaedics and Related Research, 1994;303:67-78.
80. Levitt RL, Malinin T. Allograft reconstruction of the anterior ligament reconstruction of the anterior cruciate ligament—A 6 year experience with 300 allografts. Arthroscopy (abstract), 1991;7:318-9.
81. Lewis D, Davis C. Experimental direct transplantation of tendon and fascia. Journal of American Medical Association, 1911;57:540-6.
82. Lubowitz JH. All-inside ACL: Reconstruction controversies. Sports Medicine Arthroscopy Review, 2010;18:20-6.
83. Lysholm J, Gillquist J. Evaluation of knee ligament surgery results with special emphasis on use of scoring scale. American Journal of Sports Medicine, 1982;10:150-4.
84. Macnicol MF, Penny ID, Sheppard L. Early results of the Leeds-Keio Anterior cruciate ligament replacement. Journal of Bone and Joint Surgery (Br), 1991;73:377-80.
85. Maeda A, Shino K, Horibe S, et al. Anterior cruciate ligament reconstruction with multistranded autogenous semitendinosus tendon. American Journal of Sports Medicine, 1996;24:504-9.

86. Malinin T. Processing and storage of viable human tissues. Public Health Service, US. Washington, DC: Government Printing Office; 1966.
87. Malinin T. Principles of musculoskeletal tissue banking In: Musculoskeletal Tissue Regeneration. In: WS Pietrzak (Ed): Humana Press; 2009. pp. 483-508.
88. Malinin T, Latta L, Milne T, Temple H, Flores A. Optimizing biomechanics of allograft reconstruction for anterior cruciate ligament deficiency. San Francisco: Proceedings of AAOS Annual Meeting, 2001. p. 654.
89. Malinin TI. Allografts for the reconstruction of the cruciate ligaments of the knee: procurement, sterilization and storage. Sports Medicine Arthroscopy Review, 1993;1:31-41.
90. Malinin T. Testimony before the Advisory Committee on Transmissible Spongiform Encephalitis. Bethesda, MD: FDA; 1997.
91. Malinin TI, Levitt RL, Bashore C, Temple HT, Mnaymneh W. A study of retrieved allografts used to replace anterior cruciate ligaments. The Journal of Arthroscopic and Related Surgery, 2002;18(2):163-70.
92. Marder RA, Rasking JR, Carroll M. Prospective evaluation of arthroscopically assisted anterior cruciate ligament reconstruction, patellar tendon versus semitendinosus and gracilus tendons. American Journal of Sports Medicine, 1991;35: 2148-58.
93. Masson ML, Allen HS. The rate of healing of tendons, Annals of Surgery, 1941;113:424-59.
94. Mc Allister DR, Joyce MJ, Mann BJ, et al. Allograft update. The current status of tissue regulation, procurement, processing and sterilization. American Journal of Sports Medicine, 2007;35:2148-58.
95. McGarvey KA, Lee JM, Boughner DR. Mechanical suitability of glycerol-preserved human dura mater for construction of prosthetic cardiac valves. Biomaterials, 1984;5:109-17.
96. McGuire OA, Hendricks SD. Allograft tissue in ACL reconstruction. Sports Medicine and Arthritis Review, 2009;17: 224-33.
97. Meijer R, Huiskes R, Kauer J. Mechanische Belastung in Sports. In: R Huiskes (Ed). Biomechanica-Aspekte des Bewegungsapparats. Brussels, Alphen aan den Rijn: Stafleu; 1987. pp. S9-27.
98. Melberg MW, Indelicato PA. Fresh-frozen allograft experience. Sports Medicine Arthroscopy Review, 1993;1:77-88.
99. Myers JF. Allograft reconstruction of the anterior cruciate ligament. Clinical Sports Medicine, 1991;10:487-548.
100. Noyes F, Barber-Westin S. Reconstruction of the anterior cruciate ligament with human allograft: comparison of early and later results. Journal of Bone and Joint Surgery(Am), 1996;78:524-37.
101. Noyes FR, Barber SD, Mangine RE. Bone-patella ligament tendon-bone and fascia lata allografts for reconstruction of the anterior cruciate ligament. Journal of Bone and Joint Surgery (Am), 1990;72:1125-36.
102. Noyes FR, Barber SD. The effect of an extra-articular procedure on allograft reconstructions for chronic ruptures of the anterior cruciate ligament. Journal of Bone and Joint Surgery (Am), 1991;73:882-92.
103. Noyes FR, Butler DL, Grood ES, et al. Biomechanical analysis of human ligament grafts used in knee-ligament repairs and reconstructions. Journal of Bone and Joint Surgery (Am), 1984;66:344-52.
104. Noyes FR, Grood ES. The strength of the anterior cruciate ligament in humans and rhesus monkeys. Journal of Bone and Joint Surgery (Am), 1976;58:1074-82.
105. Orlando F, Weiss JS, Beyer-Machule CK, et al. Histopathologic condition of fascia lata 42 years after ptosis repair. Archives of Ophthalmology, 1985;103:1518-9.
106. Oster G. Testimony before the Advisory Committee on Transmissible Spongiform Encephalitis. Bethesda, MD: FDA; 1997.
107. Otero AL, Hutcheson L. A comparison of the doubled semitendinous/gracilis and central third patellar tendon autografts in arthrascopic anterior cruciate ligament reconstruction. Anthroscopy, 1993;9:143-8.
108. Paget J. Lectures on surgical pathology, Longmans, Philadelphia, 1853.
109. Peer LA. Transplantation of Tissues. Baltimore: The Williams and Wilkins Co, 1955. pp. 243-77.
110. Pirogoff NI. In: Fried MF In: The Crimean Campaign, 1854-1855. New York Academy of Medicine. New York Society for Medical History, 1840;31.
111. Pritchard JC, Drez D, Moss M, et al. Long-term follow-up of anterior cruciate ligament reconstruction using freeze-dried fascia lata allografts. American Journal of Sports Medicine, 1995;23:593-6.
112. Prolo DJ, Pedrotti PW, White DJ. Ethylene oxide sterilization of bone, dura mater and fascia lata for human transplantation. Neurosurgery, 1980;6:529-39.
113. Puig LB, Brandão CM, Kawabe L, et al. Dura mater mitral and tricuspid bioprosthesis: 30 years of follow-up. Review of 291 cases treated at the Hospital das Clinicas da Faculdade de Medicina da Universidade de São Paulo, 2003;58(3):163-8.
114. Quilici PJ, Vieta JO, Privitera L. The use of dura mater allograft in the surgical repair of large defects of the abdominal wall. Surgery, Gynaecology and Obstetrics, 1985;161:47-8.
115. Rappé M, Horodyski MB, Meister K, et al. Non-irradiated versus irradiated Achilles tendon allograft: An in vivo failure comparison. American Journal of Sports Medicine, 2007;35:1651-8.
116. Re L, Weiss R, Rintz K. Incidence of anterior knee pain after treatment of anterior cruciate ligament rupture. Orthopaedic Transactions, 1993;17:1005-11.
117. Rehn E. Free transplantation of tendons. Neue Deutsche Chirurgie, 1924;26:370.
118. Rehn E. Die homoplastische Sehnentransplantation im Tier experiment. Beitrage zur Klinische Chirurgie, 1910;68:471.
119. Rosomoff HL, Malinin TI. Freeze-dried dura mater—20 year's experience. Transplantation Proceedings, 1976;8(2):133-8.
120. Rothschild. Über functionalle Heilung der Cucularis-Lehmung mittels freier Fascienplastik. Zentralblatt für Chirurgie, 1910;45:1441.
121. Rougraff B, Shelbourne KD, Garth PK, et al. Arthroscopic and histologic analysis of human patellar tendon auto-

grafts used for anterior cruciate ligament reconstruction. American Journal of Sports Medicine, 1993;21:477-84.

122. Runza M, Pietrabissa R, Montero S, et al. Lumbar dura mater biomechanics: experimental characterization and scanning electron microscopy observations. Anesthesia and Analgesia, 1999;88(6):1317-21.
123. Sacks MC, Himenez Hamann MC, Otano-Lata SE, et al. Local mechanical anisotropy in human cranial dura mater. Journal of Biomechanical Engineering, 1998;120:541-4.
124. Schatzman L, Brunner P, Staubli HU. The effect of cyclic preconditioning on the tensile properties of human quadriceps tendons and patella ligaments. Knee Surgery, Sports Traumatology, Arthroscopy, 1998;6:556-61.
125. Schwartz HE, Matava MJ, Proch FS, et al. The effect of gamma irradiation on anterior cruciate allograft biomechanical and biochemical properties in the caprine model at time zero and at 6 months after surgery. American Journal of Sports Medicine, 2006;34:1747-55.
126. Seggel R. Histologische Untersuchungen über die Heilung von Sehnenwunden und Sehnendefected. Beitrage zur Klinische Chirurgie, 1903;37:342.
127. Sevin B-U, Malinin TI. Dura mater allograft in gynecologic reconstruction. In: PG Knapstein, V Friedberg, B-U Sevin (Eds). Reconstructive Surgery in Gynecology. New York: Thieme Medical Publishers Inc, 1990. pp. 151-65.
128. Shelbourne K, Trumper R. Preventing anterior knee pain after anterior cruciate ligament reconstruction. American Journal of Sports Medicine, 1997;25:41-7.
129. Shelton WR, Papendick L, Dukes AD. Autograft versus allograft anterior cruciate ligament reconstruction. Arthroscopy, 1997;13:446-9.
130. Shino K, Inoue M, Horibe S, et al. Reconstruction of the anterior cruciate ligament using allogeneic tendon: Long-term follow-up. American Journal of Sports Medicine, 1990;18:457-65.
131. Shino K, Kimura T, Hirose H, et al. Reconstruction of the anterior cruciate ligament by allogeneic tendon graft. An operation for chronic ligamentous insufficiency. Journal of Bone and Joint Surgery (Br), 1986;68:739-46.
132. Shino K, Kimura T, Horibe S, et al. Maturation of allograft tendons transplanted into the knee: an arthroscopic and histologic study. Journal of Bone and Joint Surgery (Br), 1988;70:556-60.
133. Sidwell RW, Dixon GJ, Westbrook L, et al. Procedure for the evaluation of the virucidal effectiveness of an ethylene oxide gas sterilizer. Applied Microbiology, 1969;17:790-6.
134. Siebold R, Baelow JU, Bos L, et al. Primary ACL reconstruction with fresh-frozen patellar versus Achilles tendon allografts. Archives of Orthopaedic and Trauma Surgery, 2003;123:180-5.
135. Silvagio VJ, Fu F. Anterior cruciate ligament replacement materials: allograft and synthetic ligaments. In: EW Ewing (Ed). Articular Cartilage and Knee Functions. New York: Raven Press, 1990. pp. 273-9.
136. Simonds RJ, Holmberg SD, Hurwitz RL, et al. Transmission of human immunodeficiency virus type I from seronegative organ and tissue donor. New England Journal of Medicine, 1992;326:726-32.
137. Simonian PT, Harrison CD, Cooley VJ, et al. Assessment of morbidity of semitendinosus and gracilis tendon harvest for ACL reconstruction. American Journal of Knee Surgery, 1997;10:54-9.
138. Singhal MC, Gardiner JR, Johnson DL. Failure of primary anterior cruciate ligament surgery using anterior tibialis allograft. Arthroscopy, 2007;23:469-75.
139. Smith DJ, Jones CS, Kleinert HE. Evaluation of gluataraldehyde-treated tendon xenografts. Journal of Hand Surgery, 1986;11(1):97-106.
140. Strocchi R, de Pasquale V, Gubellini P, et al. The human anterior cruciate ligament: histological and ultrastructural observations. Journal of Anatomy, 1992;180:515-9.
141. Sun K, Tian SQ, Zhang JH, et al. ACL reconstruction with BPTB autograft and irradiated fresh frozen allograft. Journal of Zhejiang University-SCIENCE B (Biomedicine and Biotechnology), 2009;10:306-16.
142. Takahashi M, Ono K, Wakakuwa R, et al. The use of human dura mater allograft for the repair of a contaminated abdominal wall defect. Surgery Today, 1994;24(5):468-72.
143. Tateishi J. Testimony before the Advisory Committee on Transmissible Spongiform Encephalitis. Bethesda, MD: FDA, 1997.
144. Tauro JC, Parsons JR, Ricci J, et al. Comparison of bovine collagen xenografts to autografts in the rabbit. Clinical Orthopaedics and Related Research, 1991;266:271-84.
145. Thatcher HV. Repair of flexor tendons with fascia lata. Northwest Medicine, 1949;48:848-52.
146. Thomas ED, Gresham RB. Comparative tensile strength study of fresh frozen and freeze-dried fascia lata. Surgical Forum, 1964;14:442-3.
147. Tarandy DI, Williamson JS, Bach BR. An unusual intrinsic complication of a patellar tendon allograft and recommendations for tissue banking. Arthroscopy, 1995;11:475-7.
148. Upshur FC. Use of lyophilized homografts of dura mater in repair of incisional hernia. AMA Archives of Surgery, 1958;76:58-61.
149. Vandenabeele F, Creemers J, Lambrichts I. Ultrastructure of the human spinal arachnoid mater and dura mater. Journal of Anatomy, 1996;189:417-30.
150. van Dijk R, Huiskes R, Selvig G. Roentgen stereophotogrammetic methods for the evaluation of the three-dimensional kinematic behavior and cruciate ligament length patterns of the human knee joint. Journal of Biomechanics, 1979;12:727-31.
151. van Noort RM, Black MM, Martin TRP, et al. A study of the uniaxial mechanical properties of human dura mater preserved in glycerol. Biomaterials, 1981;2:41-5.
152. van Rens T, van den Berg A, Huiskes R, et al. Substitution of the anterior cruciate ligament: A long-term histologic and biomechanical study with autogenous pedicled grafts of the iliotibial band in dogs. Journal of Anthrascopic and Related Surgery, 1986;2:139-54.

153. Veillette CJ, Cunningham KD, Hart DA, et al. Localization and characterization of porcine patellar tendon xenograft antigens in a rabbit model of medial collateral ligament replacement. Transplantation, 1998;65:486-93.
154. Viering W. Experimentelle Untersuchung über die Regeneration des Sehnengewebes. Archiv für Pathologische Anatomie, 1891;125:252-86.
155. Vulpius O. Operationspläne für schnenüberpflanzung. Zentralschrift für Orthopaedische Chirurgie, 1923;44:57-71.
156. Wallace MP, Howell SM, Hull ML. *In vivo* tensile behavior of a four bundle hamstring graft as a replacement for the anterior cruciate ligament. Journal of Orthopaedic Research, 1997;15:539-45.
157. Weiner RA, Clark TJ, Poehling GG. Arthroscopic reconstruction of the anterior cruciate ligament using allograft tendon. Arthroscopy, 1988;4:199-205.
158. Wolfinbarger L, Zhang Y, Adam BT, et al. Biomechanical aspects on rehydrated freeze-dried allograft dura mater tissues. Journal of Applied Biomaterials, 1994;5:265-70.
159. Woo SL, Orlando CA, Camp JF, et al. Effects of postmortem storage by freezing on ligament tensile behavior. Journal of Biochemistry, 1986;19:399-404.
160. Wright EJ, Iselin CE, Webster GD. Pubovaginal sling using cadaveric allograft fascia for the treatment of intrinsic sphincter. Journal of Urology, 1998;160:759-62.
161. Yasuna E. Use of prepared fascia lata in correction of ptosis. American Journal of Ophthalmology, 1962;54:1097-103.
162. Zimmerman MC, Cantiliano JH, Parsons JR, et al. The biomechanics and histopathology of chemically processed patellar tendon allografts for anterior cruciate replacement. American Journal of Sports Medicine, 1994;22:378-86.
163. Zoltan DJ, Reinecke C, Indelicato PA. Synthetic and allograft anterior cruciate ligament reconstruction. Clinical Sports Medicine, 1988;7:773-8.

Chapter 11

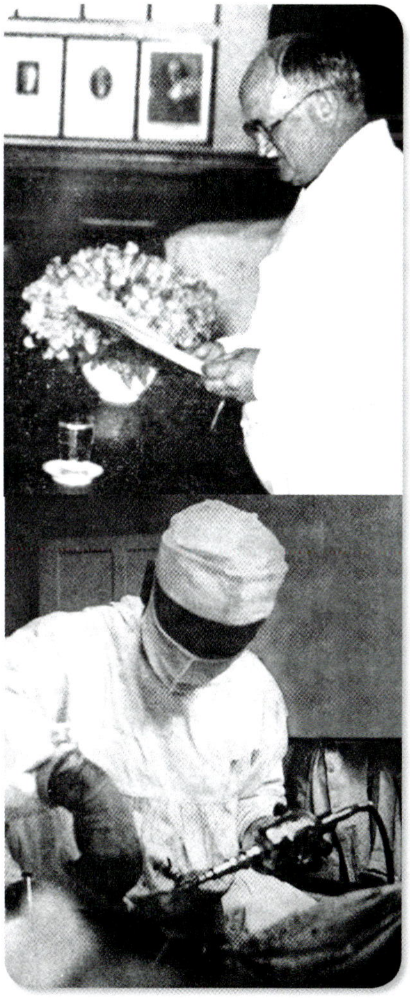

Fred H Albee MD (1876–1948), Professor of Orthopaedic Surgery, Columbia University and Consulting Surgeon, Broad Street Lutheran Hospital, New York, USA, Founder of Florida Medical Center, Venice, Florida. Dr Albee devised the operations and instruments for spine fusions. The photographs show Dr Albee lecturing and excising an autograft for a spinal fusion.

(*Reference:* Abee FH. A Surgeon's Fight to Rebuild Men. EP Dutton & Co., Inc. New York 1943.)

Bone Grafts in Spine Stabilization

INTRODUCTION

The replacement of segments of long bones of the extremities represents one of the most spectacular applications of cadaver bone allografts. In somewhat less remarkable but no less important circumstances, allogeneic bone has been used for achieving arthrodesis of the spine (Albee 1944).

The wide variety of sizes and shapes of bone allografts makes them the most versatile grafts in spinal surgery. For the past several decades, dowel or wedge-shaped bone allografts have been used routinely in many institutions for spine interbody fusions (Malinin and Brown 1981).

Nowhere in orthopedic surgery is the loss of bone fraught with so many serious ramifications as in the spinal column. Methyl methacrylate has not proved as useful as anticipated in stabilizing the spine (Dunn 1983). A most desirable alternative would be a graft that could replace the missing structure, and then become incorporated into the spinal column. Bone autografts serve this purpose. However, they have distinct anatomic and physiologic limitations. Likewise, the excision of large autografts results in considerable morbidity and longstanding discomfort. It also prolongs the operation which, in salvage surgery of the spine, by itself, is long enough. It increases blood loss and results in many cases in donor site pain and morbidity.

Readily available bone allografts obviate these problems and in many circumstances have proven to be valuable adjuncts to the surgery of the spine. Anatomically sound bone allografts provide immediate support of the spinal column and augment the use of internal stabilization devices, although lately there is a trend to use allografts in combination with metallic screws and plates (Ehler and Vaccaro 2000).

The prime concern of surgeons contemplating the use of banked bone grafts is the delayed union or nonunion of the graft with the host bone, and the possibility of transmission of an infection. The host's own fresh bone offers the best healing rate. However, as we said, it is not always available in the sufficient quantity or quality necessary in many clinical circumstances. In addition, in the anterior cervical spine arthrodesis and the fusions for scoliosis, the difference in the healing rate is not significant enough to offset the morbidity resulting from the removal of the patient's own bone. Particles between 100 µm and 300 µm are optimal for bone healing and are equivalent to autologous bone.

The risk of transmitting disease by tissues from cadavers is often of concern to surgeons who contemplate using cadaver tissue grafts. However, this potential threat is practically nonexistent if bone allografts are taken from selected donors under aseptic conditions and are subjected to standard forms of processing and adequate microbiologic monitoring. Careful review of the medical history, interpretation of an exhaustive serological panel, adequate microbiologic monitoring of blood and tissue samples and a complete autopsy result in assurance that the tissues are safe for transplantation.

The transplantation of freeze-dried and frozen bone allografts has definitely produced an advancement in reconstructive surgery of the axial skeleton. Prevention and correction of spinal deformity and replacement of bone loss have been the major indications for the use of these grafts.

Spinal fusion is effected by a series of events which depend on the biological and biomechanical conditions of the host. The presence of three components of equal importance is required for new bone formation at the fusion site. These are the presence of osteoprogenitor cells derived from mesenchymal stem cells, osteoinductive factors which bring about differentiation of osteoprogenitor cells and scaffolding to provide support for new bone formation. The efficacy of bone graft material depends in a large measure on the environment provided by the host. Therefore, the understanding of various surgical techniques used to achieve spinal fusion is essential when determining the type of graft to be used in a particular patient.

Posterolateral fusion is still the most commonly performed fusion in the lumbar spine (Bosworth 1948). Decortication of posterior surfaces of the vertebral bodies before placing bone graft on the denuded surfaces results in the formation of new bone which fuses adjacent transverse processes. These fusions are frequently stabilized with bone plates secured in place with pedicle screws. The biomechanical properties of the posterior spine are unique because the motion segment is subjected to tensile rather than compression forces. Therefore the graft used to produce osseous fusion of the posterior spine does not need to provide structural support, but most important, osteogenesis. If the latter is not the case, the fusion fails. Multiple comparative studies have reported failure rates between 40% and 60% in non-instrumental posterolateral fusions regardless of the type of graft used (Simpson et al 2008). Because of these poor results, supplemental internal fixation gained acceptance, as did the interbody fusion techniques.

Bone Autografts for Spinal Fusions

Although in spinal fusions employing various surgical techniques, fresh human autogenous grafts produce the most favorable rate of healing and induction of new bone formation, donor sites for these are limited to the ilium, tibia, and fibula. The drawbacks to one's own bones for heterotopic grafting are the inherent limitations of size, shape and quality of graft material. For example, the correction and fusion of a paralytic scoliosis curve of the spine may require banked bone because there is usually an insufficient quantity and quality of bone available from the patients with disuse atrophy of the pelvis. Not infrequently, a spinal fusion, such as an interbody fusion in the lumbar spine at more than one level, requires a large volume of cortico-cancellous bone graft fashioned to provide biomechanical stability of the interspace. To obtain the needed volume of properly shaped bone graft required for each interspace would necessitate removing large portions of the patient's own ilium on both sides. This is a major operative procedure with a distinct incidence of morbidity. Graft-site infections, chronic pain, neuroma formation, stress fractures and hematomas are listed among the complications encountered from obtaining autogenous bone (Kim et al 2009). Although graft site complications are not serious, there is significant and persistent pain and morbidity after elective spine surgery that impacts on the patient's postoperative course.

Spine fusions in children with tuberculosis of the spine were performed by Albee as early as 1911 (Albee 1911). For this purpose, he used autologous bone taken from the tibia. Tibial bone autografts are still used, but fortunately infrequently. Corticocancellous graft from the ilium is now the preferred bone autograft. If placed into a recipient's bed in the spinal column with an adequate blood supply, these grafts not only induce osteogenesis, but also act as scaffolding for the ingrowth of new bone, thus producing an arthrodesis.

The use of autografts in spinal fusions is limited by the quantity and dimensions of the graft and by the quality of the bone. If an autograft is not osteopenic and anatomically normal, it is biologically the gold standard to which other grafts are compared in achieving spinal arthrodesis.

Autologous bone for spinal fusions can be taken from the ilium in a variety of shapes and configurations. Straight corticocancellous strips can be obtained from the lateral aspect of the ilium. Curved tricortical struts are obtained from the iliac crest, as are wedges for the Smith-Robinson procedure. Large amounts of cancellous bone can be taken from the superior and posterior portions of the iliac crest.

Longer and structurally strong grafts can be made from sections of the ilium, but the latter procedure is fraught with considerable morbidity and the possibility of fracture as well as nerve injury.

Another source of strong and long autogenous bone is the mid-portion of the fibula. Ribs offer another source of corticocancellous bone. They can also serve as struts. However, ribs are structurally weak and may not provide as much resistance to axial compression as might be needed. Ribs, as well as sections of fibulas, have been used with vascular pedicles, particularly in the reconstruction of thoracic and lumbar spines. Viable bone marrow may contribute mesenchymal cells to the healing site, but whether or not these cells participate significantly in bone formation is unclear. If they do so, care must be taken to preserve these cells within the autograft tissue. In bone marrow aspirates, most viable "stem cells" are present in the first few cc's of aspirate (Muschler et al 2005).

Exposure to air and drying results in the loss of cell viability, as do prolonged bathing in saline, heating and exposure to chemicals. Bone autograft cells will survive 6 hours or so in blood-soaked gauze (Marx et al 1979).

The biologic differences in the cortical and cancellous bone favor cancellous bone. Cancellous bone is replaced uniformly throughout its thickness, while the replacement process in the cortical bone is incomplete.

Despite long standing continuous employment of autografts in spinal surgery, little is known of the biomechanical properties of these transplants or the fusion mass they produce. At times the fusion mass continues to enlarge, i.e., the new bone in the fusion mass continues to proliferate. This can also occur with allografts (**Figs 1A and B**).

In general, throughout the spine the union rate decreases with the increase in number of segments being fused. A known factor associated with a poor fusion rate is placement of the graft on the tension which is the posterior side of the spine. Host factors play a significant role in the quality and rate of fusion.

Complications encountered with autografts used for spinal fusions include pseudarthrosis, graft resorption, spinal stenosis, collapse of the graft under compression and bone mass overgrowth. Spinal stenosis may develop at the site of decompression or pseudarthrosis (Winter et al 1973).

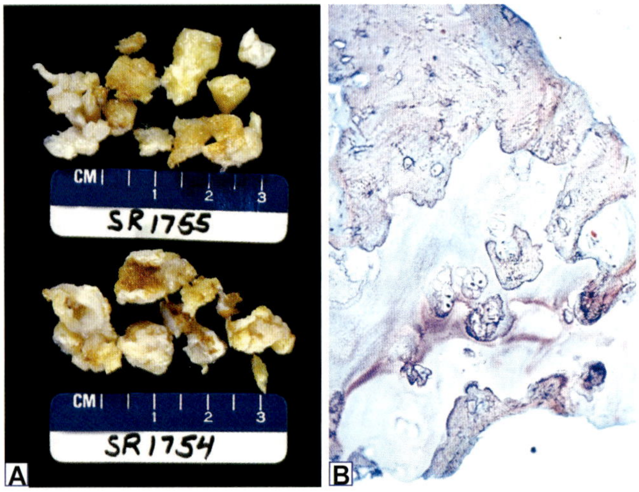

Figures 1A and B Overgrowth of an allograft (SR 1755) top and an autograft (SR 1754) bottom from the same patient. (A) The allograft had been actively replaced with host bone; (B) It might well be that grafted bone overgrowth is due to the host reaction rather than the type of graft used (Romanowski-Giemsa stain × 25)

Surgical technique obviously influences the outcome of spinal fusions with autografts (Benzel et al 2008). However, this is difficult to establish because of the inability and inadvisability of conducting human studies. Despite this limitation, it would seem obvious that careful optimization of bone fusion beds as well as contact between the graft surfaces and the recipient's tissue would be important. These considerations come into play when dealing with various surgical methods of spinal fusion.

Bone Allografts for Spinal Fusion

Bone allografts used to effect spinal arthrodeses are usually either frozen or freeze-dried. The proliferation of bone banks has made these allografts readily available (Malinin 1992). There are, however, differences between autografts and allografts which should not be overlooked. With allografts, new bone formation and vascular penetration proceeds more slowly than with autografts. Bone allografts can be frozen and stored at different temperatures. The storage temperature and the duration of storage affect the biological and biomechanical properties of allografts. Allografts stored at around -20° C for 1 year have been associated with a high failure rate (Wilson 1951). On the other hand, Brown et al (Brown et al 1976) reported satisfactory incorporation of spinal grafts stored at the same temperature for 6 months. Minus seventy degrees Celsius (dry ice temperature) provides for a longer storage, but for reliable long-term storage, cryogenic gas temperatures below -120°C are recommended (Malinin et al 1986).

Freeze-drying of spinal allografts is more complicated than freezing them, but the end result is a biomechanically acceptable graft which can be stored at room temperature. A variety of cancellous, cortical and combined allografts can be prepared and preserved by freeze-drying. Freeze-drying has been shown to attenuate the immunologic response to allografts.

In a weight-bearing interbody position, freeze-dried allografts perform well. In the lumbar spine, interbody fusions can be performed posteriorly (PLIF), anteriorly (ALIF) or transforaminally. In these procedures, the intervertebral disc is excised and replaced with a bone allograft spacer. Apposition and compression are critically important in optimizing fusion. Adjunctive posterior transpedicular rigid fixation may result in higher fusion rates (Madan et al 2003).

Although interbody fusions were first introduced in the lumbar spine, they were popularized in the cervical spine. Anterior cervical discectomy and fusion is now a frequently performed procedure involving the cervical spine. The operation of the vertebral body fusion following subtotal removal of the intervertebral disc has been performed by Cloward since 1943 (Cloward 1953). A detailed description of the operation and its results with 47 patients were presented in 1958 by the Harvey Cushing Society, and the use of banked bone for vertebral interbody fusion was described by Cloward in 1952. Published in the same year, Robinson and Smith described their methods before the John Hopkins Medical Society in 1955 (Robinson and Smith 1955). A definitive paper on the technique and results was published in 1958 (Smith and Robinson 1958). The preference between the Smith-Robinson and the Cloward technique rests with individual surgeons (Malinin and Brown 1996). The difference in the methods of fusion depends upon the shape of the grafts as shown in **Figures 2 and 3**. Either technique produces satisfactory arthrodesis (Cloward 1952, Robinson et al 1962). Bailey and Badgleys (1960) technique of placing the graft anterior without obliterating the interbody space never achieved popularity.

Cervical spine arthrodesis was based on the premise that the symptoms resulting from intervertebral disc degeneration might be relieved through immobilization of the motion segment by fusion of an interspace (Robinson et al 1962).

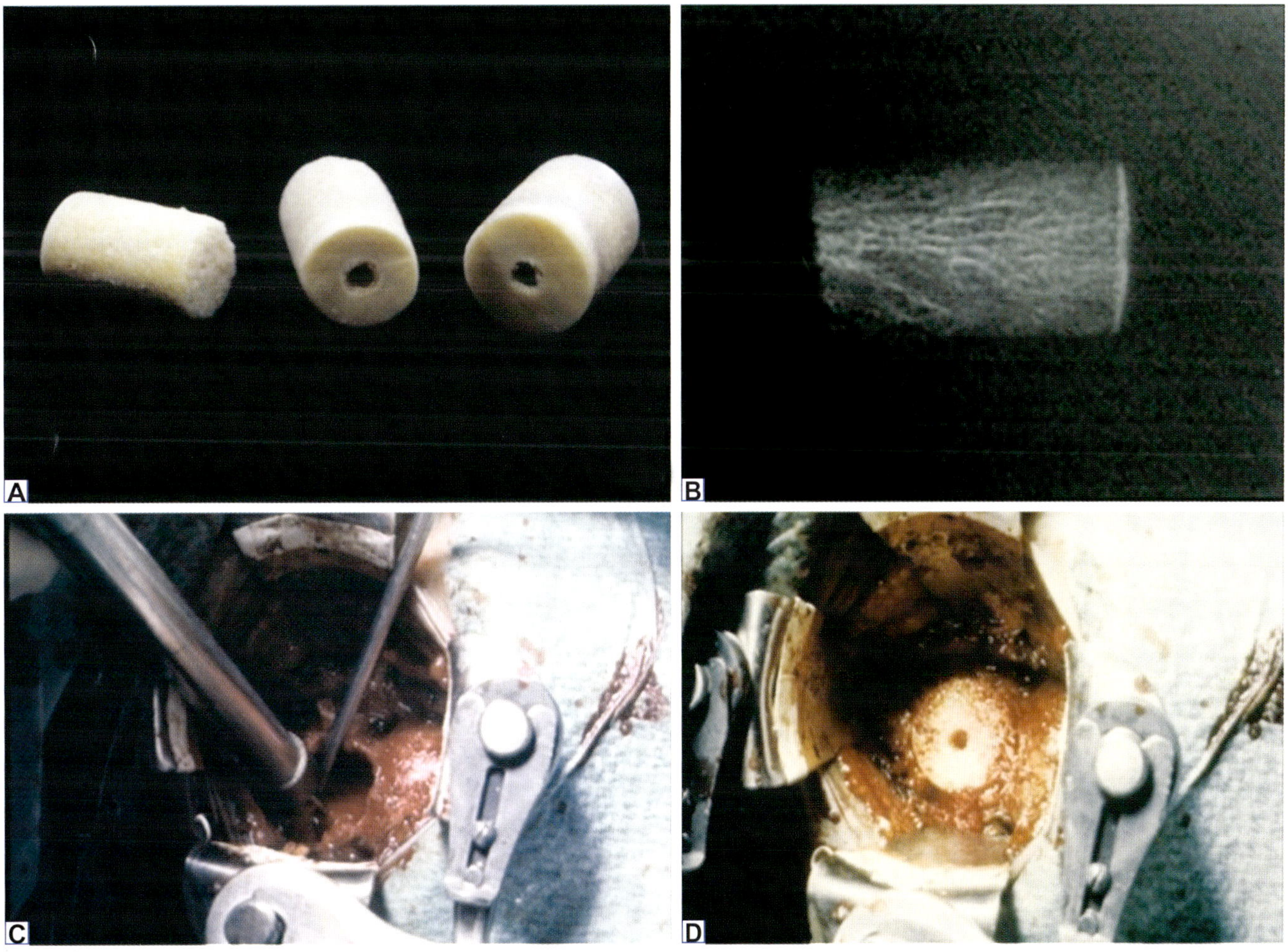

Figures 2A to D These photographs of historical interest show Cloward plugs being prepared with various diameters, usually 10, 12, 14, 16 and 18 mm. (A) The cylinders are composed of dense cancellous bone; (B) The grafts are inserted into a predrilled matching cavity; (C) The grafts do not require cortical plates on their proximal ends; (D) Although the technique is no longer used frequently, it shows the versatility of bone allografts.

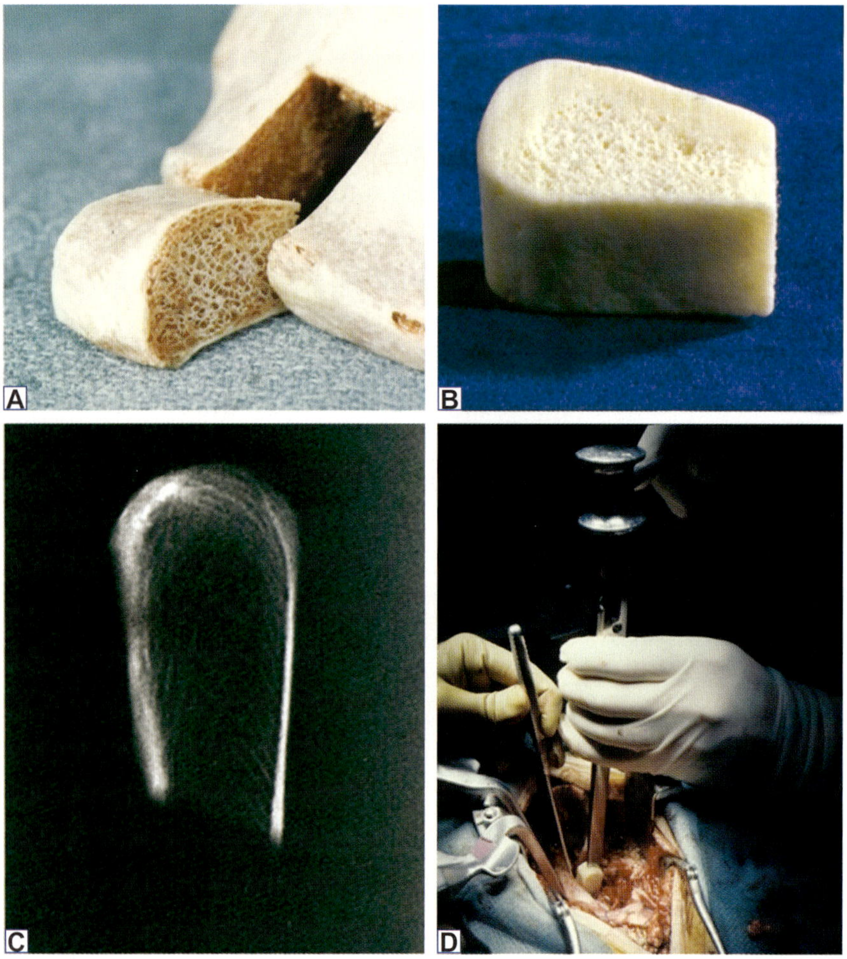

Figures 3A to D Freeze-dried bone allograft for an anterior cervical fusion according to Smith-Robinson technique. (A) Allograft is obtained from the crest of the ilium; (B) it is then washed and freeze-dried; (C) typically iliac crest wedge shows cortical bone surrounding finally trabeculated cancellous bone; (D) The allograft must be impacted tightly at its insertion. Therefore the graft must be strong enough to withstand the force of impaction

Initially, interbody fusions were achieved with autologous bone grafts. Later, frozen cadaver bone grafts were introduced, followed by freeze-dried bone allografts (Malinin et al 1977).

In comparing allografts with autografts in the cervical spine, regardless of whether wedge or dowel-shaped grafts are used, little difference in the overall rate of fusion is noted. Our experience with freeze-dried bone allografts in this operation exceeds 1,000 patients.

In 1976, Rish et al (1976) presented the results of a comparative study of freeze-dried allogeneic and autogenous bone grafts used in anterior cervical fusion in patients with arthritis and chronic debilitating neck pain. There were no significant differences between the two in either clinical or radiographic results. The fate of frozen allografts was compared with that of fresh autografts in anterior cervical spine fusions by the Smith-Robinson method (Brown et al 1976). Little difference in the fusion rates was noted. Similar results were reported by Schneider and Bright (1976). Malinin and associates (1977) described freeze-dried dowel-shaped Cloward grafts prepared from the femoral heads and condyles and not from iliums. The results with these grafts were comparable to autografts. Likewise, freeze-dried grafts performed as frozen grafts. Good quality freeze-dried bone allografts may well constitute the best grafting material for interbody fusions of the cervical spine as well as for spine stabilization (Rossier et al 1977).

These grafts are stable until healing occurs between the grafted bone and host bone of the adjacent vertebral bodies. Properly shaped grafts are assured with banked bone, but this is not true of autogenous bone. Although immunologic rejection and, at times, sensitization may occur with banked bone, this does not seem to be clinically significant if the bone grafts are subjected to compression (**Fig. 4**). In these circumstances, any graft, be it autogenous or allogenic, is a nonliving bioresorbable spacer upon which a scaffolding of new bone is formed. Both autografts and allografts, if properly prepared, contain cytokines which participate in the process of bone healing.

With the advent of instrumentation used in conjunction with anterior cervical interbody fusion, cortical or combined machined corticocancellous allografts came into vogue (**Fig. 5**). Although cortical bone, incorporates more slowly than cancellous bone, it provides stability with the slowness of incorporation overcome by metallic fixation. Biomechanically machined cortical bone allografts are strong. Compression force exceeding 20,000 Newtons (N)

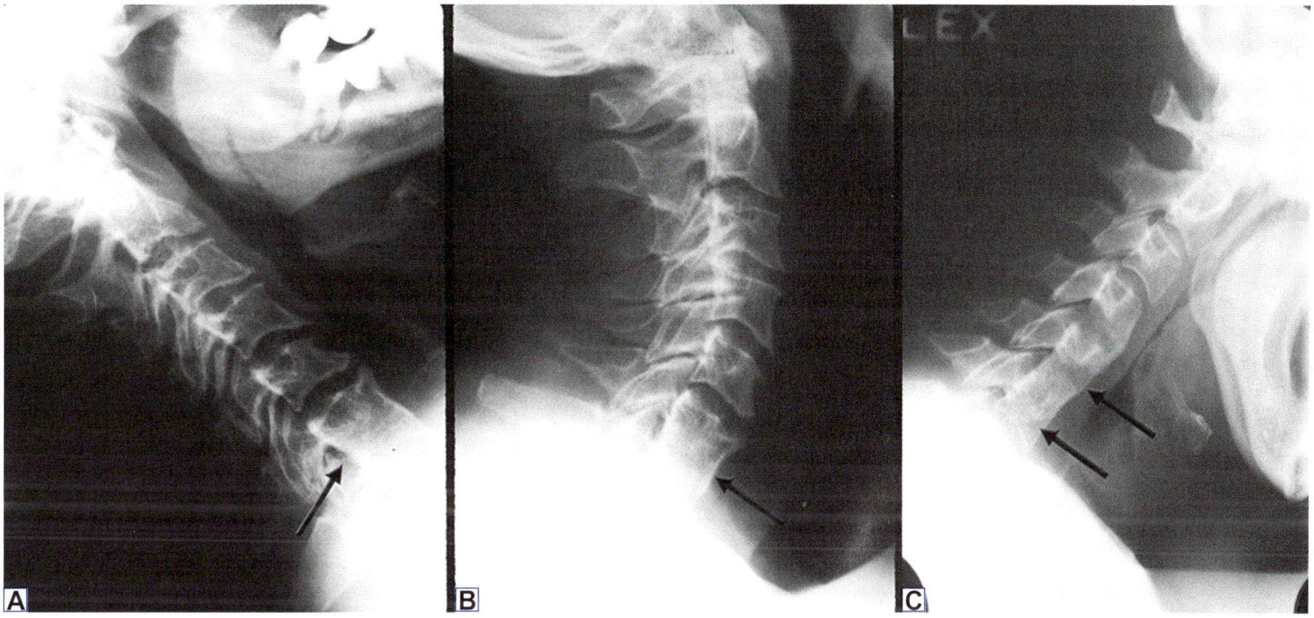

Figures 4A to C (A) Cervical spine fusion with a freeze-dried Cloward allograft 8 years post-transplantation and; (B) 10 years post-transplantation; (C) Cervical spine fusion with frozen Smith-Robinson wedge grafts, 1 year post-transplantation

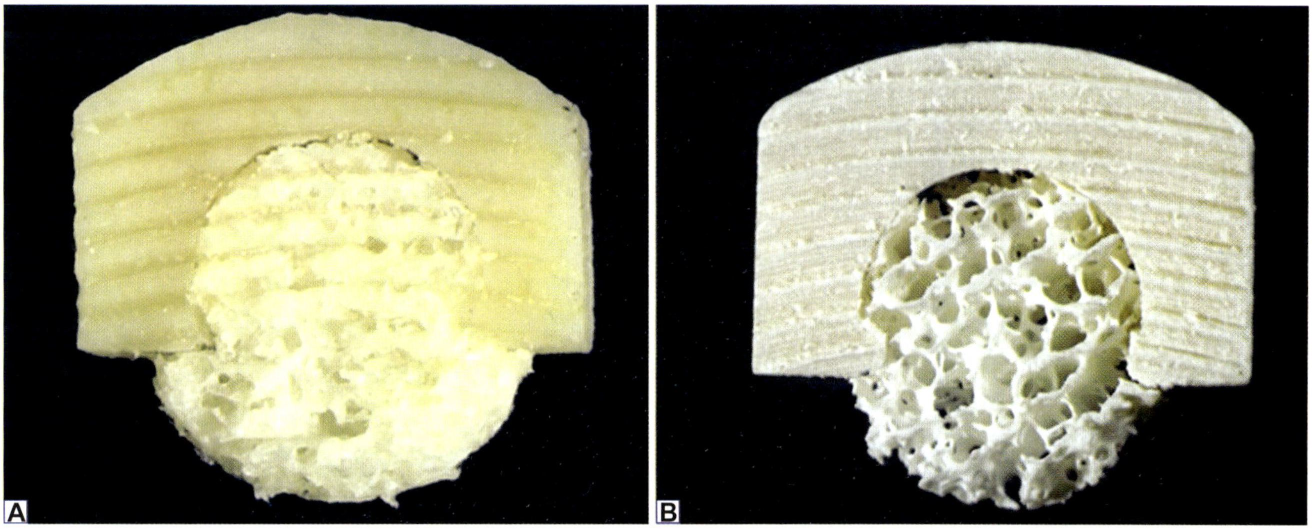

Figures 5A and B Machined interbody fusion allografts with cortical bone combined with cancellous bone. (A) Composite allograft bleached with Dakin solution. (B) Bleaching provides "cosmetic" color as well as disinfection. Bleaching with hydrogen peroxide should be avoided since hydrogen peroxide decreases or abolishes osteoinductive properties of the allograft, depending on the time of exposure is needed to produce failure in machined cortical spacers. Compact cancellous bone is much weaker. The force needed to produce failure in these is around 3,000 N. Several configurations of machined interbody fusion of allografts are available. Two examples of these are illustrated in **Figures 5 and 6**.

Freeze-dried and frozen bone allograft segments have been used in the anterior interbody fusions at one, two and three levels in patients with cervical spine myelopathy. With the use of these grafts, 92% union rate is usually obtained by 6 months after surgery. Machined cortical allografts fulfill a similar function. The replacement of vertebral bodies destroyed by trauma, tumors and infection can be accomplished through an anterior approach. In these circumstances the use of allografts becomes very attractive, as these are readily available in appropriate sizes and shapes.

Although these grafts, as advocated by Bailey and Badgley (1960), are effective they have not gained in popularity. A maltese cross shaped graft is shown in **Figure 7**.

To replace bone loss from tumors, allografts of the tibia, femur diaphysis or ilium blocks have been used. Fibular struts as well as notched cortical allografts have also been employed with success for the replacement of vertebral bodies after partial or complete corpectomies.

Negative results with allografts as compared to autografts in anterior cervical spine fusions were reported in only one study (Zdeblick and Ducker 1991). In this report, with frozen allografts, a higher rate of non-unions and graft collapse were reported in two level fusions, but not in one level fusion.

In the cervical spine, aseptically processed allografts, when placed in an interbody position, produce fusion rates

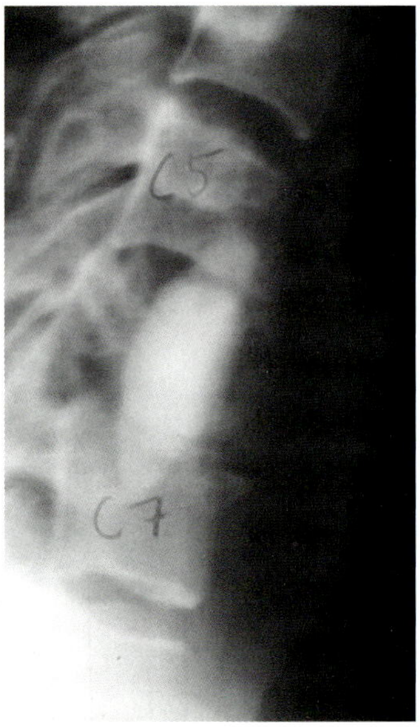

Figure 7 A cortical "maltese" cross shaped allograft placed to stabilize the spine after removal of a vertebral body 1 year post-transplantation. Utilizing the principle advocated by Bailey and Badgley (1960) allograft fused with the vertebral below and above the gap. Although these grafts worked well, the technique never became popular

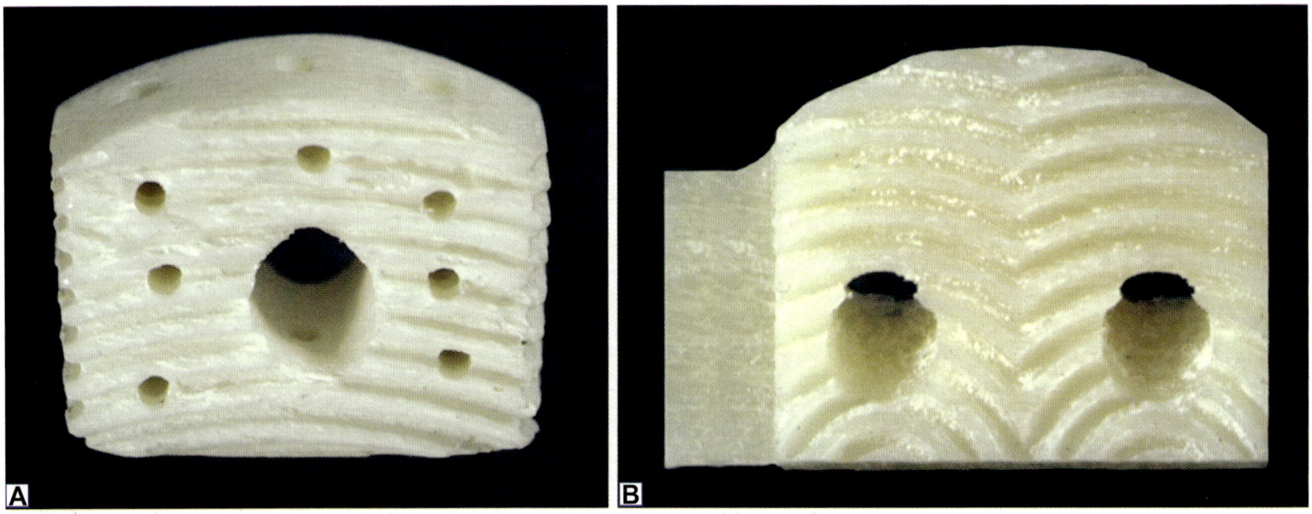

Figures 6A and B Two machined bone allografts for the vertebral interbody fusion. Allograft on the left has small drill holes to accommodate ingrowth of recipient's bone, in addition to larger apertures which can be packed with autologous, growth factors, etc. Allograft, on the right has a shaft which accommodates forceps for ease of insertion. After the insertion shelf is broken off. Gull wing grooves prevent the graft from being displaced by slippage very similar to those obtained with autografts (Brown et al 1976, Cloward 1962, Rish et al 1976, Schneider and Bright 1976). Experience with dowel-shaped or tricortical iliac crest allografts show certain advantages over autografts obtained from the ilium. Dowel-shaped bone allografts are prepared from the femoral head, femoral condyles or tibial plateaus.

In posterior cervical fusions the indications for bone allografts are not as clear cut as they are for the anterior cervical fusions. Although posterior cervical fusions have been a cornerstone for the stabilization of the cervical spine, posterior interbody fusion in the cervical spine is technically not feasible. Hence, strut grafts have usually been employed. Most frequently these were autografts, as complications with allografts, particularly in the upper cervical spine, have been reported (Stabler et al 1985). This might be explained in part by poor performance of cortical bone plates and struts under tension (White and Panjabi 1984). At present, autogenous bone grafts from the ilium or the fibula are generally preferred to allografts for the posterior cervical spine fusions. Cancellous bone from the ilium incorporates into the fusion mass more rapidly than does cortical bone. However, this sentiment does not apply to all cases, as in a number of patients, posterior fibular bone allografts have performed well in stabilizing the cervical spine. This is particularly true in the lower cervical spine, as is shown in **Figure 8**. Therefore the decision regarding the use of allografts in stabilization of the cervical spine posteriorly must be made individually for each patient. Several factors will influence the decision, not least of which is the availability and quality of autologous bone.

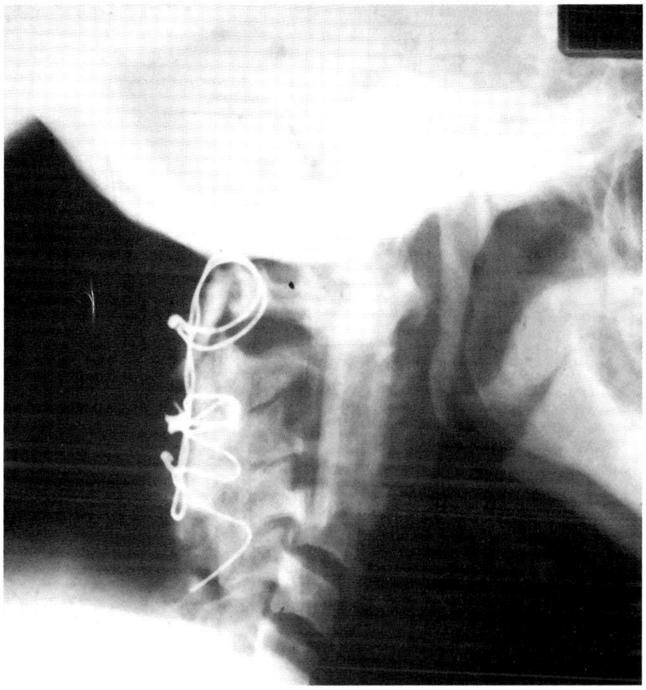

Figure 8 The spine stabilized with autograft fibula anteriorly and a freeze-dried cortical allograft posteriorly in a patient with metastatic thyroid carcinoma. By 1 year solid fusion was produced by the autograft as well as the allograft

The above statements likewise do not apply to posterior cervical laminoplasties. Both the rib and the fibula allografts perform well in these operations (**Figs 9A and B**).

Thoracolumbar Spine

In the lumbar spine, indications for the use of freeze-dried allogeneic bone grafts are more complex than in the cervical spine. Certainly, in the face of significant bone loss, surgeons have little choice but to use a bone spacer. One clinical

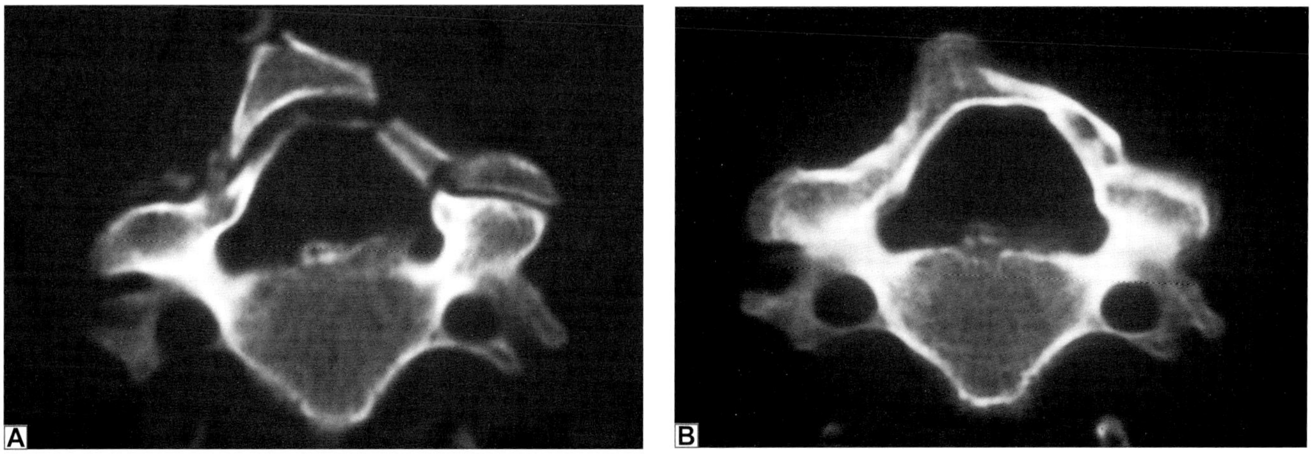

Figures 9A and B (A) Freeze-dried fibula strut allograft in the cervical spine laminoplasty, postoperatively; (B) 1 year post-transplantation

observation made by us with respect to fusion of the spine following the application of freeze-dried bone allografts is that the rate of union of grafts under compression load, i.e., interbody or strut grafts, does not appear to be delayed or hampered by the low-grade immunologic response. Freeze-dried allografts placed under tension posterior to the spine without internal fixation do not fare as well. This seems to be particularly true with cortical bone grafts. A high incidence of graft resorption, leading to nonunion with cortical strips of freeze-dried bone wired to the facets, as well as with intertransverse lateral lumbar fusions with corticocancellous matchsticks, have been noted. However, a large number of patients with allogeneic iliac strip (corticocancellous) grafts and internal fixation progressed to satisfactory fusion.

In salvage spinal surgery, banked bone with and without supplemental internal fixation is useful in replacing vertebral bodies destroyed by infection, trauma, congenital deformity or tumor. Since instituting the use of allografts, the use of methyl methacrylate in the stabilization and replacement of portions of the spine for this purpose has been abandoned except in cases of extensive spinal metastatic disease in patients with limited life expectancy. Even in these patients, the use of modern cage devices and allograft particulate bone are preferred.

Arthrodesis of the spine is performed to achieve permanent fixation of the spinal column. Rigid arthrodesis can be produced by bone grafts interposed between the vertebrae or by grafts superimposed on the exposed bone of the vertebral column. Rigid arthrodesis is difficult to achieve with grafts placed on the transverse processes.

A posterolateral fusion is the most common type of fusion performed in the lumbar spine (Sandhu HS et al 1999). Decortication of the posterior surfaces of the spine with superimposed autograft bone results in the formation of bridging bone across adjacent transverse processes. The fusion is often enhanced by supplementary metallic fixation, usually consisting of rods and pedicle screws. The biomechanics of the posterolateral spine are unique, since the motion segment is subjected to tensile forces rather than to compression. Therefore graft material used for this type of fusion does not have to provide mechanical support. Because posterolateral fusions do not correct anterior deficiency of the spinal column, interbody fusions are used in a number of circumstances alone or in combination with autografts (Ehrler and Vaccaro 2000). These exhibit higher fusion rates than the posterolateral fusions. Unlike the posterior spine, the anterior vertebral column is under compression. Thus any bone graft placed into intervertebral space must withstand these considerable loads. For this reason, cortical bone allografts became popular for this application.

Posterolateral fusion without instrumentation: Fred H Albee is credited with performing the first direct posterior fusion in patients with tuberculosis of the spine. He performed arthrodeses with autologous bone taken from the tibia and placed it on denuded spinous processes and laminae. In Albee's technique, the fusion was obtained by uniting the spinous processes into one continuous bony ridge. The technique gained popularity and the indications were expanded to include degenerative diseases of the spine. Autograft bone functioned well in these applications, as it did with Hibbs' technique, and became the gold standard for posterior spinal fusions. Exactly a century later it still is a commonly performed procedure. However, the placement of the grafts gradually changed toward posterolateral arthrodesis (Glasser et al 2004).

In the method developed by Hibbs, fusion of the neural arches is produced by overlapping small osseous flaps from spinous processes, contiguous laminae and articular facets. Posterolateral arthrodesis has an advantage over posterior arthrodesis because of the extensive blood supply from the transverse processes, and the ease of soft tissue coverage (Bosworth 1948).

Ever since Albee introduced the technique of spinal fusion, discussions have centered on the type of arthrodesis and on the use of autologous and allogeneic bone grafts. The question of spinal instrumentation had been settled in favor of the latter, particularly with respect to the employment of pedicle screws (Roy-Camille et al 1993, Louis 1986, Steffee and Sitkowski 1988, Bridwell et al 1993). However consideration of spinal instrumentation, important as it might be, falls outside of the scope of this chapter. With respect to the grafts used for posterolateral fusions, autografts, as pioneered by Albee, still remain the first choice when available.

The use of allografts for posterolateral fusion has been reported in several publications dating back to the middle of the last century. No unanimity of opinion has been reached on the subject. The data is difficult to interpret since allografts prepared by different methods had been used in these studies and compared with autologous bone grafts. Bosworth and his colleagues fused spines with both autologous and allogeneic bone in patients with tuberculosis of the spine, as did Albee (Bosworth 1948). The specific type of bone allograft used in each case was not recorded. The pseudarthrosis rate was 32.7% with banked bone and only 13.3% with autografts. However, the authors noted: "To claim that the high percentage of failures with bank bone was due to the fact that the bone was not always of the best quality (that is cortex rather than cancellous) is pointless, since, when one uses bank bone, one takes what is on hand or goes without." This problem has been largely eliminated by the modern tissue banking technology.

As happens not infrequently, the observations of Bosworth and colleagues, as well as the experimental studies of Campbell and his co-workers, were not accepted without reservations. Dr Robert D Ray (1953) noted: "Dr Bosworth and his associates are certain to be highly commended for their well controlled clinical study, as are Dr Campbell and his co-workers for their timely experiments.

Both groups have concluded that fresh autogenous bone grafts are superior to other types of bone grafting materials. However, many of those here today also attended the recent meeting of the American College of Surgeons in San Francisco, where another group of investigators reported on a series of similar experiments from which they drew conclusions diametrically opposed to those just presented."

The problem on the divergence of opinions and results relating to the performance of freeze-dried or frozen bone allografts in the posterolateral spine fusions, regretfully, has not been resolved with time. To this day there is still a divergence of opinions on the subject. The method of graft preparation is diverse and many banks rely on secondary sterilization techniques that have been shown to have inferior inductive properties compared to aseptically processed bone. The matter is further complicated by the use of particulate bone allografts to achieve posterolateral fusions without instrumentation.

In summary, for posterolateral fusions of the lumbar spine, without instrumentation, autologous bone is preferable to allogeneic bone. Cancellous bone incorporates into the fusion mass more rapidly than does cortical bone. Therefore, strips of ilium are the most frequently employed grafts. Most authors report significantly higher fusion rates with autografts as compared to allografts (An et al 1995, Nasca and Whelchel 1987). Good performance of bone allografts in the cervical spine or the thoracolumbar spine in adolescents cannot be extrapolated to lumbar spine in adults with degenerative disease (Boden 1994, Malinin and Brown 1996).

Arthrodesis of the thoracic and lumbar spine with supplemental instrumentation: For the last several decades discussions regarding indications for spinal instrumentation have dominated orthopedic literature. Although some spinal instrumentation has been available for years (Hadra 1975, Bosworth 1948), currently the insertion of screws through the pedicles or vertebral bodies attached to rods or plates respectively has been most prominent. The statement, of course, does not apply to instrumentation employed for the correction of scoliosis, kyphosis and lordosis which has been around since Harrington introduced distraction rods for correction of scoliosis in the 1960s.

Papers have been published arguing for and against supplemental instrumentation with posterior lumbar arthrodesis. The goals of instrumented versus non-instrumented posterolateral fusion are to augment the fusion, increase the rate and quality of the fusion, and hasten recovery by decreasing the time of postoperative immobilization. Unfortunately, papers dealing with pedicle screw fixation and comparison to non-instrumented fusions do not focus on the types of grafts used to effect the fusions in patients included in the studies. Presumably most of the fusions employing structural grafts were autografts. A similar assumption cannot be made with regard to particulate grafts. Particulate allograft with particle size of 160–250 μm produced excellent results when implanted into bone defects (Temple and Malinin 2008).

Arthrodesis of the lumbar and thoracic spine is a commonly performed operation for scoliosis. Instrumentation used to realign the spine must be supplemented with spinal fusion to avoid breakage of metallic fixation devices.

In these circumstances the fusion technique involves the destruction of articular cartilage, removal of soft tissue attachments and decortication of host bone surfaces. Unlike the experience of posterolateral fusions in adults performed for spondylolisthesis and other degenerative conditions, the indications for the use of allograft bone in surgery for the correction of idiopathic adolescent scoliosis is clear-cut. There is a general agreement among the authors on the adequacy of fusions effected by allogeneic bone graft in this patient population (Aurori et al 1985, McCarthy et al 1986, Dodd 1988, Recht et al 1993, Farby 1991, Stricker 1997, Knapp et al 2005). Dodd and co-workers compared patients receiving autografts from the iliac crest with those receiving frozen corticocancellous allogeneic bone grafts. In a blind radiographic assessment of bone graft fusion, no mass difference was demonstrated between the two groups. The maintenance of the curve correction was also similar. McCarthy et al reported similar complication rates and fusion rates similar or better in patients receiving allografts versus those receiving autologous bone. Montgomery et al found freeze-dried bone allografts to be safe and effective substitutes for autologous bone in patients with cerebral palsy and scoliosis (Montgomery et al 1990). Stricker and Sher (1997) demonstrated the efficacy of freeze-dried crushed cortical allograft in promoting fusion in patients with idiopathic and other types of scoliosis. Farby critically evaluated autograft versus allograft bone using multivariate statistical analysis. The subjects of his study were two groups of patients matched for age, angle of curve, and length of fusion. The group receiving autologous bone consisted of 88 patients, whereas the allograft group had 99 patients. A significant reduction in the operating time and blood loss were recorded with allografts versus autografts. One year postoperatively there were no significant differences in the correction of the curve. The authors recommended that, given the morbidity with autografts (Summers and Eisenstein 1989) as well as the discomfort at the autograft site, the use of bone allograft was preferred in scoliosis surgery.

The follow-ups in most of the above cited reports were relatively short, with McCarthy et al and Stricker and Sher observing patients for 2.8–3.7 years. Others had even shorter periods of observation. Recognizing this deficiency, Knapp and co-workers performed a retrospective multi-institutional review of 111 patients with a follow-up time of 5–15 years (average 6 years). They confirmed the results of previous studies (Knapp et al 2005). The average loss of correction with segmental instrumentation such as Cotrel-Dubousset instrumentation and Texas Scottish Rite Hospital instrumentation was 6.3%, while loss with the Wisconsin technique (Drummond et al 1984) was at 4.4%. Thus the overall average rate of 5.9% was superior to the 7–14% figures quoted in the literature for autograft fusions. All authors in this study used aseptically excised and processed corticocancellous bone allograft.

In view of the cumulative results with allografts, the reticence of some spinal surgeons to use allograft bone in idiopathic scoliosis is puzzling. The criticism of the procedure of using allograft in spinal fusion by Czitrom is based on neither scientific nor clinical evidence (Czitrom 1998). Therefore it is unwarranted.

The data accumulated thus far tips the scale in favor of bone allografts in posterior spinal fusions used for patients with idiopathic adolescent and neuromuscular scoliosis. In these patients, the use of bone allografts not only matches the results with autografts, but also reduces operative time and blood loss. It eliminates morbidity and discomfort resulting from graft harvest. Such being the case, the use of adequately processed bone allograft is indicated even in the presence of an adequate iliac crest. Because of its biological properties, aseptically excised and processed banked bone is preferable to irradiated allografts or bone secondarily sterilized by other means. Many surgeons prefer freeze-dried bone over frozen bone, not only because of its convenience, but also because of the reduction of antigenicity in freeze-dried bone allografts (Malinin and Brown 1996). Now there remains the question of an optimal size of particulate bone allograft. Recent studies indicate that particle sizes in the range of 100–300 microns produce optimal results. These can be either cortical or cancellous. Both types of grafts are osteogenic bone and accelerate the healing process by direct rather than endochondral ossification (Temple and Malinin, 2008).

Posterior Lumbar Interbody Fusion

Posterior lumbar interbody fusion, as originally described by Cloward, was subsequently modified (Lin et al 1983). The placement of four iliac crest grafts into the intervertebral space is difficult, but the success of the graft incorporation depends on adequacy of alignment and close match of these grafts. Cloward had used allografts for this procedure almost from its inception and reported good results in 623 posterior lumbar interbody fusions (Cloward 1963). However, he cautioned, the graft must be placed in the prepared interspace so that the cortical surface of the graft lies in a vertical plane and the cancellous bone is in contact with the vertebral bodies as shown in **Figure 4**. Subsequently, posterior lumbar interbody fusions utilizing two larger grafts rather than four thin ones have been performed with success. The fusion rates with this procedure have been roughly the same regardless of whether the patients' own bone or allografts were used. Steffee combined posterior interbody fusions with segmental instrumentation and devised rectangular cancellous block grafts which are placed in the intervertebral body position (Steffee and Sitkowski 1988). He claimed a predictable fusion outcome utilizing this technique. Several other configurations of bone allografts for this purpose have been designed since then.

Anterior Lumbar Interbody Fusion

Since the original description of the anterior lumbar interbody fusion in 1933, different success rates have been described with this procedure. Allografts and autografts have been used interchangeably with equal success (Loquidice 1988), although Crock, who described the use of ALIF for internal disc derangement, favored autografts (Crock 1982). Few spinal fusion operations theoretically appear as ideal and as unpredictable and difficult as the anterior lumbar interbody fusion, hence there are a number of discrepancies in the reports dealing with it. Watkins et al in a review of 82 patients with ALIFs recorded a 60% union rate for autografts and 64% for iliac crest block allografts, and noted that the fusion rate did not increase with additional operative time required for autograft patients (Watkins et al 1985).

On the other hand, Kozak et al recorded a 97% fusion rate with composite allografts consisting of femoral rings and crushed cancellous bone (Kozak et al 1994). However with the usage of cortical bone allografts prepared from fibulas, the fusion rate was only 58% (Wetzel et al 1993).

This suggests that ALIFs cortical bone allografts do not fare as well as cancellous allografts. To overcome the biological deficiencies of solid cortical bone grafts, machined grafts combining cortical and cancellous bone have been devised **(Fig. 5)**.

Anterior interbody fusion procedures are used widely with considerable success in patients undergoing lumbar spine surgery. The use of allograft bone is an effective adjunct when used in proper clinical circumstances (Ehler and Vaccaro 2000).

If used posteriorly in interbody fusion, the allografts fare well.

On the other hand, posterolateral fusions with allograft only resulted in poor fusion rates. This can be explained by the lack of compression in the posterolateral grafts.

Anteriorly or posteriorly placed interbody spinal allografts are in an optimum position for structural reconstruction. These allografts have an advantage of providing immediate support and being available in adequate quantities. Anterior interbody spine allografts produce fusion rates comparable with autografts **(Fig. 10)**.

Since interbody allografts are called upon to provide structural support immediately after transplantation, their biomechanical properties are of importance. Although it is well known that cancellous bone density is age related, the choice of the grafts based on age alone is not a reasonable procedure. Some individuals below 55 years may have bone that is structurally weak, while some over this age may have dense, structurally sound bone. Therefore, the suitability of the allografts should be determined by inspection and mechanical testing. This does not apply to cortical bone allografts. When tested under the same conditions, the femoral ring allografts were stiffest in compression and flexion over the threaded interbody fusion constructs. However, the latter were stiffer in the lateral bending and axial torsion. The stiffest construct in flexion, extension and lateral bending was the femoral ring allograft combined with screw-rod constructs (anterior) (Glazer et al 1996).

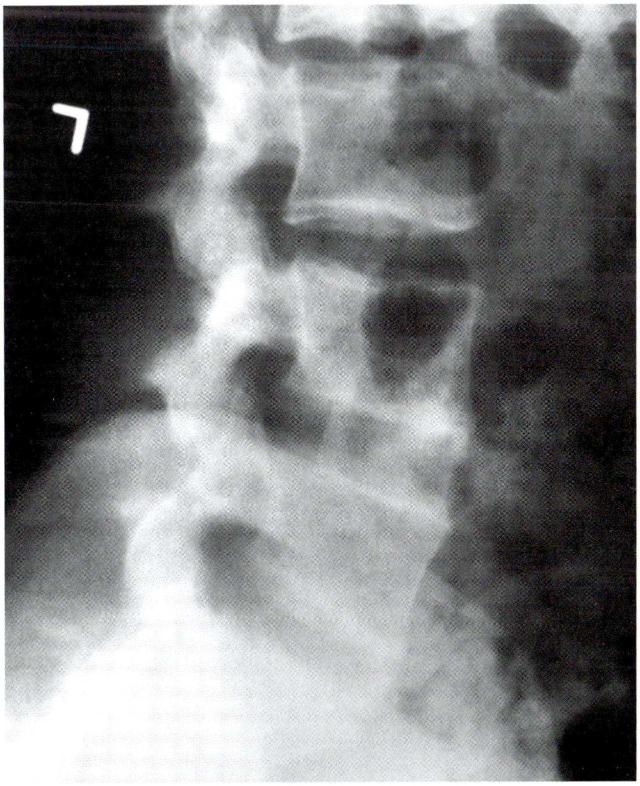

Figure 10 Anterior interbody fusion with frozen allograft 1 year post-transplantation. The radiograph shows integration and healing between the allografts and host vertebrae

Extensive Bone Loss

Stabilization of the spinal column in the presence of extensive bone loss usually cannot be accomplished with autogenous bone. There is simply not enough material to fill the gap. Therefore, bone allografts or an allograft-autograft combination are used for the purpose. Strut allografts spanning one or several spaces work well to stabilize the spine in the presence of sizeable defects caused by trauma, infection or tumor. The fibula provides the best anatomical construct for this purpose, but other bone struts can also be used.

In the lumbar spine, tibial segments are the graft of choice, because the flat medial surface can be abutted against the spinal canal. Segments of tibial diaphysis can support the spinal column biomechanically (**Figs 11A and B**).

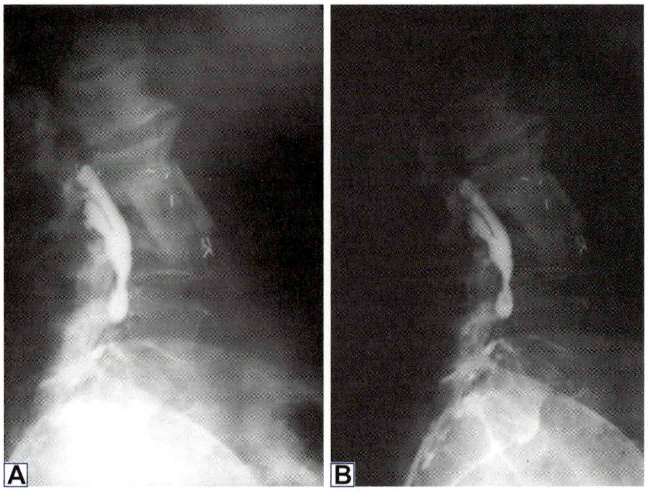

Figures 11A and B Corpectomy performed for chronic osteomyelitis. The defect was obliterated with the freeze-dried segment of the tibia. (A) Tibia graft is suitable for vertebral body replacement because one surface of it is flat. (B) One year later the allograft is united with the vertebral bodies and supports the continuity of the vertebral column (Note the myelogram dye is posterior to the allografts)

Allografts sterilized with ionizing irradiation cannot be recommended for posterior stabilization of the spine. Doses of irradiation required to sterilize bone allografts significantly reduce their osteogenic potential, and hence prevent speedy fusion of the graft needed to achieve stabilization of the affected segments of the spinal column (Buring and Urist 1967) (**Fig. 12**).

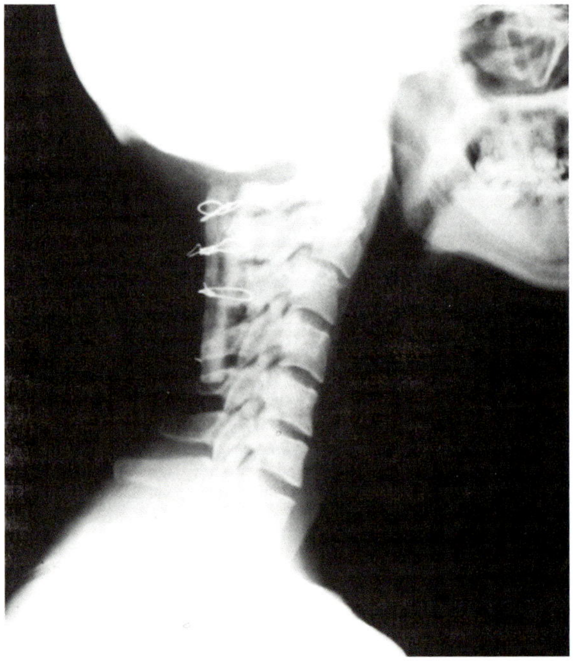

Figure 12 Irradiated freeze-dried allograft used for posterior cervical spine fusion. The graft had not fused three and a half years post-transplantation

Xenografts generally have likewise not met with success in spine fusion. Treated bovine xenografts are biologically inert. Deproteinized bovine bone (DBB) has architectural geometry that is similar to human cancellous bone and can theoretically support tissue ingrowth, osteoprogenitor cell attachment and new bone formation. However DBB is mechanically weak and prone to fracture. It cannot be used in load bearing applications and is therefore unsuitable for spinal fusions in humans (Stavropoulos 2008). A xenograft used for spine fusion that years later, remain biologically inert is shown in **Figure 13**.

Bone Morphogenetic Proteins

Bone morphogenetic proteins (BMPs) are essentially responsible for bioactivity in bone healing. Utilizing recombinant gene technology, large quantities of recombinant BMPs (rhBMPs) can be produced, albeit expensively.

Recombinant human BMP-2 and recombinant human BMP-7 have been used to induce bone formation in spinal fusions. These molecules, also known as osteogenic protein 1 or OP-1, have been used in a number of clinical studies, which resulted in approval of these products for spinal fusions (Grauer et al 2000, Blatert et al 2002, Cunningham et al 2002, Benzel et al 2008). Bone morphogenetic protein had been placed into spines alone in combination with spinal instrumentation and in spinal cages. Clinical studies evaluating BMP in humans have suggested replacement of

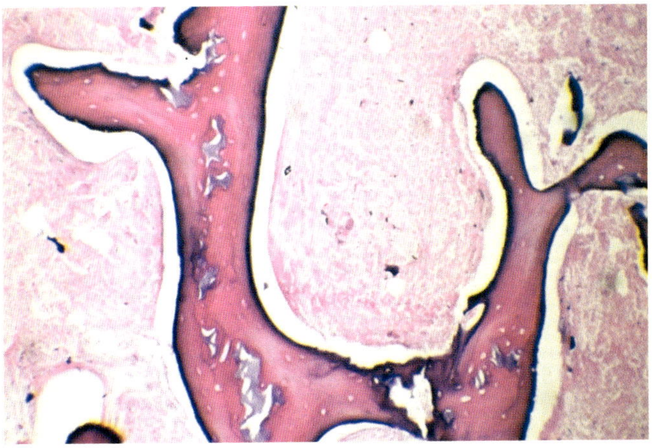

Figure 13 A bovine xenograft used to fuse lumbar spine. Fifteen years later the xenograft remains virtually unchanged. The graft is surrounded by fibrous connective tissue. Bone marrow is lacking in the trabecular spaces (H&E × 250)

viable and non-viable bone grafts with these compounds. Bone morphogenetic proteins were shown to promote bone growth and remodeling. The use of recombinant BMP was approved by the US Food and Drug Administration in 2002 for surgery of anterior lumbar spine to promote bone fusion. However, the current rates and patterns of BMP use since the clinical introduction 6 years ago are not known. No population-based data are available. Higher complication rates and greater inpatient charges in patients receiving BMP, compared to those in whom BMP was not used, have been recorded (Cahill et al 2009). In the posterolateral spine, Glassman reviewed 1,037 patients who received rhBMP-2 as an adjunct to posterolateral lumbar spine fusion. They noted a low complication rate directly attributable to the addition of rhBMP-2, (0.6%) (Glassman et al 2008, 2011). The use of BMPs as an adjunct to cervical spine fusion have been abandoned due to the resultant pretracheal edema and reports of airway compromise (Yaremchuk et al 2010).

Since the results with fusions in which BMP was employed do not differ appreciably from those with bone grafts, the use of BMPs in the spinal fusion should be approached with great caution, particularly in view of higher costs (Friedman et al 2002).

Alternatives to Bone Grafts

Composites which are made up of ceramics, demineralized bone matrices (DBM), synthetic polymers and others are osteoconductive, but with the exception of DBM, these are not osteoinductive compounds. Bone graft substitutes such as calcium sulfate, hydroxyapatite, tricalcium phosphate and beta-tricalcium phosphate provide three dimensional structures which are resorbable. These are not biologically active, but facilitate remodeling. No definitive clinical data on spine fusions using these compounds is currently available. Demineralized bone matrices are biologically active, but they provide no structural support.

Introduction of the transforaminal lumbar interbody fusion technique (TLIF) accelerated the use of spinal cages containing bone grafts as well as synthetic polymers such as polyglycolic acid (PGA). The latter provides scaffolding for bone ingrowth. Long term results with these techniques are not available.

DISCUSSION

The ready availability of bone allografts has led to several advances in reconstructive surgery of the spine. Correction and prevention of spinal deformity, as well as the replacement of bone loss, have been accomplished with the aid of these grafts. However, orthopedic and neurologic surgeons contemplating transplanting allogeneic, preserved bone must be aware of the advantages and disadvantages of these transplants.

Despite long-standing interest by these surgeons in achieving arthrodesis of the spine with bone from other individuals, systematic gathering of information for these efforts is relatively spotty. By now, accumulation of clinical experience suggests there is an indication for allografting in spinal surgery. Understanding the biological behavior of bone transplants and the dependence of their success on osteoinduction is of paramount importance. Clearly, allografts subjected to irradiation as well as other treatments which diminish or abolish their osteoinductive capacity will not function as well as their counterparts in which such activity is preserved. Curiously, in spinal surgery, allografts function well in some anatomical sites but not in others. Failures, for instance, have only been reported with allografts used in non-instrumental intertransverse fusion of the lumbar spine (An et al 1995, Brown et al 1976). Allografts are not autografts. They do not contribute living cells, and cannot be expected always to induce bone formation from the host bone. A fundamental difference between autografts and allografts is that the latter will not produce heterotrophic bone in a muscle bed. This must be borne in mind in determining where the allografts are to be placed.

The attainment of a solid spine fusion is multifactorial. There is a race between bone resorption and new bone formation in creating a fusion mass. The new bone mass must eventually withstand mechanical loads. Standard methods of bone allograft preparation are now widely accepted. However, the behavior of bone allografts cannot always be predicted. At this point, the use of bone allografts in spinal surgery can be recommended where its effectiveness has been clearly established. Hopefully future research will result in bone grafting material and methodology which will fulfill all of the demands of reconstructive surgery of the spine. At present we do not have such materials. Allografts have their limitations. Their use after the failure of autologous bone must be approached with caution.

Back pain is still the leading cause of disability worldwide and is one of the most common reasons for seeking help from a physician. Although non-surgical treatment remains the first line of defense, many patients eventually come to surgery and undergo spinal fusion. To achieve fusion, development of various surgical techniques, as well as a variety of bone grafts and subsequently the development of spinal instrumentation have evolved. Later still biologics and bone substitutes came into play.

Bone grafts, including bone allografts remain the keystone of spinal arthrodeses. A variety of these are available to the spinal surgeon, but the knowledge of how these are obtained and produced eludes many spine surgeons who obtain the grafts through the hospitals where they work, frequently on the recommendation of orthopedic appliance sales persons. Clearly this is not an optimum arrangement which benefits the patients.

Choice of spinal allografts on the basis of bleached color obtained by bleaching or subjection to a proprietary process during processing are but few examples of travesty. Such being the case, the authors urge spinal surgeons to become familiar with the technology of allograft preparation. It is our hope this chapter might be of help in this regard.

Among biologicals, in the last 6 years, rhBMPs have come into prominence. The usage of BMP in the US has increased from 0.69% of all fusions in 2002 to about 26% of all fusions in 2006. The figures for other countries are not available. The economic implications of BMP use in spinal fusion had little formal analysis, but BMP use is associated with higher total hospital charges in all spinal fusions (Friedman et al 2002). This is likely related to implant charges for cases that use BMP (Glassman et al 2008).

The clinical results with BMP in posterolateral fusions are comparable to those with microparticulate bone allografts, but the complication rate with BMP is higher (Cahill et al 2009). In view of high cost, higher complication rates and comparable clinical results, the choice of BMP versus bone allograft is not justified at this time.

BIBLIOGRAPHY

1. Albee FH. Evolution of bone grafts surgery. American Journal of Surgery, 1944;63:421-36.
2. Albee FH. Transplantation of a portion of the tibia into the spine for Pott's disease. Journal of American Medical Association, 1911;57:885-7.
3. An HS, Lynch K, Toth J. Prospective comparison of autograft versus allograft for adult posterolateral lumbar spine fusion: differences among freeze-dried, frozen and mixed grafts. Journal of Spinal Disorders, 1995;8:131-5.
4. Aurori BF, Weierman RJ, Lowell HA, et al. Pseudoarthrosis after spinal fusion for scoliosis: a comparison of autogenic and allogenic bone. Clinical Orthopaedics and Related Research, 1985;199:153-8.
5. Bailey RW, Badgley CE. Stabilization of the cervical spine by anterior fusion. Journal of Bone and Joint Surgery (Am), 1960;42:565-94.
6. Benzel EC, Gilbertson L, Mericle RA. Enhancing spinal fusion. Clinical Neurosurgery, 2008;50:63-71.
7. Boden S. Point of view. Spine, 1994;18:2053
8. Blattert TR, Delling G, Dalal PS, et al. Successful transpedicular lumbar interbody fusion by means of a composite of osteogenic protein-1 (rhBMP-7) and hydroxyapatite carrier: a comparison with autograft and hydroxyapatite in the sheep spine. Spine, 2002;27:2697-705.
9. Bosworth DM. Technique of spinal fusion. AAOS Instructional Course Lectures, 1948;3:295-313.
10. Bridwell KH, Anderson PA, Boden SD, et al. What's new in spine surgery. Journal of Bone and Joint Surgery (Am), 2010;92:2017-28.
11. Bridwell KH, Sedgewick TA, O'Brien MF, et al. The role of fusion and instrumentation in the treatment of degenerative spondylolisthesis with spinal stenosis. Journal of Spinal Disorders, 1993;6:461-72.
12. Brown MD, Malinin TI, Davis PB. A roentgenographic evaluation of the rate of fusion of allografts versus autografts in anterior cervical spine fusions. Clinical Orthopaedics and Related Research, 1976;119:231-8.
13. Buring K, Urist MR. Effect of ionizing radiation on the bone induction principle in the matrix of bone implants. Clinical Orthopaedics and Related Research, 1967;55:225-34.

14. Cahill KS, Chi JH, Day A, et al. Prevalence, complications and hospital charges associated with use of bone-morphogenetic proteins in spinal fusion procedures. Journal of American Medical Association, 2009;302(1):58-66.
15. Cloward RB. The treatment of ruptured lumbar intervertebral disc by vertebral body fusion. I Indications, operative technique, after care. Journal of Neurosurgery, 1953;10:1954-68.
16. Cloward RB. The treatment of ruptured lumbar intervertebral disc by vertebral body fusion. III Method of use of banked bone. Annals of Surgery, 1952;136:987-92.
17. Cloward RB. Lesions of intervertebral disc and their treatment by interbody fusion methods. Clinical Orthopaedics and Related Research, 1963;27:51-77.
18. Cloward RB. The anterior approach for removal of ruptured cervical discs. Journal of Neurosurgery, 1962;15:602-17.
19. Cloward RB. The treatment of ruptured intervertebral disc by vertebral body fusion. Annals of Surgery, 1952;136:967-71.
20. Crock HV. Anterior lumbar interbody fusion. Clinical Orthopaedics and Related Research, 1982;165:157-63.
21. Cunningham BW, Shimamoto N, Sefter JC et al. Osseointegration of autograph versus osteogenic protein-1 in posterolateral spinal arthrodesis: emphasis on the comparative mechanisms of bone induction. Spine, 2002, 2:11-24.
22. Czitrom AA. Freeze-dried cortical allografts. A letter to the editor. Orthopaedics, 1998;21:242
23. Dodd CA, Fergusson CM, Freedman L, et al. Allograft versus autograft bone in scoliosis surgery. Journal of Bone and Joint Surgery (Br), 1988;70:431-4.
24. Drummond DS, Guadangni J, Keene JS, et al. Interspinous process segmental spinal instrumentation. Journal of Pediatric Orthopaedics, 1984;4:397-404.
25. Dunn HK. Spinal instrumentation. In: Evart CM (Ed). AAOS Instructional Course Lectures. Mosby, S.V: St. Louis; 1983. pp. 209-18.
26. Ehler DM, Vaccaro AR. The use of allograft bone in lumbar spine surgery. Clinical Orthopaedics and Related Research, 2000;371:38-45.
27. Farby G. Allograft versus autograft bone in idiopathic scoliosis surgery: a multivariate statistical analysis. Journal of Pediatric Orthopaedics, 1991;11:465-8.
28. Friedman B, De La Mare J, Andrews R, et al. Practical option for estimating cost of hospital inpatient stays. Journal of Healthcare Finance, 2002;29(1):1-13.
29. Glasser JA, Bernhardt M, Found EM, et al. Lumbar arthrodesis for degenerative conditions. AAOS Instructional Course Lectures, 2004;53:325-40.
30. Glassman SD, Carreon LY, Campbell MJ, et al. The perioperative cost of infuse bone graft in posterolateral lumbar spine fusion. Spine Journal, 2008;8(3):443-8.
31. Glassman SD, Howard J, Dimar J, et al. Complications with recombinant human bone morphogenic protein-2 in posterolateral spine fusion: a consecutive series of 1037 cases. Spine, 2011;36(22):1849-54.
32. Glazer PA, Colliou O, Lotz JC, et al. Biomechanical analysis of lumbosacral fixation. Spine, 1996;21:1211-22.
33. Grauer JN, Patel TC, Erulkar JS, et al. 2000 Young Investigator Research Award winner. Evaluation of OP-1 as a graft substitute for intertransverse process lumbar fusion. Spine, 2001;26:127-33.
34. Hadra BE. The classic: wiring of the vertebrae as a means of immobilization in fracture and Potts' disease: Berthold E. Hadra: Med Times and Register, May 23, 1891. Clinical Orthopaedics and Related Research, 1975;112:4-8.
35. Kim DH, Rhim R, Li L, et al. Prospective study of iliac crest bone graft harvest site pain and morbidity. Spine Journal, 2009;9:886-92.
36. Knapp DR, Jones ET, Blanko JS, et al. Allograft bone in spinal fusion for adolescent idiopathic scoliosis. Journal of Spinal Disorder Technology, 2005;18 (Sept 1):S73-6.
37. Kozak JA, Heilman AE, O'Brien JP. Anterior lumbar fusion options. Technique and graft material. Clinical Orthopaedics and Related Research, 1994;300:45-51.
38. Lin PM, Cautilli RA, Joyce MF. Posterior lumbar interbody fusion. Clinical Orthopaedics and Related Research, 1983;180:154-68.
39. Loquidice VA, Johnson RG, Guyer RD, et al. Anterior lumbar interbody fusion. Spine, 1988;13:366-9.
40. Louis R. Fusion of the lumbar and sacral spine by internal fixation with screw plates. Clinical Orthopaedics and Related Research, 1986;203:18-33.
41. Madan SS, Harley JM, Boeree NR. Anterior lumbar interbody fusion: does stable anterior fixation matter? European Spine Journal, 2003;12:386-92.
42. Malinin T, Brown MD. Arthrodesis of the spine with bone allografts. In: AA Czitrom, H Winkler (Eds). Orthopaedic Allograft Surgery: Springer Verlag; 1996. pp. 193-201.
43. Malinin TI, Brown MD. Bone allografts in spinal surgery. Clinical Orthopaedics and Related Research, 1981;154:68.
44. Malinin TI, Eismont FJ, Brown MD. Materials used in spine stabilization. In: Dunster M, et al. (Eds). The Unstable Spine. Philadelphia: W.B Saunders; 1986. pp. 25-43.
45. Malinin TI, Rosomoff HL, Sutton CH. Human cadaver femoral head homografts for anterior cervical spine fusions. Surgical Neurology, 1977;7:249-51.
46. Malinin TI. Acquisition and banking of bone allografts. In: HB Habal, AH Reddi (Eds). Bone Grafts and Bone Substitutes. Philadelphia: WB Saunders Co; 1992. pp. 206-25.
47. Marx RE, Snyder RM, Kline SN. Cellular survival of marrow during placement of marrow cancellous bone graft. Journal of Oral Surgery, 1979;37:712-8.
48. McCarthy RE, Peek RD, Morrissy RT, et al. Allograft bone in spinal fusion for paralytic scoliosis. Journal of Bone and Joint Surgery (Am), 1986;68:370-5.
49. Montgomery OM, Aronson DD, Lee CL, et al. Posterior spinal fusion: Allograft versus autograft bone. Journal of Spinal Disorders, 1990;3:370-5.
50. Muschler GG, Matsukura Y, Nitto H, et al. Selective retention of bone marrow derived cells to enhance spinal fusion. Clinical Orthopaedics and Related Research, 2005;432: 242-51.

51. Nasca RJ, Whelchel JD. Use of cryopreserved bone in spinal surgery. Spine, 1987;3:222-7.
52. Ray RD. Discussion. Journal of Bone and Joint Surgery (Am), 1953;35:343-4.
53. Recht J, Bayard F, Delloye C, et al. Freeze-dried allograft versus autograft bone in scoliosis surgery. European Spine, 1993;2:235-8.
54. Rish BL, McFadden JT, Penix JO. Anterior cervical spine fusion using homologous bone grafts: A comparative study. Surgical Neurology, 1976;5:119-21.
55. Robinson RA, Smith GW. Anterolateral cervical disc removal and interbody fusion for cervical disc syndrome. Bulletins of Johns Hopkins Hospital, 1955;96:223-9.
56. Robinson RA, Walker AE, Ferlic DC, et al. The results of anterior interbody fusion of the cervical spine. Journal of Bone and Joint Surgery (Am), 1962;44:1569-86.
57. Rossier AB, Hussey RW, Kenzora JE. Anterior fibular interbody fusion in the treatment of cervical spine cord injuries. Surgical Neurology, 1977;7:55-60.
58. Roy-Camille R, Benazet JP, Desauge JP, et al. Lumbosacral fusion with pedicular screw plating instrumentation: A 10-year follow-up. Acta Orthopaedica Scandinavica Supplementum, 1993;251:100-4.
59. Sandhu HS, Grewal HS, Parvataneni H. Bone grafting for spinal fusion. Clinical Orthopaedics and Related Research, 1999;30:685-98.
60. Schneider JR, Bright RW. Anterior cervical fusion using freeze-dried bone allografts. Transplantation Proceedings, 1976;8 (Suppl 1):73-6.
61. Simpson AK, Whang PG, Grauer JN. Biological Approaches to Spinal Fusion. In: Orthopaedic Biology and Medicine: Musculoskeletal Tissue Regeneration, Biologic Materials and Methods (WS Pietrzak ed) Humana Press, Totowa, NJ 2008. pp. 247-58.
62. Smith GW, Robinson RA. The treatment of certain cervical spine disorders by anterior removal of the intervertebral disc and interbody fusion. Journal of Bone and Joint Surgery (Am), 1958;40:607-24.
63. Stabler CL, Eismont FJ, Brown MD, et al. Failure of posterior cervical fusion using cadaveric bone graft in children. Journal of Bone and Joint Surgery (Am), 1985;67:370-5.
64. Stavropoulos A. Deproteinized bovine bone xenograft In: WS Pietrzak (Ed). Musculoskeletal Tissue Regeneration: Humana Press; 2008. pp. 119-51.
65. Steffee AD, Sitkowski DJ. Posterior lumbar interbody fusion and plates. Clinical Orthopaedics and Related Research, 1988;227:99-102.
66. Stricker SJ, Sher JS. Freeze-dried cortical allograft in posterior spinal arthrodesis: Use with segmental instrumentation for connection of idiopathic adolescent scoliosis. Orthopaedics, 1997;20:1039-43.
67. Summers BN, Eisenstein SM. Donor site pain from the ilium, a complication of the lumbar spine fusion. Journal of Bone and Joint Surgery (Br), 1989;71:677-80.
68. Temple HT, Malinin TI. Microparticulate cortical allografts. An alternative to autograft in the treatment of osseous defects. The Open Orthopaedics Journal, 2008;2:91-6.
69. Watkins RG, Springer D, Wiltse L, et al. Anterior interbody fusion of the lumbar spine—a review. Orthopaedic Transaction, 1985;9:430.
70. Wetzel FT, Hoffman MA, Arcieri RR. Freeze-dried fibular allograft in anterior spinal surgery: cervical and lumbar applications. Yale Journal of Biology and Medicine, 1993; 66:263-75.
71. White AA, Panjabi MM. The role of stabilization in the treatment of the cervical spine injuries. Spine, 1984;9:512-22.
72. Wilson PD. Follow-up study of the use of refrigerated homologous bone transplants in orthopedic operations. Journal of Bone and Joint Surgery (Am), 1951;37:307-26.
73. Winter RB, Moe JH, Wang JF. Congenital kyphosis. Journal of Bone and Joint Surgery (Am), 1973;55:223-56.
74. Yaremchuk KL, Toma MS, Somers ML, et al. Acute airway obstruction in cervical spine procedure with bone morphogenic protein. Laryngoscope, 2010;1954-7.
75. Zdeblick TA, Ducker TB. The use of freeze-dried allograft bone for anterior cervical spine fusion. Spine, 1991;16:726-9.

Chapter 12

Walid Mnaymneh MD (1931), Professor of Orthopaedic Surgery, University of Miami, Miami, Florida. Dr Mnaymneh was chief of Orthopaedic Oncology at the University of Miami, Jackson Memorial Medical Center from 1977–1996. A proponent of massive allografts for reconstruction of extremities following resection of tumors, he developed a number of innovative surgical techniques. Among others he described resections and reconstructions of scapulas and hemi-pelvises.

(*Reference:* Mnaymneh W, Malinin T, *Massive allografts in surgery of bone tumors*. Orthopaedic Clinics of North America 1989;20: 455-67.)

Limb Saving Procedures

Bone allografts, both structural and particulate have played an important clinical role in bone repair for segmental and partial defects in bone.

HISTORY

The history of human homograft (allograft) replacement is long and interesting. The first depiction of allograft transplantation was of the saints Damien and Cosmos in the 3rd century AD (**Fig. 1**).

The first report of successful allograft transplantation in the English literature was by MacEwen in Scotland whereby a large segmental defect in the humeral diaphysis of a child was healed with allograft bone (MacEwen 1881). The work by Lexer in the early 20th century produced compelling results. He initially replaced whole joints and later hemi-joints around the knee (Lexer 1925). Subsequently, clinical bone transplantation was taken on by Ottolenghi in Argentina (Ottolenghi 1961), Parrish in the United States (Parrish 1966), and Volkov in Russia (Volkov 1970). Their empiric results were validated by the scientific observations and discoveries of doctors Hyatt and Kreutz at the National Naval Medical Center in Bethesda, Maryland (**Fig. 2**).

Figure 1 Maestro de los Balbases, Burgos, Spain, c. 1495. Painless amputation of the ulcerated leg of a Christian, and the substitution of the undiseased leg of a dead Moor

Hyatt espoused aseptic recovery and processing, and developed techniques for freeze-drying bone. Processing and distribution of cryopreserved osteochondral allografts was pioneered by the University of Miami Tissue Bank. Later scientific contributions by Malinin, Mnaymneh,

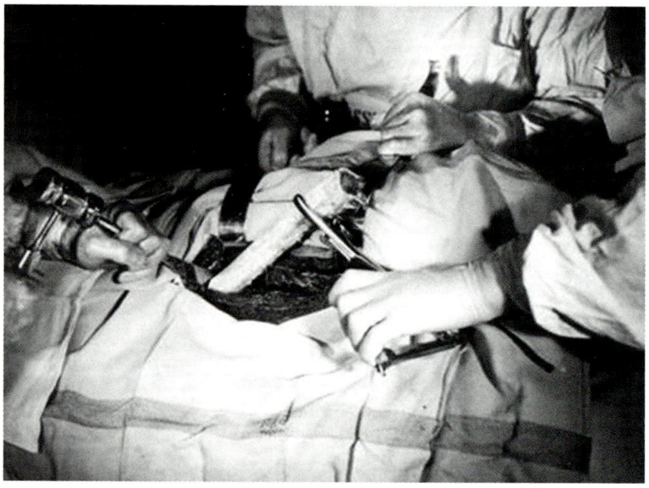

Figure 2 Aseptic recovery of bone 1960: Tissue Bank, National Naval Medical Center, Bethesda, Maryland

Mankin, Tomford, and others, as well as clinical applications of allografts in many major universities around the world, confirmed the clinical utility of bone and soft tissue allografts. (Mankin et al 1982, Mnaymneh et al 1990, 1992).

STRUCTURAL ALLOGRAFTS

These grafts are used primarily for replacement of large segments of bone and/or joints. This occurs following tumor surgery, but also in cases of bone loss in patients with failed implants due to osteolysis, trauma and to a lesser extent, bone infections (osteomyelitis). With the advent of chemotherapy for primary malignant bone tumors, most commonly in children with osteosarcoma, survival has improved markedly. Thus, limb sparing surgery became not only possible but preferred over amputation for the treatment of this disease. Subsequent studies comparing amputation and limb salvage demonstrated comparable local disease control and overall survival after administration of chemotherapy (Rougraff et al 1994). Osteoarticular allografts, aseptically excised and processed by a method of controlled rate freezing preserves articular cartilage and is especially well suited for restoring joint function. Structural allografts incorporate slowly into the host bone principally from the host tissue into the allograft at the host-allograft junction. Carefully reconstructed joints are stable and kinematically superior to endoprosthesis (Temple 2011). When properly transplanted, osteoarticular allografts can provide excellent function for many years. When the articular surfaces fail, the graft provides a substrate for further reconstruction, usually joint arthroplasty.

Allografts are especially useful in reconstructing complex defects such as those following resection of the pelvis which, in our experience, remains the best option for reconstructing acetabular defects and restoring the hip articulation (Bell et al 1997).

In growing children, preservation of the physis can avoid significant limb length discrepancies especially about the knee (Ramseier et al 2006). In these circumstances an osteoarticular allograft is useful as this technique can preserve at least one of the physeal growth plates. Careful preoperative planning allows the surgeon to estimate the remaining growth of the child's limb and thus plan to lengthen the affected limb during surgery (with a longer graft) and/or to perform an epiphysiodesis of the contralateral limb at a later time. Furthermore, with advanced imaging and improved chemotherapeutic agents, the physis adjacent to the tumor can be preserved, in selected cases, at the time of tumor resection using precise image-guided osteotomies above and below the lesion (**Figs 3A to H**).

Imaging is critical in the decision making process as this technique is limited to patients in whom an excellent tumor response to chemotherapy is observed in addition to the preservation of adequate bone to achieve adequate fixation.

Allograft reconstructions have certain disadvantages. For example, the surgery requires very careful preoperative planning and critical sizing of the allograft to satisfy the anticipated defects. Furthermore, these grafts typically come from young donors that represent a small portion of the allografts available. The problem is compounded by the fact that there are only a limited number of banks in the world that actually process and distribute cryopreserved and fresh grafts for this purpose.

Surgically, removing a tumor and sparing the limb is more difficult than an amputation. Reconstructing the defect and the joint with an allograft is much more challenging than using a metal implant. Operative times are longer, infections are more common (up to 12% of patients) than corresponding metal implants, healing is delayed, especially in patients undergoing subsequent chemotherapy and

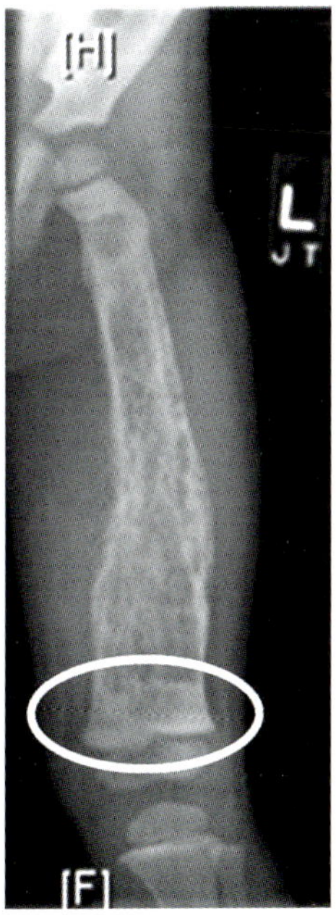

Figure 3A A 2-year-old boy with Ewing sarcoma of the left distal femur. The physis and a portion of the metaphyseal bone are uninvolved (circle)

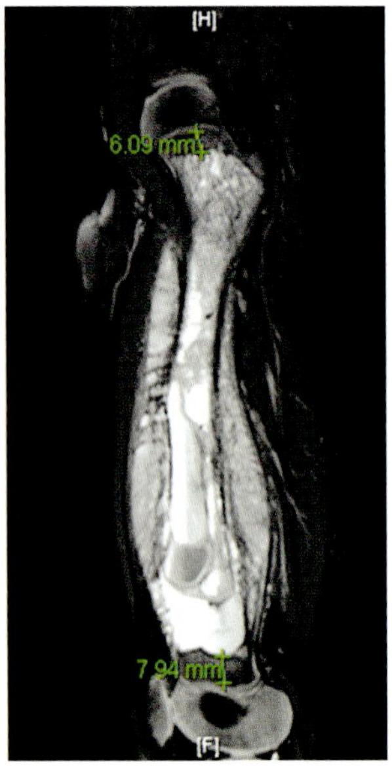

Figure 3B Sagittal MR of the left distal femur demonstrating uninvolved metaphyseal bone prior to chemotherapy

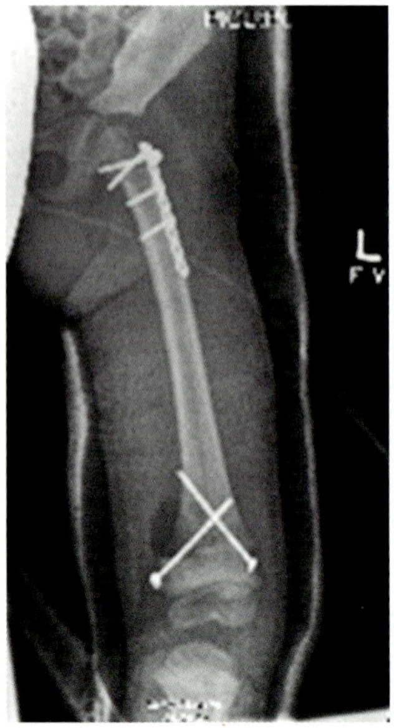

Figure 3D AP radiograph immediately after surgery demonstrating fixation of an intercalary allograft with preservation of the proximal and distal femoral physes

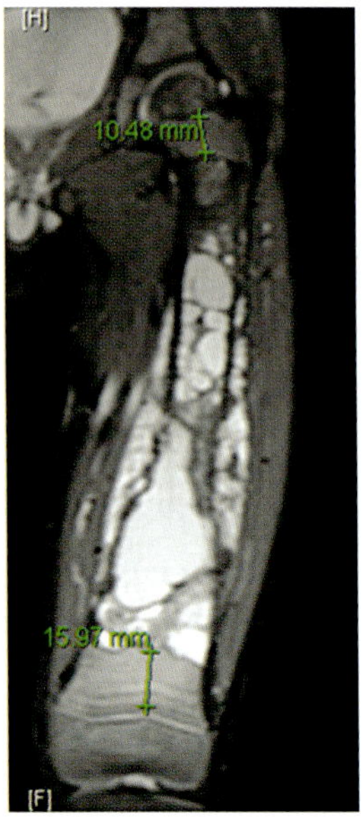

Figure 3C Coronal MR of the same patient following chemotherapy demonstrating intervening growth of the femur. The additional growth coupled with an excellent chemotherapeutic response made limb and physeal salvage possible

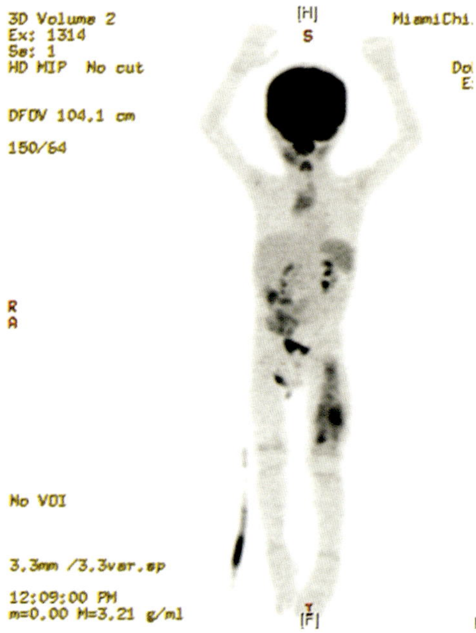

Figure 3E Fluorodeoxyglucose-PET scan prior to chemotherapy in the same patient as shown in Figure 3A

non-unions are not infrequent (up to 30% of patients) (Ortiz-Cruz et al 1997). Potential complications include fracture (in as many as 17% of the patients) (Sorger et al 2001) usually at the non-union site early on and around the joint months and years later (Berrey et al 1990), late infection, joint instability that may lead to early and advanced arthritic changes, and the remote possibility of disease transmission (Buck et al 1990).

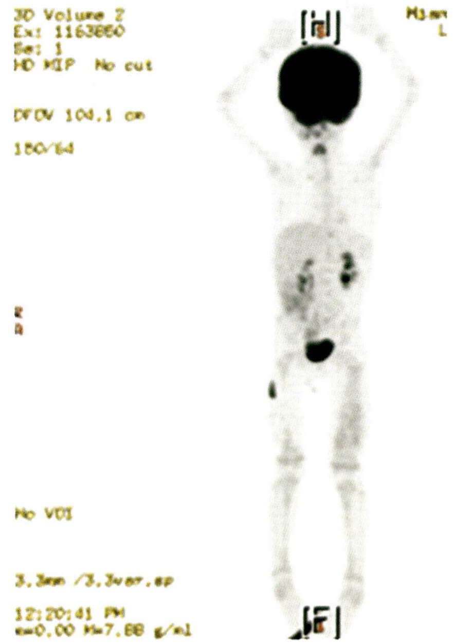

Figure 3F Fluorodeoxyglucose-PET scan in the same patient following chemotherapy administration demonstrating significantly decreased FDG uptake in the involved left femur suggesting an excellent tumor response to the treatment

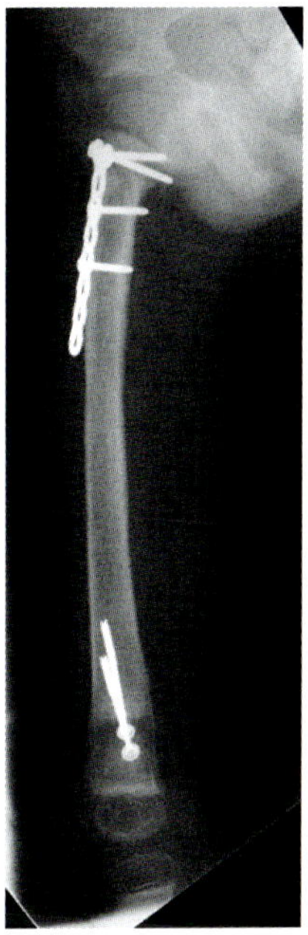

Figure 3H Lateral radiograph 6 months after surgery demonstrating growth in the distal femur

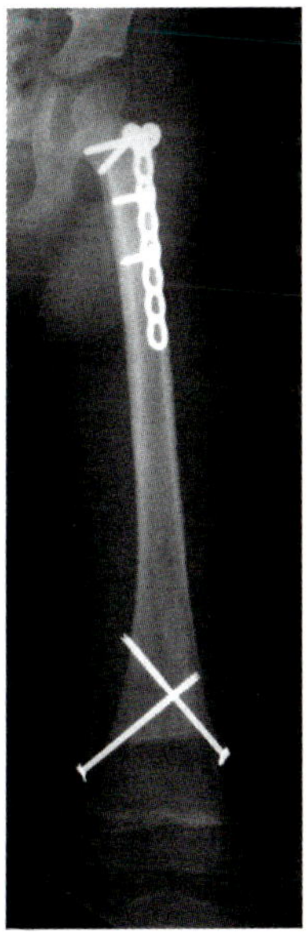

Figure 3G Anteroposterior radiograph 6 months after surgery demonstrating growth in the proximal femur

Fractures typically occur through holes drilled into the bone to affix plates with screws. Pathologic studies of allograft retrievals almost always show revascularization of the graft through these defects in the allograft. These sites act as stress risers and can result in significant weakening of the bone and subsequent fracture (**Fig. 4**). It is very important to have a stable graft host junction and this is best accomplished with two plates to achieve adequate rigid fixation. Thus, it is impossible to avoid creating holes in the graft. This problem has been mitigated in part with the advent of locking plates allowing the surgeon to achieve a stable construct with fewer holes, mostly unicortical, in the graft. Some surgeons advocate nail fixation. However in our experience, this results in inadequate stability at the allograft host junction producing a higher rate of hypertrophic non-unions. Furthermore, the use of a nail makes reconstruction around a joint more complicated as the nail may have to be removed to allow placement of stemmed implants when performing joint replacement.

The initial treatment of a fractured allograft is to repeat surgery with autogenous bone grafting and internal fixation for fractures that occur at the host-allograft junction site. Fractures that occur later, sometimes many years later, occur through the allograft itself. Depending on the degree of revascularization, some allografts are capable of healing with bone grafting and rigid fixation (Scarborough 2001) (**Fig. 5**).

For others, depending on the fracture pattern or patient age and medical condition, another allograft may be necessary, or possibly conversion to a metal implant with or without retention of the remaining graft (Wang et al 2006).

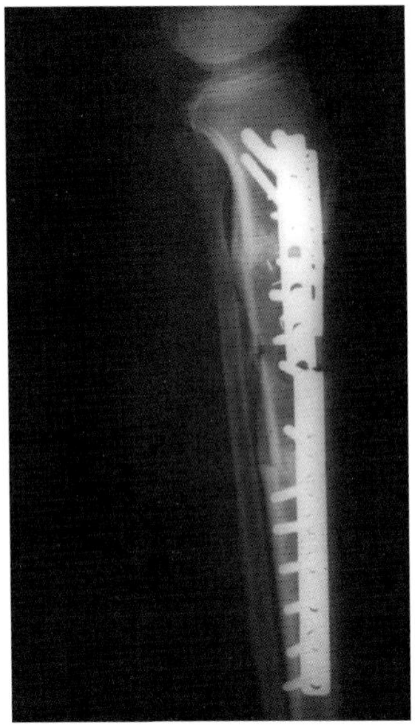

Figure 4 Lateral radiograph of left tibia with a fracture through an intercalary allograft

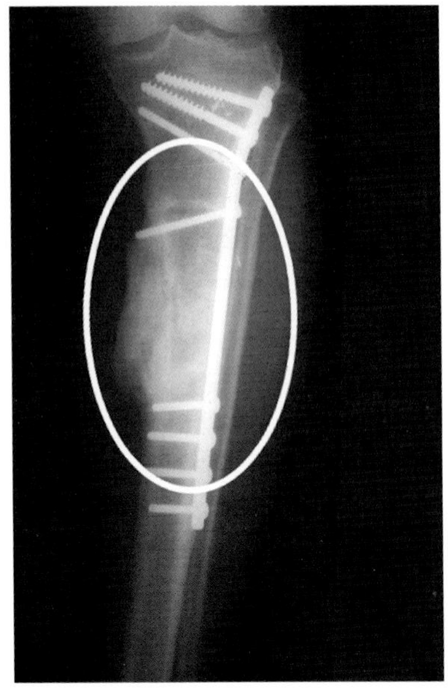

Figure 5 AP radiograph of the left tibia from the same patient following removal of the broken implant, autologous bone grafting and replanting of the tibia. Note the abundant callus (circled area)

Infections reported in one study typically occurred in the first 6 months of transplantation and were most commonly due to Staphylococcus organisms (Lord et al 1988). In our experience, both early and late infections have been observed with equal frequency. Prolonged surgery time, chemotherapy, immunocompromised patients, excessive intraoperative blood loss, poor soft tissue coverage and persistent postoperative drainage have been implicated as risk factors for infection.

Treatment of an allograft infection requires early diagnosis and prompt surgical debridement as well as prolonged parenteral antibiotic administration. Initially the antibiotic therapy is broad-based then, once a bacterial isolate is obtained, specific for the underlying infecting organism. Many patients require allograft removal, placement of antibiotic-loaded cement spacers and prolonged antibiotic therapy. The spacer should be of adequate size to maintain an adequate soft tissue envelope for future reconstruction. Furthermore, the longer the interval between implant removal and reimplantation, the better the chance of a successful outcome. Of course, this is predicated on the immunologic competency of the host, and the virulence of the organism. A late exchange reconstruction with another allograft or metal implant can be contemplated after a prolonged infection-free time interval. The wound should be completely healed without overt clinical signs of infection and the white blood count, erythrocyte sedimentation rate and C-reactive protein should be normal before considering reimplantation. At the time of reoperation, the tissue should be sampled and reviewed microscopically by a skilled pathologist to ensure that there are no acute inflammatory cells to suggest continued infection.

Of course, the best strategy is to avoid infection and to this end, the surgeon should handle soft tissues gently, ensuring appropriate antibiotic dosing at the beginning of the case and throughout the procedure if necessary. The allograft should be covered by muscle and normal skin and subcutaneous tissues then closed without tension. A gastrocnemius muscle flap is frequently used in patients undergoing proximal tibial reconstruction to provide adequate muscle coverage over the graft and to augment the extensor mechanism repair. Drains should be used liberally to avoid fluid collections and prolonged wound drainage. The optimal duration of postoperative antibiotics is controversial, but generally should continue as long as wound drains are in place. Bulky dressings and splints are used in the perioperative period to restrict joint motion for a period of 4–6 weeks to ensure adequate soft tissue healing.

Joint arthrosis generally occurs due to the lack of chondrocyte viability that results from the freezing process itself in addition to the presence of joint instability and/or incongruity. Painful symptoms though do not occur until many years after the signs of arthritis are apparent radiographically. When pain interferes with activities of daily living and diminishes the quality of life, joint replacement is an alternative. This however is, after all reasonable nonoperative strategies are exhausted. In well reconstructed knees, for example, a constrained kinematic knee results in pain relief, adequate stability and improved function. For those patients with poor ligamentous constraints, a knee replacement with a rotating hinge platform is appropriate. Stemmed and cemented implants are generally used.

Metal-Allograft Composites

The combination of a mobile prosthetic joint with an allograft is an effective modality in some anatomic locations. This mode of reconstruction combines modern endoprosthesis with allograft bone. The advantages are superior joint function, preservation of bone stock and attachment sites for soft tissues. This mode of reconstruction is especially attractive in the proximal humerus, proximal femur and proximal tibia; sites where soft tissue attachment and muscle function are critical for stability and proper joint function.

Site Specific Allograft Reconstruction

Shoulder

Soft tissue reconstruction is critical to normal shoulder function and stability. The most common complication following shoulder reconstruction after tumor resection is mechanical instability due to the reliance on soft tissue restraint over osseous conformity in this anatomic location (Henderson et al 2011). In those patients in whom a movable joint is contemplated, soft tissues are critical for adequate functional results. For this reason, the osteoarticular allograft is ideally suited to reconstructing this joint after tumor resection. Ogilvie reported on 20 patients who underwent tumor resection of the proximal humerus in which nine patients sustained fractures, five had progressive arthrosis, four a non-union and two, an infection (Ogilvie et al 2009). Sixty percent of allografts were removed at a mean of 5.2 years. Despite these results, the authors felt that allograft replacement of the proximal humerus was comparable to metal implantation.

In a review by Potter, osteoarticular allografts resulted in acceptable stability (Potter et al 2009). However they were prone to fracture through the surgical neck, a complication observed in 50% of such reconstructions. For this reason, the allograft prosthetic composite graft has been employed to mitigate the high incidence of fractures observed in the osteoarticular grafts and to provide soft tissue attachments to improve shoulder stability. The problem though is the technical difficulty in balancing the rotator cuff muscles and reproducing a normal rotator interval. Over time, despite meticulous capsule reconstruction and soft tissue reattachment, the implant tends to migrate proximally and anteriorly causing impingement, loss of shoulder motion and pain.

These problems are magnified with the use of modular segmental metal implants where maintaining soft tissue attachments to the implant are challenging. Most of these reconstructions act as nothing more than a spacer with limited shoulder range of motion and strength. Although aseptic loosening is a long-term potential complication of the procedure, implant fracture is infrequently observed. Instability, however, is the chief cause of failure in patients undergoing reconstruction with this modality.

An alternative to a mobile joint is the use of an allograft with or without a vascularized fibula as an intercalary graft to create an arthrodesis. In one study, this method of reconstruction produced both stability and better strength than comparative reconstructions whereby joint motion was preserved (Wang et al 2011). The downside to this type of reconstruction is the high incidence of pseudarthrosis at both junction sites and the risk of fracture.

Finally, the use of a reverse shoulder joint reconstruction is attractive as both a primary reconstruction and salvage procedure for functionally debilitating instability. For this type of reconstruction the proximal humerus is replaced with an allograft with a cup component while the glenoid has a ball configuration. This articulation is stable and does not rely on the rotator cuff musculature for stability or motion. An intact deltoid and axillary nerve is necessary for a successful outcome. Glenoid malpositioning is a potential complication that over time leads to loosening. Long term results of this type of reconstruction are not available at this time.

Scapula

Although malignant tumors of the scapula are relatively rare, enough information has been accumulated to conclude that scapulectomy results in a flail, unstable shoulder with an unsightly appearance (Enneking 1983). To this end, the use of scapular allografts produces very satisfactory results (Mnaymneh et al 1991) and cosmetically, the procedure avoids shortening and narrowing of the shoulder. However, the procedure is technically rather difficult and requires a close anatomic match between the allograft and the recipient (Mnaymneh et al 2002). Other authors also noted that the scapular allograft reconstruction following tumor resection can be performed with satisfactory functional, cosmetic, and oncological results. The deltoid muscle and articular capsule contribute significantly to shoulder function, stability, and contour. Thus, their preservation is important. The rotator cuff should be reconstructed despite the inherent difficulties with its intraoperative reattachment (Zhang et al 2008). Finally, it must be remembered that, if at all possible, it is advisable to preserve the glenoid. If glenoid is preserved the resulting functional disability is usually minor (O'Connor et al 1996).

In view of the good results with scapula allografts and the availability of the latter, there seems to be little justification for using anatomically unrelated allografts to reconstruct the scapula.

Elbow

Both benign and malignant tumors in the elbow region are uncommon. However, when they occur, primary and metastatic tumors may result in a significant periarticular bone loss and joint destruction that requires reconstruction. Allografts are best suited to reconstruct hemi-condylar resections of the distal humerus. Accurate matching of the allograft to the articular defect, as well as collateral ligament reconstruction is necessary to produce optimal results. For this purpose, both a fresh or cryopreserved graft may be used depending on the size of the defect and the availability of a suitable size-matched graft.

Wrist

Benign, aggressive and malignant tumors of the distal radius may result in significant osseous and joint destruction, and occasionally require structural allograft reconstruction. The chief problem with osteoarticular allograft reconstruction in this anatomic location is instability (Bianchi et al 2005). In our experience, meticulous size matching as well as ligament and capsular reconstruction may result in acceptable motion and strength **(Fig. 6)**. A more reproducible result is achieved with arthrodesis of the residual radius to the carpus fusing the wrist in neutral rotation and approximately 20° of dorsiflexion to maximize grip power.

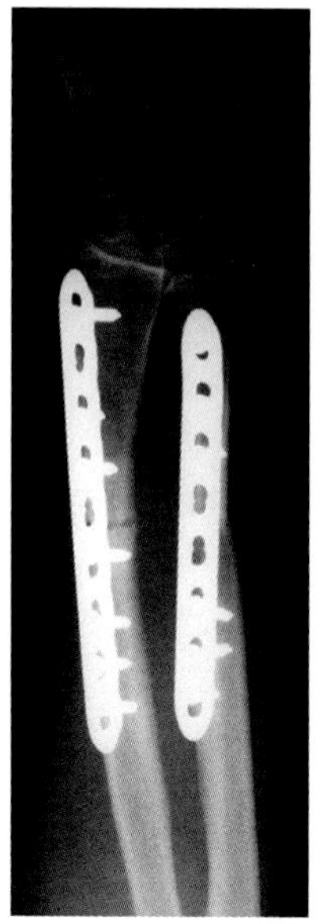

Figure 6 A 15-year-old girl with a high grade osteosarcoma of the right distal radius who underwent excision of the distal radius and ulna with osteoarticular allograft replacement

Tumors of the distal ulna are also rare and reconstruction is not always necessary. Reconstructing the triangular fibrocartilage complex, however, results in superior functional results and seems to improve wrist stability.

Pelvis

Pelvic reconstructions following tumor resections are the most challenging and frequently result in unsatisfactory function and significant complications. Early results employing arthrodesis techniques of the proximal femur to the ischium or residual ilium resulted in pseudarthrosis and significant shortening respectively. Custom metal and "saddle" type reconstructions are fraught with problems related to loosening implant dissociation and instability. Long term observations of patients undergoing saddle reconstruction are disappointing due to proximal migration of the implant.

Many surgeons have abandoned reconstruction altogether and attempt to rehabilitate patients with a flail extremity. Although perhaps 20–30% of these patients are independently functional, most require assistive devices and braces to ambulate and they all have significantly shortened extremities. Many develop proximal migration of the femur that results in lumbosacral and sciatic nerve irritation that too can be significantly debilitating.

For this reason, reconstruction of the pelvis with an allograft is an acceptable alternative (Bell et al 1997). This reconstruction requires careful preoperative planning to match the anticipated defect as closely as possible with an allograft. Due to gender differences in pelvic anatomy, gender matching is necessary. Most patients with this type of reconstruction undergo resection of all or part of the acetabulum. Thus, in addition to the graft, a femoral implant is used to restore the hip joint. Preserving and maintaining the patient's own femoral head in the acetabulum of the allograft results in rapidly progressive arthrosis generally requiring early arthroplasty. A bipolar implant is preferred over a total hip construct since it is more stable and easier to revise.

The procedure is generally done through a Mercedes type incision with the horizontal limb starting at the pubic symphysis and extending along the ilioinguinal line up to the pelvic brim. A second vertical incision is made from the anterior superior iliac spine to the tip of the greater trochanter. An extended iliofemoral approach is an alternative.

In reconstructing the pelvis with an allograft, it is important to achieve osseous stability using plates and screws (**Fig. 7**). When the entire ilium is removed, fixation to the sacrum with iliosacral screws or anterior plates is preferred. Despite stable fixation, proximal migration of the graft is often observed.

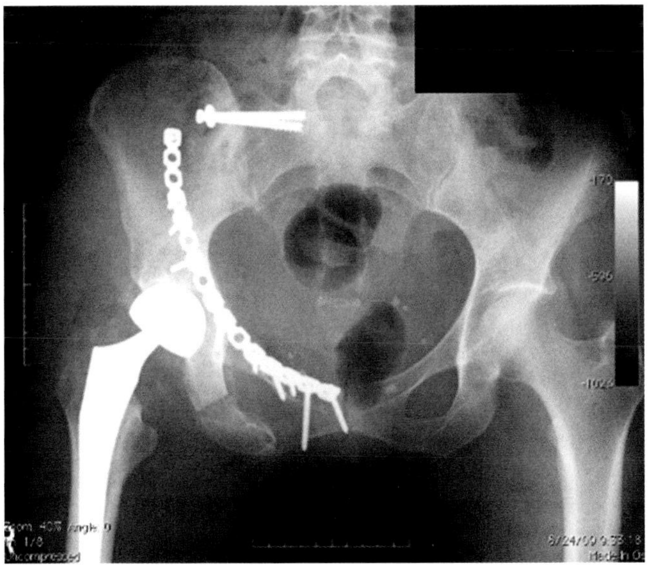

Figure 7 A 53-year-old woman with a grade I chondrosarcoma of the ilium and acetabulum who underwent a Type I/II pelvic resection and reconstruction with a pelvic allograft

The advantages of this type of reconstruction following tumor resection that involves the acetabulum are reconstitution of a stable hip joint, maintenance of limb length and restoration of abductor muscle function.

This mode of reconstruction is associated with significant potential complications. Early in the postoperative period, infection occurred in one study in as many as 8 of 22 patients (36%) (Ozaki et al 1996). Early deep infections almost always require graft removal. Late infections are also observed and can sometimes be treated with multiple debridements and suppressive antibiotics.

In the intermediate term, fracture, often through the acetabulum is noted in a significant number of patients. Although the femur may migrate proximally resulting in lower extremity shortening, pain is not a prominent symptom and often patients continue to be functional with the use of a cane or other assistive device.

The functional results of pelvic allograft reconstruction are highly variable and depend on a number of factors including patient age, premorbid functional status, preservation of the abductor muscles, extent of bone resection and concomitant complications occurring at the time of resection (neurologic injury, urologic injury, blood loss and operative time). Most patients require assistive devices for ambulation although some walk independently.

Patients with protrusio defects of the acetabulum resulting from metastatic disease and failed implant reconstructions can benefit from structural allograft augmentation of the pelvis in conjunction with a hip arthroplasty (**Figures 8 to 11**).

For very large defects, the use of an acetabular cage may be necessary to recreate a stable acetabulum, (**Fig. 12**).

Proximal Femur

Although osteoarticular allograft reconstructions of the proximal femur have been largely abandoned, the use of a structural allograft with soft tissue attachments and a long stem prosthesis (alloprosthesis) results in hip stability and is best suited for re-establishing near normal abductor muscle function. This approach also preserves bone stock should be necessary for future reconstructions. Pseudarthrosis of the allograft to host bone is uncommon. Roque reported on the

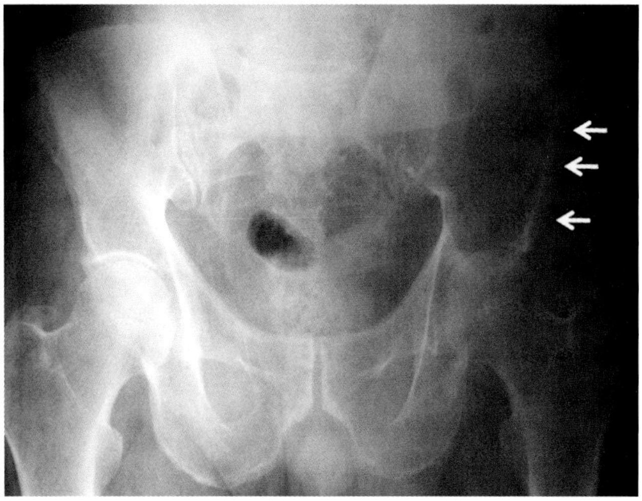

Figure 8 A 69-year-old male with a large plasmacytoma in the left superior acetabulum. The subchondral bone was intact, but most of the weight bearing dome was destroyed by tumor

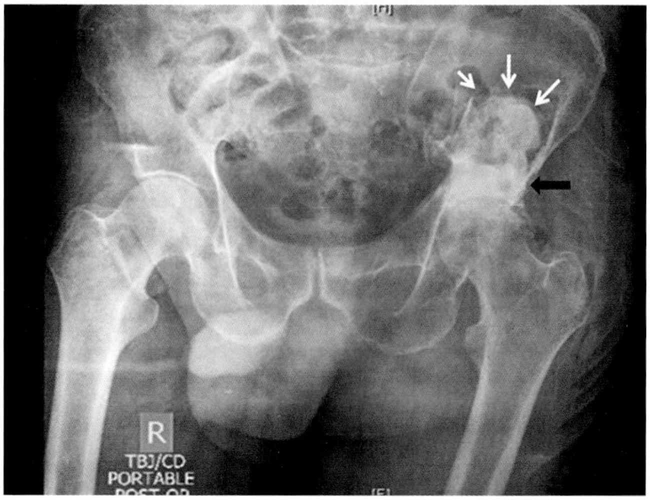

Figure 9 Following radiation and embolization, the patient underwent structural bone grafting in the superior aspect of the defect (white arrows) and microparticulate bone grafting in the subchondral area of the hip joint (solid black arrow)

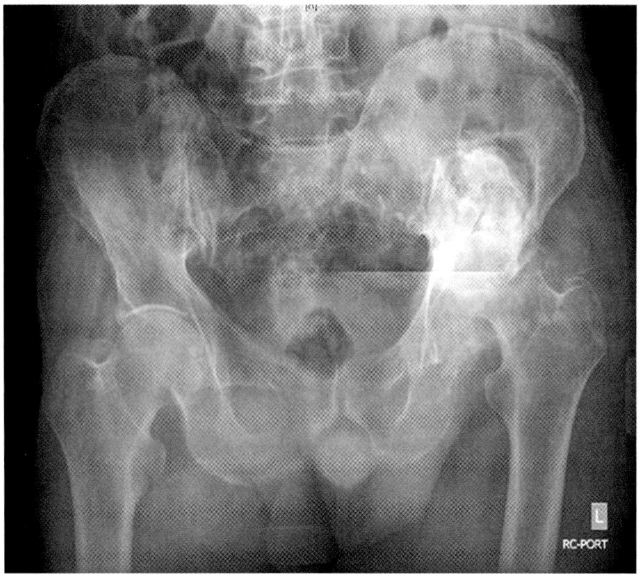

Figure 10 Before bone consolidation could occur, the patient began weight bearing and sustained a fracture through the dome of the acetabulum with superior migration of the femoral head

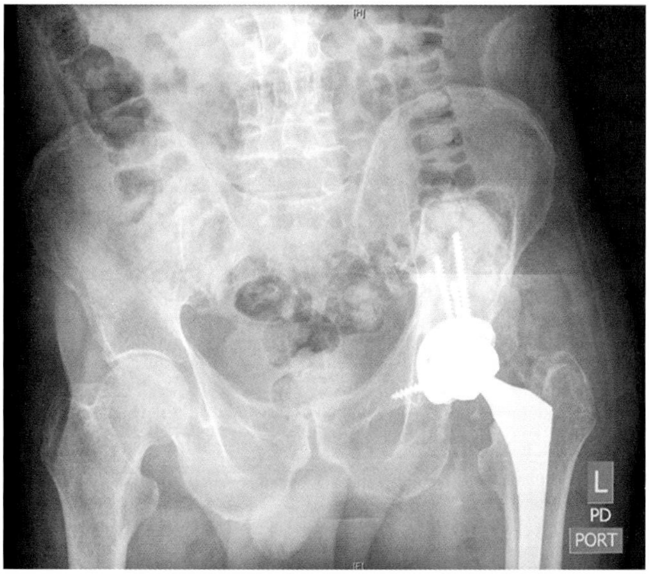

Figure 11 A second structural graft was placed in the acetabular defect and an acetabular implant was stabilized with multiple screws. A total hip arthroplasty was successful in restoring normal function

Massachusetts General Hospital experience involving 150 patients who underwent resection and allograft transplantation of the proximal femur. Most resections were for malignant bone tumors. Four types of allografts were used: osteoarticular (46 patients), allograft-prosthesis (73 patients), intercalary (20 patients), and allograft-arthrodesis (5 patients). Only 16% of the patients died of disease and 3% required amputation. The overall success rate for the series was 77% with the best results for the allograft prosthetic (82%) and intercalary procedures (87%). Graft infection (15 patients), allograft fracture (26 patients), and local recurrence (11 patients) most markedly affected outcome (Roque et al 2010).

The principal problem with alloprosthetic reconstruction in our experience is resorption of the graft at the insertion of the abductors into the graft **(Fig. 13)**. Bone loss may progress to the point where structural bone grafting is necessary.

Distal Femur

Osteoarticular allograft reconstruction of the distal femur is restricted to patients that are near skeletal maturity with malignant bone tumors, most commonly osteosarcomas. This method preserves the physis in the tibia to allow for further growth; epiphysiodesis is generally performed on the contralateral lower extremity at the appropriate time to allow for equal limb lengths. This is a technically challenging procedure in which the cruciate and collateral ligaments must be repaired or reconstructed to create a stable joint. Once the soft tissues are repaired or reconstructed, the allograft is attached to the host bone, generally with plates and screws to create a stable osteosynthesis and optimize osseous healing at the site. As noted in previous chapters, the bone heals from the endosteum of the host bone into the allograft over several weeks and months. Over time, remodeling occurs to the extent that the junction site is barely visible (**Fig. 14**).

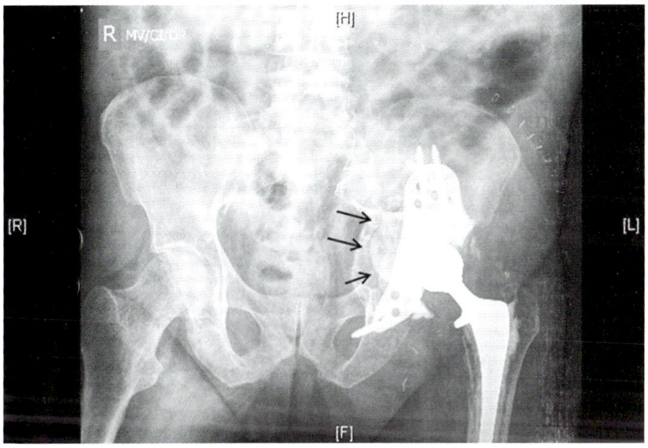

Figure 12 Postoperative AP radiograph of the pelvis augmented with a structural femoral head allograft (arrows) and a protrusio cage

Preoperative planning is essential to select a graft of appropriate dimensions in both the anteroposterior and mediolateral planes. The grafts are prepared aseptically and subjected to treatment with glycerol or dimethyl sulfoxide prior to controlled-rate freezing to preserve the articular cartilage. However, despite initial satisfactory cartilage preservation, cartilage is subjected to latent injury which eventually results in slow but progressive degeneration (Malinin et al 2007). Prior to implantation, the frozen graft is rapidly thawed in warm lactated Ringer's solution with antibiotics. The marrow contents are removed to the extent possible to prevent a potential host immunologic response.

Campanacci et al reported on long-term results in 13 patients with distal femoral allograft replacement, three of which died in the first 2 years of follow-up; of the remaining 10 patients, four grafts failed. All the failures were related to the graft fracture. The overall rate of allograft survival

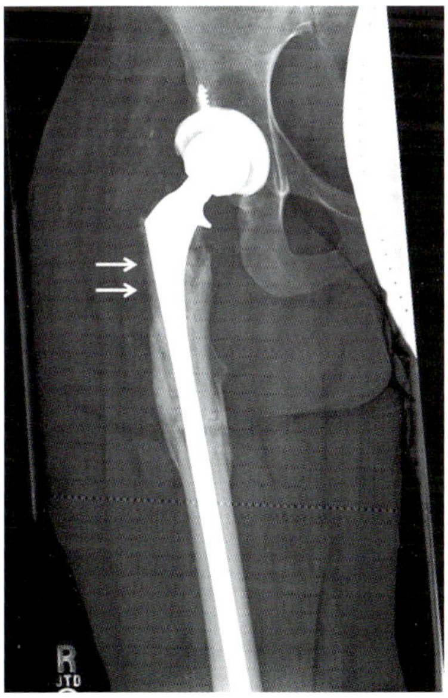

Figure 13 Significant resorption of the proximal femoral allograft

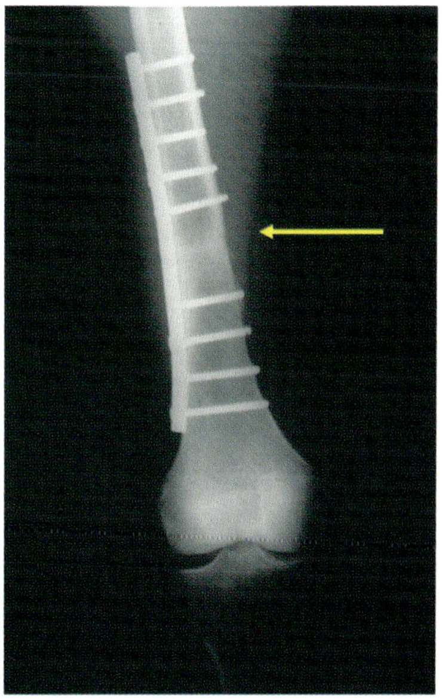

Figure 14 A 34-year-old woman with a distal femoral allograft following resection of the distal femur for a giant cell tumor. The graft is completely healed and the allograft host junction site is barely visible (arrow)

was 70% at 5 years and 58% at 10 years (Campanacci et al 2010). In a similar long-term retrospective study, Toy et al reviewed 26 patients who had osteoarticular allograft reconstructions of the distal femur after resection of a malignant or aggressive benign tumor of bone. At the last follow-up, 16 of the 26 original allografts were still in place. The overall 5-year and 10-year allograft survival rates were 69% and 63%, respectively (Toy et al 2010). Mnaymneh et al described 96 patients with distal femoral allografts and a follow-up of 2–15 years. These authors noted considerable difference in allograft survival between patients receiving and not receiving adjuvant chemotherapy. In patients not receiving chemotherapy, good to excellent results were obtained in 70% of the patients, but only in 53% of patients receiving chemotherapy (Mnaymneh and Malinin 1989, 1994). Perhaps the most encouraging results were reported by Musculo et al reporting on 80 reconstructions with a distal femoral osteoarticular allograft following resection of a bone tumor in 76 patients. Fourteen grafts failed for a variety of reasons: infection (6), local recurrence (4), massive resorption (1), fracture (3). At the time of final follow-up, at a mean of 125 months, 48 allografts were still in place. The overall rate of allograft survival was 78% at both 5 years and 10 years, and the rate of allograft survival without the need for resurfacing with a knee prosthesis was 71% at 5 years and 10 years (Musculo et al 2008).

More commonly, osteoarticular grafts in this location are used to replace a hemicondyle or portion of a hemicondyle following tumor excision or as a result of significant trauma that precludes adequate fracture repair. Unicondylar osteoarticular allografts of the knee are mainly used after bone tumor resections for benign aggressive tumors or relatively small malignant tumors with clearly defined margins. In a study by Bianchi et al (2009), 12 unicondylar osteoarticular reconstructions were performed. One allograft was removed after 29 months because of an intra-articular displaced fracture, and substituted with a new osteoarticular graft. One patient died of metastatic disease at 24 months. Of the remaining 10 patients, three had excellent results, five good and three fair. Arthrosis seemed to be the principal cause of failure over time occurring in 8 of the 10 patients in this series (Bianchi et al 2009). Again, careful preoperative planning and graft selection are critical to optimizing results. These grafts result in a stable knee when the collateral ligaments are anatomically repaired; they incorporate rapidly and result in durable and high-level joint function. In 40 cases, Musculo et al reported very good results with this technique in the distal femur and proximal tibia. There were only six failures in that series that were salvaged with repeat grafting (Musculo et al 2006).

Both fractured and infected allografts can be salvaged with segmental metal replacements (Wang et al 2006).

Proximal Tibia

The indications for osteoarticular allografts of the proximal tibia are similar to those for grafts in the distal femur (**Figs 20 and 21**). The advantage of this graft in the proximal tibia, as in other long bones is preservation of bone stock and the ability to reconstruct the extensor mechanism. Due to the subcutaneous nature of the tibia, both infection and fracture appear to be more common than allograft reconstructions in other anatomic sites. Compared to reconstruction of the proximal tibia using metal implants, extensor mechanics are superior in allograft patients. Both infection and fracture are common causes of graft failure. Musculo et al reported on 52 patients (58 reconstructions including six repeat reconstructions) who underwent osteoarticular proximal tibia allograft reconstructions after tumor resection. There were 46 surviving patients with a mean follow-up of 72 months. At last follow-up, 32 of the 52 remaining allografts were still in place; 20 failed due to infections, local recurrences, or fractures. Overall allograft survival was 65% at 5 years and 10 years, with an average Musculoskeletal Tumor Society functional score of 26 points and an average radiographic result of 87% (Musculo et al 2010).

Like the distal femur, arthrosis is common after many years requiring total joint arthroplasty. This too can generally be accomplished with constrained implants (DeGroot and Mankin 2000). Those patients with collateral ligament insufficiency may require a rotating hinged implant. Hornicek reviewed 38 patients who underwent proximal tibial allograft reconstruction after wide resection of benign and malignant tumors. One or more complications were observed in 21 patients (55%). Of the 26 patients who received chemotherapy, 15 (58%) experienced one or more complications, whereas of the 12 patients who did not receive chemotherapy, six (50%) experienced one or more complications. In the chemotherapy group, there were 12 (46%) fractures, four (15%) infections, three (12%) non-unions, and four (15%) instabilities. In the non-chemotherapy group there were three (25%) infections, two (17%) fractures, one (8%) instability, and one (8%) non-union. These complications were managed adequately with multiple subsequent surgical procedures. Three patients underwent amputations for deep wound infections. Twelve (32%) patients underwent removal of the allograft, and the limb was salvaged by another allograft or by total knee arthroplasty. The results of both groups were 66% (25 of 38 patients) satisfactory (good or excellent).The chemotherapy group had a significantly higher incidence of fractures. All other complication rates and functional outcomes were not significantly different between these groups (Hornicek et al 1998).

Because of the high rate of fracture complications with osteoarticular allografts, many surgeons prefer allograft prosthetic implant composite reconstructions. This has the advantage of a durable joint replacement, reconstitution of proximal tibial bone and soft tissue attachments for the extensor mechanism. Gilbert et al reported results of allograft prosthetic reconstruction in 12 patients; nine had no extensor lag, and three patients had 5–15 degrees of extensor lag. The mean Musculoskeletal Tumor Society score was 24.3 (81%) of a maximum of 30. Complete bone union occurred in nine patients, and partial union occurred in three patients (Gilbert et al 2009).

Hemi-resections of the proximal tibia plateaus, like hemi-resections of the distal femoral condyles, result in excellent and durable function.

Distal Tibia

Tumors of the distal tibia are rare. In addition, most traumatic injuries of the tibial plafond result in tibiotalar fusion. Thus, osteoarticular allograft reconstruction in this site is uncommon. In most cases, these patients undergo allograft arthrodesis with or without a vascularized fibula graft to promote early and durable healing. In those unusual cases where the tumor can be resected because of its small size or robust response to systemic chemotherapy, osteoarticular reconstruction can be considered.

The principal problems are the selection and availability of a suitably sized graft and associated wound complications in this anatomic location. Although osteoarticular allograft reconstruction can be supplemented with a vascularized fibula, in our experience, this has not been necessary.

The short-term complications of allograft reconstruction of the distal tibia are principally related to wound complications. Minor complications can be managed with local wound care and antibiotics. Skin slough and dehiscence must be managed aggressively with debridement and free flap coverage.

We have reported good results with this type of reconstruction (Balsamo et al 2007) but ultimately, arthrosis supervenes and is best managed with tibiotalar arthrodesis. Revision surgery using prosthetic ankle implants is not recommended.

Intercalary Allografts

With better chemotherapeutic agents and advanced imaging techniques to measure tumor response, coupled with better implant fixation, the indication for intercalary allografts has increased. Using pre-and postcontrast MRI, the intramedullary extent of tumor and the relationship of the lesion to the proximal and distal physis can be accurately assessed. For patients with robust responses to neoadjuvant chemotherapy and uninvolved metaphyseal bone, adequate in length to support fixation, the physis can be preserved in a young patient. The resulting segmental defect following tumor resection can be satisfied with an intercalary allograft. Growth of the affected extremity progresses normally and the allograft incorporates and remodels over time. The process of graft incorporation is slowed by continued chemotherapeutic administration; in fact, chemotherapy is the most significant risk factor for non-union at the allograft host junction site (Hazan et al 2001). Following cessation of therapy, healing can accelerate.

Joint sparing in young patients is preferable to resection and reconstruction using either prosthetic components or osteoarticular allografts. In older patients with primary and metastatic tumors of bone, intercalary allograft reconstruction can restore patients to near-normal function.

Adult patients with metadiaphyseal tumors are best served by this technique following segmental resection. Although the incidence of delayed union and non-union of the allograft-host junction sites are relatively high (up to 30% per junction in one series) (Hornicek et al 2001), these problems can be mitigated in part by percutaneous or open autologous bone grafting techniques. Generally, practitioners have employed the use of microparticulate allograft combined with a percutaneous bone marrow aspirate for this purpose with excellent results and minimal morbidity.

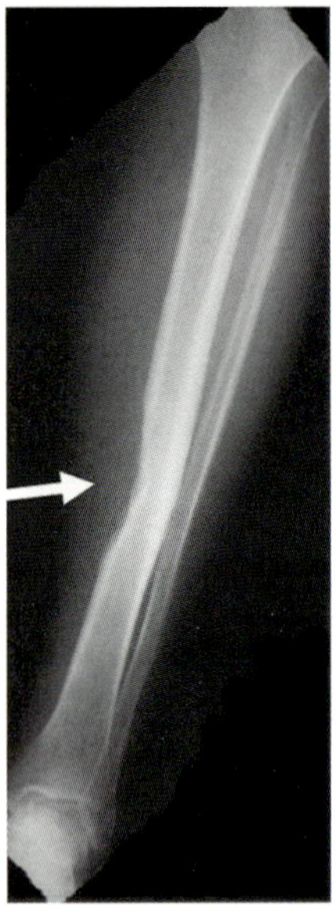

Figure 15 A 65-year-old male with a solitary renal cell metastasis in the mid tibial diaphysis. Note periosteal scalloping (arrow)

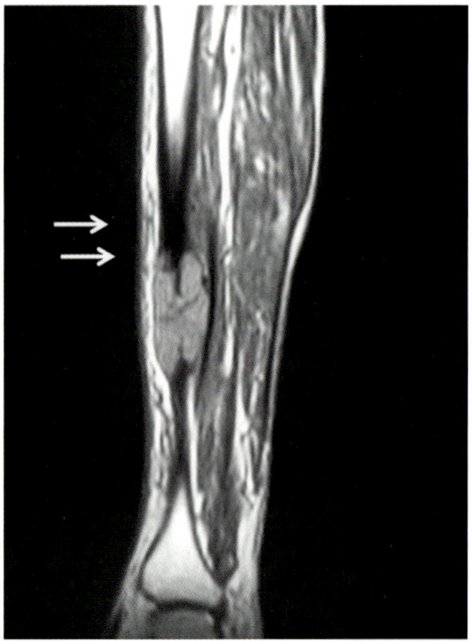

Figure 16 T2 weighted MR image demonstrating the intraosseous tumor extent and soft tissue extension (double arrow)

Selected patients with relatively radioresistant tumors such as renal cell carcinoma may also be good candidates for aggressive resection of the solitary osseous metastasis and replacement with an intercalary allograft. These patients have a reasonable life expectancy (especially those that have undergone nephrectomy, developed a metachronous and solitary metastasis and exhibit high performance preoperative function) (**Fig. 15**).

These allografts are aseptically recovered and cleaned. They may be prepared and stored frozen at -150°C or 70°C, or freeze dried. For the fresh frozen grafts, it is important to remove any residual marrow prior to implantation. In our experience, the freeze dried graft preparations heal as effectively, if not better than the frozen grafts and are easier to ship and store. The freeze dried graft must be rehydrated and therefore should be removed prior to beginning the procedure and allowed to soak in lactated Ringer's solution with antibiotics. There are no scientific or validated studies to suggest optimal rehydration time and in the authors' experience there have been no adverse outcomes with graft fracture following screw insertion. Self-tapping locking unicortical screws are recommended for graft fixation. Bicortical screws may be used in the host bone (**Figs 16 to 19**).

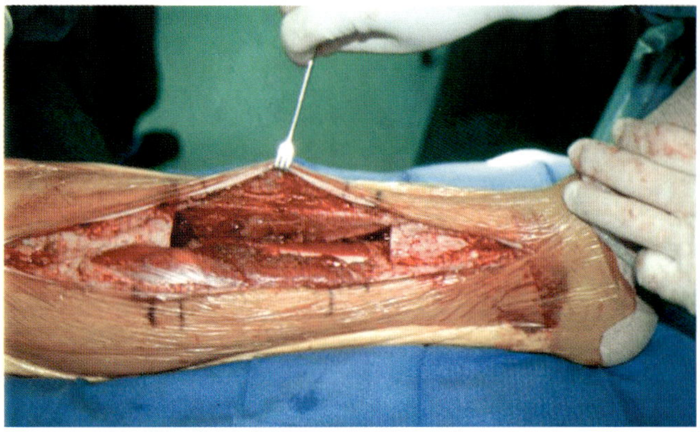

Figure 17 Intraoperative photograph demonstrating the large segmental diaphyseal defect in the tibia following tumor resection

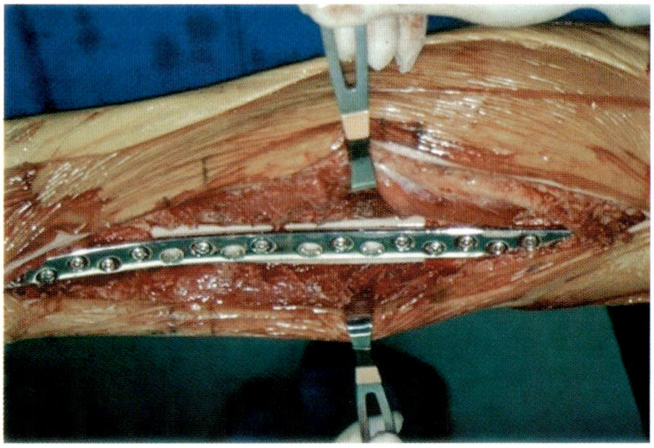

Figure 18 Intraoperative photograph showing the intercalary allograft fixed above and below with a 4.5 plate and screws

Hemicortical Allografts

Large hemicortical defects in bone are well suited to allograft replacement. The large cancellous surfaces on the host bone side are ideal for rapid and complete allograft healing. Graft fitting is generally done in the operating theater after tumor resection. Depending on the size of the defect, fixation may require screws only or plates and screws (**Figs 18, 19 and 22**).

Particulate Grafts

Particulate allografts are generally prepared by freeze drying methods and vary considerably in size and shape. They may be cortical, cancellous or a combination of both. They are

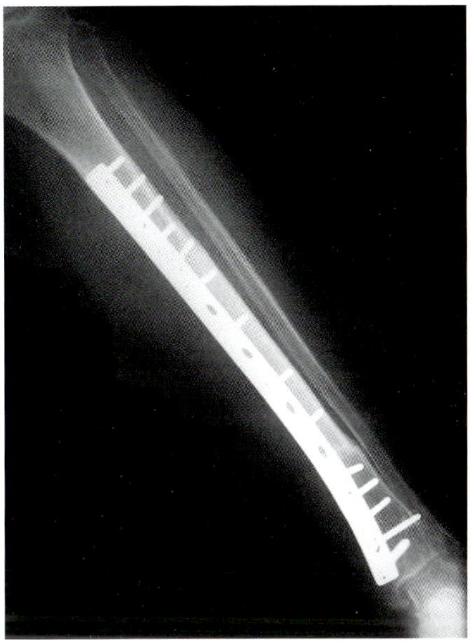

Figure 19 AP radiograph taken 7 years later showing complete incorporation of the allograft and no obvious tumor recurrence

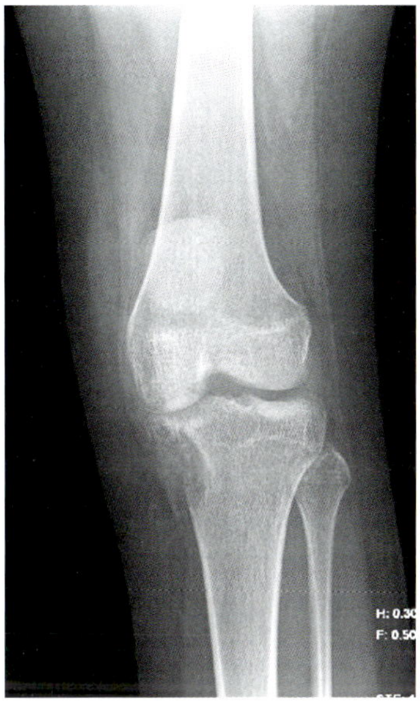

Figure 20 A 16-year-old girl with a radiolucent abnormality of the medial tibial metaphysis that on biopsy was a high grade osteosarcoma

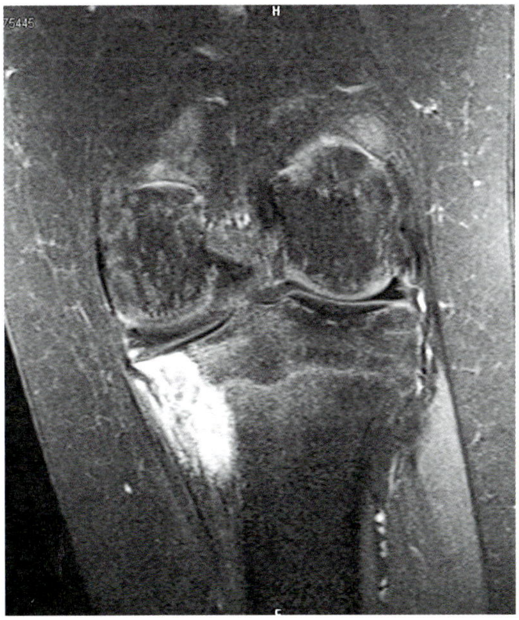

Figure 21 Corresponding coronal T1 STIR MR with gadolinium showing enhancement and limited to the metaepiphyseal portion of the medial tibial plateau

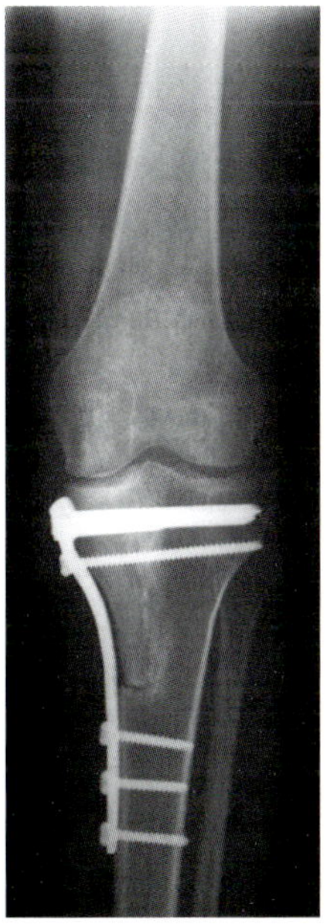

Figure 22 AP radiograph 6 months after wide excision and hemi-osteoarticular allograft reconstruction. The patient had a normal gait and near normal knee range of motion

generally used to fill defects following intralesional tumor resections (Temple, Malinin 2008), traumatic injuries with bone loss and to induce bone formation to achieve spine fusion. These particles have been combined with antibiotics and used to fill defects following bone debridement for patients with osteomyelitis.

Studies performed to assess the effect of particle size on osseous integration indicated that particles in a relatively narrow size range (100–300 μ) were optimal. Smaller sizes demonstrated limited capacity to incorporate in the time interval observed whereas larger particles were slow to heal. The reasons for improved osseous integration using small particles are unknown (Malinin et al 2008, Temple, Malinin 2008). Presumably the 100–300 μ particles maintain the morphologic features necessary for osteoinduction coupled with preservation and optimal bioavailability of molecules that are involved in the complex cascade of events that initiate mesenchymal stem cell recruitment and sustain the process of osteointegration.

In the same study, assessing optimal particle sizes in non-human primates, demineralized grafts of the same "optimal size" (100–300 μ), produced robust sclerotic bone in the periphery of the defect but remained largely unconsolidated with a cystic-appearing center. In a large clinical series of patients with large closed segment defects in bone, the use of fully mineralized (100–300 μ) particles produced results strikingly similar to those observed in the non-human primates, that is, complete defect filling, rapid consolidation and resorption of bone particles in the soft tissues.

The following case is illustrative in demonstrating how these particles stimulate mesenchymal cells, induce new bone formation and remodel.

A 17-year-old woman presented with a large giant cell tumor of the right proximal femur. She underwent curetting, burring and packing of the defect with (100–200 μ) bone particles. She relapsed in 8 months and underwent recuretting and bone grafting. Within 6 months, she recurred again. This time, the tumor recurred in the peripheral margin of the bone defect in addition to the adjacent soft tissues. For this she underwent wide excision of the proximal femur and proximal femoral replacement (**Fig. 23**). This allowed a rare opportunity to analyze the microscopic features of the previous bone graft.

Figure 24, shows histologic sections progressing from the graft-host interface, there is a transition from cortical

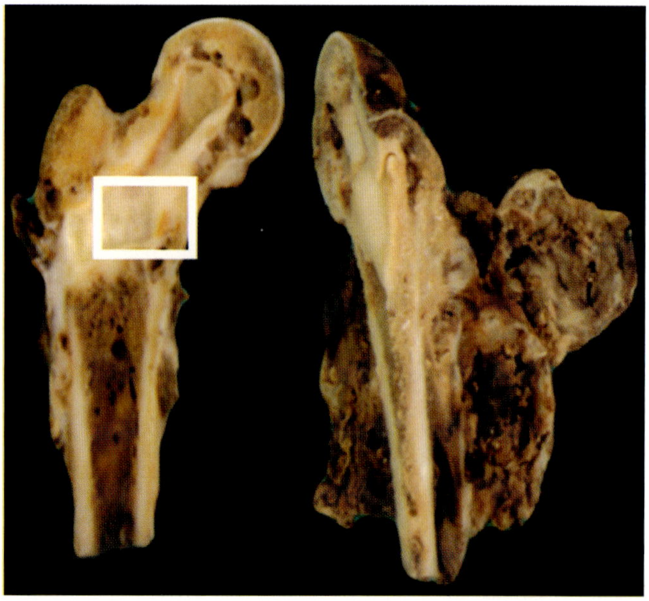

Figure 23 Specimen of the resected proximal femur. The area inside the square represents the bone host-bone graft interface that was studied microscopically

to cancellous bone, the marrow is relatively quiescent and there are small capillaries. The more central sections show recapitulation of the cancellous bone architecture and mature fibrous stroma. More centrally, there is only partial incorporation of the graft and abundant immature fibrous stroma. Finally in the center of the grafted defect, there are unincorporated particles and no stroma. These changes are summarized in **Figure 25**.

Interestingly, when these particles are placed in continuous flow osteogenic media with mesenchymal stem cells, a similar phenomenon is observed. The sharp angles are rounded and the particles coalesce. These particles are surrounded by the mesenchymal cells that form bridges between the particles. This is demonstrated in **Figure 26**.

Finally, the particles coalesce after approximately 3 weeks in culture (**Fig. 27**).

These observations are informative and suggest that there is an optimal particle size for bone induction which allows for rapid healing of bone cavities after tumor curettage or resection. Furthermore, the process follows a logical sequence *in vitro* and *in vivo*. That is mesenchymal recruitment, partial resorption of the sharp angulations, bridging of particles by cells and finally coalescence.

In summary, bone allografts, both structural and particulate constitute an important repertoire in surgical reconstruction of skeletal defects resulting from the tumor resection. To

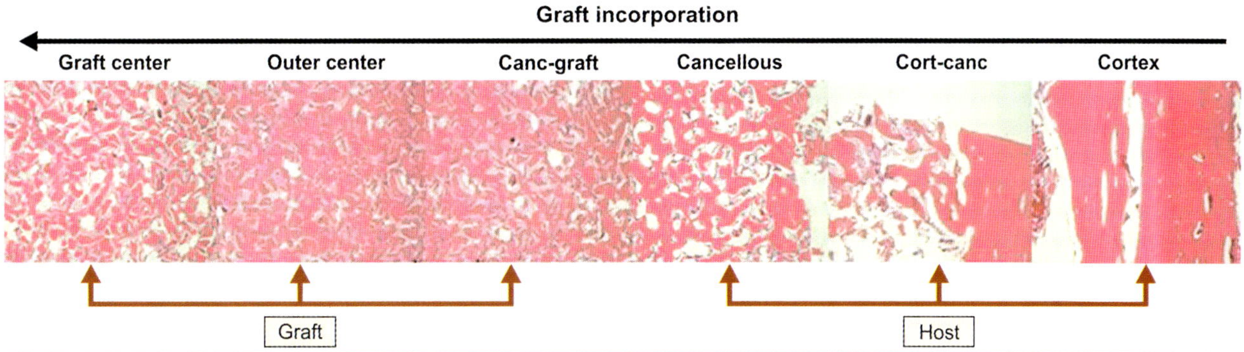

Central zone of the graft unaffected by repopulation from host precursors	Intermediate zone of the graft. Initial repopulation from host precursors	Peripheral zone of the graft populated by adjacent host precursors	Cancellous bone of the host. Source of precursors and singals for engraftment	Host-cortical-canclous interphase	Host cortex
Unincorporated particles	Partial incorporation	Recapitulation of cancellous architecture	Normal three-dimensional structure	Transition between cancellous and cortical areas	Normal cortex
No interparticular stroma	Inmature fibrous stroma	Mature fibrous stroma	Quiescent marrow stroma	Quiescent marrow stroma	No marrow stroma
No bone marrow elements	No bone marrow elements	Occasional bone marrow elements at the interphase with host cancellous bone	Few bone marrow elements	Few bone marrow elements	No bone marrow elements
No vascularity	Small capillaries	Small capillaries	Capillaries developed	Small capillaries	Haversian system
No particle fusion	Partial particle fusion	Maturel particle fusion			

Figure 24 The process of healing starts with significant mesenchymal cell recruitment that can be best seen in panel one. The bone graft particles are sharp and pointed. This is how they appear following grinding and freeze drying. Later, there is a significant fibrous stroma between the bone fragments

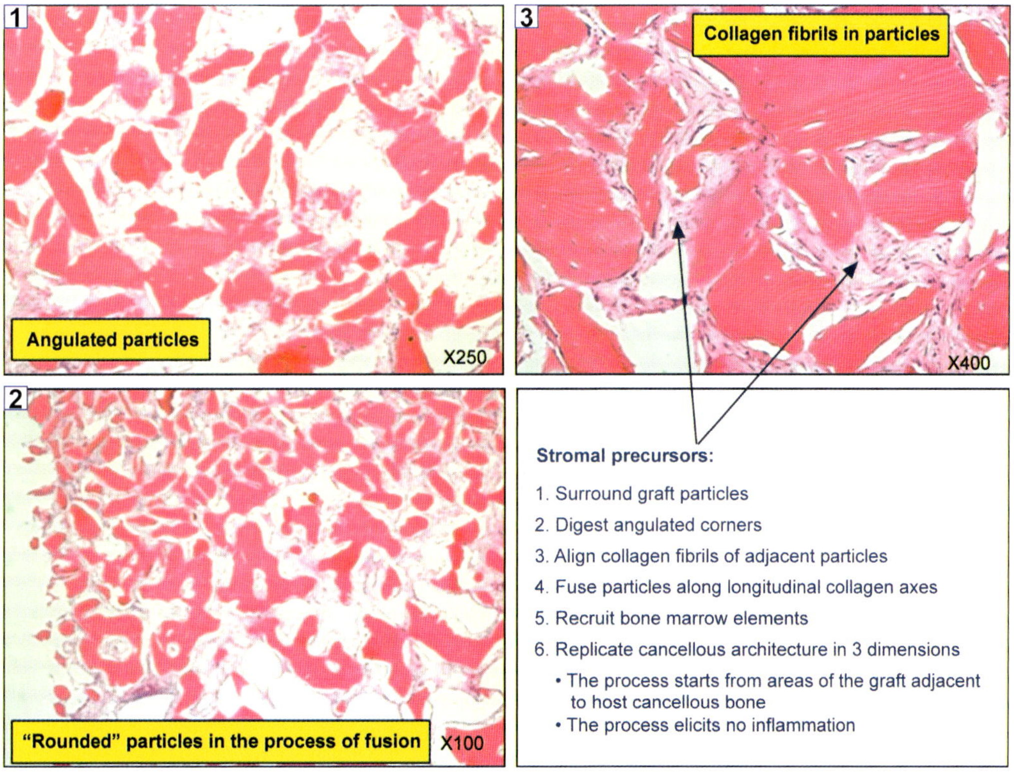

Figure 25 The sharp edges have been resorbed and rounded off. At this point, the particles begin to coalesce. In panel three, the three dimensional architecture is restored and there are mature appearing bone marrow elements. An inflammatory component is not evident at any time during this process

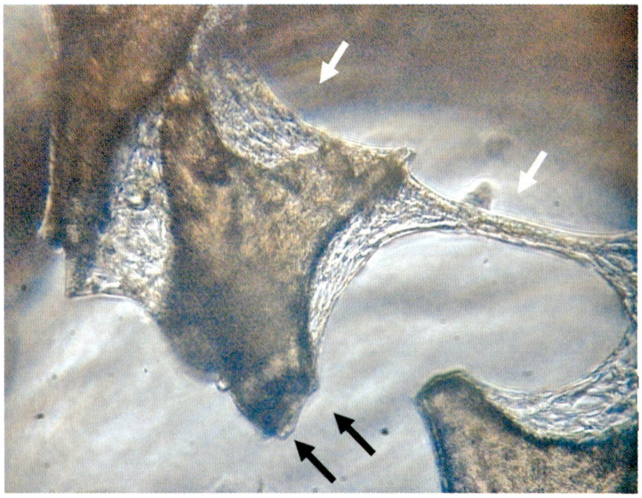

Figure 26 The black arrows show the rounded edge of the particle while the white arrows depict the cells forming around the particle and bridging it to other particles

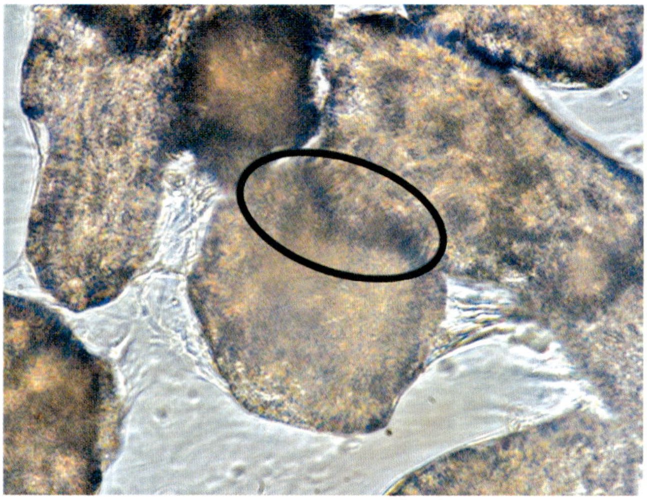

Figure 27 The dark circle denotes the coalescence of particles

optimize graft success, aseptic processing and minimal tissue manipulation is essential. In addition to processing and selection of the appropriate sized graft, surgical technique is critical in achieving successful outcomes. This involves strict asepsis, antibiotic coverage before and during the procedure, rigid internal fixation for structural grafts and meticulous soft tissue handling and muscle flap coverage. The chances of HIV transmission with adequately monitored allografts are remote (Carlson et al 1995).

For the foreseeable future, traditional grafts will continue to play an important role in the skeletal repair however, the combination of microparticles with other scaffolds, both natural and synthetic, and the measured and scientifically-based use of mesenchymal stem cells might be important technologies in the future.

BIBLIOGRAPHY

1. Balsamo LH, Malinin TI, Temple HT. Distal tibial osteoarticular allografts. Clinical Orthopaedics and Related Research, 2007;459:92-5.
2. Bell RS, Davis AM, Wunder JS, et al. Allograft reconstruction of the acetabulum after resection of stage-IIB sarcoma. Intermediate-term results. Journal of Bone and Joint Surgery (Am), 1997;79(11):1663-74.
3. Berrey BH, Lord CF, Gebhardt MC, et al. Fractures of allografts. Frequency, treatment, and end-results. Journal of Bone and Joint Surgery (Am), 1990;72:825-33.
4. Bianchi G, Donati D, Staals EL, et al. Osteoarticular allograft reconstruction of the distal radius after bone tumor resection. Journal of Hand Surgery (Br), 2005;30:369-73.
5. Bianchi G, Staals EL, Donati D, et al. The use of unicondylar osteoarticular allografts in reconstructions around the knee. Knee, 2009;16(1):1-5.
6. Buck BE, Resnick L, Shah SM, et al. Human immunodeficiency virus cultured from bone. Implications for transplantation. Clinical Orthopaedics and Related Research, 1990;251:249-53.
7. Campanacci L, Manfrini M, Colangeli M, et al. Long-term results in children with massive bone osteoarticular allografts of the knee for high-grade osteosarcoma. Journal of Paediatric Orthopaedics, 2010;30:919-27.
8. Carlson ER, Marx RE, Buck BE. The potential for HIV transmission through allogeneic bone. A review of risks and safety. Oral Surgery, Oral Medicine, Oral Pathology, Oral Radiology and Endodontics, 1995;80:17-23.
9. DeGroot H, Mankin H. Total knee arthroplasty in patients who have massive osteoarticular allografts. Clinical Orthopaedics and Related Research, 2000;(373):62-72.
10. Enneking WF. Musculoskeletal Tumor Surgery. New York: Churchill Livingston; 1983. pp. 355-69.
11. Gilbert NF, Yasko AW, Oates SD, et al. Allograft-prosthetic composite reconstruction of the proximal part of the tibia. An analysis of the early results. Journal of Bone and Joint Surgery (Am), 2009;91(7):1646-56.
12. Hazan EJ, Hornicek FJ, Tomford W, et al. The effect of adjuvant chemotherapy on osteoarticular allografts. Clinical Orthopaedics and Related Research, 2001;385:176-81.
13. Henderson ER, Groundland JS, Pala E, et al. Failure mode classification for tumor endoprostheses: retrospective review

of five institutions and a literature review. Journal of Bone and Joint Surgery (Am), 2011;93:418-29.
14. Hornicek FJ, Mnaymneh W, Lackman RD, et al. Limb salvage with osteoarticular allografts after resection of proximal tibia bone tumors. Clinical Orthopaedics and Related Research, 1998;(352):179-86.
15. Hornicek FJ, Gebhardt MC, Tomford WW, et al. Factors affecting non-union of the allograft-host junction. Clinical Orthopaedics and Related Research, 2001;382:87-98.
16. Lexer E. Joint transplantation and arthroplasty. Surgery, Gynaecology and Obstetrics, 1925;40:782-809.
17. Lord CF, Gebhardt MC, Tomford WW, et al. Infection in bone allografts. Incidence, nature, and treatment. Journal of Bone and Joint Surgery (Am), 1988;70:369-76.
18. MacEwen W. Observations concerning transplantation of bone. Proceedings of the Royal Society London, 1881;32:232-4.
19. Malinin TI, Carpenter EM, Temple HT. Particulate bone allograft incorporation in regeneration of osseous defects: importance of particle sizes. Open Orthopaedics Journal, 2007;1:19-24.
20. Malinin TI, Mnaymneh W, Lo HK, et al. Cryopreservation of articular cartilage. Ultrastructural observations and long-term results of experimental distal femoral transplantation. Clinical Orthopaedics and Related Research, 1994;303:18-32
21. Mankin HJ, Doppelt SH, Sullivan TR, et al. Osteoarticular and intercalary allograft transplantation in the management of malignant tumors of bone. Cancer, 1982;50:613-30.
22. Mnaymneh W, Malinin T, Mnaymneh L, et al. Pelvic allograft: a case report with a follow-up evaluation of 5.5 years. Clinical Orthopaedics and Related Research, 1990;255:128-32.
23. Mnaymneh W, Malinin TI, Lackman RD, et al. Massive distal femoral osteoarticular allografts after resection of bone tumors. Clinical Orthopaedics and Related Research, 1994;303:103-15.
24. Mnaymneh W, Malinin T, Mnaymneh LG, et al. Scapula allografts: a report of two cases. Clinical Orthopaedics and Related Research, 1991;262:124-8.
25. Mnaymneh WA, Malinin TI. Massive allografts in surgery of bone tumors. Orthopedic Clinics of North America (Am), 1989;20:455-67.
26. Mnaymneh WA, Temple HT, Malinin TI. Allograft reconstruction after resection of malignant tumors of the scapula. Clinical Orthopaedics and Related Research, 2002;405:223-339.
27. Musculo DL, Ayerza MA, Aponte-Tinao LA, et al. Unicondylar osteoarticular allografts of the knee. Surgical technique. Journal of Bone and Joint Surgery (Am), 2008;90(suppl 2 Pt 2):206-17.
28. Musculo DL, Ayerza MA, Aponte-Tinao LA, et al. Use of distal femoral osteoarticular allografts in limb salvage surgery. Surgical technique. Journal of Bone and Joint Surgery (Am), 2006;88(suppl 1 Pt 2):305-21.
29. Musculo DL, Ayerza MA, Farfalli G, et al. Proximal tibia osteoarticular allografts in tumor limb salvage surgery. Clinical Orthopaedics and Related Research 2010;468(5):1396-404.
30. O' Connor HI, Sim FH, Chao EYS. Limb salvage for neoplasms of the shoulder girdle. Intermediate reconstructive and functional results. Journal of Bone and Joint Surgery (Am), 1996;782:1872-88.
31. Ogilvie CM, Crawford EA, Hosalkar HS, et al. Long-term results for limb salvage with osteoarticular allograft reconstruction. Clinical Orthopaedics and Related Research, 2009;467:2685-90.
32. Ortiz-Cruz E, Gebhardt MC, Jennings LC, et al. The results of transplantation of intercalary allografts after resection of tumors. A long-term follow-up study. Journal of Bone and Joint Surgery (Am), 1997;79(1):97-106.
33. Ottolenghi CE. Massive osteo and osteo-articular bone grafts. Technic and results of 62 cases. Clinical Orthopaedics and Related Research, 1972;87:156-64.
34. Ottolenghi CE. Massive osteoarticular bone grafts. Transplant of the whole femur. Journal of Bone and Joint Surgery (Br), 1961;48,646-59.
35. Ozaki T, Hillman A, Bettin D, et al. High complication rates with pelvic allografts: experience of 22 sarcoma patients. Acta Orthopaedica Scandinavica, 1996;67:333-8.
36. Parrish FF. Treatment of bone tumors by local excision and replacement of massive autologous and homologous grafts. Journal of Bone and Joint Surgery, 1966;48A:968-90.
37. Parrish FF. Allograft replacement of part of the end of a long bone following excision of a tumor: report of twenty-one cases. Journal of Bone and Joint Surgery, 1973;55A:1-22.
38. Potter BK, Adams SC, Pitcher JD, et al. Proximal humerus reconstructions for tumors. Clinical Orthopaedics and Related Research, 2009;467:1035-41.
39. Ramseier LE, Malinin TI, Temple HT, et al. Allograft reconstruction for bone sarcoma of the tibia in the growing child. Journal of Bone and Joint Surgery (Br), 2006;88(1):95-9.
40. Roque PJ, Mankin HJ, Malchau H. Proximal femoral allograft: prognostic indicators. Journal of Arthroplasty, 2010;25(7):1028-33.
41. Rougraff BT, Simon MA, Kneisl JS, et al. Limb salvage compared with amputation for osteosarcoma of the distal end of the femur. A long-term oncological, functional, and quality-of-life study. Journal of Bone and Joint Surgery (Am), 1994;76(5):649-56.
42. Scarborough MT. Allograft-allograft healing? Salvage of massive allografts after fracture. Clinical Orthopaedics and Related Research, 2001;382:28-33.
43. Sorger JI, Hornicek FJ, Zavatta M, et al. Allograft fractures revisited. Clinical Orthopaedics and Related Research, 2001;382:66-74.
44. Temple HT, Malinin TI. Microparticulate cortical allograft: an alternative to autograft in the treatment of osseous defects. Open Orthopaedics Journal, 2008;14(2):91-6.

45. Temple HT. Gait analysis in patients with distal femoral resection comparing osteoarticular allograft reconstruction to metal endoprostheses. Unpublished data 2011.
46. Tomford WW, Mankin HJ. Bone Banking: update on methods and materials. Clinical Orthopaedics and Related Research, 1999;30:553-65.
47. Toy PC, White JR, Scarborough MT, et al. Distal femoral osteoarticular allografts: long-term survival, but frequent complications. Clinical Orthopaedics and Related Research, 2010;46:2914-23.
48. Volkov M. Allotransplantation of joints. Journal of Bone and Join Surgery, 1970;52B:49-53.
49. Wang J, Shen J, Dickinson IC. Functional outcome of arthrodesis with a vascularized fibular graft and a rotational latissimus dorsi flap after proximal humerus sarcoma resection. Annals of Surgical Oncology, 2011;18:1852-9.
50. Wang J, Temple HT, Pitcher JD, et al. Salvage of failed massive allograft reconstruction with endoprosthesis. Clinical Orthopaedics and Related Research, 2006;443:296-301.
51. Zhang K, Duan H, Xiang Z, et al. Journal of Experimental and Clinical Cancer Research. 2008;28:45 (Open Access).

Chapter 13

George W Hyatt MD (1920–1993), Dr Hyatt was the founder and the first director of the US Navy Tissue Bank. He made many contributions to the study of bone allografts and allograft transplantation. After leaving the Navy he became Professor and Chairman of the Department of Orthopaedic Surgery, Georgetown University School of Medicine Washington DC.

(*Reference:* Hyatt GW, Butler HC. Bone grafting. The procurement, storage and clinical use of bone allografts. Instructonal Course Lectures. American Academy of Orthopaedic Surgeons. 1957;14:343-73.)

Tissue Banking

INTRODUCTION

Transplantation of bone and tissue allografts is now a commonly performed surgical procedure. The current widespread use of bone and tissue allografts can be attributed largely to the ready availability and relative safety of these tissues (Malinin et al 1985, Malinin 1992, Mankin and Friedlaender 1983). This has been brought about by the development of well organized tissue banks capable of supplying surgeons with a variety of tissue allografts for transplantation (Kreutz et al 1951, Malinin 1977, Malinin and Mnaymneh 1985). However, bone and tissue banks vary in their complexity and scope of operation. Some tissue banks provide virtually all types of bone and soft tissue allografts employed in reconstructive procedures of the musculoskeletal system (Malinin et al 1985), while others provide only a limited variety of allografts to surgeons in their own institutions (Tomford and Mankin 1999, Doppelt 1981, Hart et al 1986). Some banks are involved only in tissue recovery whereas others are dedicated to processing and distribution, still others do both.

Bone allografts have been and continue to be prepared in numerous ways (Elliot and Scott 1951). Some tissue banks rely on complex methods of selecting donors, preparing, and storing the allografts and follow standards and guidelines adapted by various professional organizations (AATB 2005). On the other hand, bone for transplantation has also been obtained from donors about whom little was known, and prepared in the simplest of ways. Thus, techniques of bone allograft preparation have varied tremendously (Tomford et al 1987). Virtually everything has been tried. To date, bone allografts have been excised and processed either under rigorous aseptic conditions (Malinin et al 1985, Kreutz et al 1951) or without aseptic technique (Cloward 1980).

Bone allografts excised without rigorous aseptic precautions must be subjected to secondary sterilization, either chemical, irradiation or other methods. The wide divergence in tissue recovery and acquisition underscores the obvious differences between the allografts and explains, in many instances, the disparity in clinical results often observed with these grafts. Thus, the results of "bone allografts" can be misleading and, in retrospective studies, difficult to interpret. In some instances, only after considerable scrutiny, it is possible to determine that clinical results reported were obtained with secondarily sterilized allografts. The divergence of techniques in tissue banking is not limited to recovery, processing and storage of bone allografts. It also encompasses the degree and thoroughness with which donors are investigated for potential transmissible diseases (Malinin 1973).

When using bone allografts, surgeons must be familiar with the methods of their preparation (Ghormley 1942). All bone allografts are not alike! The differences between bone allografts prepared by various methods is worth knowing as these may account equally for safeguarding the recipients and for attaining the desired clinical results (Henry 1948).

It was estimated in 1977 that more than 100,000 bone grafting procedures were performed each year in the United States. This was a sizeable increase since 1948 when Alldredge suggested that successful use of homogenous bone would be far reaching (Alldredge 1948). The numbers have increased steadily and today, bone is the most frequently transplanted tissue. The number of bone grafts performed currently is many multiples of that in 1977, but exact numbers of bone transplantation performed annually are not known. More than 500,000 bone graft procedures are performed in the US each year and approximately 2.2 million are performed worldwide (Giannoudis et al 2005). In an aging population, diseases that affect the skeleton like osteoporosis, diabetes and cancer will increase and thus increase demand for tissues to effectively treat patients with these conditions. Historically, autologous bone was standard for bone grafting. As early as 1981, a survey showed that allografts were preferred over autografts; allografts were preferred by as much as 55% in some institutions (Miller 1984).Widespread acceptance of bone allografts suddenly increased their demand. This in turn resulted in the promulgation of shortcuts and materials that were less costly and perhaps less effective in achieving the desired clinical results. When using allografts, which may behave differently from what a surgeon would expect, a thorough understanding of bone allograft biology and the methods of their preparation is very important (Jackson 1987).

Transplantation of bone allografts in skeletal reconstructions is commonplace. Although cadaver bone allografts have been transplanted successfully since the beginning of the last century (Baschkirzew et al 1912), and in large numbers for several decades, the correct postulate that autografts are biologically superior to allografts still prevails (Rodrigo et al 1978). However, generalizations add little to the comfort and confidence of surgeons who use bone allografts to reconstruct skeletal defects. For large defects autografts are simply not available. Furthermore donor site morbidity, prolonged hospitalization, recovery and delayed return to function are serious drawbacks. Thus, the decision making process regarding the transplantation of bone allografts or autografts is a complex one and should be based on the understanding of the biology, and respective advantages and limitations of both types of grafts (Goldberg and Heiple 1983, Goldberg and Lance 1972). Once familiarity with the subject is gained by the surgeon, he or she will be in a position to answer the question as to whether or not transplantation of bone allograft will be beneficial for a particular patient. Today, allografts are the preferred method of bone grafting for spinal fusion, tumors and selected trauma reconstructions.

The question of safety and efficacy is another matter. Some allografts may be safe and effective, some may be either effective or safe, and others unfortunately may be neither. There are numbers of ways by which bone allografts can be obtained and prepared. Some tissue banks rely on complex methods of donor selection, excluding those donors whose medical profile and social behavior constitute increased risk of potential disease transmission. The risk of transmitting disease from the donor to the recipient is slight if allografts are taken from adequately studied and selected cadaver donors (Buck and Malinin 1994). On the other hand, bone can be obtained from donors about whom little is known, and are not scrutinized by postmortem examination (autopsy).In these cases safety hinges on secondary sterilization.

The general acceptance of bone and tissue allografting in the United States was brought about by the availability of allografts from a network of tissue banks (Kozak et al 1994, Malinin 1992, Mankin et al 1983, Emerson et al 1992). The ready availability of bone allografts, at times, led to the transplantation of the graft with surgeons knowing little about where the grafts came from and how they were prepared. Unfortunately, this is still true today (Lavernia et al 2004).

Graft demand seems to have trumped the principle of recovering grafts in an aseptic environment espoused by the pioneers of tissue banking such as Kreutz (1951), Hyatt and Butler (1957) and their associates and successors.

Although clinical transplantation of bone was not new, the development of tissue banks is of recent vintage. In the United States, the first full-service tissue bank was organized by the US Navy in response to the need for bone allografts for reconstructive surgery of skeletal structures in the casualties of the Korean conflict (Hyatt and Butler 1957). Full-scale tissue banks in the United States were developed subsequently and used the experience of the US Navy Tissue Bank as the basis of their operation (Malinin 1992).

Once clinical allograft transplantation became accepted, the demand for bone allografts resulted in the proliferation of tissue banks across the US and other parts of the world. These varied in the techniques of allograft excision and preparation. Attempts to standardize tissue banking by voluntary membership organizations have, by and large, resulted in failure since the standards were broad. Some tissue banks excised allografts under strict aseptic conditions whereas others chose to secondarily sterilize tissue after recovery or at some later time during processing and packaging. Criteria for selection of donors are broad and vary from one bank to another. There is a finite balance between safety and efficacy. The basic tenets of tissue banking stress donor selection and evaluation. Strict donor selection, careful postmortem examination and extensive cultures of bone, marrow, blood and tissues, in addition to comprehensive serologic analysis, constitute reasonable efforts to ensure safety. Careful selection and intense testing minimize the risk of transmitting disease from the donor to the recipient. On the other hand, allografts obtained from donors about whom little is known present a problem. In the latter cases, secondary sterilization is relied upon to achieve graft safety. Under these conditions confidence in graft safety, was illusionary and a number of patients have acquired diseases through allografts that could have been avoided. The most notorious was the transmission of Jakob Creutzfeldt disease with dura mater allografts. Sterilization with irradiation was not effective in these cases. Failures in donor selection, microbial surveillance and process failures in manipulating and processing allografts have resulted in transmission of several bacterial and viral pathogens.

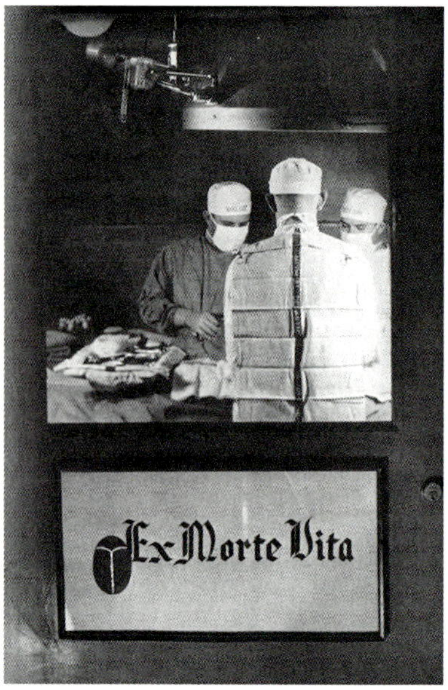

Figure 1 A view through the door leading to the sterile recovery room of the US Navy Tissue Bank. The sign "Ex Morte Vita" (from death, life) adorns the door. Aseptic removal of tissues for transplantation is in progress
Source: Official US Navy photograph

SELECTION OF THE DONORS AND STUDIES DESIGNED TO SAFEGUARD THE RECIPIENTS

Safeguarding the transplant recipients from the possible transmission of disease is dependent upon the screening and selection of donors of tissue allografts. Transmissible infectious diseases preclude tissue donation.

Donor selection is the first and perhaps most important step in acquiring tissues for transplantation (Friedlaender 1983). This process involves not only medical parameters on which the selection or exclusion of donors is based, but also includes medicolegal and social behavioral factors, ethical considerations, and public acceptance of the concept of transplanting tissues from deceased individuals to help the living. The US Navy Tissue Bank succinctly summarized the concept of the dead helping the living in its motto "Ex Morte Vita"(**Fig. 1**).

If one accepts the premise that transplantation of human cadaver tissues to replace diseased or injured parts of the musculoskeletal system is beneficial and desirable, then ethical questions regarding the circumstances under which these tissues can be obtained must be dealt with. The ethical issue of postmortem tissue donation is a complex one. Of most importance and perhaps most paradoxical is the freedom of choice of the next-of-kin of the donor. Equally important are the wishes of the donors themselves which are often expressed quite clearly in antemortem instructions, public declaration (motor vehicle registration) or legal documents. These decisions become even more difficult when the decedent's wishes are not shared by his relatives. Balanced against these factors are the needs of the members of society to acquire such tissues to improve their quality of life.

Permission to excise tissues from cadaver donors is usually given by the decedent's next-of-kin. Although donor cards and driver's license instructions are valid, and in fact the law in some states, tissue is often not recovered when living relatives object to donation. This creates a real conflict between the law and will of living relatives. In obtaining permission for excision of bone and tissues, it is important to discuss funeral arrangements with the family. Open casket funerals are not precluded by the excision of bone, provided the surgical excisions are carefully planned and executed and the body is adequately reconstructed. Dialog between funeral directors and tissue recovery agencies is very important in offering families a unified front that is positive and constructive to the donation process.

In our experience, most next-of-kin support and grant permission for tissue recovery. In the last three decades the University of Miami Tissue Bank (UMTB) has received permission for tissue donation from the families of over 10,000 donors. Seventy percent of those asked, consented to donation. It is important to note that donor qualification is highly selective, thus limiting the total number of potential musculoskeletal donors.

Sources of bone allografts include both cadavers and living donors (mostly femoral heads, osteotomy wedges). Most tissue banks rely on cadaver donors since a large amount of bone can be removed from a single donor and the body can be subjected to an extensive laboratory and anatomic examination needed to establish transplant suitability. However, bone banks of surgically excised tissues, i.e., bone obtained from the living donors, have in the past provided a large number of allografts to recipients of bone transplants (Hart et al 1986, Tomford et al 1985). The trend has now decreased.

Unlike vascular organ transplantation, in which only a short ischemic period can be tolerated, bone allografts can be excised from cadavers as long as 24 hours postmortem providing the donor's body has been refrigerated. If the cadaver has not been refrigerated, tissues must be excised within 12 hours postmortem. If excised within these limits, the allograft remains "biologically useful" (Friedlaender 1983). This time constraint is based on two considerations: The first has to do with tissue and cell viability, while the second main concern is the possibility of postmortem microbial contamination, especially with clostridia. Many studies have demonstrated the survival of human cells after cessation of cardiac activity by cell culture techniques, supravital staining, biochemical activity and other methods (Malinin 1966). However, recovery of viable cells from various tissues decreases with increased postmortem intervals, and reaches relatively low levels by 24 hours; hence, a justifiable but arbitrary time limit (Malinin 1966). It is obvious that postmortem changes commence almost immediately after death and proceed relatively rapidly. With bone, however, these changes are more gradual and within the above mentioned time frame, do not appreciably alter its osteogenic potential. Viable bone marrow cells (i.e., capable of growing in cell culture) have been recovered as long as 48 hours postmortem.

The second reason for imposing a time limit on recovery of bone from cadaver donors is bacterial contamination. The incidence of postmortem bacterial contamination increases precipitously at around 24 hours, since the rate of purifications from nonpathogenic microorganisms is a function of time and temperature. In our own studies, we have identified the length of the postmortem interval to be the most predictive of clostridia contamination at the time of recovery (Malinin et al 2003).

Proper donor screening is one of the most important aspects of tissue banking. Screening eliminates, at least partially, donors with potentially transmissible diseases and allows for the selection of allografts that are biologically, biomechanically and anatomically suited for their intended application.

Since the appearance of the AIDS epidemic in the US, the most frequently asked questions by surgeons regarding bone allografts is whether the human immunodeficiency virus (HIV) can be transmitted by bone implants. Two single instances of the transmission of HIV with a bone and tissue allograft have heightened this concern (CDC 1988, Simonds et al 1992).

Of more significance was the study of the recovery of HIV from the bone of patients with AIDS (Buck et al 1989). The study demonstrated, by a virological culture method, that HIV resided in bone-marrow rich tissue. In subsequent studies, HIV was recovered from the bone of five of twelve patients. Thus, HIV has been isolated from bone and is a legitimate cause for concern. After freezing, the virus could no longer be recovered from some initially positive bone specimens. However, those specimens that were positive for virus post-freezing continued to yield the virus after additional washings and even following freeze-drying. This demonstrated clearly that under many circumstances HIV was not inactivated by freezing, washing, or freeze-drying.

After documented transmissions of HIV by transplanted skin, kidneys and freeze-dried factor VIII (CDC 1984), it could be predicted that bone allografts would also transmit HIV. The above cited incidents of HIV transmission with bone allografts and the virological studies demonstrating the presence of HIV in bone are cases in point. However, if the presence of the virus in bone at the time of its excision was the sole factor responsible for the development of HIV infections in the recipients, one would expect more than a few cases of HIV infection among at least 200,000 patients estimated to have received bone transplants since the epidemic began in the US, and before screening programs designed to eliminate potential HIV carriers from the donor pool were instituted (Buck et al 1989). Since then, most concerns about the safety of allograft focused on human immunodeficiency virus/acquired immunodeficiency syndrome (HIV/AIDS) (AAOS 1989). Using available methods in a multiphasic screening and testing sequence, it was estimated in 1989 that the risk of obtaining an allograft from an unrecognized HIV infected donor was approximately one in 1.6 million (Buck et al 1989). Since then, the epidemic has grown somewhat, but concomitantly the polymerase chain reaction for HIV (PCR-HIV) has been added to the armamentarium of laboratory tests for HIV. Therefore, currently the risk is less than it was in 1989, provided all available precautions are taken. Comprehensive multiphasic screening and testing sequence involves the following steps:

Investigation of a medical and social history; direct physical examination of the donor by a trained physician; serologic tests for HIV-I/II antibodies, HIV-antigen, PCR-HIV, Hepatitis B surface antigen, Hepatitis B surface antibodies, Hepatitis B core antibodies, Hepatitis C antibodies, human T-cell lymphotrophic virus-III, 5 (HTLV-I/II) antibodies, and the rapid plasma regain (RPR) test for syphilis; autopsy with a separate study of lymph nodes; if available, follow-up of organ recipients may also provide additional information on donor tissue safety.

Perhaps the best way to address the issue of allograft safety is to review the reasons for post-excision exclusion of potential donors from a large donor pool in a single tissue bank. In our tissue bank in a sample of 1,000 consecutive-donors, 187 (18.7%) were excluded. Other than bacterial contamination (85 cases), hepatitis detected by either serologic or morphologic methods, or usually both, was the single largest reason for eliminating 68 donors. One of these was excluded on histological findings alone, all serological markers being negative (CDC 2003).

Laboratory documented or suspected HIV infections accounted for 10 exclusions. Three may have been false positive HIV-Ag tests. Four of the ten had other positive serologic findings, illustrating the value of other serologic tests as surrogates for detecting individuals with HIV infections. One had tuberculosis detected only at autopsy. The remaining two cases were excluded for morphological reasons, although the serology was negative. In one, there were classic nonspecific changes consistent with early HIV infection in the lymph nodes only. The other one demonstrated classic findings of recreational drug abuse with birefringent material within granulomas in the liver and lungs. Granulomatous diseases accounted for seven exclusions. Three were tuberculosis (includes the case noted above), one was a drug abuse case, and the other was inactive histoplasmosis. The latter may be particularly dangerous because of its propensity to "reactivate" in times of "stress". The two other cases were due to granulomas in the liver or lungs without an etiology being identified in either.

Four unsuspected malignancies were also recognized in potential donors as were two cases of myocarditis, conditions clearly exclusionary for transplantation. Both patients were clinically thought to have died of myocardial infarction.

Other morphologic or serologic exclusions were of one case only and included an unsuspected Alzheimer's disease.

Safeguarding recipients against disease transmission remains a primary responsibility of institutions preparing bone allografts. This is particularly true in view of the fact that most musculoskeletal allografts are life enhancing, and therefore elective, rather than lifesaving, as are their vascularized organ counterparts.

The incident of HIV transmission with bone and tissue allografts (so-called Life Net incident) created a great deal of notoriety and resulted in the FDA becoming involved in regulating tissue banks (Simonds et al 1992). The donor tested negative for HIV antibody, and was felt to have a negative medical and social history, although the latter might be questionable. The donation included liver, kidneys and heart as well as bone and soft tissues. Three recipients of organs developed AIDS and died. Three of four frozen bone allograft recipients likewise developed HIV infection. Thirty-two recipients of the freeze-dried allografts did not develop infections. This information was used to suggest that freeze-drying and accompanying graft treatments inactivated HIV in bone. However, laboratory studies cited above demonstrate that, in some circumstances, neither washing nor freeze-drying inactivates HIV in bone. The investigation of the reported donor stated that there were no known risk factors for HIV. Since the test performed for HIV antibodies was negative, a conclusion was reached that the donor was in the "window" period of infection. The report only peripherally mentioned what could be another possibility: that of a false negative and inaccurate test for HTLV-III antibodies.

Since the publication of the original reports of this incident, some additional information has come to light. Public records revealed the donor was incarcerated at least once, had an association with a known intravenous drug abuser and females with multiple sex partners. This case was instructive for many reasons. Detailed medical and social histories should have excluded these types of donors but there are significant limitations to the so-called next-of-kin medical social histories. Sometimes the designated next-of-kin actually knows very little about the donor's medical and social history, as they live separate lives often times in other states and countries. Even intimate spouses, for example, are unable to relate salient and potentially important components of the donor's medical history (Young and Wilkins 1995). As clinicians, we are often surprised by the limited insight some patients have in their own medical history, let alone that of their spouse, step mother, father or child! This highlights and underscores the importance of reviewing any and all available medical records prior to making an informed decision about donor qualification.

Laboratory tests by themselves, interpreted in a vacuum, can lead to a disastrous decision. Most tissue banks now do not rely upon a single laboratory test. Most sensitive, currently performed HIV-Ab, HIV-Ag and PCR tests now available would have likely identified this HIV infection (Strong et al 2005). Beyond a test or panel of tests for specific diseases, however, there is no substitute for clinical and pathologic correlation. For example, histopathologic examination of a donor's lymphoid tissue reveals abnormalities which in turn may question donor suitability. On the positive side, the incident prompted communication between organ transplant organizations, regulatory agencies, advocacy groups and tissue banks. Currently, if an organ recipient develops an HIV infection, a full investigation of all blood donors and the organ donor is launched. In view of these additional safeguards and adherence to basic tissue banking principles, it is unlikely, yet not impossible, that a similar maloccurrence will be repeated.

MICROBIOLOGIC STUDIES OF THE DONOR AND BONE ALLOGRAFTS

Thorough microbiologic analysis of the donor and the excised tissue is necessary not only to ascertain the adequacy of aseptic technique, but also to determine the presence or absence of microorganisms which might pose a threat to graft recipients (CDC 2003). Thus, microbiologic studies of cadaver musculoskeletal tissue donors are an essential component in assessing the suitability of allografts for transplantation.

Since bone for transplantation is removed postmortem, one cannot rely on the phagocytic action of leukocytes to remove even the most fastidious microogranisms. For this reason, every possible precaution must be taken to prevent the possibility of contamination. Identification of a particular organism does not preclude continued processing and future transplantation, but certainly alters the way in which the tissue is treated from the time bacterial contamination is identified. To this end, it is just as important to determine the source of the contamination as it is the contaminant itself (Malinin 1966).

Since tissues from cadaver donors may harbor microorganisms, allografts excised from these donors cannot be assumed to be sterile, even if strict antiseptic precautions are taken. If microorganisms are present, it is essential to know what they are and where they were found. To prevent contamination of the entire donor, segregating the recovery into defined (anatomic) zones is important. For example, if the tibia on one side is contaminated but segregated from the rest of the tissue, processing does not need to be altered nor does the donor necessarily have to be discarded. This knowledge enables one to determine the suitability or unsuitability of tissue for transplantation, as well as its potential for secondary sterilization.

Blood and bone marrow cultures are helpful in predicting contamination of musculoskeletal tissues. Positive blood and bone marrow cultures with similar microbial species correlate with a higher rate of positive cultures from bone (30%) as compared with positive blood (15%) or marrow cultures (11%) alone (Martinez et al 2003). The origin of microorganisms in cadaver blood is unclear. Whether blood contamination from the GI tract, the respiratory tree or the mucus membranes occurs concurrently with the events leading to death, or whether it takes place after death, has not been established (DuMolin 1992). The recovery of pathogens from autopsy specimens and the reported discrepancy between antemortem evidence of infection and postmortem culture results lead to the belief that visceral microflora may be subject to agonal and postmortem dissemination. The time and sequence of such dissemination has not been defined (Koneman and Davis 1974). That postmortem microbial contamination does occur within 2 or more days is unquestionable (Roberts 1998). Postmortem dissemination within 24 hours is supported by the study, which shows that the only statistically significant difference between clinically contaminated donors from which clostridia were isolated and those not harboring these organisms was the interval between death and the excision of musculoskeletal tissue. Therefore, the adoption of limiting tissue excision to 24 hours or less by the AATB as a standard is reasonable.

Postmortem multiplication of bacteria of normal intestinal and other flora leads to eventual breakdown of tissues. However, the process is a gradual one, and to be present in the normally sterile tissues within 12 hours of death implies antemortem infection, i.e., while the heart was still beating and the circulation maintained. The rate of putrification from and multiplication of non-pathogenic bacteria is a function of time and temperature. After the cessation of circulation, some tissues become non-viable very rapidly while others maintain viability for extended time periods (Lewis and McCoy 1922, Dolan 1964, Perry et al 1965, Malinin 1969). After tissues and cells become non-viable, they no longer serve as physiologic barriers to microorganisms. With tissue degradation, bacterial dissemination proceeds at a rapid rate. The first gross signs of this process are usually seen around the intestinal tract, and manifest themselves as a greenish discoloration of tissues.

In a living host, microorganisms gain entrance to the tissues or the bloodstream either directly, i.e., via local invasion through a wound, or by dissemination through blood or lymphatic vessels or by a combination thereof (Malinin 1966). With the exception of dissemination through the bloodstream and lymphatic channels, the same holds true after death. This has been demonstrated by repeated negative cultures of postmortem blood, bone marrow, mesenteric lymph nodes and other viscera (Martinez et al 2003, Minckler et al 1966). Microorganisms do not reach distant tissues and organs unless they are introduced there mechanically or until the tissue barriers begin to break down. This allows for the recovery of sterile postmortem tissues within the 12–24 hours period alluded to earlier.

In interpreting the results of microbiologic cultures on cadaver bone donors, it is important to differentiate between antemortem infections or contamination at tissue recovery and postmortem dissemination of microorganisms.

Reports in the literature on transmission of bacterial infection with bone allografts are relatively scant. Although some authors report higher infection rates with massive allografts, there are several explanations other than allograft contamination for these infections (Mankin and Friedlaender 1983). These include the extent of resection, operating time, adjuvant chemotherapy and radiation therapy, and seeding of the non-viable allograft with normally circulating bacteria (Zierd 1983).

The concern with patient safety and potential transmission of microbial infections naturally stimulated interest in the study of microbiology in cadaver donors. To this end, several studies were conducted by the authors and associates (Malinin 1992, Martinez et al 2003, Malinin et al 2003, Malinin et al 1985). Final results were derived from 1,747 consecutive donors. From 41,434 samples collected, 4,631 (11%) were positive. However, pathogenic microorganisms were recovered from only 2,245 (5.4%) of total samples. A single species pathogenic microorganism was from 83% of these. The remaining 17% yielded two or more pathogenic species.

The groups of pathogenic microorganisms recovered from musculoskeletal cadaver donors included: β-hemolytic Streptococci, *Clostridium sporogenes*, Enterococci, group D Streptococci, gram-negative anaerobic microorganisms, *Staphylococcus aureus* and *Streptococcus viridans* group. One or more positive cultures of skin flora and environmental organisms of low pathogenic potential were recovered from 46% of all donors. Entirely negative cultures were obtained from 29% of donors. These data also show a statistically significant increase in postmortem microbial contamination with the increase of postmortem interval.

Contamination of musculoskeletal allografts with Clostridium species deserves special attention because of the reports of *Clostridium sordellii* infections associated with allograft transplantation (CDC 2002). Extensive microbiological studies of cadaver donors indicate contamination with *C. sordellii* is not as rare as once thought. Clostridial contamination occurs in a significant percentage of tissue donors (8.1%), with *C. sordellii* being the most commonly isolated organism (Malinin et al 2003).

Clostridial infections transmitted with allografts have emphasized the need for adequate microbiologic assessment of donors (Eastlund 2006). This requires multiple samples, including those of blood and bone marrow. Since the percentage of pathogenic microorganisms is relatively small, these can be detected reliably only when an adequate number of tissue, blood and marrow samples are obtained. Otherwise, the extent of tissue contamination with dangerous microorganisms can be underestimated, or missed altogether (CDC 2006).

In the authors' experience, bone excised within 24 hours generally has a reasonable chance of not harboring microorganisms acquired through postmortem contamination. However, the incidence of postmortem bacterial contamination appears to increase from 18 hours onwards. Postmortem contamination manifests itself by the recovery from the tissues and the blood of several microbial species, particularly enteric organisms. In paired anaerobic and aerobic cultures of blood taken from the inferior vena cava of bone donors, three or more microbial species isolates were recovered from only seven cases (2.3%). Two bacterial species were recovered from 25 cases (8.3%) and one isolated from either an aerobic or an anaerobic culture was found in 64 cases (21.3%).

Microbiologic data gained from bone allograft donors emphasize the importance of thorough microbiologic analysis of the donors. In the authors' experience, reliance on a single bacteriologic culture could be misleading. Interpretation of microbiologic data can be best made by examining the results of multiple cultures and multiple culture sites. Blood and bone marrow samples from an ilium taken before bone excision should be incubated in aerobic and anaerobic blood culture bottles. After each allograft is excised, three tissue samples are submitted to the laboratory. These are bone marrow, a sample from the surface of the graft, and soft tissue from the joint. These samples are obtained before the allografts are placed into antibiotic-containing solutions. After the allografts are washed in antibiotic-containing solutions, they are rinsed in antibiotic-free fluid and again sampled in an identical manner for microbiologic studies. Whenever possible, actual samples of bone or other tissue rather than swabs are submitted for bacteriologic examination (**Figs 2A and B**). Recent investigations have demonstrated increased sensitivity with liquid cultures compared to standard swab cultures. For this reason, we rely on liquid cultures for recovered tissues in an effort to improve detection and identification of potential pathogens (Dennis et al 2011).

AUTOPSY

Autopsy remains an accurate diagnostic modality for defining pathologic changes in the deceased, including those that might preclude transplantation of tissues. Numerous studies report major discrepancies between clinical diagnoses and postmortem findings (Thurlbeck 1981, Chacon et al 1997, Friedrici and Sebastian 1984). Most studies report discrepancies of about 12%.The figure has not improved significantly, despite advances in imaging and other diagnostic techniques.

The sentiment that an autopsy is not necessary for the welfare of tissue allograft recipients is emotive and a very efficient way of promoting the pecuniary interests of tissue banks. In reality, lack of information which might be gained from the procedure places the allograft recipients at risk. A misclassification of cause of death, which we have seen, is common in the absence of a postmortem examination that promotes the acceptance of donors that should have been

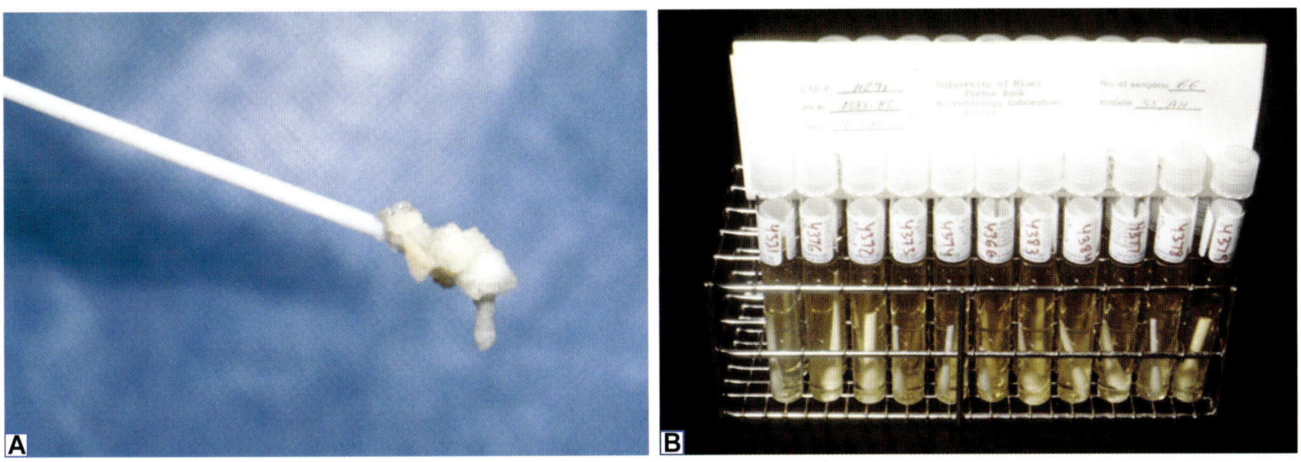

Figures 2A and B (A) A sample of tissue submitted for microbiological assessment; (B) Thioglycolate tubes with microbiologic samples obtained at the time of bone and tissue allograft excision

excluded. This is particularly true of donors with malignancies. The risk of a recipient dying from any cancer and its possible relationship to receiving a transplant from donors harboring malignancies is unknown. However, it cannot be dismissed since development of cancers related to possible viral etiologies has taken years! On the other hand, exclusion of donors from the donor pool whether on a hearsay history or on a past medical history is wasteful and unjustified, although promoted by the FDA. An autopsy can provide clear cut evidence of the absence of potentially transmissible disease with much greater accuracy than any diagnostic strategy. Our review of autopsy findings on 4,575 consecutive musculoskeletal donors showed major discrepancies between clinical diagnoses and autopsy findings in 3% of musculoskeletal donors. These included unsuspected thyroid, renal, lung, larynx and colon carcinomas with metastases, as well as cases of myocarditis, granulomatous disease, lymphoma, active tuberculosis, active histoplasmosis and Alzheimer's disease, etc. Tissues from 135 donors with serious diseases would have been transplanted if donors were not excluded by autopsy findings. We suspect most surgeons and tissue recipients would find these risks unacceptable.

Autopsies performed on tissue donors have unique features, as do autopsies performed for medicolegal reasons, teaching purposes and research. While medical examiners are masters at determining the sequence of events leading to a person's demise, and anatomic pathologists at unraveling the pathophysiology of an illness, the prosector of transplantation donors is responsible for preventing the transmission of disease from a donor to a recipient. It follows, therefore, that such an autopsy must be an orderly search for both obvious and subtle evidence of disease. The task is important for two reasons: First, the number of recipients of grafts from a multiple organ, cornea, bone and tissue donor may exceed 100. Second, recognizing potentially dangerous diseases as well as the lack thereof rests with the pathologist. Clinicians and/or coordinators involved in the evaluation of donors screen out unacceptable donors as best as they can. However, many potential donors may have disease such as viral myocarditis, a wide range of cancers, a host of neurologic diseases and diseases of unknown etiology. Mostly these may become apparent only upon microscopic examination. Conversely, the presence of conditions suggested by a medical history can be ruled out by the negative findings at autopsy.

Another advantage of autopsy is the discovery of disease processes that may help living family members of the decedents. There have been a number of donors whose health was unrelated to an underlying genetic abnormality that was discovered on postmortem examination and disclosed to family members. Furthermore, infectious diseases uncovered in postmortem discovery were revealed to appropriate public health authorities who, in turn, use this information to screen donor contacts and improve public health and safety. We feel that it is incumbent on tissue banks to help the living, not only by providing safe and effective grafts for transplantation, but also through sharing medical discovery with living family members. In the future, directed genomic screening of living relatives, based on careful and comprehensive postmortem examination of donor relatives may improve and extend the lives of family members. Why should not tissue banks offer this valuable service to families who unselfishly offer loved ones for tissue donation?

Excision of Tissues

The excision of tissues, especially bone, from cadaver donors is indeed a large-scale surgical procedure requiring constant diligence and surgical discipline. Knowledge of surgical anatomy is important. During the excision, medical findings often dictate what types of grafts can be best made from particular tissues. Initial transections of bones, tendons, ligaments, fascia latas, etc. are made accordingly. Several anatomic findings may exclude donors, and the decision to do so is best made as early as possible.

Several methods for the excision of bone allografts have been described in the medical literature (Kreutz et al 1951, US Navy Tissue Bank 1958, Hyatt and Butler 1957, Malinin 1992). Ideally, bone is best excised under aseptic conditions in an operating room. In this chapter descriptions of the methods and techniques of excision of bone allografts are based on the cumulative experience of obtaining bone allografts from over 4,500 cadaver donors over 40 years. It is natural, in that time, several techniques and procedures have been tried, discarded and retried. The following description of techniques is based on several trial and error attempts to excise bone allografts with the utmost efficiency and with as little contamination as possible.

The University of Miami Tissue Bank utilizes a specifically designed operating room suite for excision of bone allografts. Because of convenient proximity to the donors and other logistical considerations, the suite is located in the Miami-Dade County Medical Examiner's Office (**Figs 3A to C**).

This highly sophisticated concept has been reproduced in a self-contained facility that is designed for recovery, postmortem examination, imaging, processing and distribution.

Upon arrival of a donor body in the Tissue Bank, the extremities and the trunk are prepared for surgery by shaving and thorough scrubbing with the povidone-iodine soap. However, other antiseptic detergents can be used just as effectively. To avoid bringing a contaminated stretcher into the operating room the donor's body can be transferred to the OR table through a window.

In the operating room, the body is prepped with povidone-iodine prep. Other surgical preparations can also be used. This can be followed with a 70% alcohol soak. If alcohol is not used, iodine solution is best left on the body. The operative site is then draped as for major surgery. The peroneum is sealed with a towel and plastic adhesive. Figure 3B shows a prepped donor body. The extremities are prepped on the anterior and posterior aspects while being held up by a technician or by a mechanical device.

After the body is prepped and draped, it is covered with a sterile disposable sheet which is cut to allow for skin incisions. After the initial skin incisions, the edges of the skin are sprayed with an antiseptic solution and the area is isolated from the surrounding skin with towels and towel clips in a standard manner. If this is not practical, sterile adhesive sheets can be used, but these generally do not adhere to the edges of the wound.

The first incision is made in the groin over the greater saphenous vein. The vein is isolated and a catheter is passed through it into the inferior vena cava for collection of blood samples. Once these have been obtained, the catheter is withdrawn and the vein ligated.

A stab incision is then made over the iliac crest to facilitate collection of a bone marrow sample. Bone marrow can be collected and obtained by aspiration or by a method described by Fedotenkov et al (Fedotenkov et al 1963). The latter procedure consists of driving two Jamshidi needles of unequal length into the ilium. A sterile wash-out solution is infused through a longer needle and bone marrow aspirate is collected through a shorter needle. Bone marrow effluent is injected immediately into an anaerobic and an aerobic blood culture bottle. Upon completion of blood and bone marrow collection, instruments are discarded and gloves changed.

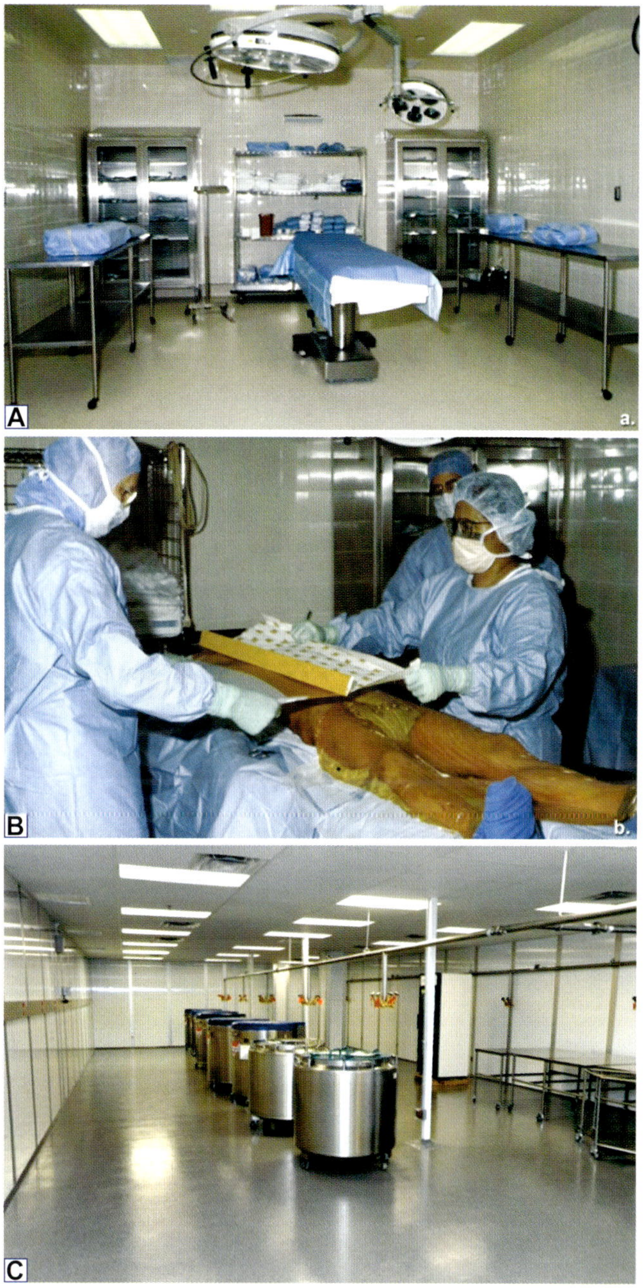

Figures 3A to C (A) Operating room for the excision of bone and tissue allografts, University of Miami Tissue Bank (UMTB); (B) Excision of allografts under aseptic conditions; (C) Liquid nitrogen storage facility at UMTB

The femurs, tibias and the knees are removed through anterior incisions over the thighs, which curve medially to the patella and continue downward parallel to the tibial crest. If circumstances permit, the incision may be continued cephalad over the hip joint using an extended iliofemoral approach.

This facilitates the dissection around the proximal portion of the femur and the subsequent removal of the iliums.

Excision of the tissues and bones of the lower extremities should be performed in layers. First, fascia lata is identified, dissected free and excised (**Fig. 4**).

Next the quadriceps tendon is identified and divided at its origin in the rectus femoris muscle.

The suprapatellar bursa is incised and the extensor mechanism is dissected free and reflected downward to expose the knee joint. Dissection is then carried downward between the rectus femoris and the vastus lateralis to expose the diaphysis of the femur. The vastus intermedius is divided and the femoral shaft is subperiosteally dissected free of soft tissue (**Fig. 5**).

The femur may be divided with an osteotomy at a level dictated by the intended use of the graft. If necessary, the femur may also be removed *in toto* with or without tibia.

After the femur is osteotomized or the hip joint disarticulated, it is reflected anteriorly and mobilized. The tibia is then dissected subperiosteally, divided at a predetermined level and also reflected anteriorly. The tibio-fibular joint is then identified, incised and divided. The knee joint is then dissected and freed together with the joint capsule. The distal femur and the proximal tibia can be removed together with the knee joint or separately (**Figs 6A and B**).

It is easier to separate the knee joint in the processing room rather than at the time of its excision. To minimize the chance of injury to popliteal vessels which facilitate the

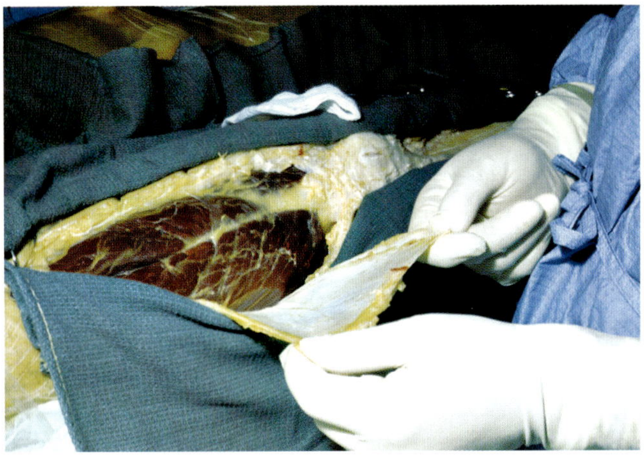

Figure 4 Tissues in the extremity are dissected in layers. Fascia lata is removed first

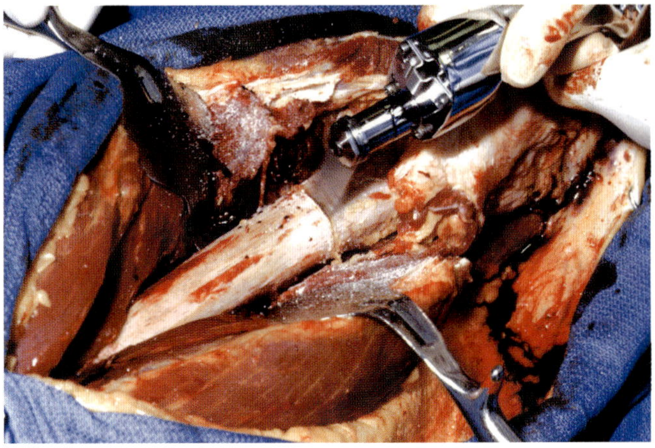

Figure 5 Exposed femur diaphysis may be osteotomized or removed *in toto*

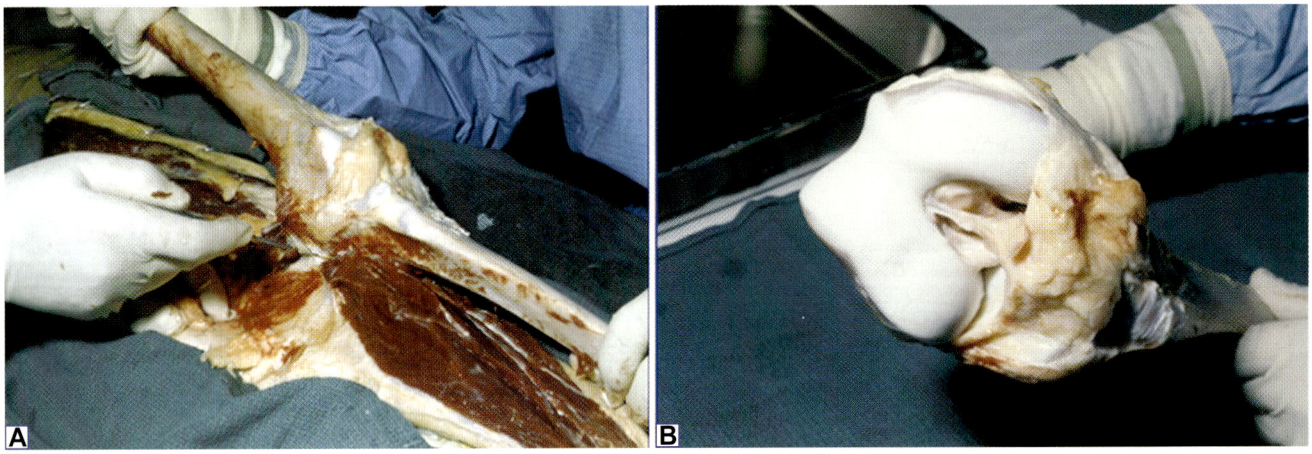

Figures 6A and B (A) The femur and the tibia can be removed together with the knee joint; (B) Separation of the knee joint, either at the time of the excision or in the processing room, facilitates the inspection of articular surfaces

embalming process, it is helpful to dissect the knee with the extremity flexed.

The proximal femur may be removed either with or without the femur head. If it is desired to retain the femur head, the joint capsule is incised circumferentially, the femur is dislocated and the ligamentum teres is cut.

To remove the entire hemipelvis, the muscles attached to the anterior and inferior iliac spines, the iliac crests and the iliums are detached and the vessels are identified. Peritoneum is detached from the medial surface of the ilium with curved osteotomes. The skin and soft tissue are retracted to expose the pubic symphysis which is osteotomized. An osteotomy is then performed across the SI joint and the pelvis is mobilized by sharply dividing the sacrospinous and sacrotuberous ligaments and the remaining tendinous insertions of the adductors, abductors, and hamstrings muscles.

The other alternative is to divide the neck of the femur after incising the capsule, removing the head of the femur separately or *in toto* with the acetabulum. Next the gloves are changed and the ilium subjected to subperiosteal dissection. This is best accomplished with curved osteotomes. It is important to divide the sacroiliac ligaments during the dissection. The sacroiliac joint is then identified and divided with curved osteotomes. If only the ilium is to be removed, it is osteotomized with straight osteotomes immediately below the inferior anterior iliac spine and with curved osteotomes at the sacroiliac juncture. Osteotomized ilium is grasped with Lambotte forceps and rotated. The ilium is then removed (**Fig. 7**).

Bones of the upper extremity are excised after changing the gloves and the instruments. These are approached through a standard deltopectoral incision which can be extended inferiorly to facilitate removal of the elbow and the radius. The deltoid muscle is detached anteriorly from the acromion to facilitate exposure of the shoulder joint. The biceps tendon is divided distally and reflected superiorly, the subscapularis is divided and retracted medially and the tendons of the latissimus and pectoris muscles are divided medially. The deltoid insertion is detached, leaving a generous cuff on the humerus followed by an osteotomy of the distal humeral diaphysis. The bone is then reflected anteriorly and the soft tissue attachments are divided sharply. The capsule is the last structure divided, leaving a small cuff of tissue on the glenoid if excision of the scapula is anticipated.

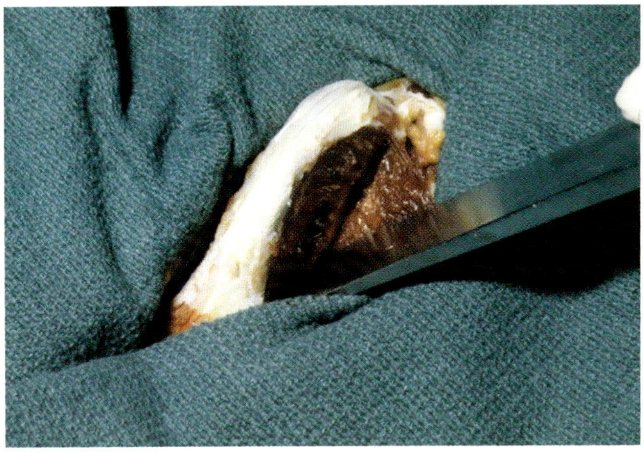

Figure 7 After osteotomies, the ilium is grasped with bone (Lambotte) forceps, rotated and removed after severing the ligaments

In selected cases and depending on family wishes and funeral arrangements, the elbow can be removed *in toto* with the distal humerus and the proximal ulna and radius. The distal radius is removed by extending the incision to about 3 cm above the radial styloid. Further extension of the incision into the wrist may expose it at the funeral.

The ribs are removed last because, of all bones, ribs have the highest incidence of microbial contamination. This can probably be explained by the close proximity of the ribs to the skin and underlying lung pleura. Ribs are removed subperiosteally through a standard autopsy incision.

Separate incisions can be used for the removal of any additional bone in the body, using standard aseptic precautions. The only bone which cannot be excised aseptically is the mandible; typically this bone is either irradiated or treated with ethylene oxide following excision and cleaning.

Historically, each bone was immediately sampled for microbiologic studies as described above. More recently, in light of our findings comparing liquid cultures to swabs, liquid cultures are obtained from collective bone samples segregated by zone then transferred to enriched culture media and incubated in special incubators with CO_2 sensors (Bac T Alert™).This results in sensitivity for microbes and decreases detection time. After obtaining samples for microbiologic cultures, the bones are individually wrapped and transported to the processing facilities. Osteoarticular allografts that are designated to be fresh or cryopreserved are best placed into tissue culture medium such as CMRL-1415, RPMI-62 and other similar tissue culture media.

The practice of placing osteoarticular allografts in lactated Ringer's solution is dictated by the FDA (2004) approval of the latter. However, lactated Ringer's solution has a relatively low osmolarity (256) compared to other balanced salt solutions used in cell culture. It also lacks nutrient additives (Malinin and Wolcott 1973). Other non-osteoarticular grafts are placed in saline. Antibiotics are added to these solutions to prevent contamination of the grafts from the environment and to inactivate possible nonpathogenic organisms which may have been placed on the surface of the allografts during excision.

Reconstruction of the Donor's Body

Skeletal reconstruction is of utmost importance in maintaining the dignity of the donor's body. Unless otherwise specified, the reconstruction of the body is left to the funeral directors who will prepare and display cosmetically acceptable remains. In addition, after the excision of bone allografts, the body is turned over for an autopsy. A reconstructed body facilitates postmortem examination in as much as it allows the pathologist to ascertain the gross appearance of the body.

Reconstruction of the extremities is accomplished by replacing excised bone with wooden dowels, hinged plastic or metal rods, or similar devices, packing the cavities with gauze and cotton, and tightly closing the incisions.

Processing of Bone Allografts

Bone and tissue allografts are ideally processed in a distinct and separate operation in separate facilities. Typically, tissue processing occurs in rooms that are better rated for air flow and quality than the standard operating rooms. A dedicated team of specialized technicians are familiar with the varied processes and end products. Separating the two functions also prevents potential contamination. Temporally separating processing from recovery allows for preliminary culture and serology findings to be interpreted so that tissue can be appropriately chosen for specific types of processing and special grafts. For example, separate procedures are used for processing osteoarticular cryopreserved allografts, fresh articular allografts and frozen and freeze-dried preparations.

Osteoarticular Allografts

After storage at 4–6°C, osteoarticular allografts are brought into the processing room in sealed and doubly wrapped containers. The solution into which the allografts have been placed in our institution is culture medium CMRL-1415 to which 100 mg/L of streptomycin sulfate and 80 mg/L of gentamicin are added. However, other tissue culture media, balanced salt solutions, as well as lactated Ringer's solution have been used. After the containers with the allografts are brought into the processing room they are transferred to a room-temperature antibiotic-free tissue culture medium, washed, and cleaned of unwanted soft-tissue attachments and bone marrow. After at least 1 hour of washing in antibiotic-free solution and cleaning, a second set of tissue samples is obtained for microbiologic culturing. Again, at least three samples from each allograft are submitted for culturing.

Osteochondral allografts are prepared in a way that preserves chondrocytes to sustain joint function after transplantation. Femurs are usually osteotomized at the proximal or midportion of the diaphysis, humeruses, and tibias in the distal one-third. A portion of the knee capsule, along with the collateral and cruciate ligaments, is preserved with the distal femur. The patellar ligament, menisci, and cruciate and collateral ligaments must also be preserved with proximal tibias. Thus, from a given donor, either the distal femur or proximal tibia from each lower extremity is used as an osteochondral allograft.

With the exception of soft-tissue needed for joint reconstruction, the periosteum and all other soft-tissue attachments are meticulously removed. The bone marrow is likewise reamed and washed out of the shaft of the graft. However, bone marrow and cancellous bone are retained in the epiphysis. Photographs of distal femur allografts are shown in **Figures 8A and B**.

The rotator cuff and the capsule are preserved on the humerus **(Fig. 9)**.

The elbow is preserved with the capsule and ligaments. The distal radius is preserved with as much capsule as possible.

Hemipelvic allografts include the entire ilium, acetabulum, pubis and ischium. The hip joint capsule is preserved, only if the recipient's femur head is retained (an uncommon situation). A bipolar hemiarthroplasty is often used in reconstruction following tumor resection for acetabular lesions. In

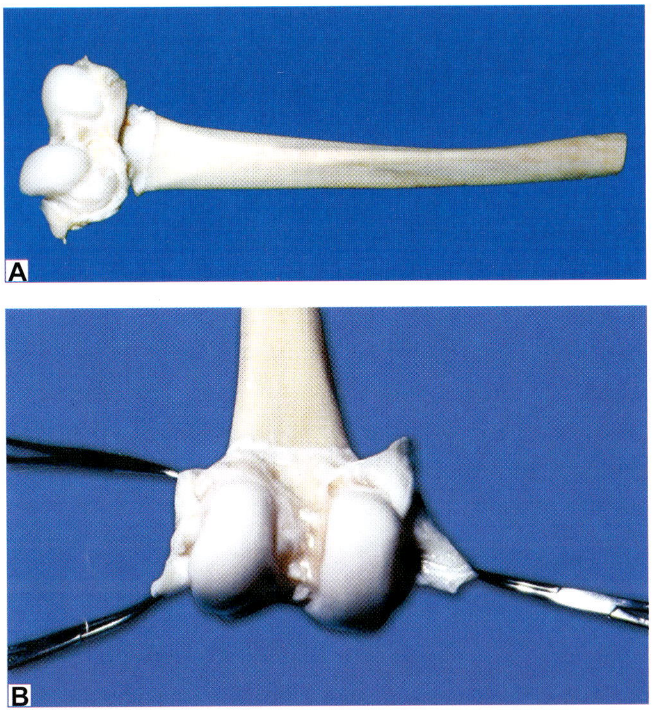

Figures 8A and B (A) Distal femur allograft used for the extremity reconstruction following tumor resection. The ligaments and the joint capsule are preserved for reattachment to those of the recipient to restore normal joint kinematics; (B) The length of the diaphysis is maintained to match that of the resected femur

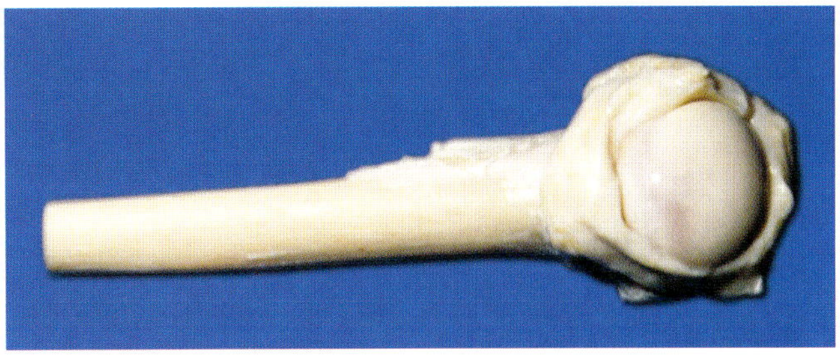

Figure 9 Proximal humerus allograft with the capsule, ligament and tendons

Figure 10 Liquid nitrogen freezers provide convenient means of storing allografts at cryogeneic temperatures

these cases, it might be reasonable to cryopreserve articular cartilage. If total joint arthroplasty is planned, cryopreservation is not necessary.

Less frequently used osteoarticular allografts such as phalanges, carpal bones, distal ulnas, distal tibias, tarsal and metatarsal bones, as well as scapulas, are prepared and cryopreserved in the same manner as more frequently used allografts.

During the entire period of processing, osteochondral allografts are kept in tissue culture solution to prevent articular cartilage dessication and maintain viability. After the allograft is prepared, it is washed in cold tissue culture solution and then transferred into 15% glycerol in CMRL-1415 or other tissue culture medium for 30-40 minutes. After exposure to glycerol, the grafts are placed in two plastic bags, double wrapped in cloth towels, and placed in a chamber for liquid nitrogen vapor, in a control-rate freezing apparatus precooled to 4°C. Thermocouples are placed into a control bone. The allografts are cooled to -40° to -60°C at rates from 0.5° to 2.0° per minute. They are then brought down rapidly to -100°C, and are then transferred into the vapor phase of liquid nitrogen or a mechanical freezer cooled to -126°C or below until the day they are used or shipped. Liquid nitrogen freezers used for storing bone allografts are shown in **Figure 10**.

It is advisable to photograph each allograft after it is processed. These photographs and unmagnified or digital radiographs of these allografts are used for recipient matching. The photographs provide the surgeon with information regarding the length of the tendons and ligaments, the size of the capsule, and the general state of the allograft. Inspection of these photographs and images before the start of the surgical operation allows for extremity reconstruction. Interactive web-based programs are ideal in providing end users with accurate updated imaging and real time dialog to ensure that both the bank and surgeon are well informed.

Fresh Allografts

Transplantation of fresh hemijoint allografts has been described previously (Lexer 1985). Currently, fresh allografts are used primarily when cartilage with a thin bone shelf is transplanted, generally, for focal articular defects. These are stored and shipped at refrigerator temperatures (4–6°C) in tissue culture solutions without serum. The acceptable limit for the storage of cartilage allografts at these temperatures is 5–7 days, although experimental studies indicate that cartilage stored in CMRL-1415 remain viable for over 10 days (Malinin et al 1985, Malinin et al 2006). Storage of fresh cartilage allografts in saline is not recommended, as prolonged exposure of chondrocytes to hypotonic saline results in a loss of cell viability. Lactated Ringer's solution has been used by several investigators, but it is inferior to tissue culture solutions (Czitrom et al. 1986, 1990). Fresh allografts are prepared in the same manner as cryopreserved osteoarticular allografts, except for preservation of capsules, ligaments and tendons.

Preparation of Allograft to be Preserved by Freezing or Freeze-Drying

After removal and cleaning, bone may be cut into any desired shapes and sizes in anticipation of specific surgical applications. After the excised bone is brought into the processing room, it should be placed into antibiotic-free saline and recultured in a manner identical to that employed in the operating room. Bone is then cleaned of all soft tissue attachments and periosteum. To clean bone well is a tedious and laborious process. First, periosteum is removed with an osteotome or with specially designed power instruments. Ligamentous and tendinous attachments are excised with a scalpel. The bone is then cleaned with a steel brush or any other device. Cleaned bone is washed in saline and then cut. Copious amounts of fluid are used during cleaning.

All visible marrow is removed from the tubular bones. After each step, samples for bacteriologic cultures are obtained.

Bone allografts are then cut to prepare allografts for various surgical applications. Iliums are cut to prepare wedges for anterior cervical and lumbar fusions, preserving both cortices. Large and long strips are made from the iliac crests for mandibular replacements. The remainder of the iliac bone may be crushed and made into various particulate sizes.

Cloward dowels for anterior cervical fusions can be made from the heads of femurs and femoral condyles. Patellar tendons and Achilles tendons for replacement of anterior cruciate ligaments are prepared with bone blocks. Cancellous blocks of bone made to fit a variety of instrumentation systems are cut from femur condyles and proximal tibias.

Cortical plates are prepared either from femurs or tibias, as is crushed cortical bone. Diaphyseal segments and the entire diaphyses of tibias, femurs, fibulas and humeri are preserved either in segments or *in toto*.

Femur heads, distal femurs, femur condyles and proximal tibias to be used in revision hip arthroplasties can be either stripped of cartilage or preserved with intact cartilage. Proximal femurs are cleaned of soft tissue attachments and preserved *in toto*, as are the acetabulums.

Cut allografts are subjected to extensive washing and agitation in washing solutions to remove the bone marrow. The reasons for doing this are several. First, the antigenicity of bone allografts has been attributed to the so-called antigen presenting cells which reside in the bone marrow (Czitrom and Axelrod 1985). Second, sensitization to Rh factor by marrow-containing bone allografts have been reported (Johnson and Urist 1998). Third, dead allogeneic bone marrow cells do not contribute to the incorporation of the allograft. These have to be removed before vascularization of the allograft takes place.

Preservation of Bone Allografts by Freezing

Bone allografts have been stored at temperatures below 0°C since Ollier's time (Ollier 1867). With the advancement of freezing technology, a notion emerged that the lower the storage temperature the longer the biologic properties of the frozen material would be maintained. This sentiment was proven to be correct. However, all too often, bone allografts were simply frozen at random, the selection of storage temperature being based not on the knowledge of the phenomenon associated with freezing, but on the availability of a freezer. The results were often not satisfactory (Wilson 1951), but the blame was often placed on the inadequacy of the allografts, not their preparation. To achieve reproducible results, the selection of temperatures for storage of bone allografts must be based on understanding the change produced in tissues as they are being cooled, stored at subzero temperatures, and warmed through the freezing point (Malinin 1966).

A frequently asked question related to the temperatures of -15°, -20°, -30° and -40°C referred to in several tissue banking manuals, guidelines and other publications, is why were these temperatures singled out? The answer is simple. These were the temperatures achieved in the then available commercially produced mechanical freezers and the household freezers (Brown et al 1976).

Freezing is prefaced by the reduction of temperatures from the normothermic to a hypothermic range. Hypothermia decreases the metabolic rate, but cooling by itself does not kill cells. Irreversible injury to frozen tissues is caused by a combination of physical and chemical phenomena, many of which are still poorly understood.

To achieve complete cessation of enzymatic activity and to arrest restructuring of ice crystals, it is necessary to store tissues well below 0°C. In the historical development of freeze-preservation of tissues, many investigators have stored various biological materials at -79°C, the melting point of easily obtainable dry ice. Later, mechanical freezers were constructed which would maintain temperatures between -60° and -80°C. Storage of frozen bone allografts in mechanical freezers in these temperature ranges are still common. However, biologic materials cannot be stored indefinitely even at these temperatures (Malinin and Perry 1967). Regretfully, data which would define exactly the temperature limits of storage of allografts is not available. Suggestions of not keeping bone allografts at -60 to -80°C for more than a year have been made (Brown and Cruess 1982). The suggestions were made on the basis of empirical observations, and the application of bone to observations made on other tissues and cells (Malinin and Perry 1967). In the review paper on bone and cartilage preservation, Brown and Cruess (1982) state "even at low temperature bone can only be kept for as long as a year", citing a paper in which no mention is made on the time limitations of storage of bone at "low temperatures". This illustrates the confusion which exists with regard to low temperatures. The confusion is natural, as the term has been used across the board without an exact agreement as to what is meant by it.

Ice crystal growth ceases only at temperatures below -120°C, temperatures easily attainable by cryogenic gases and by recently developed ultra-low temperature mechanical freezers. Because of the relationship between ice crystal formation and growth and temperature, it seems reasonable to use the term low temperature almost synonymously with cryogenic temperatures, and to refer to temperatures above -100°C as intermediate low temperatures, dry-ice temperatures and the like.

Liquid nitrogen, which boils at -196°C at atmospheric pressure, is the most popular and readily accessible cryogenic gas. It provides ideal storage temperatures in its vapor phase (around -150 to -130°C) at a reasonable cost. A major property of importance of this gas is its inertness. Unlike carbon dioxide and other solvents used as cooling media, nitrogen does not react with what it comes in contact with and when warmed, vaporizes without leaving a residue. It has no direct effect on the pH of frozen preparations, unlike carbon dioxide. Because of the reliability of liquid nitrogen and the simplicity of liquid nitrogen freezers, liquid nitrogen freezing and storage has been employed routinely by tissue banks. With the advent of ultra-low temperature mechanical freezers, these too have been used, but their reliability (frequent break-downs), cost and limited storage space per unit, make them considerably less effective than cryogenic gas freezers. The advantage of mechanical freezers is that no gas vapors obscure vision in the freezing compartments.

Cryopreservation of Osteoarticular Allografts and of Cartilage

The term cryopreservation is used here in a specific, rather than generic sense, although in the true sense of the word this is not accurate. The common usage of the term, which emerged in the last few decades, implies the use of cryoprotective agents and controlled rate cooling in conjunction with the term. To understand the principles, advantages, and limitations of cryopreservation procedures, it is necessary to consider and review events associated with cooling, freezing, storage at subzero temperatures and thawing of tissues. Attention must be directed not only to heat transfer, but also to various other phenomena associated with freezing.

The change from the colloidal to the solid state does not occur instantaneously at a particular point in the cooling curve. It may take place within minutes or hours depending on the volume of the object being frozen and the temperature differential between the specimen and the cooling environment.

Temperature shock: When tissues are injured by cooling to near 0°C, they are considered sensitive to temperature shock. Thermal shock, cold shock and constitutional cold death are terms used interchangeably. The deleterious effects of rapid cooling are manifested in the deterioration of physiological function of different cell types (Sherman 1964). It has been postulated that most mammalian cells are to some degree susceptible to cold shock. Chondrocytes, the preservation of which constitutes a major effort of tissue bankers, are apparently not very sensitive to thermal shock. Isolated animal chondrocytes recover from cooling (Tomford et al 1985) as do cartilage slices (Brighton et al 1984). Cartilage on osteoarticular allografts survives cooling hypothermic storage and freezing, but the survival is time dependent and not permanent (Malinin et al 1985).

Freezing damage and cryoprotection: With lowering temperature, molecular kinetic energy decreases and produces changes in chemical reaction rates, solubility, diffusion rates, and viscosity. These changes can cause irreversible injury, but they are far less damaging than changes which take place when tissues are taken below their freezing point. If conclusions with regard to mammalian tissues can be drawn from studies with yeasts, the biochemical changes caused by temperature drop alone do not produce cell death. Cell death is associated with freezing of water, i.e. transformation of available water into ice crystals. Formation of interstitial and intracellular ice subjects the cells to the direct physical effects of crystal formation and exposes them to high concentration of solutes in yet unfrozen fluids. Cells and tissues can be damaged by either mechanism. When 90% of the freezable water is converted to ice, the concentration of sodium chloride increases almost tenfold. A concentration of sodium chloride of about 8.5% has a denaturing effect on the cell membrane lipoprotein components (Lovelock 1954, 1957). Based on these findings, a theory of death from the effects of high solute concentration has evolved. The injury mechanism is applicable to solid osseous tissue as well. It would follow that the longer the cells are exposed to harmful temperatures, the greater the injury. However, in microorganisms, longer exposure to cold either reduced injury or had no affect whatsoever (Mazur 1962). Slow cooling produced longer exposures to concentrated solutes, but it also produced better survival than rapid cooling. The presence of electrolytes in the surrounding medium actually reduced the damage from low temperatures. On the basis of these observations, it was (Mazur 1962) who concluded that cell death was associated with the direct effects of ice crystals rather than with the effect of high solute concentration. This conclusion may be substantiated by studies on human and animal tissues *in vitro.* Morosow (Morosow 1930) dehydrated embryonic chicken hearts and human hearts to a water loss of 75% by weight. The preparations, placed in Ringer's solution for 1 hour to several hours, were cultured at 37°C. Pulsations began at once or after a time lapse of up to 7 days. In another series of experiments, pieces of brain, heart, spleen, testicle, and skin were dried with a water loss of up to 80% by weight. Cells from these tissues grew out in the usual manner in explant cultures (Morosow 1928). The concentration of solutes was presumably increased several-fold by dehydration, but it did not appear to inhibit the growth of rehydrated cells.

Cell death in mammalian tissues may be simply the result of failure to survive crystallization of freezing. Ice crystals may be formed both extracellularly and intracellularly. Whether extracellular or intracellular, ice crystal formation is a function of the rate of freezing. Ice formation is essentially extracellular when the drop in temperature is relatively slow. During ultra rapid freezing, (many degrees per second) microscopic ice crystals form simultaneously within the cells and throughout the suspending fluid. Crystal size is also a function of the rate of cooling. After the solid state has been reached, recrystallization growth of crystals may occur, depending on the temperature of storage. Ice crystal growth does not appear below -120°C, thus the justification for low temperature storage as previously discussed.

As water undergoes crystallization, energy is released in the form of so-called "latent heat of fusion" (Luyet 1966, Meryman 1963). Much was made of this phenomenon in designing control rate freezing equipment which compensates for the release of latent heat of fusion by rapid cooling of the freezing chamber. This is done to avoid the rise in temperature in the specimen being frozen and thus preventing its rewarming and reaching melting point again. The importance of the phenomenon is greatly exaggerated. Release of the latent heat of fusion is a phenomenon manifested most readily in aqueous solutions. However, the recorded amount of heat released is dependent on the placement of thermocouples. If a 5 ml tube with saline thermocouples is placed near the wall of the vessel, it will demonstrate a transient rise in temperature as freezing proceeds. Thermocouples placed in the center will register only slight temperature elevations (Luyet 1966). Therefore, the latent heat of fusion might be a laboratory phenomenon. In practice it might not be as important as previously thought.

Tomford achieved the best results with preservation of chondrocytes by cooling the cells to -40°C without compensation for the release of the latent heat of fusion, maintaining cell suspensions at this temperature for 5 months and then transferring frozen suspensions into a -80°C freezer (Tomford 1985). However, freezing of chondrocyte suspension is not identical to cryopreservation of organized cartilage. Articular cartilage matrix is high in water content (Linn and Sokoloff 1965). Since about 75% of the wet weight of cartilage is water, alteration of the water component during freezing would cause a major disruption of the cartilage (Pegg et al 2006).

Water is a major component of articular cartilage and other tissues. The transition of water from a liquid to a solid phase with crystal formation and subsequent crystal structure alteration is responsible for freeze injury and physical destruction of the tissue. It is important to understand this phenomenon if injury is to be avoided.

During slow cooling, the temperature of the material being frozen remains near or at the melting point of the solution. All freezable extracellular water will become immobilized by large water crystals. During this process, intracellular water is gradually withdrawn into extracellular space and incorporated into extracellular crystals. This creates a physiological condition akin to edema, except the extracellular water is immobilized. It is estimated that about 90% of cellular water is freezable, the remainder being bound or bulk water. Removal of the chemically bound water would result in irreversible chemical alterations in the cell.

If, on the other hand, the cooling is rapid, the temperature of the specimen will fall until sufficient numbers of water crystals are formed to release the heat. Consequently, rapid cooling produces many small crystals. There is no time for translocation of water between the compartments and ice crystals formed within the cells and in the interstitial fluid. The crystals are uniform and are evenly distributed throughout the tissues.

Many attempts have been made to control ice crystal formation by altering the cooling rates. Parkes applied a two-stage freezing technique (Parkes 1958). It consisted of plunging the ampules with tissue into freezing mixtures at -20° or -30°C. These experiments demonstrate the importance of the rate of freezing as well as that of the critical temperature below which tissues do not survive except under unusual circumstances. Critical temperatures vary between tissues.

Controlled cooling velocity was devised to minimize the destructive effects of crystallization of water. It was empirically determined that the rate of cooling of 1°C per minute yielded the best results.

As usual, an isolated phenomenon in biology does not stand on its own. In addition to cooling rates, thawing rates must also be considered. Early in the development of cryobiology, Taylor (1960) noted the best results were obtained when an ampule of frozen tissue was thawed by placing it directly into a +45°C bath. An oscillographic record showed that for small fragments of tissue, warming from -196°C to 0°C occurred in 0.25 seconds. If the same tissue was thawed in a +20°C bath, the thawing time was increased threefold. Thawing of tissues in a +40–45°C solution is a procedure of choice.

Cryoprotective agents: In the early phases of low temperature studies, it became apparent that controlling the rate of freezing alone and staying above the critical temperature was not enough to avoid freezing damage. To this end, a search was made for chemical protective agents. The first such agent which, when added to tissues or cells prior to freezing, protected them from the freezing damage, was glycerol, also known as glycerin, 1, 2, 3 propanetriol, or trihydroxypropane. This colorless, syrupy, sweet-tasting fluid is hydroscopic and miscible with water. Its boiling point is +290°C, and it solidifies forming shiny orthorhombotic crystals on cooling at 0°C. Its cryoprotective effect on spermatozoa was discovered by Polge, Smith and Parkes in 1949 (Polge et al 1949). Somewhat later, their observations were extended to red blood cells. These investigations show that 10-15% glycerin permitted recovery of slowly frozen erythrocytes.

Glycerol lowers the freezing point of water and, through its hydroscopic properties, facilitates the movement of water out of the cell. One mole of glycerin is capable of binding approximately 3 moles of water (Lovelock 1953). The glycerin-water complex is an effective cryoprotective agent. To be effective, glycerin has to penetrate into the cell. Paradoxically, the way in which it penetrates the cell has caused most of the problems associated with its use and much discussion on its mode of action. The rate of cooling of 1°C/min is optimal with glycerolized cells because it is presumably slow enough to permit adequate diffusion of glycerin through the cell membranes (Sherman 1964). On the other hand, the same rate of freezing is equally as effective with dimethyl sulfoxide, an exceedingly rapid penetrating agent.

The most convincing evidence on the site of action of glycerol was produced by Lovelock, who showed intracellular glycerol to be essential to the survival of human erythrocytes and suggested glycerol acted as a "salt buffer" (Lovelock 1953). This led to the assumption that living cells can be protected by glycerol only when they are completely permeated by the substance. This assumption has been questioned on the basis of data showing detectable amounts of intracellular glycerol was not essential, but rather detrimental to the maximum survival of unfertilized mouse eggs. This suggested the extracellular protective action of glycerol is achieved, in part, by cellular dehydration (Sherman 1964).

Since the advent of glycerol as a cryoprotective agent, many other compounds have been tested for their ability to prevent freeze-thaw damage in tissues. This list includes polyvalent alcohols, sugars, sera, ethylene glycol, egg yolk, pyridine N-oxide, dimethyl sulfoxide, and many others. With the exception of dimethyl sulfoxide, none of the tested compounds were found to be as effective as glycerol. DMSO also possesses radio-protective properties (Ashwood-Smith 1962, 1967).

Pure dimethyl sulfoxide (DMSO) (methyl sulfoxide, methanesulfonic acid CH_3SOCH_3) is a very hygroscopic, colorless fluid with practically no odor and a slightly bitter taste. It is very soluble in water because of proton bonding of water molecules to the oxygen atom (Wu and Malinin 1980). It also has systemic effect in animals (Panuska et al 1966). Lovelock and Bishop thought dimethyl sulfoxide was, in certain circumstances, a better cryopreservative than glycerol (Lovelock and Bishop 1959). Soon afterward Ashwood-Smith began detailed studies of this compound in bone marrow preservation (Ashwood-Smith 1961). The outstanding virtue of dimethyl sulfoxide was that, unlike glycerin, it appeared to pass freely through cell membranes. However, this might not have been the case in organized tissues (Malinin and Fontana 1969). Since its introduction, dimethyl sulfoxide has been used in freezing virtually all animal tissues and, with the exception of spermatozoa, has been found to be equally effective as glycerin (Sherman 1964).

In the study with transplantation of osteochondral allografts in animals, little difference was found between glycerol and dimethyl sulfoxide (Malinin et al 1985). In fact, one study suggested glycerol was a better cryopreservative for cartilage than was DMSO. For this reason we continued the use of glycerol as cryoprotective agent in the preservation of osteoarticular allografts. Before proceeding with the discussion of methods used for the preservation of osteoarticular allografts, it must be noted that cryopreservation of cartilage has achieved only partial success (Pegg et al 2006). Although isolated chondrocytes can be cryopreserved with success, such is not the case with intact articular cartilage (Tomford 1985). With the best methods available, cryopreservation of cartilage results in latent injury which leads eventually to transformation of hyaline cartilage into fibrocartilage (Malinin et al 1985). However, cryopreserved cartilage maintains articular surfaces longer than cartilage frozen without cryopreservatives and without regard to freezing rate. To ascertain the extent of the latent injury, long-term observations are necessary. In baboons with transplanted cryopreserved distal femurs, normal joint spaces were present for the first 3 years of observation. This would be equivalent to about 9–12 years of human life. However, by 5 years (15–20 years in human life) the cartilage underwent degenerative changes and was transformed into fibrocartilage (Malinin et al 1989). It is clear that present cryopreservation techniques are not perfect, but it is also clear that cryopreserved articular cartilage functions longer than randomly frozen cartilage. Because of this, and until better methods of cartilage preservation are found, the use of the present methods of preservation of osteoarticular allografts appears justified.

Preservation and Storage of Osteoarticular Allografts

Osteoarticular allografts are kept in tissue culture solution during the entire period of allograft processing (Malinin et al 1985). This prevents desiccation of cartilage and exposes it to nutrients in the medium. After allograft preparation is complete, it is washed in fresh cold tissue culture solution. The allografts are placed for 30–40 minutes into 15% glycerol in tissue culture media. After exposure to glycerol, the grafts are sampled for microbiologic studies, placed into plastic bags, heat sealed, doubly wrapped in towels, taped and labeled. The allografts are then placed into the chamber of liquid nitrogen control-rate freezing apparatus precooled to 4°C (**Fig. 11**). When transplanted into experimental animals the results with cartilaginous allografts were similar to those obtained with osteochondral allografts.

Figure 11 The chamber of the controlled velocity apparatus is cooled by injections of liquid nitrogen vapor

Thermocouples are placed into a control bone package wrapped in an identical manner to the allografts being frozen. Osteochondral allografts are cooled at a rate of 2°C per minute to -40°C. They are then cooled at 10°C/min to -100°C. Once the allografts are frozen, they are transferred to the liquid nitrogen freezers. To avoid seepage of nitrogen into the packages, osteoarticular allografts are stored in the vapor phase of liquid nitrogen. The allografts remain in liquid nitrogen storage until the day they are used or shipped to another facility. Allografts are transported to the operating rooms on dry ice. In the operating room, osteoarticular allografts are thawed rapidly in a 40–45°C solution. The solution used for thawing should be the same solution as that used in the preparation of the grafts. However, from a practical standpoint, lactated Ringer's solution is frequently used because it is readily available in most hospitals. To avoid additional injury to the cartilage, thawing in saline is not recommended. If balanced salt solutions or tissue culture media are not available, lactated Ringer's solution is preferable over saline.

Dehydrated Bone and Tissue Allografts

Currently, most dehydrated bone and tissue allografts are freeze-dried. As mentioned previously, freeze-drying has been reported to decrease the antigenicity of bone (Burwell et al 1985, Friedlaender 1983). Combined with decreased antigenicity and ease of storage, freeze-drying and other methods of dehydration increase the utility of bone grafts (Bonfiglio et al 1955, Bull et al 1928).

Freeze-drying, although a more recent development than freeze-preservation, represents one of several endeavors created in the search of better techniques for the storage of bone and tissue allografts.

Freeze-drying is a precise technique embodying three requirements. The product to be freeze-dried has to be solidly frozen, the condenser has to be colder than the product and the chamber must have vacuum. If any of these three conditions are not met, the process is something other than freeze-drying.

The technique of freeze-drying tissues developed by Greaves and Adair (1959), Flosdorf and Hyatt (1952), Gresham (1964) and others became a reality in the 1950s. However, it is evident that many transplanted allegedly freeze-dried allografts were not freeze-dried at all, at least not by the US Navy Tissue Bank (Gresham 1964).

One suspects that other tissue banks followed the same procedures, i.e., transplanted tissues that were simply dried at hypothermia and not freeze-dried. Nevertheless, bone allografts prepared in this manner and transplanted into thousands of recipients were by and large successful.

According to Flosdorf (1949), biophysical advantages of freeze-drying depend upon maintenance of the solid state of tissue fluids during dehydration. This was not the case in many instances. The drying was not accomplished from the solid ice, but from a liquid phase of water. Bone allografts subjected to chemical dehydration (Tutoplast Process) have also been used, but reports on clinical results, in peer reviewed journals, are scarce (Gorschewsky et al 2005).

As has been the case with many surgical techniques, freeze-drying of human tissues has enjoyed wide popularity. At one time or another, virtually all tissues which could be transplanted have been freeze-dried (Malinin et al 1978). However, the technique did not work well for the preservation of some tissues, particularly vascular conduits, heart valves, corneas and skin. Consequently, for a while, the technique of freeze-drying tissue allografts fell into disrepute. However, later on it re-emerged as a firmly established method of perseving bone and tissue allografts. By now, extensive clinical and experimental data on freeze-dried bone allografts has been accumulated. Evaluation of this information shows considerable variation in the preparation of bone allografts by freeze-drying, the process of freeze-drying itself, and most importantly the end results.

It is apparent that the term "freeze-dried bone" means different things to different authors. Because of this, the entire subject of freeze-drying warrants attention. Some degree in the standardization of methods of assessing freeze-dried bone is desirable. Such an endeavor would have to include both the biologic and clinical assessment of the freeze-dried allografts, and determination of residual water in freeze-dried bone, a principal criterion by which the success or failure of freeze-drying is judged.

By necessity, transplanted freeze-dried bone allografts have been evaluated by comparing them to fresh autologous and frozen allogeneic bone grafts transplanted into identical anatomic sites (Malinin 1993). Although it is difficult to avoid, such an evaluation scheme may not be appropriate in all circumstances. The grafts are different anatomically and biologically. Bone marrow is removed from allografts, but retained in autografts. Autografts are also limited in availability. Their recovery is associated with morbidity and the graft is often structurally inadequate to subserve immediate function. When the clinical utility and and biomechanical adequacy of freeze-dried allografts is viewed in this context, their advantages become much more apparent (Heo et al 1992).

History of Freeze-Dried Bone Allografts

Freeze-drying of bone allografts is a relatively new development in tissue banking. Although the principles of freeze-drying were known at the turn of last century, and the process was described before World War II, it was not applied to bone until some 40 years later.

Freeze-drying found its practical applications, at first, in the preservation of serum and plasma (Flosdorf 1949, Greaves 1968). The mechanically refrigerated condenser and controlled heating in a freeze-drying apparatus was described by Greaves and Adair (Greaves and Adair 1939). Material freeze-dried in this apparatus was maintained at temperatures below 0°C in a sterile chamber, and ice was transferred to the condenser. It was then brought to ambient temperature for further desiccation.

Thus, the process was not freeze-drying in the true sense of the word. In the initial stages of the development of freeze-drying, it was necessary to establish an adequate surface relationship between the evaporating area of the liquids being dried and the area for the absorption of the heat of sublimation. For this purpose, a solid carbon dioxide condenser was attached to a manifold connected to the chamber. An attempt to increase the capacity of the freeze-drying apparatus was made by providing a manifold to which glass flasks with frozen material were attached. When surrounded with dry-ice, the manifold which was connected to a vacuum pump functioned as a condenser. Freeze-dryers of this configuration, although simple and inexpensive, are no longer in use. Their principal disadvantage was a relatively small capacity and inability to maintain sufficiently low temperatures of the product being dried, again compromising freeze-drying principles. An improved design which incorporated a mechanically refrigerated condenser and controlled temperature of the product served as a basis of modern freeze-dryers (Greaves and Adair 1939).

The US Navy Tissue Bank was the first to systematically preserve bone allografts by freeze-drying (Kreutz et al 1951, Carr and Hyatt 1955). At first, bone was embedded in a block of ice by building up successive layers of ice around a previously frozen specimen. Bone in a block of ice was then put into the freeze-dryer. Obviously, this placed unnecessary demands on the freeze-dryer. Consequently, soon thereafter the ice block was eliminated and frozen bone in wide-mouth glass containers was subjected to drying. Following a 14 day freeze-drying cycle, residual moisture (residual water) was reportedly reduced to between 5% and 10% as the temperature of the bone reached zero (Hyatt and Butler 1957). The 14 day freeze-drying cycle was later replaced by 7 day cycles with a step-wise increase in the shelf temperature. This was done during the first 3 days of the cycle and produced bone allografts of adequate dryness. However, no data on the actual water content of grafts dried in this fashion has been published.

Subsequently, newer freeze-dryers with external condensers further reduced the drying cycle (**Fig. 12**). Greshem, comparing the regimens for freeze-drying of bone allografts used by the US Navy Tissue Bank, came to the conclusion that in most instances bone was not, in fact, dried from a frozen state. These products were not strictly freeze-dried because in the early phases of the 14 day freeze-drying cycle used for nearly 10 years, bone was maintained at temperatures above the freezing point of saline.

Freeze-Drying Techniques

Freeze-drying and vacuum packing allows storage and shipment at room temperature, a clear advantage over storage in cumbersome tanks and shipment in special containers with ice or nutrient media. The process of freeze-drying, despite being time consuming and cumbersome, is one of the most satisfactory methods of preservation, storage and shipment of human bone tissues. However, as in other tissues, changes produced in freeze-dried bone are not insignificant. Freeze-drying, as practiced now, renders all tissues and cells non-viable. The only organisms that can survive the process are microorganisms; however, microbial survival depends on the species and the freeze-drying regimens employed (Nei 1964).

The process of freeze-drying consists of removing water from the frozen material by sublimating *in vacuo*. In other words, freeze-drying is achieved through dehydration by removing water vapor from ice without melting it. The water vapor is again converted into ice crystals on a condenser which is significantly colder than the material being freeze-dried.

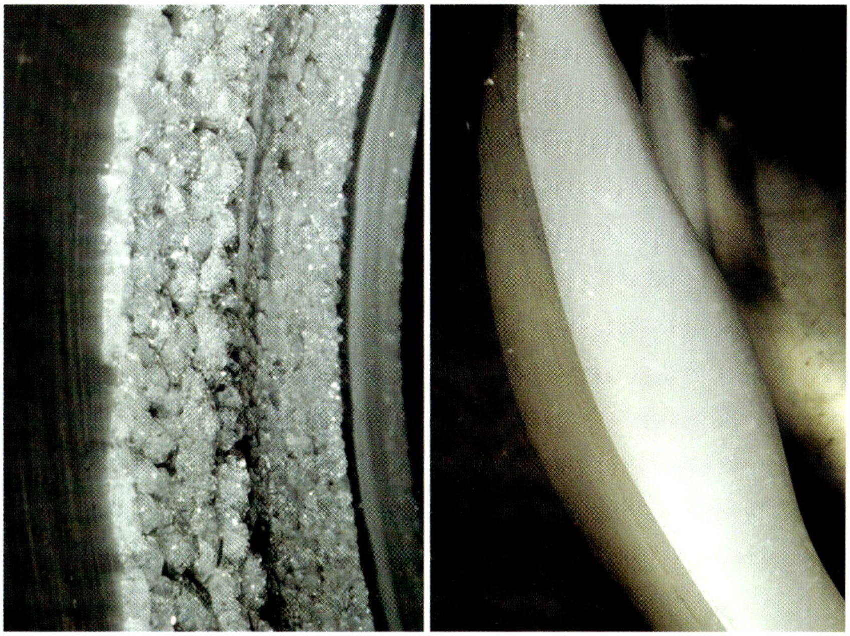

Figure 12 Ice crystals deposited on the condensers of freeze-driers will vary in their structure and size, depending on the condenser temperatures and the speed of sublimation. The same discrepancy in ice crystal formation occurs in tissues with concomitant destruction of the tissues by larger crystals

To understand the principles of freeze-drying, we must recall that at about -20°C and below 2,000 atm pressure there is only one solid phase in the water system, namely ordinary ice. The solid ice, liquid water and water vapor constitute the three possible single phases in this system. These phases may be involved in three possibly two-phase equilibria, namely: liquid-vapor, solid-vapor, and solid-liquid. The deposition of the translocated ice may assume several forms, manifested by the different structures of ice crystals (**Fig. 12**).

Freeze-drying should depend entirely on the solid-vapor equilibrium, and the sublimation curve of ice. Since the equilibrium between water, ice and water vapor is possible only at 0.0098°C and 4.58 mm pressure, and the equilibrium between ice and liquid water at temperatures above 0°C, freeze-drying is conducted at lower temperatures and pressures. However, this is not always possible and in many applications tissue is simply dried and not freeze-dried.

The process of freeze-drying had been divided into two successive phases (Rey 1964). The first phase, termed the sublimation or desiccation phase, encompasses extraction of free water frozen to ice. This phase is readily defined since its end is marked by the disappearance of the visible ice crystals from the product and their reappearance on a condenser. The second phase includes the removal of the so-called bound or structured water. This process can be demonstrated only indirectly by physicochemical measurements. Bound water is removed much slower than the free water, as can be judged by the decreasing water content of tissues commensurate with the progression of freeze-drying. Complete removal of bound water cannot be achieved without marked morphologic alterations of bone (Malinin et al 1983).

The methods of freeze-drying bone allografts described here are based on the experience of freeze-drying over 500,000 bone allografts. The present methods have been developed by the empirical means accomplished by varying the techniques to obtain a product of sufficient dryness (Greiff 1973).

Most freeze-dryers have several features in common. These include control of the shelf temperature, sterilization of the chamber, capacity for the chamber aeration, high efficiency vacuum systems, and translucent door or port holes (**Figs 13A and B**).

There are several variations in freeze-drying techniques. In accordance with one technique, bone allografts are first frozen to below -130°C and are then placed into the chambers of the freeze-dryers. The condensers of the freeze-dryers are set to temperatures between -60°C and -70°C. The shelves are cooled to around -40°C. The doors of the freeze-dryers are closed and the vacuum pumps are turned on. Since the bone and the containers in which it is stored are initially colder than the condensers, no drying occurs at the start. However, the bone equilibrates rapidly with the shelf and within 2 hours the product-condenser temperature differential is sufficient to commence freeze-drying.

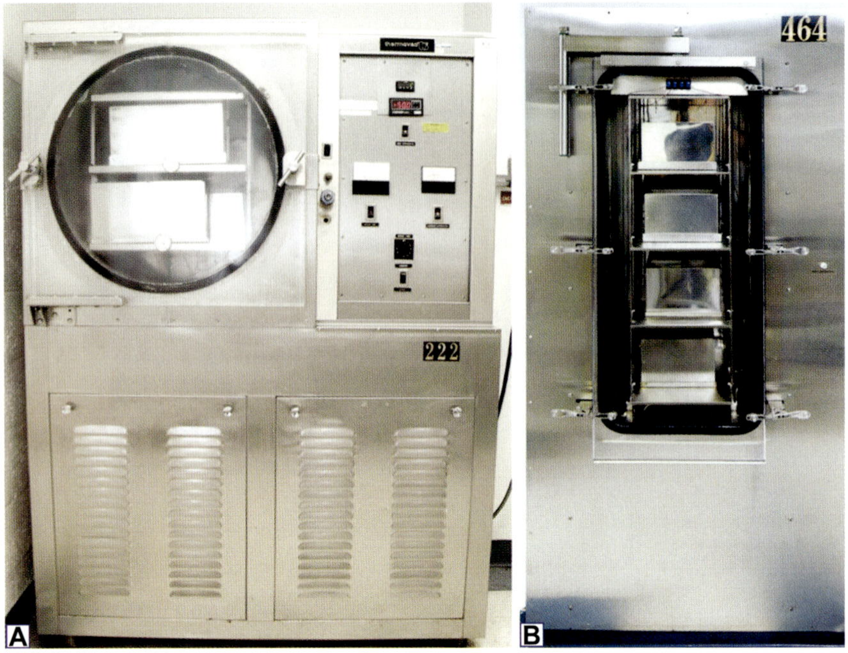

Figures 13A and B (A) Older model freeze-dryer with an internal condenser; (B) Recent vintage freeze-dryer with an external condenser. Both types of freeze-dryers have distinct advantages and disadvantages

In performing studies with freeze-dried bone, it was learned that the removal of water in phase one of the freeze-drying cycle depended primarily on the vacuum in the chamber. However, the removal of total water (free plus bound water) is dependent on the length of the freeze-dried cycle.

The vacuum in the freeze-dryers is maintained below 100 millitorr after the first 2 days of the cycle. The length of the freeze-drying cycles can vary from 2–3 days to 20 days. The length of the cycles, of course, depends on the amount of bone placed into a freeze-dryer, the smaller the amount of bone the shorter the cycle.

At the end of the freeze-drying cycle, after all ice has been sublimed, the shelf temperature is raised to +30°C, and the vacuum in the chamber is released by allowing air to enter the chamber. To maintain sterility, air in the chamber is passed through a filter.

The changes in the properties of freeze-dried tissues have been attributed to two chemical phenomena: in one case, by changing the protein configuration caused by underdrying and in the other, saturation or blocking of hydrophilic sites of proteins by oxidation associated with overdrying (Malinin 1966). However, in bone allografts, alterations associated with freeze-drying may represent a desirable feature since they may be responsible for the reduction of antigenicity of freeze-dried tissue.

Residual Moisture (Water) Determinations

The water not removed from the tissues by freeze-drying is generally referred to as "residual moisture" or "residual water". The residual water is determined to assess the efficacy of freeze-drying, but the most frequently used methods employed to measure residual water are indirect measurements and are subject to variations. Nevertheless, the quantity of water remaining in the freeze-dried tissue has a direct bearing on the characteristics of the preserved tissue. The matter becomes rather complicated when structural and biological peculiarities of the tissue being freeze-dried are considered. It becomes obvious that to ensure optimum preservation of bone and soft tissue allografts, it is important to accurately determine the levels of residual water for freeze-dried bone, and to adjust the freeze-drying regimens accordingly. Of course, water is not the only factor influencing the stability of freeze-dried tissue. Oxygen also contributes to the changes in freeze-dried substances. For this reason, freeze-dried materials are best preserved either in a vacuum or in an atmosphere of inert gas.

A certain amount of bound water will remain in bone after freeze-drying regardless of the total length of the process and the operating conditions. However, the levels of residual water correspond to the sensitivity of freeze-dried bone to environmental conditions.

At first glance, the problem of residual water determination would appear to be a simple one, but in practice the measurement of water in freeze-dried bone is not as simple and reproducible as it might appear. This is attested to by the existence of several physical and chemical methods for determining water in freeze-dried material. In order to interpret the data correctly, the limitations and advantages of each technique employed to determine residual water must be understood. As mentioned earlier, a considerable volume of literature is devoted to the nature of water in biologic systems. To avoid controversy, and for a matter of simplicity and consistency, we will continue to refer to water in two categories free water and bound or structured water. Free water is the basic component of all biologic fluids. It maintains in solution the minerals, sugars, amino acids, lipids and proteins. At low temperatures it is crystallized as ice.

The bound water represents water in which molecules participate in the macromolecular structures. It may be slightly or tightly bound to intracellular structures or to matrix. Although some of this water may be expelled by physical change in the external environment, some of it cannot leave the cell without causing profound chemical and morphologic alterations. Measurements of the residual water of both varieties must be performed to determine the efficacy of freeze-drying.

Gravimetric Method

Of all methods currently employed, the gravimetric (or weight for equilibrium) method is the most popular, the simplest and the cheapest. However, if not carefully performed, it is also subject to significant errors. Gravimetric determination of residual water is based on the assumption that in a freeze-dried tissue there is a certain amount of water for a given mass of the substance. If a sample mass is placed in an air-tight container with a drying agent (e.g. P_2O_5, silica gels), heated to a given temperature, and weighed periodically, it will eventually reach a constant weight. However, the heating temperature must be adequate to drive the water vapor off, but not high enough to char the tissue. The residual water (RH_2O) is calculated on the basis of weight loss that represents water (m) while the initial weight of the sample represents the mass (M). The following formula is used to calculate the grams of water per 100 grams of bone:

$$RH_2O\% = m \times 100/M\%$$

This method is relatively simple and inexpensive. In many countries it is the official method for determining residual moisture in freeze-dried biologic materials. However, it is based on the assumption that when the weight of the sample becomes constant, the sample no longer contains water. This is an incorrect assumption. The method gives only an assumed value for the "water" present, making it an indirect determination. Erroneous low residual water levels may be obtained if the drying is done at insufficient temperatures to drive off the water vapor. Misleading results may also be obtained if an insufficient number of consecutive weighings are recorded. If these errors of omission are made, incorrect low residual moisture figures are obtained (Burchard et al 1978).

Several instruments are available for automatic residual moisture determination. These instruments operate on the thermogravimetric principle. A halogen radiator or other heating device dry tissue samples while a precision balance concomitantly measures the sample weight. The total weight loss is interpreted as moisture content as it does in the manual method. The advantage of halogen radiator is that it affords shorter measuring time with results comparing favorably with other drying methods. Halogen radiator drying is a modification of infrared drying method. It allows for achieving full equilibrium state where vapor pressure of the sustance measured is equal to the vapor pressure of the surrounding atmosphere. The thermogravimetric methods are classical methods. For this reason they are often a part of governmental regulations dealing with biological products.

Karl Fisher Method

The Karl Fisher method is based on a selective reaction of water with a solution of iodine in a mixture of methyl alcohol, anhydrous pyridine, and sulfurous anhydride. Iodine, which is inactive in the anhydrous solution, oxidizes SO_2 to SO_3 while being reduced to hydroiodic acid which fixes pyridine. The color reaction is based on the fading of the reagent solution, which is initially dark brown. Although the Karl Fisher reaction is simple in principle and has some definite advantages, such as technical rapidity, precision, reduced number of manipulations, and direct dissociation of the sample in the reagent, it is difficult to standardize. Furthermore, since the ability of alcohol to penetrate some tissue is poor, there may be a problem of incomplete extraction. It also gives only a "general" value for existing water. For this reason this technique had not been frequently employed. However, recently modifications of the technique and the construction of specialized apparatus had been made. These have greatly simplified the procedure and improved reliability of results.

Proton Magnetic Resonance (Nuclear Magnetic Resonance) Method

The proton or nuclear magnetic resonance (NMR) method provides a direct measurement of water. This method, based on magnetic resonance spectroscopy, is widely employed in the study of chemical structures. However, the instrumentation required for NMR spectroscopy is expensive and cumbersome. The magnetic field provided by the NMR apparatus is either steady or pulsed. For determination of water, steady-state NMR spectroscopy is used. NMR measurements are based on the fact that most nuclei have an angular momentum in addition to a magnetic moment. When the orthogonal radio frequency is identical to the processional frequency of the nucleus being examined, absorption of energy becomes possible and produces a transition from one energy level to another. At certain frequencies this phenomenon, referred to as resonance, can be detected independent of other atomic species present in the sample. The intensity of the absorption that follows resonance is proportional to the nuclear concentration of the atomic species, whose resonance is being measured, thus allowing for quantification. At the magnetic field strength of 14,100 gauss, the protons of the water molecule have an intense absorption at 60 MHz. Their detection in an apparatus tuned to a specific frequency presents no difficulty. Since every proton of every water molecule in the field is reactive, NMR provides the only absolute method for water determination (Hazlewood 1972).

Electrical Resistance Method

An electrical resistance method for determining the moisture content of porous material is a highly accurate technique. However, it has not been applied to tissue banking except on experimental basis.

Results of Freeze-Drying of Bone Allografts

During the initial phase of the freeze-drying cycle, large amounts of water are removed from frozen bone specimens as is evident by an accumulation of ice on the condenser. Of course, ice insulates the condenser, reducing the temperature of the condenser by nearly 20°C.

The level of the vacuum in the chamber of the freeze-dryer bears a direct relationship to the degree of freeze-drying. The higher was the vacuum, the faster the freeze-drying.

Nuclear magnetic resonance analysis appears to provide the most accurate measurement of the total water in freeze-dried bone. The values obtained by NMR exceeded those obtained by gravimetric methods by 3-4 gm percent. The additional NMR measured water probably represents water that is tightly bound to the macromolecular structures (bound water). It can be extracted via the D_2O:H_2O exchange process but cannot be driven off by conventional drying procedures. However, if the samples are charred, the NMR and gravimetric values become very close.

At this time, it is not likely that NMR residual water determination, accurate as it may be, will replace the time-honored gravimetric method. NMR equipment and deuterium oxide are expensive and are not readily available in most laboratories. Furthermore, NMR spectroscopy requires continuous calibration.

It is apparent that in freeze-dried bone, levels of residual water below 5% are attainable with conventional freeze-driers. The limit to freeze-drying of bone seems to lie in this range. However, misleading low residual water values may be reported if bone is subjected to gravimetric analysis in a 50°C convection oven. Gravimetric measurements with bone must continue until true weight equilibrium is achieved. This may take close to 200 hours.

Freeze-dried human tissue allografts prepared according to the methods described herein function quite satisfactorily in a variety of clinical circumstances, including in particulate allograft preparations (Malinin and Temple 2004, 2007, Temple and Malinin 2009, Sloof 1996). It is difficult to assess whether the variation of freeze-drying techniques alter the behavior of allografts in patients.

Before implantation, freeze-dried grafts must be rehydrated. Rehydration is mandatory in grafts which may require cutting, drilling or compression. Reconstitution of freeze-dried allografts is simple and dependent on the size and configuration of the graft; all it requires is the immersion of the specimen in saline. The process of rehydration may be accelerated in a vacuum-sealed pouch attached to a fluid chamber by a one way valve and is available in the packaging of some commercial freeze-dried products (Malinin et al 1981).

Biomechanical Properties of Freeze-Dried Bone Allografts

Biomechanical properties of freeze-dried bone allografts have received unfair criticism over the years, all on account of rodent experiments performed 30 years ago (Pelker et al 1984). This study showed a significant decrement in the torsional strengths of freeze-dried long bones. Therefore, it was concluded, and generally accepted, that freeze-dried bone was much weaker than its fresh or frozen counterparts. There is a possible explanation for these results. Femur specimens were freeze-dried to less than 3% residual moisture. This means they were possibly over-dried. Over-drying, of course, makes bones brittle. Although Triantafyllou (1995) and associates also observed weakening of the freeze-dried and irradiated bone in a three point fracture testing, other more recent animal studies showed little difference between frozen and freeze-dried cortical bone (Kang and Kim 1995, Heo et al 2009).

Our own testing of the biomechanical properties of human femoral cross sections showed no significant decrease in the bending strength following freeze-drying and rehydration (Malinin 1989). Subsequently studies using three point bending tests exhibited similar patterns of failure in both frozen and freeze-dried specimens. The results of these studies indicate that adequately dried, but not over-dried freeze-dried diaphyseal bone allografts possess strengths similar to fresh and frozen bone counterparts. Therefore it should provide safe applications as a structural allograft in reconstructive surgery. Washing and impaction of particulate grafts improve their shear stress (Dunlop et al 2003).

There is little in the literature giving technical details and providing scientific data on bone and tissue allografts treated by proprietary processes such as Tutoplast™ process, Allowash™ and Biocleanse™.Therefore, we cannot comment on preservation of bone osteoinduction in grafts treated by these proprietary methods.

Fresh Cartilage with Subchondral Bone Allografts

Allotransplantation of articular cartilage for the treatment of isolated articular cartilage defects has gained in popularity during the last decade (Bugbee 2001). To this end, either fresh allografts or allografts stored in tissue culture media or salt solutions at hypothermic temperatures are employed. There has been some controversy regarding the length of storage at hypothermia, and retention of viability of cartilage grafts. Initially, investigators who pioneered the procedure transplanted cartilage within 48 hours of its excision from cadaver donors (Gross et al 1983, Meyers 1989). Subsequently, for logistical reasons, some tissue banks extended the storage time for over 3–4 weeks. However, the results of laboratory and animal experiments show time-dependent deterioration of cartilage stored in tissue culture media with or without serum. Because of these findings and the findings by other investigators, it seems unwise to transplant cartilage stored at hypothermia for more than 2 weeks (Ball et al 2004).

Sterilization of Bone and Tissue Allografts

A number of methods and procedures designed to sterilize the allografts were tried in the past. A simple method of sterilizing bone allografts would be by immersion into a liquid. Many chemical solutions including ethyl, methyl, and isopropyl alcohols as well as paraphenol have been investigated but proved to be entirely unsatisfactory, failing the test of time (Sabanas et al 1955). Acetone and formaldehyde have also been tried without success, but acetone is still used as a dehydrating agent (Carnevale 1951, Christophe 1926). Alcohol-fixed bone implanted into rodents became absorbed, although a few layers of new bone appeared in and around it. Osteoclasts were absent at the periphery of alcohol-fixed grafts. Thus, it became evident that unlike boiling or autoclaving, the osteoinductive principle of the graft was not destroyed altogether but was partially extracted by alcohol (Lacroix 1951). Osteogenesis was induced by injection of the alcohol into rabbits in about one-third of the animals.

The boiled and alcohol-fixed grafts are mentioned here to demonstrate the differences of the recipient's response to bone transplants prepared by different methods. For better or for worse, these allografts, as well as those sterilized by immersion in other sterilizing solutions, are still in use. There is scant information on the behavior of secondarily sterilized bone allografts transplanted into humans. This has resulted in the development of methods for bone allograft preparation that have proceeded empirically, but devoid of scientific evidence. Virtually everything was tried, but the only two accepted methods of secondary sterilization which have withstood the test of time and are still used by many tissue banks, irradiation and sterilization with ethylene oxide gas.

Irradiation

The only reason for irradiating bone and soft tissue allografts is the fear of infections, including those caused by HIV. Irradiation is ineffective against prions, and thus cannot prevent transmission of Creutzfeld-Jacob disease. A relatively high dose of irradiation is needed to inactivate HIV in bone, but the actual dose estimates vary. Although a 15–25 KGy dose is commonly used, Conway found that 15 KGy would not reliably inactivate HIV in bone allografts (Conway et al 1990). To overcome this deficiency, doubling the dose to 30 KGy may be necessary, but irradiation in this dose range alters the biomechanical integrity of the graft and reduces its osteoinductive potential (Buring and Urist 1967, Urist and Hernandez 1974). It also diminishes the mechanical stress of the grafts (Gibbons et al 1991). Therefore a choice as to whether or not to implant irradiated allografts is a personal one. It must be balanced between the reduction of a chance for possible transmission of an infection and the undesirable effects of irradiation on the biointegrity of an allograft (Brooks et al 1963). Sterilization of allograft bone by irradiation may be effective at certain doses, but it alters the biologic potential of the bone (Maeda et al 1988, Loty et al 1990, Anderson et al 1992). It also reduces the osteoinductive capacity of the grafts (Skinner 1990).

Sterilization with Ethylene Oxide

Ethylene oxide gas penetrates many types of material such as paper, cloth and cellophane. It also penetrates tissues and bone. It has been shown to possess bactericidal, fungicidal and virucidal properties (Sidwell et al 1969, Prolo et al 1980 Klaienbeek and van Torngen 1954). However, variation in resistance to ethylene oxide among spore-forming organisms has also been noted (Kerulek et al 1970).

When tissues are sterilized with ethylene oxide, its secondary by-products, ethylene glycol (EG) and ethylene chlorohydrins (ECH) remain in the tissue and other materials. For this reason, the FDA has published a limit for the quantity of these compounds allowed on the implantable devices (Gardener 1978).

Residuals of ethylene oxide sterilization are toxic and hence must be removed before treated grafts can be implanted (Arizona et al 1994).

Bone morphogenic proteins are resistant to degradation by ethylene oxide sterilization (Moore et al 1990, Urist 1980, 1980a). Ethylene oxide gas is very soluble in water as well as in fat (Chichester 1968). For this reason, some authors recommend defatting bone before sterilization with ETO. Although the FDA has set the standards for ETO and its residuals, Arizona et al (1995) found that ETO adversely affected fibroblasts even below FDA established levels. This may explain a report of the failure of spinal fusions with ETO-sterilized bone allografts (Herron and Newman 1989). Experimentally, Thoren and Aspenberg (1995) reported inhibition of bone formation by ethylene oxide gas. However, ETO sterilized allografts have been used clinically with success (Cloward 1953, 1980, Prolo et al 1980). Consistent bone induction with ETO sterilized bone has also been reported in experimental animals (Singholm et al 1992, Syftestad and Urist 1982). Ethylene-oxide treated materials have a toxic effect on fibroblasts in culture (Kudryk et al 1992). This may have accounted for persistent synovitis reported by Jackson and his associates 1990. However, the lesion described in humans could not be reproduced in animals. All in all, despite the above cited complications, thousands of ethylene-oxide sterilized bone and tissue allografts have been implanted into many patients without complications. Therefore, it appears that ethylene oxide sterilization is a safe and effective procedure for sterilization of allografts if adequate precautions are exercised to remove ethylene oxide residuals from the grafts (Rosomoff and Malinin 1976).

Hydrogen Peroxide

Bone allografts are frequently placed into hydrogen peroxide solution primarily for cosmetic reasons. Hydrogen peroxide bleaches the bone white, and in addition has some disinfectant effects. Ideally, exposure to any sterilizing or bleaching agent should not adversely affect graft healing, but the available data indicates hydrogen peroxide affects osteoinductivity of bone allografts (Boyce et al 1999, DePaula et al 2005). The effect on osteoinductivity is time-dependent. While exposure to hydrogen peroxide for short periods of less than 1 hour might not affect osteoinductivity of allografts to a significant degree (Beebe et al 2009), longer exposures abolish osteoinductivity altogether (Carpenter et al 2006).

Demineralized Bone

Although demineralized bone is actively promoted, its use is limited primarily to ground cortical bone and cancellous bone chips (Wolfinbarger et al 2008). These do not provide structural support, i.e., they are not osteoconductive. A notable exception is demineralized perforated bone of Gendler (Gendler 1986).

The demineralization of bone allegedly uncovers BMPs and other bone growth promoting factors. It also promotes differentiation of primitive mesenchymal cells of the host into osteoprogenitor cells. However, so does non-demineralized bone of equal particle sizes. (Malinin et al 2009). The advent of demineralized bone is not new. Senn (1889) used demineralized bone as a vehicle for the delivery of antiseptics. The first definitive demonstration of heterotropic bone formation in rodents induced by decalcified bone was provided by Urist (1965). However, he did not employ demineralized bone matrix (DBM) as it is known today, but surface decalcified bone, later termed AAA (Chemosterilized antigen-extracted surface demineralized autolyzed allogeneic bone) (Urist 1965, 1983).

There are considerable differences between demineralized bone matrix (DBM) and demineralized, or more precisely, partially decalcified bone. The methods of preparation of these allografts are distinct, as are their biological properties. DBM is prepared by simply demineralizing bone in hydrochloric acid until the calcium content is reduced to less than 2%. Since DBM is prepared from particulate bone, the preparation frequently entails freeze-drying, grinding, demineralizing and refreeze-drying. Thus, it is unlike other freeze-dried bone allografts. DBM is freeze-dried twice. DBM is grossly amorphous, soft and provides no structural support.

Demineralization is thought to increase availability of bone matrix-associated bone morphogenetic proteins (BMPs) rendering these grafts osteoinductive. However, it must be noted, non-demineralized particulate grafts are also osteoinductive. There seems to be little difference in clinical results between demineralized and non-diminineralized particulate allografts (Cammack et al 2005). Thus, extra effort and expense of using demineralized bone matrix might not be justified. Although demineralization with hydrochloric acid releases BMP from the bone, it also facilitates BMP elution and loss into the acid bath (Pietrzak et al 2009). Likewise it may alter intraosseous lipids needed for BMP transport (Urist et al 1997). Finally, other bone antigens are exposed during demineralization that may induce an immunologic reaction. This is evident in non-human primate studies described in Chapter VI.

Partially demineralized, partially decalcified or surface decalcified bone is exemplified by Urist's chemosterilized, antigen-extracted allogeneic (AAA) bone (Urist 1983). Preparation of this allograft is complex and time-consuming, but its clinical efficacy is worth the effort (Urist et al 1975).

The calcium content of AAA bone is in the neighborhood of 10–15%. The calcium content of surface demineralized bone allograft is usually the same. Both of these types of preparations maintain osteoconductive properties.

Bleached Bone

A number of tissue and commercial bone allograft distributors promote allografts on the basis of their whiteness. As everyone knows, normal bone is not white bone, but grayish, brownish or yellowish. Chalk white bone is bleached, as are bones left under the sun in a desert. There is no reason to bleach bone allografts other than for cosmetic or marketing purposes. Bleaching is usually achieved by bathing allografts in a solution of hydrogen peroxide (H_2O_2). Exposure to H_2O_2 diminishes or abolishes osteoinductivity, and as noted, the effect is time-dependent. This is balanced against the disinfecting activity of the compound. According to Beebe et al (2009) 1 hour exposure to H_2O_2 does not have a profound effect on osteoinduction by bone allografts. However, prolonged exposure of bone to H_2O_2 abolishes its osteogenecity (Carpenter et al 2006).

DISCUSSION

This chapter, reviewing tissue banking, was intended to familiarize the reader with the developments that led to tissue banking as it is practiced today as well as past practices. In order to know where we are now and where we are going in the future, it is important to have some perspective on where we have been. The current practices evolved from trial and error, observation, laboratory studies and clinical results. Regrettably, by necessity, these investigations were conducted neither in a systematic manner nor in a logical sequence. Allograft demand and commercial interest trumped scientific investigation. Thus, pieces of information from disjointed laboratory studies and clinical experiences, when these became available, influenced the modification and development of new techniques. Several tissue banks adopted proprietary techniques and methodology for allograft preparation. These were promoted mainly through advertising without adequate data published in peer-reviewed scientific literature. This is not to imply that these products and techniques are inferior and without merit; some have proven to be valuable and beneficial.

In writing this chapter, references to unsubstantiated claims and generalizations have been avoided. Instead, sound, well-established principles and findings were summarized and brought to the reader's attention in hope that this information would help guide further investigation and discovery regarding the preparation, properties and transplantation of musculoskeletal tissue allografts.

Laboratory investigations of bone allografts helped to predict the behavior of allografts transplanted into patients (Küttner 1913, Hutchison 1949). Experiments with intercalary allografts clearly showed the importance of rigid internal fixation. Press-fitted allografts simply do not heal, even with external immobilization. Incorporation of internally fixed allografts relies on the external callus and endosteal replacement of the graft, whereas vascularization of the cortical allografts is associated with biomechanical weakness. This has clinical relevance.

Laboratory studies with onlay cortical plate allografts paved the way for their success in revision surgery for proximal femoral deficiencies. Studies involving articular cartilage allografts demonstrated latent injury of cryopreserved chondrocytes in intact cartilage. Although viable cells were present in intact cryopreserved cartilage, as demonstrated by tissue culture experiments, morphology and histochemistry, the survival of chondrocytes in the intact frozen cartilage was not achieved in quantities sufficient to maintain normal cartilage structure and function over time. Retrieval studies and direct observation have consistently shown that articular hyaline cartilage in cryopreserved osteochondral allografts undergoes fibrocartilaginous transformation and subsequent joint degeneration.

Limitation of cold storage time for fresh osteoarticular grafts applicable to human transplants has been demonstrated in controlled laboratory investigations. This information has influenced commercial practices and improved the clinical results in patients receiving these grafts.

These are but a few examples that underscore the principles of tissue banking. Hopefully, these principles will be the foundation upon which future scientific inquiry and discovery advance the field of musculoskeletal transplantation.

BIBLIOGRAPHY

1. Alldredge RH. Discussion. Journal of Bone and Joint Surgery (Am), 1948;30:75.
2. American Association of Orthopaedic Surgeons Task Force on AIDS and Orthopaedic Surgery, American Association of Orthopaedic Surgery, Park Ridge, IL; 1989.
3. American Association of Tissue Banks. Standards for Tissue Banking. McLean, VA; 2005.
4. Anderson MJ, Keyak JH, Skinner HB. Compressive mechanical properties of human cancellous bone after gamma irradiation. Journal of Bone and Joint Surgery (Am), 1992;74(5):747-52.
5. Arizona T, Iwamoto Y, Okayama K, et al. Ethylene oxide sterilization of bone grafts. Acta Orthopaedica Scandinavica, 1994;65:640-2.
6. Ashwood-Smith MJ. Radioprotective effect of combinations of AET or cysteamine with dimethyl sulphoxide. International Journal of Radiation Biology, 1962;5:201-2.
7. Ashwood-Smith MJ. Radioprotective and cryoprotective properties of dimethyl sulfoxide in cellular systems. Annals of the New York Academy of Sciences, 1967;141(1):45-62.
8. Ashwood-Smith MJ. The radioprotective action of dimethyl sulphoxide and various other sulphoxides. International Journal of Radiation Biology, 1961;3:41-8.
9. Ball ST, Amiel D, Williams SK, et al. The effect of storage on fresh human osteochondral allografts. Clinical Orthopaedics and Related Research, 2004;418:246-52.
10. Baschkirzew NJ, Petrow NN. Beitrage zum Knochenuberpflanzung. Deutsch Zeitschrift für Chirurgie, 1912;113:490.
11. Beebe KS, Benevenia J, Tuy BE, et al. Effect of new allograft processing procedure on graft healing in canine model. Clinical Orthopaedics and Related Research, 2009;467:273-80.
12. Bonfiglio M, Jeter WS, Smith CL. The immune concepts: Its relation to bone transplantation. Annals of the New York Academy of Sciences, 1955;59:417-33.
13. Boyce T, Edwards J, Scarborough N. Allograft bone: the influence of processing on safety and performance. Clinical Orthopaedics and Related Research, 1999;30:571-81.
14. Brighton CT, Lane JM, Kohl JK. In vitro rabbit articular cartilage organ model.11 35S incorporation in various oxygen tensions. Arthritis and Rheumatism, 1984;17:245-51.
15. Brooks OB, Heiple KG, Herndon CH, et al. Immunological factors in homogenous bone transplantation. IV. The effects of various methods of preparation and radiation on antigenicity. Journal of Bone and Joint Surgery (Am), 1963;45:1617-26.
16. Brown KLB, Cruess RL. Bone and cartilage transplantation in orthopaedic surgery. Journal of Bone and Joint Surgery (Am), 1982;64:270-9.
17. Brown MD, Malinin TI, Davis PB. A roentgenographic evaluation of frozen allografts versus autografts in anterior cervical spine fusions. Clinical Orthopaedics and Related Research, 1976;119:231-6.
18. Buck BE, Malinin TI, Brown MD. Bone transplantation and human immunodeficiency syndrome (AIDS). Clinical Orthopaedics and Related Research, 1989;240:129-36.
19. Buck BE, Malinin TI. Human bone and tissue allografts; preparation and safety. Clinical Orthopaedics and Related Research, 1994;303:8-17.
20. Bugbee WD. Alternatives to arthroplasty of the knee: biologic resurfacing. Current Opinion in Orthopaedics, 2001;12:1-7.
21. Bull CR. Experimentelle Studien uber Knochentransplantation und Knochenregeneration. Norske Videnskaps-Akademie; Oslo; 1Maten-Naturv; 1928; Klasse No. 9:105.

22. Burchard H, Jones H, Glowczewskie F, et al. Freeze-dried allogenic segmental cortical bone grafts in dogs. Journal of Bone and Joint Surgery (Am), 1978;60:1082-90.
23. Buring K, Urist MR. Effect of ionizing radiation on the bone induction principle in the matrix of bone implants. Clinical Orthopaedics and Related Research, 1967;55:225-34.
24. Burwell RG. Marrow in incorporation of bone graft. Clinical Orthopaedics and Related Research, 1985;200:125-41.
25. Cammack GV II, Nevins M, Clem DS III. et al. Histologic evaluation of mineralized and demineralized freeze-dried bone allograft for ridge and sinus augmentation. International Journal of Peridontics and Restorative Dentisry, 2005; 25:231-37.
26. Carnevale V. A proposito de injecto oseo conservado en parofenol (Communication previa) Soc. Argent. De Ortopedia y Trauma, 1951;16:121.
27. Carpenter EM, Gendler E, Malinin TI, et al. Effect of hydrogen peroxide on osteoinduction by demineralized bone. American Journal of Orthopaedics, 2006;35:562-7.
28. Carr CR, Hyatt GW. Clinical evaluation of freeze-dried gone grafts. Journal of Bone and Joint Surgery (Am), 1955; 37:549-66.
29. Centers for Disease Control and Prevention. Update: Allograft associated bacterial infections. Morbidity and Mortality Weekly Report, 2002;5:207-10.
30. Centers for Disease Control and Prevention. Hepatitis C virus transmission in an antibody-negative organ and tissue donor, US 2000-2002. Morbidity and Mortality Weekly Report, 2003;52:273-4.
31. Centers for Disease Control and Prevention. Invasive Streptococcus pyogenes infection after allograft implantation-Colorado. Morbidity and Mortality Weekly Report, 2003;52:1174-6.
32. Centers for Disease Control and Prevention. Transmission of HIV through bone transplantation. Case report and public health recommendations. Morbidity and Mortality Weekly Report, 1988;37:597.
33. Centers for Disease Control and Prevention. Update: Acquired immunodeficiency syndrome (AIDS) in persons with hemophilia. Review of Medical Child Abuse, 1984;33:589-91.
34. Centers for Disease Control and Prevention. Brief Report: Investigation into recalled human tissue for transplantation. United States, 2005-2006, Morbidity and Mortality Weekly Report, 2006;55(20):564-6.
35. Chacon M, Gazitua R, Paebla C. Clinical correlation between the premortem study and autopsy. Review of Medical Child, 1997;125(10):1173-6.
36. Chichester DF. Antimicrobial food additives. In: Furia TE (Ed). Handbook of Food Additives. The Chemical Rubber Co, Cleveland; 1968. pp. 137-207.
37. Christophe L. Recherses sur les greffes d'os fixe a l'alcohol et sur le mechanism de l'osteogenese. Arch Fr-Belg Chir, 1926;26:13.
38. Cloward RB. The treatment of ruptured lumbar intervertebral disc by vertebral body fusion: Indications, operative technique, aftercare. Journal of Neurosurgery, 1953;10:154-67.
39. Cloward RB. Gas sterilized cadaver bone grafts for spinal fusion operations. A simplified bone bone bank. Spine, 1980; 5:4-10.
40. Conway B, Tomford WW, Hirsch MS, et al. Effects of gamma radiation on HIV-1 in a bone allograft model. Transaction of 36th Annual Meeting. Orthopaedic Research Society, 1990;15:225.
41. Czitrom AA, Axelrod FB. Antigen presenting cells in allotransplantation. Clinical Orthopaedics and Related Research, 1985;197:27-31.
42. Czitrom AA, Keating S, Gross AE. The viability of articular cartilage in fresh osteochondral allografts after clinical transplantation. Journal of Bone and Joint Surgery (Am), 1990; 72:574-81.
43. Czitrom AA, Langer F, McKee N, et al. Bone and cartilage allotransplantation. Clinical Orthopaedics and Related Research, 1986;208:141-5.
44. Dennis J, Martinez OV, Landry DC, et al. A comparison of two methods used in aseptic processing of musculoskeletal allograft tissues. Cell and Tissue Banking, 2011;12:45-60.
45. De Paula CA, Truncale KG, Gertzman AA, et al. Effects of hydrogen peroxide cleaning procedures on bone graft osteoinductivity and mechanical properties. Cell and Tissue Banking, 2005;6:287-98.
46. Doppelt SH, Tomford WW, Lucas AD et al. Operational and financial aspect a hospital bone bank. Journal of Bone and Joint Surgery (Am), 1981; 63:1472-78.
47. DuMolin G, Love W. The value of autopsy microbiology. Clinics of Microbiology Newsletter, 1982;10:165-7.
48. Dunlop DG, Brewster NT, Madabhushi SP, et al. Techniques to improve the shear strength of impacted gone graft: The effect of particle size and washing of the graft. Journal of Bone and Joint Surgery (Am), 2003;85:639-46.
49. Eastlund T. Bacterial infection transmitted by human tissue allografts. Cell and Tissue Banking, 2006;7(3):147-66.
50. Elliott H, Scott HJ. The bone bank in neurosurgery. British Journal of Surgery, 1951;39:31-4.
51. Emerson RH, Malinin TI, Cuellar AD, et al. Cortical strut allografts in the reconstruction of the femur in revision total hip arthroplasty. A basic science and clinical study. Clinical Orthopaedics and Related Research, 1992;285:35-44.
52. Fedotenkov AG, Danilova LP, Diuhkant IP, et al. A new method of processing cadaver bone marrow designed for transplantation. Problemy Gematologii, 1962;2:28-32.
53. Flosdorf EW, Hyatt GW. The preservation of bone grafts by freeze-drying. Surgery, 1952;31:716-9.
54. Flosdorf EW. Freeze-Drying. New York: Reinhold Publishing Co; 1949. pp. 1-67.
55. Friedlaender GE. Immune responses to osteochondral allografts. Clinical Orthopaedics and Related Research, 1983;174:58-68.
56. Friedrici HH, Sebastian M. Autopsies in a modern teaching hospital. A review of 2,537 cases. Archives of Pathology and Laboratory Medicines, 1984;108(6):518-21.
57. Gardener S. Ethylene oxide, ethylene oxide, ethylene chlorohydrins and ethylene glycol. Proposed maximum residue limits and maximum levels of exposure. Federal Register, 1978;43:27474-83.
58. Gendler E. Perforated demineralized bone matrix: A new form of osteoinductive material. Journal of Biomedical Materials Research, 1986;20:687-97.
59. Ghormley RK. Choice of graft methods in bone and joint surgery. Annal of Surgery, 1942;115:427-34.

60. Giannoudis PV, Dinopoulos H, Tsiridis E. Bone substitutes, an update. Injury, 2005;36(Suppl 3):520-7.
61. Gibbons MJ, Butler JH, Grood ES, et al. Effects of gamma irradiation on the initial mechanical and material properties of goat bone-patella tendon-bone allografts. Journal of Orthopaedic Research, 1991;9:209-18.
62. Goldberg VM, Heiple KG. Experimental hemijoint and whole joint transplantation. Clinical Orthopaedics and Related Research, 1983;174:43-53.
63. Goldberg VM, Lance EM. Revascularization and accretion in transplantation. Journal of Bone and Joint Surgery (Am), 1972;54:807-16.
64. Gorschewsky O, Klakow A, Riechert K, et al. Clinical comparison of the Tutoplast allograft and autologous patellar tendon (bone-patella tendon-bone). American Journal of Sports Medicine, 2005;33:1202-9.
65. Greaves RIN, Adair ME. High vacuum condensation drying of protein from frozen state. Journal of Hygiene, 1939;39:417-26.
66. Greaves RIN. Serum plasma preservation Cryobiology, 1968;5:76-86.
67. Greiff D. The important variable in the long-term stability of viruses dried by sublimation of ice in vacuo. Progress in refrigeration science and technology. Proceedings of the XIII International Congress of Refrigeration. Westbort: AVI Publishing Co.; 1973. pp. 657-62.
68. Gresham RB. Freeze-drying of human tissues for clinical use. Cryobiology, 1964;1:150-6.
69. Gross AE, McKee NH, Pritzker KPH, et al. Reconstruction of skeletal deficits at the knee. A Comprehensive osteoarticular transplantation program. Clinical Orthopaedics and Related Research, 1983;174:96-106.
70. Hart MM, Campbell ED, Kartub MG. Bone banking: a cost effective method for establishing a community hospital bone bank. Clinical Orthopaedics and Related Research, 1986;206:295-300.
71. Hazlewood CF, Chang DC, Rorschach HE, et al. Cellular water and macromolecules. In: Reversability of cellular injury due to inadequate perfusion. (T. Malinin et al, eds.) Charles C Thomas Publisher, Springfield, 1972. pp22-38.
72. Henry M. Homografts in orthopaedic surgery. Journal of Bone and Joint Surgery (Am), 1948;30:70-6.
73. Heo SY, Lee HB, Kim MS, et al. Biomechanical assessment of freeze-dried allograft cortical bone plate graft in canine bone defect model. Veterinari Medicina, 2009;54:183-90.
74. Herron LD, Newman MH. The failure of ethylene oxide gas-sterilized freeze-dried bone graft for thoracic and lumbar spine fusion. Spine, 1989;14(5):496-500.
75. Hutchison J. Observations of bone transplants in the anterior chamber of the eye. Glasgow Medical Journal, 1949;39:552.
76. Hyatt G, Butler MC. Bone grafting: The procurement, storage and clinical use of bone allografts. AAOS, Instructional Course Lectures, 1957;14:343-73.
77. Jackson JB. Bone banking: an overview. Laboratory Medicine, 1987;18:830-3.
78. Jackson DW, Windler GE, Simon TM. Interarticular reaction associated with the use of freeze-dried, ethylene-oxide sterilized bone-patellar tendon-bone allograft in the reconstruction of anterior cruciate ligament. American Journal of Sports Medicine, 1990;18:1-11.
79. Johnson EE, Urist MR. One-stage lengthening of femoral nonunion augmented with human bone morphogenetic protein. Clinical Orthopaedics and Related Research, 1998;347:105-16.
80. Kang JS, Kim NH. The biomechanical properties of deep freezing and freeze-drying bones and their biomechanical changes after in vivo allograft. Journal of Yonsel Medicine, 1995;36:332-5.
81. Kerulek K, Gammon RA, Lloyd RS. Microbiological aspects of ethylene oxide sterilization. II. Microbial resistance to ethylene oxide. Applied Microbiology, 1970;19:152-6.
82. Klaienbeek A, Van Torngen HAE. Virucidal action of ethylene oxide. Journal of Hygiene, 1954;52:525-8.
83. Koneman E, Davis M. Postmorem bacteriology 3. The significance of microorganisms recovered at autopsy. American Journal of Clinical Pathology, 1974;61:28-40.
84. Kozak JA, Heilman AE, O'Brien JP. Anterior lumbar fusion options: techniques and graft materials. Clinical Orthopaedics and Related Research, 1994;200:45-51.
85. Kruetz FP, Hyatt GW, Turner TC, et al. The preservation and clinical use of freeze-dried bone. Journal of Bone and Joint Surgery (Am), 1951;33:863-72.
86. Kudryk VL, Scheidt MJ, McQuade MJ, et al. Toxic effects of ethylene-oxide sterilized freeze-dried bone allograft on human gingival fibroblasts. Journal of Biomedical Materials Research, 1992;26:1477-88.
87. Küttner H. Einige Dauerresuctete der Transplantation aus der Leiche und aus der Affen. Verhandlung Deutsche Gesels-chaft Fue Chirurgie, 1913;11:353.
88. Lacroix P. The Organization of Bones. London: J & A Churchill; 1951.
89. Lavernia CI, Malinin TI, Temple HT, et al. Bone and tissue allograft use by orthopaedic surgeons. Journal of Arthroscopy, 2004;19:430-3.
90. Lewis WH, McCoy CC. The survival of cells after death of the organism. Bull Johns Hopkins Hospital, 1922;33:284-90.
91. Lexer E. The classic: joint Joint transplantation. Clinical Orthopaedics and Related Research, 1985;197:4-10.
92. Linn FC, Sokoloff L. Movement and composition of interstitial fluid of cartilage. Arthritis and Rheumatism, 1965;8:481.
93. Loty B, Courpied JP, Tomeno B, et al. Bone allografts sterilized by irradiation. Biological properties, procurement and results of 150 massive allografts. International Orthopaedics, 1990;14(3):237-42.
94. Lovelock JE, Bishop MWH. Prevention of freezing damage in living cells by dimethyl sulfoxide. Nature, 1959;183:1394-5.
95. Lovelock JE. The denaturation of lipid-protein complexes as a cause of damage by freezing. Proceedings of the Royal Society London Biological Sciences, 1957;1:274-84.
96. Lovelock JE. Biophysical aspects of freezing and thawing of living cells. Proceedings of the Royal Society of Medicine, 1954;47:60.
97. Lovelock JE. The mechanism of the protective action of glycerol against hemolysis by freezing and thawing. Biochimica et Biophysica Acta, 1953;11:28.
98. Luyet BJ. An attempt at systematic analysis of the notion of freezing rates and evaluation of the main contributory factors. Cryobiology, 1966;2:198-205.

99. Maeda M, Bryant MH, Yamagata M, et al. Effects of irradiation on cortical bone and their time-related changes. A biomechanical and histomorphological study. Journal of Bone and Joint Surgery (Am), 1988;70(3):392-9.
100. Malinin GI, Fontana DJ, Baumgart DC. Distribution of C14 labeled dimethyl sulfoxide in tissues of intact animals. Cryobiology, 1969;5:328-35.
101. Malinin TI, Buck BE, Temple HT, et al. Incidence of clostridial infection in donors' musculoskeletal tissues. Journal of Bone and Joint Surgery (Br), 2003;85:1051-4.
102. Malinin TI, Latta L, Milne E, et al. Effects of freeze-drying on the femoral strength and notch sensitivity of human femoral diaphysis. Orthopaedic Research Society Annual Meeting, 1989;6-9.
103. Malinin TI, Latta LL, Wagner JL, et al. Freeze-dried cortical plate allografts in fracture healing. Orthopaedic Transactions, 1985;9:284.
104. Malinin TI, Martinez OV, Brown MD. Banking of massive osteoarticular and intercalary bone allografts-12 years experience. Clinical Orthopaedics and Related Research, 1985;197:44-57.
105. Malinin TI, Mnaymneh W, Wagner JL, et al. Healing of internally fixed intercalary canine allografts of freeze-dried bone. Orthopaedic Transactions, 1985;9:339.
106. Malinin TI, Mnaymneh W. Bone banking and transplantation of bone allografts. In: Harry Goldmith's practice of surgery. Philadelphia: Harper & Row; 1985.
107. Malinin TI, Perry VP. A review of tissue and organ viability assays. Cryobiology, 1967;4:104.
108. Malinin TI, Temple HT, Buck BE. Transplantation of osteochondral allografts after cold storage. Journal of Bone and Joint Surgery (Am), 2006;88:762-70.
109. Malinin TI, Temple HT. Importance of particle size on healing of bone. Proc.71 Annual meeting. American Academy of Orthopaedic Surgery, 2004;5:633.
110. Malinin TI, Thompson CB, Brown MD. Freeze-dried tissue allografts in surgery. In: AM Karow, DE Peggs (Eds). Organ Preservation for Transplantation. New York: Marcel Dekker Inc; 1981. pp. 677-89.
111. Malinin TI, Wagner JL, Pita JC, et al. Hypothermic storage and cryopreservation of cartilage. Clinical Orthopaedics and Related Research, 1985;197:15-26.
112. Malinin TI, Wolcott MW. Infusion of cell-free fluids into rabbits in hemorrhagic shock. Archives of Surgery, 1973;106:86-9
113. Malinin TI, Wu NM, Flores A. Freeze-drying of bone for allotransplantation. In: GE Friedlaender, HJ Mankin, KW Sells (Eds). Osteochondral Allografts. Boston/Toronto: Little Brown & Co; 1983. pp. 181-92.
114. Malinin TI. Organization of a tissue bank-University of Miami experience. Procedings of American Association of Tissue Banking. Rockville, 1977;79-86.
115. Malinin TI. Acquisition and banking of bone allografts. In: M Habal, H Reddi (Eds). Bone Grafts and Bone Substitutes. Philadelphia: WB Saunders Co; 1992. pp. 206-25.
116. Malinin TI. Allografts for the reconstruction of the cruciate ligaments of the knee: Procurement, sterilization and storage. Sports Medicine and Arthroscopy Review, 1993;1:31-41.
117. Malinin TI. Processing and storage of viable human tissue. Public Health Service Publication No 1442. Washington, DC: US Government Printing Office; 1966.
118. Malinin TI, Carpenter EM, Temple HT. Particulate bone allograft incorporation in regeneration of osseous defects; importance of particle sizes. The Open Orthopaedics Journal, 2007;1:19-24.
119. Malinin TI, Temple HT, Garg A. Cancellous and cortical microparticulate allograft for dental implantation: an experimental study in non-human primates. Implant Dentistry, 2009;18:420-6.
120. Malinin TI, Claflin AJ, Brown MD. Experience with bone and dura mater allografts at the University of Miami Tissue Bank. Acta Medica Polona, 1978;19:215-228.
121. Mankin HJ, Doppelt S, Tomford WW. Clinical experience with allograft implantation. Clinical Orthopaedics and Related Research, 1983;174:69-72.
122. Mankin HJ, Friedlaender GE. Perspectives of bone allograft histology. In: GE Friedlaender, HJ Mankin, FW Sells (Eds), Osteochondral Allografts. Boston/Toronto: Little, Brown & Co.; 1983. pp. 4-5.
123. Martinez OV, Buck BE, Hernandez M, et al. 2003 Blood and marrow cultures as indicators of bone contamination in cadaver donors. Clinical Orthopaedics and Related Research, 409, 317-24.
124. Mazur P. Mechanisms of injury at subzero temperatures. National Cancer Institute Monographs, 1962;7:7-20.
125. Meryman HT. Preservation of living cells. Federation Proceedings, 1963;22:781-9.
126. Miller F. Sussman MD, Stamp WG. The use of bone allograft: A survery of current practice. Journal of Pediatric Orthopaedics, 1984;4:353.
127. Minckler JM, Newell GR, O'Toole WF, et al. Microbiology experience in collection of human tissue. American Journal of Clinical Pathology, 1966;45:85-92.
128. Moore TM, Artal R, Arenas M, et al. Influence of postmortem time and temperature on osteoinductive activity of demineralized microperforated ethylene oxide-sterilized syngeneic bone implant in the rat. Clinical Orthopaedics and Related Research, 1990;259:239-44.
129. Morosow BO. Explanation of chicken and human hearts resuscitated after drying. Munchener Medizinische Wochenschrift, 1928;75:1713.
130. Morosow BO. Weitere explanatationverduche auf Zetrocknetan Geweben der Wirbetiere (Gehirn, Herz, Mielz, Testikel und Haut). Arch Exp Zellforsch Des Gewebzuckt, 1930;10:157.
131. Nei T. Freezing and freeze-drying of microorganisms. Cryobiology, 1964;1:87-93.
132. Oilier L. Traite experimental et clinique de la regeneration des os et de la production artificielle du tissue osseux Victor Masson et Fils. Paris; 1867.
133. Panuska JA, Malinin TI, Mentz RJ. Effect of dimethyl sulfoxide on cooling rates of unrestrained rats. Cryobiology, 1966;2:345-50.
134. Parkes AS. Factors affecting the viability of frozen ovarian tissue. Journal of Endocrinology, 1958;17:337-43.
135. Pegg DE, Wang L, Vaughn D. Cryopreservation of articular cartilage. Part 2. The mechanism of cryoinjury. Cryobiology, 2006;52:347-59.
136. Pelker RP, Friedlaender GE, Markham TC, et al. Effect of freezing and freeze-drying on the biomechanical properties of rat bone. Journal of Orthopaedic Research, 1984;1:405-11.

137. Perry VP, Kerby CC, Kowalski FJ, et al. The freezing of human kidney cell suspensions. Cryobiology. 1965; 1:274-284.
138. Pietrzak WS, Ali SN, Chitturi D, et al. (2009), BMP depletion occurs during prolonged acid demineralization of bone: characterization and implication for graft preparation:Cell and Tissue Bank. Published online, 29 Dec.
139. Polge C, Smith AV, Parkes AS. Revival of spermatozoa after vitrification and dehydration at low temperature. Nature, 1949;164-6.
140. Prolo DJ, Pedrotti PW, White DJ. Ethylene oxide sterilization of bone, dura matter and fascia lata for human transplantation. Neurosurgery, 1980;6:529-39.
141. Rey LR. L'Humidite Residuelle des Moduits Lyophilisses: Nature, Origine et Methodes d'Etude. In L.R. Rey (ed)., Aspects Theoriques et Industriels de la Lyophilization. Paris: Hermann, 1964. pp199-252.
142. Reynolds F, Oliver OR. Clinical evaluation of merthiolate bone bank: A preliminary report. Journal of Bone and Joint Surgery (Am), 1949;31:792-99.
143. Roberts FJ. Procurement, interpretation and value of postmortem culture. European Journal of Clinical Microbiology and Infectious Disease, 1998;17:821-7.
144. Rodrigo JJ, Sakovich L, Travis C, et al. Osteocartilaginous allografts as compared with autografts in the treatment of knee joint osteocartilagenous defects in dogs. Clinical Orthopaedics and Related Research, 1978;134:342-9.
145. Rosomoff HL, Malinin TI. Freeze-dried allografts of dura matter 20 years experience. Transplantation Proceedings 1976;(Suppl I), 8:133-8.
146. Sabanas AO, et al. Comparison of homologous bone grafts preserved by acetone and formaldehyde and by acetone alone; experimental studies. Proceedings of Staff Meetings of the Mayo Clinic, 1955;30:432-6.
147. Senn N. (1889), Healing of aseptic bone cavities with decalcified chips. American Journal of Medical Science, 98, 219-47.
148. Sherman JK. Low temperature research on spermatozoa and eggs. Cryobiology, 1964;1:103-29.
149. Sidwell RW, Dixon GJ, Westbrook L, et al. Procedure for the evaluation of the virucidal effectiveness of an ethylene oxide gas sterilizers. Applied Microbiology, 1969;17:790-6.
150. Simonds RJ, Holmberg SD, Hurwitz RL, et al. Transmission of human immunodeficiency virus type 1 from a seronegative organ and tissue donor. New England Journal of Medicine, 1992;326(11):726-32.
151. Singholm G, Gendler E, McKellop K, et al. Graft perforations favor osteoinduction. Studies of rabbit cortical grafts sterilized with ethylene oxide. Acta Orthopaedica Scandinavica, 1992;63:177-82.
152. Skinner HB. Alternatives in the selection of allograft bone. Orthopaedics, 1990;13(8):843-6.
153. Sloof TJ, Buma P, Schreurs BW, et al. Acetabular and femoral reconstruction with impacted graft and cement. Clinical Orthopaedics and Related Research, 1996;324:108-15.
154. Strong M, Nelson K, Pierce M, et al. Preventing disease transmission by deceased tissue donors by testing blood for viral nucleic acid. Cell and Tissue Banking, 2005;6:249-53.
155. Syftestad GT, Urist MR. Bone aging. Clinical Orthopaedics and Related Research, 1982;162:288-97.
156. Taylor AC. The physical state of transition in the freezing of living cells. Annals of the New York Academy of Sciences, 1960;85:595-609.
157. Temple HT, Malinin TI. Microparticulate cortical allograft: An alternative to autograft in the treatment of osseous defects. The Open Orthopaedics Journal, 2008;2:91-96.
158. Thoren K, Aspenberg P. Ethylene oxide sterilization impairs allograft incorporation in a conduction chamber. Clinical Orthopaedics and Related Research, 1995;318:256-64.
159. Thurlbeck WM. Accuracy of clinical diagnosis in a Canadian teaching hospital. Canadian Medical Association Journal, 1981;125:443-7.
160. Tomford WW, Duff GP, Mankin HJ. Experimental freeze preservation of chondrocytes. Clinical Orthopaedics and Related Research, 1985;197:11-4.
161. Tomford WW, Mankin HJ, Friedlaender GE, et al. Methods of banking bone and cartilage for allograft transplantation. Clinical Orthopaedics and Related Research, 1987;18:241-7.
162. Tomford WW, Mankin HJ. Bone banking, update on methods and materials. Orthopaedic Clinics of North America, 1990;30:565-70.
163. Tomford WW. Cryopreservation of articular cartilage. In: GE Friedlaender, HJ Mankin, KW Sell (Eds). Osteochondral Allografts. Boston/Toronto: Little, Brown & Co.; 1985.
164. Triantafyllou N, Phillipakis S. The effects of the radiation sterilization on human transplantable bones. Progress Report No 1041 IRB. International Atomic Energy Agency, 1971:1-7.
165. Urist MR, Benham K, Krendi F, et al. Lipids closely associated with bone morphogenetic protein (BMP) and induced heterotopic bone formation. Connective Tissue Research, 1997;36(1):9-20.
166. Urist MR, Hernandez A. Excitation transfer in bone. Archives of Surgery, 1974;119:486-93.
167. Urist MR, Mikulski A, Boyd SD. A chemosterilized antigen-extracted autodigested alloimplant for bone banks. Archives of Surgery, 1975;110:416-28.
168. Urist MR, Sato K, Brownell AG, et al. Human bone morphogenetic protein (mBMP). Proceedings of the Society for Experimental Biology and Medicine, 1983;173:194-9.
169. Urist MR. Bone transplants and implants. In: MR Urist (Ed). Fundamental and Clinical Bone Physiology. Philadelphia: J B Lippincott Co; 1980. pp. 331-93.
170. Urist MR. Chemosterilized antigen-extracted surface demineralized allogeneic bone for arthrodesis. In: G Friedlaender, H Mankin, K Sells (Eds). Osteochondral Grafts. Boston: Little Brown & Co; 1983. pp. 193-201.
171. Urist MR. Bone: formation by autoinduction. Science, 1965;150:893-9.
172. US Food and Drug Administration, 21 CFR, Part 1271; 2004.
173. US Navy Tissue Bank. Manual of Procedure for the Storage. Bethesda, MD; 1958.
174. Wilson PD. Follow-up study of the use of refrigerated homologous bone transplants in orthopaedic operations. Journal of Bone and Joint Surgery, 1951;33(A):307-23.
175. Wolfinbarger L JR, Eisenlohr LM, Ruth K, Demineralized bone matrix: Maximizing new bone formation of successful boneimplantation. In: Orthopaedic Biology and Medicine: Musculoskeletal Tissue Regeneration (W S Pietzak ed). Humana Press, Totowa, NJ, 2008. pp93-117.
176. Wu NM, Malinin TI. Nuclear magnetic resonance studies of some cation and anion exchange in dimethyl/sulfoxide-water systems. Analytical Chemistry, 1980;52:186-9.
177. Young SE, Wilkins RM. Medical/Social history questionnaires; validating the process. Proceedings of 19th Annual Meeting. American Association of Tissue Banks, Atlanta, A-38. 1995.

Chapter 14

John F Connolly MD (1936–2007), Professor and Chairman Department of Orthopaedic Surgery, University of Nebraska, Omaha, Nebraska and Chief of Orthopaedics, Orlando Medical Center, Orlando, Florida. Dr Connelly was one of the first to bring tissue engineering into being by transplanting autologous bone marrow mesenchymal cells into fractured bones. (By permission of University of Nebraska Medical Center News.)

(*Reference:* Connolly J, Guse R, Lippielo, Dehne R. Development of an osteogenic bone marrow preparation. Journal of Bone and Joint Surgery (Am). 1989;71:684-91.)

Tissue Engineering

To anticipate the future, one must be mindful of the past. Although bone and tissue transplantation have been practiced for over a century, organized tissue banking is of a more recent vintage (Mastenbroek 1953). At first glance, the acquisition and storage of bone and tissue allografts do not appear to be of great significance nor of universal interest. However, such beliefs are not in line with current mainstream practice. Bone transplantation and tissue banking are now well entrenched, and since graft preparation has a direct bearing on its behavior after transplantation, the topics are important and clinically relevant (Albee 1917).

Bone banking and transplantation, in turn, stimulated the emergence of a new scientific field, tissue engineering. The term encompasses the use of a combination of cells, together with the combination of various biomechanical (scaffolding) and physical factors to improve or replace biological structures. In a broader sense, it includes various manipulations of tissue to render it functionally suitable for replacement or regeneration of missing or damaged musculoskeletal structures.

In the context of the musculoskeletal system, the term "tissue engineering" is associated with an attempt to replace or repair portions of bone, cartilage, tendons, ligaments and various membranes, or for that matter, entire three-dimensional structures. The concept is not new. In the late 1950s, there were several attempts to regenerate limbs by transplantation of embryonic cells and tissues (Malinin 1958, 1960). Tissue engineering has also been associated with attempts to produce biomechanical function, through the placement of cells in an artificial support system. The term has been used intermittently and interchangeably with regenerative medicine, but regenerative medicine usually implies production of tissues from transplanted cells.

Although the prospect of success with these technologies is intriguing, the clinical applications have not yet materialized. To translate tissue engineering into a practical discipline, several independent entities must be fused. These include cells as building blocks, scaffolds (both organic and inorganic) and various growth-promoting substances.

The experience gained from overseeing and creating clinical and research development programs that became the University of Miami Tissue Bank provided the philosophical and practical background needed for the implementation of transplantable bone, cartilage, tendons, etc.

Bone allografts became an acceptable alternative to bone autografts because their supply is virtually unlimited and they allow for immediate structural restoration of the skeleton. Some allografts stimulate bone formation on their surface, others are incorporated by accretion (Delloye et al 2007). Many kinds of allografts are now available to surgeons of many different disciplines. A number of these allografts are osteogenic. Osteogenicity does not necessarily depend on the presence of viable cells as is the case with autografts. A situation largely analogous to autograft would be when an allograft scaffolding is enriched with cultured autologous cells. However, this technique is in its infancy. The procedure is cumbersome and expensive. At this it is neither practical nor available so as to affect large numbers of patients.

Osteogenesis does not depend on the presence of viable cells. Many allografts are osteogenic provided their osteogenicity (inherent 3-D structure or proteins within the matrix) were not altered or destroyed by various chemical or physical manipulations used in graft processing.

There are two interrelated basic concepts inherent in bone graft transplantation and incorporation. The first predicates clinical success of an autograft upon survival of a significant number of graft cells with their matrix, as well as on the osteoinductivity of the graft itself. The second predicates clinical success of an allograft solely upon the ability of the graft to induce osteogenesis by causing host stem cells to differentiate into osteoprogenitor cells. This depends on the retention of the native properties of bone in the allograft as demonstrated by LaCroix 1949. To this end, many investigators have devoted efforts to design methods of allograft preservation with retention of osteoinductive properties. The early history of bone transplantation includes many findings that suggest bone storage in an unaltered state would provide acceptable allografts for clinical transplantation. In this connection, the accomplishments of Wilson and Bush in the field of frozen storage of allografts provided the basis for the clinical status that frozen bone allografts enjoy today (Bush 1947, Wilson 1947, 1951).

However, the road to this achievement was a long and arduous one as is evident from the chronological list of some of the early work on bone transplantation, tissue preservation and banking (**Table 1**).

The techniques described in the paper by Hyatt and Butler in 1957 established freeze-drying as the principal method of tissue preservation. There have been technological and mechanical improvements in freeze-drying equipment, but the principle remains the same (Kreutz et al 1951). A definite improvement in the freeze dryer has been the introduction of the external condensers, which have accelerated the rate of drying. Although this has increased freeze-drying speed, it is not certain whether or not rapid freeze-drying results in better biological preparations.

Freezing has also undergone evolution. Storage at cryogenic temperatures is now a common practice. These can be achieved both by cryogenic gases and mechanical refrigeration systems. In considering storage of tissues in a frozen state one must remember that at the beginning of tissue banking equipment capable of reaching temperatures below -15–30°C was uncommon and expensive. The introduction of cryoprotective agents, controlled velocity freezing and storage at cryogenic temperatures made possible long-term maintenance of allografts without alterations.

Table 1 Chronological list of some of the early work on bone transplantation, tissue preservation and banking

1867	First investigative report on bone transplantation (Oilier 1867).
1887	A description of a series of successful bone heterografts utilizing "decalcified" ox bone. The use of cadaver bone proposed. The ox bone was first treated with muriatic acid, then with caustic potash. The bone was stored in a solution of sublimate in alcohol (Senn 1889).
1890	First successful human homogenous bone graft transplanted (McEwen 1912).
1899	Cadaver periosteum stored at 15°C for 168 hours produced new bone when transplanted (McWilliams 1914).
1900	The periosteum of rats proliferated after preservation for 14 days in agar or gelatin at 2–29°C (Saltykov 1900).
1910	Homogenous canine grafts successfully transplanted after 3 weeks of refrigeration (Bauer 1910).
1911	Human fat, bone and cartilage stored at ordinary ice box temperature (Tuffier 1911).
1912	Classic treatise on the practicability of storing human tissue for clinical use. Storage of tissues in either Ringers' solution or petrolatum at temperature of -1–7°C (Carrel 1912).
1912	First specifications for the preservation of bone grafts published. Bone was wrapped in sterile vaseline gauze or immersed into vaseline and stored at 4°C for 24–48 hours (Albee 1912).
1917	Suggestion for using human cadavers free of transmissible disease as a source of homogenous bone. Storage in sterile receptacle at 4°C is recommended (Hey Groves 1917).
1921	Successful transplantation of a homogenous bone graft stored in alcohol (Leriche 1922).
1923	Homograft bone transplanted into dog soft tissue. Bone had been stored at room temperature, boiled and frozen. Frozen bone was osteogenic (Haas 1923, 1925).
1937	Refrigerated and autoclaved human bone allograft transplanted (Smith 1937).
1937	Treatise on the preparation and clinical use of os purum and os novum published (Orell 1937).
1942	Human bone stored in the citrated blood of donors for 3–6 days at 37°C (Inclan 1942).

Contd...

Contd...

1947	Clinical transplantation of bone homografts reported in 67 cases (Bush 1947). These grafts were taken from living donors and stored at -25°C.
1947	Criteria taken for procurement and storage of frozen bone published.Transplantation of frozen bone homografts in 307 patients reported (Wilson 1951).
1948	Reports on the use of frozen bone procured from human cadavers within 8 hours after death. No difference detected in host acceptance between frozen cadaver bone and bone obtained from living donors (Zimbron and Velasco 1948).
1949	The operation of the first frozen heterograft bone bank for human patients described. The bone was procured from the calf; bone was frozen at -25°C and stored at -15°C (Judet and Arviset 1949).
1950	The operation of the frozen human homograft bone bank in Sweden described (Hult 1950).
1951	Publication of results of 129 human transplantation cases in which merthiolate stored bone had been used (Reynolds 1951).
1951	Animal experiments and method of freeze-drying bone homografts described (Kreutz and Hyatt 1951).
1953	Netherlands Red Cross prepares and distributes frozen bovine bone.
1953	Penicillin-streptomycin bone bank. Donor bone was stored in saline solution containing penicillin and streptomycin for 3 months at 50°F (Augustine et al 1953).
1953	Method for the preservation of "living" bone homografts published. The bone was procured from cadavers free of malignancy and transmissible disease, placed in sterile containers, covered with plasma containing sulfathiazole and penicillin. The ratio of bone:plasma was 1:4. The material was stored at +4°C for 5–6 months (Tucker 1953).
1955	Treatise on the use of cathode irradiation of bone for sterilization and the storage in a frozen state published (Bassett et al. 1955).
1957	The technique of procurement, processing and storage of bone as performed at the Tissue Bank, National Naval Medical Center, Bethesda, Maryland published. A modified system of freeze-drying was used. Bone thus prepared had been used in 1,080 patients (Hyatt and Butler 1957).

These agents have definitely allowed for the improved preservation of cells, but not of all tissues. There has been progress. Trials with cryophylactic agents other than glycerol, dimethyl sulfoxide and few others have not greatly improved cryopreservation techniques. One can only look ahead to other new agents.

In going over 100 years of progress, and the trials and errors associated with tissue preservation, one cannot but notice the completion of a full circle. Chemical preservation, particularly that involving glycerol, is coming into vogue again.

Freeze-drying has achieved a great deal, but contains its own set of drawbacks. These are intimately connected to the limitations and problems associated with freezing. Freeze-drying amplifies these until it reaches a steady state.

A newer, recently introduced technique employs drying at hypothermia. The technique preserves tissues in a dessicator without subjecting them to the damage of freezing followed by that of freeze-drying. Perhaps this technique would avoid the deleterious alterations present in freeze-dried tissues (**Fig. 1**).

Figure 1 Freeze-dried bone allografts prepared by the US Navy Tissue Bank, Circa 1958

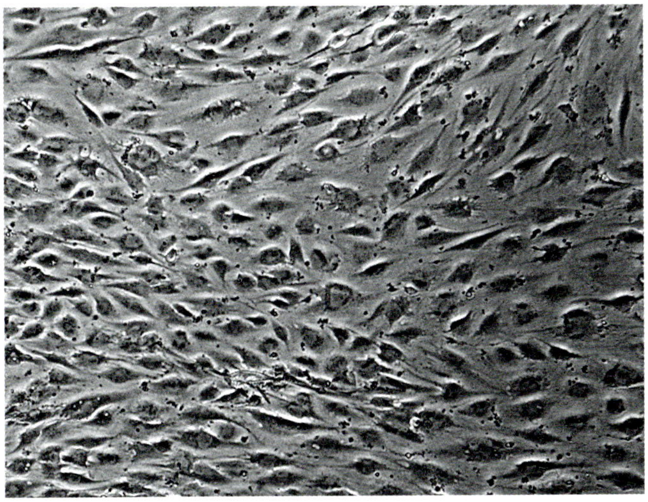

Figure 3 Phase contrast photomicrograph of human chondrocytes from explants cultures (× 25)

studies have been voluminous and should eventually lead to practical results.

Regeneration of articular cartilage defects by the introduction of autologous chondrocytes is now a well-established phenomenon. However, the technique has some serious drawbacks including the limited amount of donor tissue available, donor site morbidity and the need for expanding autologous chondrocytes in culture. On the other hand, allogeneic chondrocytes are readily available and can be grown in large quantities without the need for expansion. However, the technique is still awaiting clinical application which, at this junction, appears promising (**Fig. 3**).

In conclusion, we have defined future directions for tissue engineering. The work is still in progress, but the final outcome is difficult to predict. In the meanwhile musculoskeletal tissue transplantation is at its peak. New trends appear and disappear only to reappear again. What has been learned is very useful. A good example is osteochondral transplantation. However, successful techniques have their limits and cannot be pushed beyond these. In the meanwhile, these should be utilized judiciously until something better comes along!

BIBLIOGRAPHY

1. Albee FH. Bone graft surgery. WB Saunders Co Philadelphia; 1917.
2. Albee FH. Bone transplantation as a treatment for Pott's disease, club feet and ununited fractures. Postgraduate Medicine, 1912;27:999-1017.
3. Amin AR, Abramson SB. The role of nitric oxide in articular cartilage breakdown in osteoarthritis. Current Opinion in Rheumatology, 1998;10:263-8.
4. An HS, Simpson JW, Glover JM, et al. Comparison between allograft plus demineralized bone matrix versus autograft in anterior cervical fusion. A prospective multicenter study. Spine, 1995;20:2211-6.
5. Annersten S. Experimentelle Untersuchungen uber die Osteogenese und die Biochemie des Fracturcallus. Acta Chirurgica Scandinavica, 1940;84 (Suppl):60.
6. Augustine RW, Landmesser WE, O'Brien FD, et al. Penicillin-streptomycin bone bank. US Armed Forces Medical Journal, 1953;4:1587-91.
7. Bae HW, Zhao L, Kanim LEA, et al. Intervariability and intravariability of bone morphogenetic proteins in commercially available demineralized bone matrix products. Spine, 2006;31:1299-306.
8. Bassett, CAL, Hudgins TF, Trump JP, et al. The clinical use of cathode ray sterilized grafts of cadaver bone. Surgical Forum, 1955;6:549-53.
9. Bauer. Über Knochentransplantation. Zentralblatt für Chirurgie, 1910;37:20-1.
10. Beltelsen A. Experimental investigations into post-foetal osteogenesis. Acta Orthopaedica Scandinavica, 1944;15: 139-81.
11. Bier A. Osteoplastische Nekrotomie nebst Bemerkung über die and der Kieler chirurgischen Klinik ausgefuhrten Methoden der Nekrotomie. Archiv für Klinische Chirurgie, 1893;45:401.
12. Boden SD, Zdeblick TA, Sandhu HS, et al. The use of rhBMP-2 in interbody fusion cages. Definitive evidence of osteoinduction in humans: a preliminary report. Spine, 2000;25:376-81.
13. Burkus JK, Heim SE, Gornet MF, et al. Is Infuse bone graft superior to autograft bone? An integrated analysis of clinical trials using the LT-CAGE lumbar tapered fusion device. Journal of Spinal Disorders and Technology, 2003;16:113-22.
14. Bush LF. The use of homogenous bone grafts: a preliminary report of the bone bank. Journal of Bone and Joint Surgery, 1947;29:620-8.
15. Buttermann GR. Prospective nonrandomized comparison of an allograft with bone morphogenetic protein versus an iliac-crest autograft in anterior cervical disectomy and fusion. Spine Journal of North American Spine Society, 2008;8:426-35.
16. Carrel A. The preservation of tissue and its application in surgery. Journal of American Medical Association, 1912;59:523-7.
17. Cawston TE, Milner JM, Catteral JB, et al. Cytokine synergy, collagenases and cartilage collagen breakdown. Biochemical Society Symposia, 2003;70:125-33.
18. Chau AMT, Hobbes RJ. Bone graft substitutes in anterior cervical discectomy and fusion. European Spine Journal, 2009;18:449-64.
19. Deaver JB. Secondary bone implantation by a modification of Senn's methods. Medical News, 1889;55:714.

20. Deehan DJ, Cawston TE. The biology of integration of anterior cruciate ligament. Journal of Bone and Joint Surgery (Br), 2005;87:889-95.
21. Delloye C, Cornu O, Druez V, et al. Bone allografts. Journal of Bone and Joint Surgery (Br), 2007;89:547-9.
22. Dickerman RD, Reynolds AS, Morgan BC, et al. rhBMP-2 can be used safely in the cervical spine: dose and containment are the keys! Spine Journal of North American Spine Society, 2007;7:508-9.
23. Dmitriev AE, Castner S, Lehman RA, et al. Alterations in recovery from spinal cord injury in rats treated with recombinant human bone morphogenetic protein-2 for posterolateral arthrodesis. Journal of Bone and Joint Surgery (Am), 2011;93:1488-99.
24. Haas SL. Further observations on the survival of bone after removal from the body. Archives of Surgery, 1925;10:196.
25. Haas SL. A study of the viability of bone after removal from the body. Archives of Surgery, 1923;7:213-26.
26. Groves, Hey EW. Methods and results of transplantation of bone in the repair of defects caused by injury or disease. British Journal of Surgery, 1917;5:185.
27. Hult L. Some experience with a bone bank. Acta Orthopaedica Scandinavica, 1950;19:476-9.
28. Hyatt GW, Butler MC. The procurement, storage and clinical use of bone homografts. American Academy of Orthopaedic Surgeons Instructional Course Lectures, 1957;14:343-73.
29. Inclan A. The use of preserved bone graft in orthopedic surgery. Journal of Bone and Joint Surgery, 1942;24:81-96.
30. Judet R, et Arviset A. Banque d'os et heterogreffe. Presse Medecine, 1949;57:1007.
31. Kreutz FP, Hyatt GW, Turner TC, et al. The preservation and clinical use of freeze-dried bone. Journal of Bone and Joint Surgery (Am), 1951;33:863-72.
32. Kuklo TR, Groth AT, Anderson RC, et al. Recombinant human bone morphogenetic protein-2 for grade III open segmental tibia fractures from combat injuries in Iraq. Journal of Bone and Joint Surgery (Br), 2008;90:1068-72.
33. Lacroix P. Organizers and the growth of bone. Journal of Bone and Joint Surgery (Am), 1947;29:292-6.
34. Lee J, Green MH, Amiel D. Synergistic effect of growth factors on cell outgrowth from explants of rabbit anterior cruciate and medial collateral ligaments. Journal of Orthopaedic Research, 1995;13:435-41.
35. Leriche R. Transplantation d'os mort dans une pseudoartrose. Lyon Chirurgical, 1922;219.
36. Levander G. A study of bone regeneration. Surgery, Gynaecology and Obstetrics, 1938;67:705-14.
37. Luginbuel V, Meine L, Merkle HP, et al. Localized delivery of growth factors for bone repair. European Journal of Pharmaceutics and Biopharmaceutics, 2004;58:197-207.
38. Malinin TI. The effects of implantation of embryonic and tadpole tissues into adult frog limbs. Histological observations. The Journal of Experimental Zoology, 1960;143:1-20.
39. Malinin T, Deck JD. The effects of implantation of embryonic tadpole tissues into adult frog limbs I Regeneration after amputation. The Journal of Experimental Zoology, 1958;139(2):307-27.
40. Malinin T, Gendler E, Cvetkovic V, et al. Long-term *in vitro* maintenance of cultures of hyaline cartilage engineered from segregated chondrocytes. Houston TX: TERMIS, Annual Conference, Dec 2011;11-4.
41. Malinin T, Ouellette EA. Articulate cartilage nutrition is mediated by subchondral bone: A long-term autograft study in baboons. Osteoarthritis and Cartilage, 2000;8(6):483-91.
42. Mastenbroek G. The Frozen Bone Bank of the Netherlands Red Cross. Vox Sanguinis, 1953;3:33.
43. McKee MD, Schemitsch EW, Waddell JP, et al. The effect of human recombinant bone morphogenetic protein (rhBMP-7) on the healing of open tibial fractures: results of a multi-center, prospective, randomized clinical trial. The 18th Annual Meeting of the Orthopaedic Trauma Association. Toronto, ON, Canada 2002.
44. Mackie W. Clinical observation on the healing of aseptic bone cavities by Senn's method of implantation of antiseptic decalcified bone. Medical News, 1980;57:202.
45. Miller AG. A case of bone grafting with decalcified bone chips. Lancet, 1890;2:618.
46. Minamide A, Kawakami M, Hashizume H, et al. Evaluation of carriers of bone morphogenetic protein for spinal fusion. Spine, 2001;26:933-9.
47. Morpurgo B. The function of periosteum in bone transplants, based on four human transplantations without periosteum, and some animal experiments. Surgery, Gynecology, and Obstetrics, 1914;18:159.
48. Nagase H. Matrix metalloproteinases. In: Zink the Metalloproteinases in Health and Disease. NM Hooper, et al. (Eds). London: Taylor and Francis; 1996.
49. Orell S. Surgical bone grafting with "os purum," "os novum" and "boiled bone." Journal of Bone and Joint Surgery (Am), 1937;19:873-85.
50. Peterson B, Whang PG, Inglesias R, et al. Osteoinductivity of commercially available demineralized bone matrix. Preparation in spine fusion model. Journal of Bone and Joint Surgery (Am), 2004;86:2243-50.
51. Pietrzak WS, Ali SN, Chitturi D, et al. BMP depletion occurs during prolonged acid demineralization of bone: characterization and implications for graft preparations. Cell Tissue Bank. Published on line 29 Dec 2009.
52. Reynolds FC. Clinical evaluation of merthiolate bone bank and homogenous bone grafts. Journal of Bone and Joint Surgery (Am), 1951;33:873-83.
53. Ristiniemi J, Flinkkila T, Hyvonen P, et al. rhBMP-7 accelerates the healing in distal tibial fractures treated by external fixation. Journal of Bone and Joint Surgery (Br), 2007;89:265-72.
54. Rodeo SA, Suzuki K, Deng XH, et al. Use of recombinant human bone morphogenetic protein-2 to enhance tendon healing in bone tunnel. American Journal of Sports Medicine, 1999;27:476-88.
55. Schmitt A. Über Osteoplastik in klinischer und experimenteller Beziehung. Archiv für Klinische Chirurgie, 1893;45:401.
56. Scott J. Withdrawal of a paper. Journal of Bone and Joint Surgery (Br), 2009;91:285-6.

57. Senn N. On the healing of aseptic bone cavities by implantation of antiseptic decalcified bone. American Journal of Medical Science, 1889;98:219-43.
58. Smith AD. The use of homologous bone grafts in cases of osteogenesis imperfect. Archives of Surgery, 1937;34:687-94.
59. Smucker JD, Rhee JM, Singh K, et al. Increased swelling complications associated with off-label usage of rhBMP-2 in the anterior cervical spine. Spine, 2006;2813-9.
60. Swiatkowski MF, Aro HT, Donell S, et al. Recombinant human bone morphogenetic protein-2 in open tibial fractures. Journal Bone of Joint Surgery (Am), 2006;88: 1258-65.
61. Toriumi DM, Larrabee WF, Walikk JW, et al. Demineralized bone. Implant resorption with long-term follow-up. Archives of Otolaryngology-Head and Neck Surgery, 1990;116:676-80.
62. Tuffier T. Des greffes de cartilage et d'os humain dans les resections articulairs. Bulletins et memories, Societe de Chirurgie de Paris, 1911;37:278-86.
63. Urist MR. Bone: formation by autoinduction. Science, 1965;150(698):893-9.
64. Urist MR. Surface decalcified allogeneic bone (SDAB) implants. Clinical Orthopaedics and Related Research, 1968; 56:37-50.
65. Velasco Zimbron A, Sierra Rojas L. Banco de hueso del hospital infantile de la ciudad de Mexico: Informe preliminar. Bol Med de Hasp Inf, 1948;5:490.
66. Wilson PD. Follow-up study of the use of refrigerated homogenous bone transplants in orthopedic operations. Journal of Bone and Joint Surgery (Am), 1951;33: 307-23.
67. Wilson PD. Experience with a bone bank. Annals of Surgery, 1947;126:932-46.
68. Wolfinbarger L, Eisenlohr LM, Ruth K. Demineralized bone matrix: maximizing new bone formation for successful bone implantation. In: WS Pietrzak (Ed). Musculoskeletal Tissue Regeneration: Humana Press; 2008. pp. 93-117.

Index

Page numbers followed by *f* refer to figure and *t* refer to table